International Conference on Food Science and Engineering 2016

IOP Conference Series: Materials Science and Engineering
Volume 193

Surakarta, Indonesia
18 – 19 October 2016

ISBN: 978-1-5108-4028-7
ISSN: 1757-8981

Printed from e-media with permission by:

Curran Associates, Inc.
57 Morehouse Lane
Red Hook, NY 12571

CURRAN ASSOCIATES INC. proceedings.com

Some format issues inherent in the e-media version may also appear in this print version.

Printed by Curran Associates, Inc. (2017)

For permission requests, please contact the Institute of Physics
at the address below.

Institute of Physics
Dirac House, Temple Back
Bristol BS1 6BE UK

Phone: 44 1 17 929 7481
Fax: 44 1 17 920 0979

techtracking@iop.org

Additional copies of this publication are available from:

Curran Associates, Inc.
57 Morehouse Lane
Red Hook, NY 12571 USA
Phone: 845-758-0400
Fax: 845-758-2633
Email: curran@proceedings.com
Web: www.proceedings.com

Table of contents

Volume 193

International Conference On Food Science and Engineering 2016
18–19 October 2016, Surakarta, Indonesia

Accepted papers received: 24 April 2017
Published online: 8 May 2017

Preface

011001
OPEN ACCESS
International Conference on Food Science and Engineering 2016

011002
OPEN ACCESS
Peer review statement

Food Microbiology and Biotechnology

012001
OPEN ACCESS
Marine Biotoxins: Occurrence, Toxicity, and Detection Methods

M Asakawa.....1

012002
OPEN ACCESS
Identification of pork contamination in meatball using genetic marker mitochondrial DNA cytochrome b gene by duplex-PCR

E Novianty, L R Kartikasari, J H Lee and M Cahyadi.....17

012003
OPEN ACCESS
Development of Seaweed-based Biopolymers for Edible Films and Lectins

D Praseptiangga.....21

012004
OPEN ACCESS
Effects of different *Pediococcus halophilus* level and fermentation time on chemical properties of fermented anchovy paste "*terasi ikan*"

S D Lestari, Herpandi and G R R Simamora.....26

012005
OPEN ACCESS
Polygalacturonase production by AR2 pectinolytic bacteria through submerged fermentation of raja nangka banana peel (*Musa paradisiaca* var. *formatypica*) with variation of carbon source and pectin

R Utami, E Widowati, A Ivenaria and E Mahajoeno.....33

012006
OPEN ACCESS
Amylase production potentials of bacterial isolates obtained from the gut of *Oryctes rhinoceros* larvae

P C Aryati, A Pangastuti and S L A Sari.....40

012007
OPEN ACCESS
Prevalence of Antibiotic-resistance *Enterobacteriaceae* strains Isolated from Chicken Meat at Traditional Markets in Surabaya, Indonesia

R Yulistiani, D Praseptiangga, Supyani, Sudibya, D Raharjo and T Shirakawa.....46

012008
OPEN ACCESS
Effect of Incubation Time and Sucrose Addition on the Characteristics of Cheese Whey Yoghurt

E Nurhartadi, R Utami, A Nursiwi, A M Sari, E Widowati, A P Sanjaya and E A Esnadewi.....52

012009
OPEN ACCESS
Characteristic of Fermented Whey Beverage with Addition of Tomato Juice (*Lycopersicum esculentum*)

A Nursiwi, E Nurhartadi, R Utami, A M Sari, P W Laksono and E N Aprilia.....58

012010
OPEN ACCESS
Detection of chicken contamination in beef meatball using duplex-PCR Cyt b gene

E P Sari, L R Kartikasari and M Cahyadi.....63

012011
OPEN ACCESS
Production of Mozzarella Cheese Using Rennin Enzyme from *Mucor miehei* Grown at Rice Bran Molasses Medium

I H Rusdan and J Kusnadi.....68

012012
OPEN ACCESS
Physicochemical Properties of Dietary Fibers from *Artocarpus camansi* Fruit

V Suryanti, T Kusumaningsih and Y S Rumingtyas.....74

012013
OPEN ACCESS
Effect of Fermentation on Compositional Changes of *Cinnamomum osmophloeum* Kaneh Leaves

AD Kurniawati, TC Huang and J Kusnadi.....79

012014
OPEN ACCESS
The effect of nisin from *Lactococcus lactis subsp. lactis* on refrigerated patin fillet quality

S N Adilla, R Utami, A Nursiwi and E Nurhartadi.....86

012015
OPEN ACCESS
Isolation and Screening of Pectinolytic Fungi from Orange (*Citrus nobilis* Tan.) and Banana (*Musa acuminata* L.) Fruit Peel

K R Amilia, S L A Sari and R Setyaningsih.....93

012016
OPEN ACCESS
Preliminary Characterization of Crude Lectin Fraction of the Red Alga, *Acrocystis nana* from Wediombo Beach of the Southern Coast of Java Island, Gunung Kidul, Yogyakarta, Indonesia

C Anam, D Praseptiangga, M A Nugraheni, T Nurhayati, N D Fajarningsih, D S Zilda, E Chasanah and A Yunus.....98

012017
OPEN ACCESS
Cytotoxicity of Crude Lectins from Red Macroalgae from the Southern Coast of Java Island, Gunung Kidul Regency, Yogyakarta, Indonesia

C Anam, E Chasanah, B P Perdhana, ND Fajarningsih, N F Yusro, A M Sari, A Nursiwi, D Praseptiangga and A Yunus.....103

012018
OPEN ACCESS
Effect of temperature and pH on polygalacturonase production by pectinolytic bacteria *Bacillus licheniformis* strain GD2a in submerged medium from Raja Nangka (*Musa paradisiaca* var. *formatypica*) banana peel waste

E Widowati, R Utami, E Mahadjoeno and G P Saputro.....108

012019
OPEN ACCESS
Antibacterial Activity of Ethyl Acetate the Extract of Noni Fruit (*Morinda citrifolia* L.) Against Bacterial Spoilage in Fish

E R Nugraheni, G R Adriani and H Munawaroh.....115

Food Production System and Technology

012020
OPEN ACCESS
Food safety management and risk assessment in the fresh produce supply chain

L Jacxsens, M Uyttendaele, P Luning and A Allende.....122

012021
OPEN ACCESS
Optimization and Characterization of Cinnamon Leaves (*Cinnamomum burmannii*) Oleoresin

L.U. Khasanah, Kawiji, P. Prasetyawan, R. Utami, W. Atmaka, G.J. Manuhara and A.P. Sanjaya.....126

012022
OPEN ACCESS
The performance of Bali Cattle in transmigration area, south Kalimantan: a financial analysis

Sumanto.....133

012023
OPEN ACCESS
The role of grazing land on the buffalo population dynamics in Brebes regency

Sumanto.....138

012024
OPEN ACCESS
Effect of drying temperatures on physical characteristics of sorghum flour modified with lactic acid

G J Manuhara, B S Amanto and T A Astuti.....144

012025
OPEN ACCESS
Shelf Life Prediction for Canned Gudeg using Accelerated Shelf Life Testing (ASLT) Based on Arrhenius Method

R Nurhayati, E Rahayu NH, A Susanto and Y Khasanah.....150

012026
OPEN ACCESS
Preservation Effect of Two-Stage Cinnamon Bark (*Cinnamomum Burmanii*) Oleoresin Microcapsules On Vacuum-Packed Ground Beef During Refrigerated Storage

D Irfiana, R Utami, L U Khasanah and G J Manuhara.....155

012027
OPEN ACCESS
Exterior and interior physical quality of egg of laying hens fed diets containing different dietary purslane levels

L R Kartikasari, B S Hertanto, D Pranoto, W N Salim and A M P Nuhriawangsa.....162

012028
OPEN ACCESS
Preliminary analysis on hybrid vigor in Indonesian indigenous and crossbred cattle population using data from published studies

S Prastowo, TSM Widi and N Widyas.....168

012029
OPEN ACCESS
Characteristic of Arenga Starch-Taro (*Colocasia esculanta* L.) Flour Noodle with Addition of Beetroot Extract

E R N Herawati, D Ariani, Miftakhussolikhah, M P Elfanti and Y Pranoto.....173

012030
OPEN ACCESS
The Effect of Biduri (*Calotropis gigantean*) Latex on Meat Quality of Post Laying Hen

A M P Nuhriawangsa, W Swastike, B S Hertanto, A W and E D Pradisha.....180

012031
OPEN ACCESS
Sensory Attributes and Preliminary Characterization of Milk Chocolate Bar Enriched with Cinnamon Essential Oil

A Ilmi, D Praseptiangga and D R A Muhammad.....185

012032
OPEN ACCESS
Evaluation of Cooking Quality of Chromium Fortified – Parboiled Rice

W A Yulianto, C L Suryani and S Ulfah.....191

012033
OPEN ACCESS
Utilization of Bagasse Fly Ash to Remove the Unpleasant Odor of Stevia Extract and Soy Milk

Ashadi, M Masykuri and Haryono.....197

012034
OPEN ACCESS
Dietary Protected Feed Supplement to Increase Milk Production and Quality of Dairy Cows

A Pramono, E Handayanta, D T Widayati, P P Putro and Kustono.....201

012035
OPEN ACCESS
Starch Product of Wild Plants Species Jalawure (*Tacca leontopetaloides* L.) Kuntze as The Source of Food Security in The South Coastal West Java

Wardah, E N Sambas, Ridwan and D Ariani.....205

012036
OPEN ACCESS
Physical Characteristics of White Sweet Potato (*Ipomoea batatas* L.), Rice (*Oryza sativa* L.), and Tapioca (*Manihot esculenta)* Flours - Based Seasoning Composite Flour

NNA Alfani, D Ishartani, C Anam, D Praseptiangga and G J Manuhara.....215

012037
OPEN ACCESS
Rooms for genetic improvement in Indonesian Bali cattle population

N Widyas, T Nugroho and S Prastowo.....221

012038
OPEN ACCESS
Identification of Morphological Character and Esterase Isozyme Pattern in Second-Generation Black Rice Plant Irradiated to Gamma Rays

R S Hartanti, T A N Putri, F Zulfa, Sutarno and Suranto.....226

012039
OPEN ACCESS
Physical Characteristics of *Coleus tuberosus* Flour and Noodle in Various Arenga Starch Substitution

Miftakhussolikhah, D Ariani, M Angwar and J Kevin.....232

Physical Food Analysis and Chemistry

012040
OPEN ACCESS
Chemical changes associated with lotus and water lily *natto* production

S D Lestari, N Fatimah and R Nopianti.....237

012041
OPEN ACCESS
Physical and Chemical Characterization Of Greater Yam (*Dioscorea Alata*) And Jack Bean (*Canavalia Ensiformis*) - Based Composite Flour

D R Affandi, D Praseptiangga, F S Nirmala, B Sigit Amanto and W Atmaka.....243

012042
OPEN ACCESS
Physical Properties of Milk Cincau Curd on Different Concentrations of Green Cincau Leaf (*Cyclea barbata L.Miers*)

B S Hertanto, L R Kartikasari, Winny Swastike, M Cahyadi, A Yuliani and A M P Nuhriawangsa.....251

012043
OPEN ACCESS
Preparation and Characterization of Nano Chitosan from Crab Shell Waste by Beads-milling Method

E Rochima, S Y Azhary, R I Pratama, C Panatarani and I M Joni.....255

012044
OPEN ACCESS
The influence of composition and final pyrolysis temperature variations on global kinetics of combustion of segregated municipal solid waste

Pranoto, D A Himawanto and N A Arifin.....261

012045
OPEN ACCESS
Effect of Air Cooling and Vacuum Cooling Storage on the β-Carotene Content and Proximate Analysis (Water Content, pH, Total Protein and Content of Sugar) in Carrot

T Kusumaningsih, T Martini, K S Rini and L Okstafiyanti.....267

012046
OPEN ACCESS
Utilization of rice husk waste for Cd(II) adsorbent and its analysis using solid-phase spectrophotometry (sps)

S Saputro, M Masykuri, L Mahardiani, S Arini and A H Ramelan.....273

012047
OPEN ACCESS
The Effect of pH and Color Stability of Anthocyanin on Food Colorant

S Wahyuningsih, L Wulandari, M W Wartono, H Munawaroh and A H Ramelan.....279

012048
OPEN ACCESS
Indigo Dye Derived from *Indigofera Tinctoria* as Natural Food Colorant

S Wahyuningsih, A H Ramelan, D K Wardani, F N Aini, P L Sari, B P N Tamtama and Y R Kristiawan.....288

012049
OPEN ACCESS
Spectroscopic Study of Green Tea (*Camellia sinensis)* Leaves Extraction

A Marzuki, V Suryanti and A Virgynia.....293

012050
OPEN ACCESS
Chemical Constituents of *Luffa acutangula (L.)* Roxb Fruit

V Suryanti, S D Marliyana and I Y Astuti.....297

Miscellaneous Topics

012051
OPEN ACCESS
Barns Society as a Buffer Stock (Case in Nglaris Village Bener District Purworejo Regency)

E W Riptanti, A Qonita and Suprapti.....302

012052
OPEN ACCESS
Community food pattern and nutrition among toddlers in district of Pacitan, East Java, Indonesia

A Lestari, D Hanim, E Listyaningsih and Supriyadi.....307

012053
OPEN ACCESS
Sustainable Food Security Measurement: A Systemic Methodology

W Findiastuti, M L Singgih and M Anityasari.....313

012054
OPEN ACCESS
Soil Fertility Status, Nutrient Uptake, and Maize (*Zea mays* L.) Yield Following Organic Matters and P Fertilizer Application on Andisol

S Minardi, S Harieni, A Anasrullah and H Purwanto.....321

012055
OPEN ACCESS
Law Enforcement of Consumer Protection for Safe Food Packaging in The Decisions of Criminal Justice

F Wiryani, Herwastoeti, M Najih and A Haris.....327

012056
OPEN ACCESS
Staple Food Self-Sufficiency of Farmers Household Level in The Great Solo

Darsono.....335

012057
OPEN ACCESS
The effect of cinnamon bark (*Cinnamomum burmanii*) essential oil microcapsules on vacuumed ground beef quality

I N Brilliana, G J Manuhara, R Utami and L U Khasanah.....341

012058
OPEN ACCESS
Natrium dischargement from peripheral blood as a predominant factor influenced by the administration of banana (*Musa paradisiaca*) on elderly female hypertensive patient

A Pramono, N Noriko and S B Komara.....347

IOP Conf. Series: Materials Science and Engineering **193** (2017) 011001 doi:10.1088/1757-899X/193/1/011001

ICFSE 2016

International Conference on Food Science and Engineering

"Sustainable Development of Sciences and Technologies for Preserving Life"

IOP Conf. Series: Materials Science and Engineering **193** (2017) 011001 doi:10.1088/1757-899X/193/1/011001

PREFACE
INTERNATIONAL CONFERENCE FOOD SCIENCE AND ENGINEERING (ICFSE) 2016

The International Conference on Food Science and Engineering (ICFSE) 2016 is an international scientific forum initiated to overcome food safety and security problems as well as factors affecting them. Food safety and security problems have emerged as the main issue and challenge attracting more attention from various scientists, government, and regulatory bodies worldwide. The ICFSE 2016 was attended by leading scientists working in the field of food science and engineering. The outstanding invited speakers were:

1. Prof. Dr. Ir. Liesbeth Jacxens
   Department of Food Safety and Food Quality, Ghent University, Belgium
2. Prof. Wendy Umberger, Ph.D.
   Director of Global Food Studies, University of Adelaide, Australia
3. Prof. Dr. Agr. Sc. Ir. Vita Ratri Cahyani, M.P.
   Department of Soil Science, Sebelas Maret University, Indonesia
4. Prof. Manabu Asakawa, Ph.D.
   Department of Biofunctional Science and Technology, Hiroshima University, Japan
5. Prof. Jun Heon Lee, Ph.D.
   College of Agriculture and Life Sciences, Chungnam National University, Korea

The ICFSE 2016 aims to provide a forum for scientists, researchers, and engineers to present, discuss, share, and exchange their knowledge, experiences, and research findings on all aspects of food science and engineering. Therefore, this conference can become a platform in finding more appropriate and sustainable ways of addressing food safety and security issues to meet the need of growing population and changing lifestyle which is in line with the theme of the conference "Sustainable Development of Sciences and Technologies for Preserving Life". In this proceeding, the authors of our accepted papers come from 4 different countries, such as Indonesia, Belgium, Japan and Korea.

Date and Venue
Date: October 18-19, 2016
Venue: The Sunan Hotel Solo, Jl. Ahmad Yani 40, Surakarta, Indonesia

Venty Suryanti, S.Si., M.Phil., Ph.D.
Chairman of ICFSE 2016

IOP Conf. Series: Materials Science and Engineering **193** (2017) 011001 doi:10.1088/1757-899X/193/1/011001

IOP Conf. Series: Materials Science and Engineering **193** (2017) 011002 doi:10.1088/1757-899X/193/1/011002

Peer review statement

All papers published in this volume of *IOP Conference Series: Materials Science and Engineering* have been peer reviewed through processes administered by the proceedings Editors. Reviews were conducted by expert referees to the professional and scientific standards expected of a proceedings journal published by IOP Publishing.

IOP Conf. Series: Materials Science and Engineering **193** (2017) 012001 doi:10.1088/1757-899X/193/1/012001

Marine Biotoxins: Occurrence, Toxicity, and Detection Methods

M Asakawa[1]
Graduate School of Biosphere Science, Department of Biofunctional Science and Technology, Hiroshima University, 1-4-4, Kagamiyama Higashi-Hiroshima, Hiroshima 739-8528, Japan

E-mail : asakawa@hiroshima-u.ac.jp

Abstract. This review summarizes the role of marine organisms as vectors of marine biotoxins, and discusses the need for surveillance to protect public health and ensure the quality of seafood. I Paralytic shellfish poison (PSP) and PSP-bearing organisms-PSP is produced by toxic dinoflagellates species belonging to the genera *Alexandrium*, *Gymnodinium,* and *Pyrodinium*. Traditionally, PSP monitoring programs have only considered filter-feeding molluscs that concentrate these toxic algae, however, increasing attention is now being paid to higher-order predators that carry PSP, such as carnivorous gastropods and crustaceans. II. Tetrodotoxin (TTX) and TTX-bearing organisms - TTX is the most common natural marine toxin that causes food poisonings in Japan, and poses a serious public health risk. TTX was long believed to be present only in pufferfish. However, TTX was detected in the eggs of California newt *Taricha torosa* in 1964, and since then it has been detected in a wide variety of species belonging to several different phyla. In this study, the main toxic components in the highly toxic ribbon worm *Cephalothrix simula* and the greater blue-ringed octopus *Hapalochlaena lunulata* from Japan were purified and analysed.

Chapter I. Paralytic Shellfish Poison (PSP) and PSP-Bearing Organisms

1. Introduction

PSP is one of the most notorious and hazardous marine biotoxins known. It is mainly produced by toxic marine dinoflagellates species belonging to the genera *Alexandrium*, *Gymnodinium,* and *Pyrodinium*, and is accumulated up the food chain in many species of marine filter-feeding organisms such as bivalves [1-5]. Ingestion of PSP containing bivalves results in human intoxication. PSP blocks sodium channels distributed on the cell membrane, inhibiting the flow of sodium ions into and out of the cell, which leads to symptoms that include muscle paralysis, and often frequent death (Figure 1).

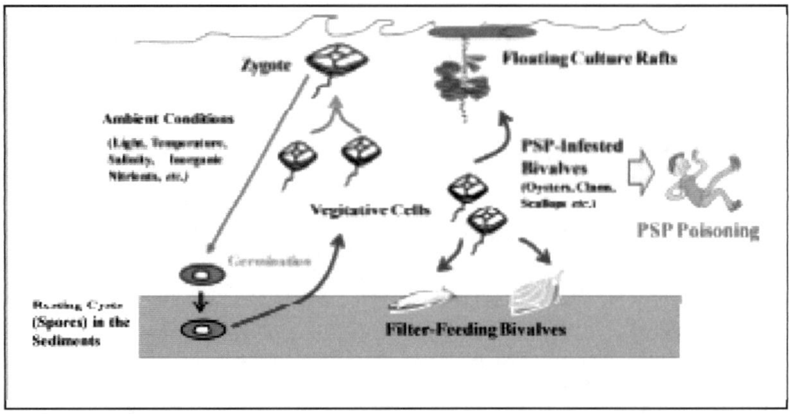

Figure 1. Generalized pathways of human intoxication with PSP via filter feeding bivalves

The PSP-producing dinoflagellates, *Alexandrium tamarense, A. catenella A. tamiyavanichii*, and *Gymnodinium catenatum* are found in Japan, where they intoxicate many edible bivalves [6-9]. *Alexandrium* spp. cause problems mainly in cold, temperate waters in the northern hemisphere, while *Pyrodinium bahamense* var. *compressum* is a tropical and subtropical species that occur in Southeast Asia [10,11]. The contamination of bivalves with PSP produced by these species has posed a serious problem to both the shellfish culture industry and public health in various parts of the world. To date, 10 PSP poisoning incidents have been officially recorded in Japan, and PSP food poisoning has also been reported in adjacent countries such as Taiwan and Korea. However, although PSP-infested bivalves are still being reported, food poisonings as a result of the ingestion of these species no longer occurs, except by accident, due to the established monitoring program for toxic shellfish and cell densities of PSP-producing dinoflagellates. In April, 1979, cultured scallops in Funka Bay, Hokkaido, were found to contain high levels of PSP and so were prohibited from being marketed [12]. However, at that time, several people collected and ate the mussels attached to the scallop-culturing rafts, three of whom were poisoned, and one of whom died. PSP is transferred from certain species of toxic dinoflagellates to shellfish by filter feeding with most of the toxins being accumulated in the digestive gland. Consequently, the digestive glands of bivalves are generally much more toxic than other tissues. For example, the adductor muscle has been found to be non-toxic to weakly toxic even in specimens whose digestive glands were extremely toxic (*c.*1000 MU/g).

PSP is a group of naturally occurring neurotoxic alkaloids that act as sodium channel-blocking agents in mammals, and causing several symptoms, including paralysis and often death. Saxitoxin (STX) was the first toxin to be isolated from PSP-infested bivalves. However, PSP is composed of at least 20 additional derivatives, all of which contain hydroxyl, carbamyl, and sulfate moieties at four sites (R1 - R4) on the backbone structure [5,13]. These PSP components have been divided into two groups, the low-toxicity group and high-toxicity group. Members of the low toxicity group are known as *N*-sulfocarbamoyl toxins (Figure 2), and are easily converted into other known toxins under acidic conditions.

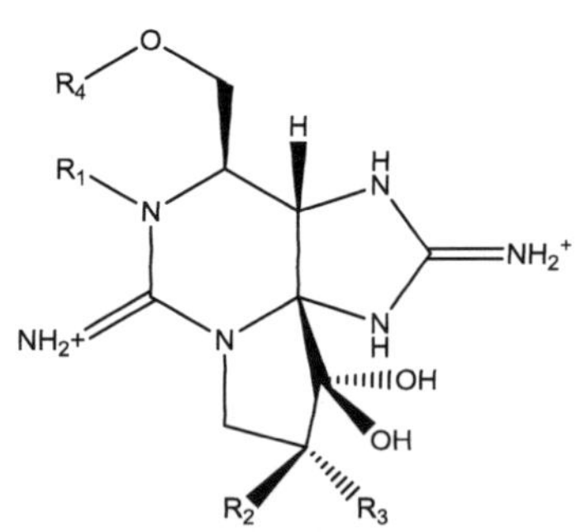

R1	R2	R3	Carbamate toxins	N-Sulfocarbamoyl toxins	Decarbamoyl toxins
				R4	
			$CONH_2$	$CONHSO_3^-$	H
H	H	H	STX (2483)	GTX5 (160)	dcSTX (1274)
OH	H	H	NeoSTX (2295)	GTX6 (180)	dcneoSTX (33)
OH	OSO_3^-	H	GTX1 (2468)	C3 (33)	dcGTX1 (1500)
H	OSO_3^-	H	GTX2 (892)	C1 (15)	dcGTX2 (1617)
H	H	OSO_3^-	GTX3 (1584)	C2 (239)	dcGTX3 (1872)
OH	H	OSO_3^-	GTX4 (1803)	C4 (143)	DcGTX4 (1080)

C = C toxins; GTX= gonyautoxin; STX= saxitoxin; dc= decarbamoyl
Value in parenthesis: specific toxicity in MU/μmol
The activity was expressed in mouse units (MU); One MU is defined as the dose of toxin required to kill a 20g ddY strain male mouse in 15 min after intraperitoneal injection.

Figure 2. Structure of the PSP analogues and their specific toxicity

For example, gonyautoxin V (GTX5) belongs to this group, and gives rise to STX when heated in dilute hydrochloric acid (HCl), increasing the toxicity sharply (mild acidic hydrolysis) [14]. These low -toxicity components are considered to be the precursors of the high toxicity carbamate toxin group. These substitutions result in analogs that vary by more than three orders of magnitude in toxicity. Some components, such as STX and GTX2, have specific toxicities comparable to tetrodotoxin (TTX). The minimum lethal dose (MLD) of PSP in humans is estimated to be 3,000 MU based mainly on fatal cases induced by this toxin [15]. PSP is roughly three times more toxic than TTX whose MLD in humans is considered to be 10,000 MU. Here, one mouse unit (MU) of PSP and TTX is defined as the amount of each toxin which can kill a 20g ddY strain male mouse in 15 and 30 min, respectively, after intraperitoneal administration.

2. Accumulation of PSP by Filter-Feeding Vectors

2.1 Toxic dinoflagellate and PSP-infested bivalves

Case study in Hiroshima Bay, Hiroshima Prefecture, Japan

Hiroshima Bay is one of the largest oyster farming areas in Japan (Figure 3). Up until 1992, there had been no records of PSP toxification of bivalves in this region. However, in April 1992, levels of PSP that substantially exceeded the quarantine limit of 4 MU/g edible parts as PSP were detected in bivalves with the appearance of *A. tamarense*, resulting in the prohibition of their harvesting and marketing by the Hiroshima Prefectural Government [1]. Fortunately, no food poisoning resulted from this outbreak. However, since the cultured oysters that are produced in Hiroshima Bay are shipped to many parts of Japan not only as fresh oysters, but also as raw materials for processed foods such as smoked oysters and oyster sauce, this posed a very serious problem to fisher-men and other workers in associated industries as well as to public health. Figure 4 shows a scanning electron microscope image of *A. tamarense* isolated from Hiroshima Bay in 1992. Subsequent monitoring of the cell density of *A. tamarense* from 1993 to 2004 showed that oysters, mussels and clams were contaminated with PSP from approximately the end of March until May each year [16].

IOP Conf. Series: Materials Science and Engineering **193** (2017) 012001 doi:10.1088/1757-899X/193/1/012001

■ : Kure Bay ▲ Kaita Bay

Figure 3. Map showing PSP-producing dinoflagellate of *A. tamarense* collecting locations in Hiroshima Bay

The location of Hiroshima bay in Japan is shown in the map to the right. The map on the left shows an enlarged image of Hiroshima Bay to pinpoint the sampling locations

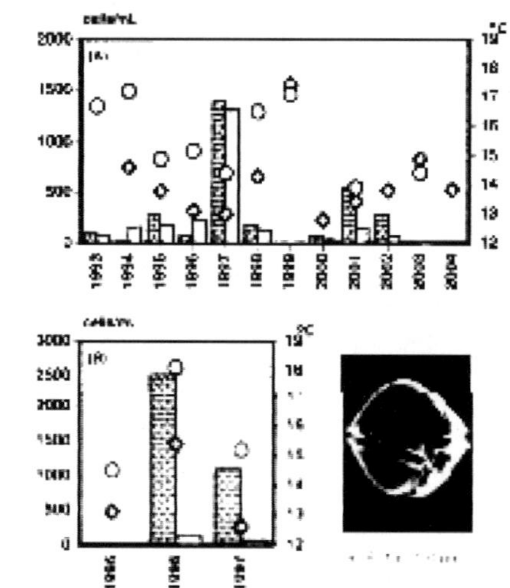

▦: Density (0m depth) ○Temperature (0m depth)
□: Density (5m depth) ◇Temperature (5m depth)

Figure 4. Maximum cells density of *A. tamarense* in Kure Bay (A) and Kaita Bay (B), Hiroshima Prefecture and scanning electron microscope image showing

For example, in Kure Bay, maximum concentrations of 1,400 and 1,300 cells/ml were observed at 0 and 5 m depths on April 21 and 24, 1997, respectively, while in Kaita Bay, remarkably high concentrations above 1,000 cells/ml were observed in two out of three years investigated. The temperature range at which the natural population of *A. tamarense* bloomed was generally between from 12°C to 16°C.

Figure 5 shows a high-performance liquid chromatography-fluorescence detection (HPLC-FLD) analysis of the GTX and STX groups of toxins contained in *A. tamarense* strain ATHS-93 isolated from Hiroshima Bay, while Table 1 shows the toxin profiles of *A. tamarense* isolated from Hiroshima Bay and the toxic bivalves.

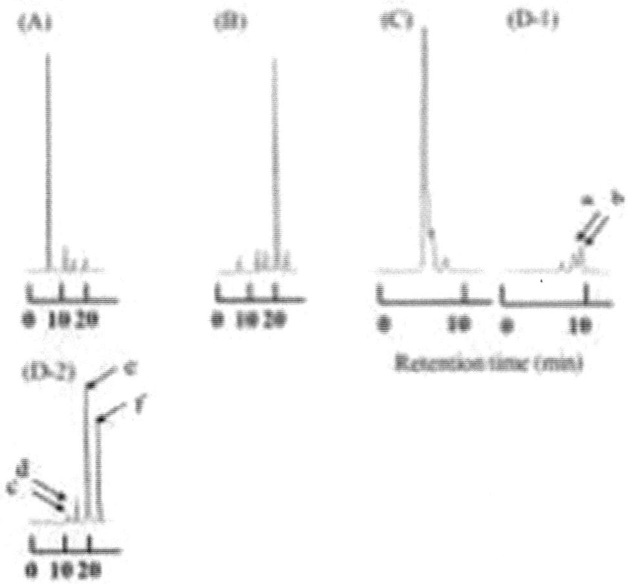

Figure 5. HPLC-FLD analysis of GTX and STX group of toxins containing in *A.tamarense* strain ATHS-93 isolated in Hiroshima Bay. (A) Purified toxins from strain ATHS-93 (B) hydrolysate of purified toxins with 0.1N HCl for 15 min. (D-1, 2) STX and GTX standards : (a, b) neoSTX, STX, (c,d,e,f) GTX4,1,3,2

Table 1. Toxin profiles of *A.tamarense* isolated in Hiroshima Bay and toxic bivalves

Chemical form	Specimens	Dinoflagellate *Alexandrium tamarense*		Short-necked clam	mussel	oyster
	Strain	ATHS-92*	ATHS-93			
α-epimer	GTX1	9.1±3.8	11.6	61.8	60.5	42.5
	GTX2	Tr.	Tr.	1.5	1.2	2.9
	C1(epi-GTX8)	0.7±0.4	1.3	0.2	0	0.9
	C3	1.7±1.3	0.2	0	0	0
	subtotal	11.5	13.1	63.5	61.7	46.3
β-epimer	GTX3	1.6±0.2	5.1	7.5	8.8	21.5
	GTX4	33.4±4.8	27.6	23.8	28.4	17
	C2(GTX8)	39.9±8.7	37.0	0.7	0	5.3
	subtotal	74.9	69.7	32	37.2	43.8
	neoSTX	13.5±5.1	17.2	5.1	1.1	9.9

All results are shown in mol%. -: not detected, Tr.: less than 0.1%
*Mean ± S.D. of five samples

Some fairly large differences were observed in the relative abundance of the toxins in the responsible dinoflagellates and the contaminated shellfish, suggesting that some *in vivo* conversion between the PSP components. The most notable difference was in the amount of C2 toxin, which is considered to be the precursor of GTX3. The two *A. tamarense* strains contained large amounts of C2 toxin (39.9 and 37.0%), whereas the bivalves only contained trace amounts of this low toxicity component, suggesting its in *vivo* conversion following ingestion. It has previously been shown that toxin composition also varies substantially between shellfish species, suggesting that there is interspecific variation in the metabolism of PSP [17-21].

2.2 PSP-infested marine mossworm

Case study in Funka Bay, Hokkaido, Japan

During the screening for paralytic potency in marine organisms around Funka Bay in Hokkaido, one of the largest scallop farming areas in Japan, a lethal potency was detected in brown seaweed, that was fouled with organisms including mossworm [22]. Figure 6 shows a scanning electron microscope image of marine mossworms that were stripped off a seaweed sample (the dried brown alga *Laminaria japonica*, known as "Kombu" in Japanese) that was harvested from Funka Bay, Hokkaido in summer of 1989. These mossworms had a lethal score of 18 MU/g dry basis, which greatly exceeded the quarantine limit of 4 MU/g. It is generally accepted that mossworms are plankton-feeders, and so, it is assumed that they were contaminated by ingesting PSP-producing toxic dinoflagellates.

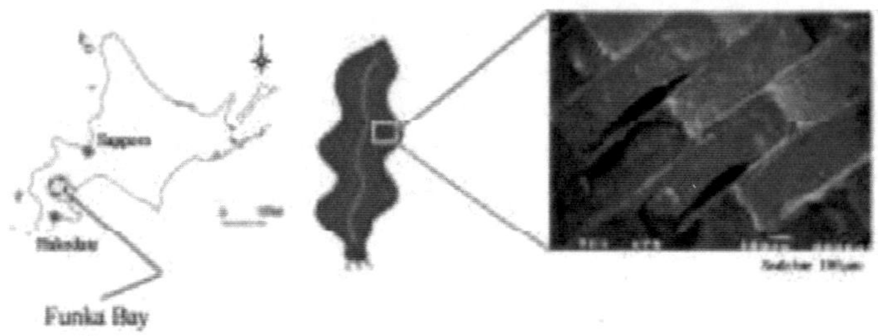

Figure 6. Scanning electron microscope image of mossworm on the brown alga *Laminaria japonica* ("Kombu" in Japanese).

3. Accumulation of PSP by Non Filter-Feeding Vectors

3.1 Echinoderms

Case study of the starfish from Hiroshima Bay, Hiroshima Prefecture, Japan

During surveillance on the toxicity of invertebrates inhabiting the coasts of Hiroshima Bay, the starfish *Asterias amurensis* was found to contain toxins that had strong paralytic action in mice. Figure 7 shows the results from the parallel monitoring on toxicity of this starfish and bivalves. Although this starfish was not found to be toxic in March to April, its toxicity increased unexpectedly n May, concurrent with an increase in the toxicity of the bivalves. This toxicity appeared to be almost exclusively present in the viscera and this toxin was unexpectedly identified as PSP by HPLC-FLD analysis [23]. Ito *et al.* also reported the occurrence of PSP in the starfish *Asterina pectinifera* collected from Kure Bay, Hiroshima Prefecture [24].

3.2 Gastropods

Case study of carnivorous gastropods from Kure Bay, Hiroshima Prefecture, Japan

Figure 8 shows a comparison of the toxicity of a gastropod (the rapa whelk, Rapana venosa) and a bivalve (the short-necked clam), along with cell densities of A. tamarense, in the Nikoh River estuary near Hiroshima Bay [25]. The period during which this carnivorous gastropod had paralytic toxicity almost completely coincided with the spring to early summer season when the toxic plankton *A. tamarense* flourished and plankton-feeding bivalves such as the short-necked clam became infested with PSP, albeit with some time lag. An analysis of the anatomical distribution of toxicity in this gastropod showed that the viscera were significantly toxic, reaching a score of 224 MU/viscera (4.2 MU/g viscera) in one specimen on May 21.

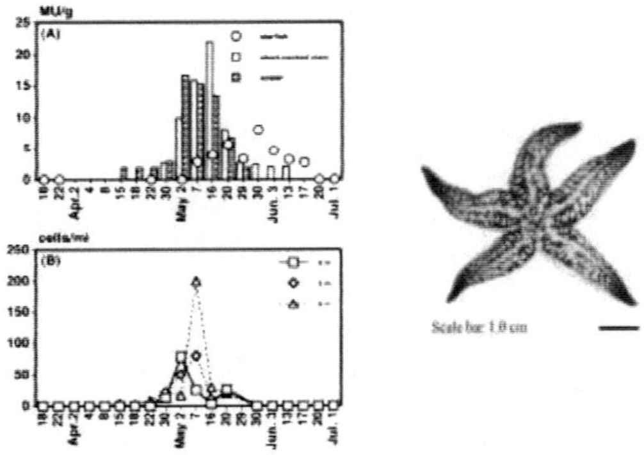

Figure 7. Starfish *Asterias amurensis* (right) and toxicities of starfish, short-necked clams and oysters in Hiroshima Bay in 1996, along with maximum cells density of *A. tamarense* (B) in each layer.

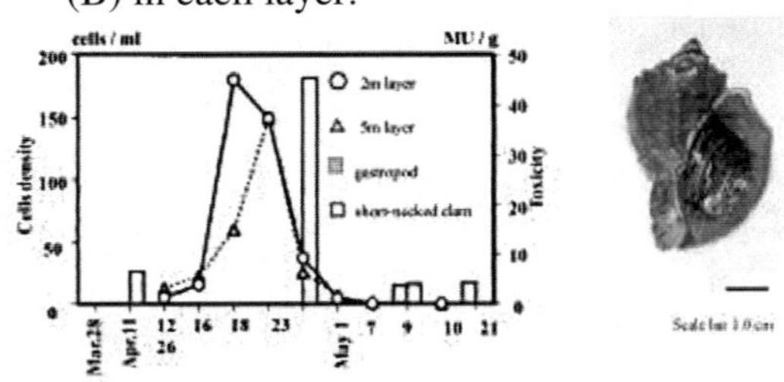

(A) (B)

Figure 8. Non-filter feeding (non-traditional) vectors of PSP (1). (A) Carnivorous species (gastropod): rapa whelk *Rapana venosa* PSP was detected in the viscera. (B) Toxicity of *R. venosa* and short-necked clam along with cells density of *A. tamarense*, in the estuary of Nikoh River in 2001.

Case study of herbivorous gastropods imported from Spain

Ormer are of great commercial value, and widely cultivated in Japan and other countries. Paralytic toxicity in excess of the quarantine limit of 4 MU/g as PSP was detected in ormer, that had been imported from Vigo, Spain, to Japan, between January and April, 1994 (Figure 9) [26, 27]. An analysis of the anatomical distribution of PSP toxicity in 10 samples showed that the muscle was the most toxic tissue in all specimens (31.9MU/g), while the viscera were less toxic (7.1MU/g). HPLCFLD analysis showed that dcSTX was the main component in all tissues, accounting for between 83 mol% (muscle) and 97 mol% (digestive gland) of all components. These results demonstrated that herbivorous marine gastropods had also accumulated PSP, and suggested that ormer possess a unique metabolic pathway for PSP. The exact source of these toxins in ormer remains unknown. Paralytic shellfish poisoning has also been detected in Haliotis tuberculata from the Galician coast, Spain, the geographical distribution of which has been investigated, along with the relationship between toxicity and the different lengths and parts of the molluscs [28].

Sample number	Toxicity (MU/g)	
	Viscera	Muscle
1	22	6.1
2	13	24
3	5.3	26
4	5	60
[illegible]	[illegible]	[illegible]
6	4.8	30
7	4.2	25
8	4.2	21
9	3.7	57
10	3.7	34
Av. ± Sd.	7.1 ± 5.9	31.9 ± 16.2

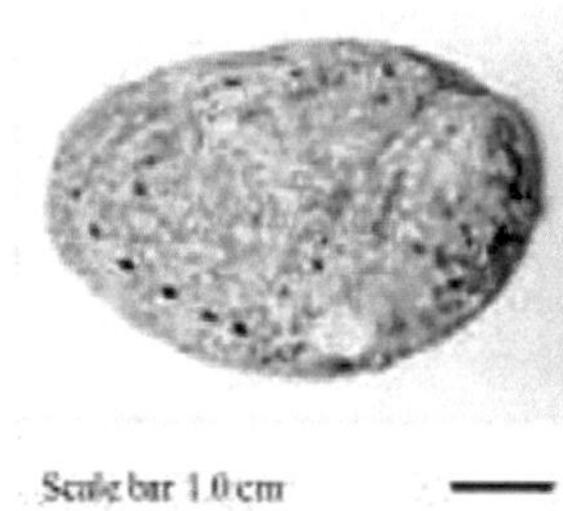

(A) (B)

Figure 9. Non-filter feeding (non-traditional) vectors of PSP (2). (A) Anatomical distribution of toxicity (B) Herbivorous species (gastropod): ormer *Haliotis tuberculata* (imported from Spain to Japan) PSP was detected in muscle and viscera.

3.3 Crustaceans

Case study of xanthid crabs from Ishigaki Island, Okinawa Prefecture, Japan, and from Camotes Island, Visayas, the Philippines

Crabs are valued as a popular seafood in many parts of the world, and are widely consumed in many ways such as by boiling, steaming, or in processed foods. While most species of crabs are edible, some are toxic to humans and other mammals.

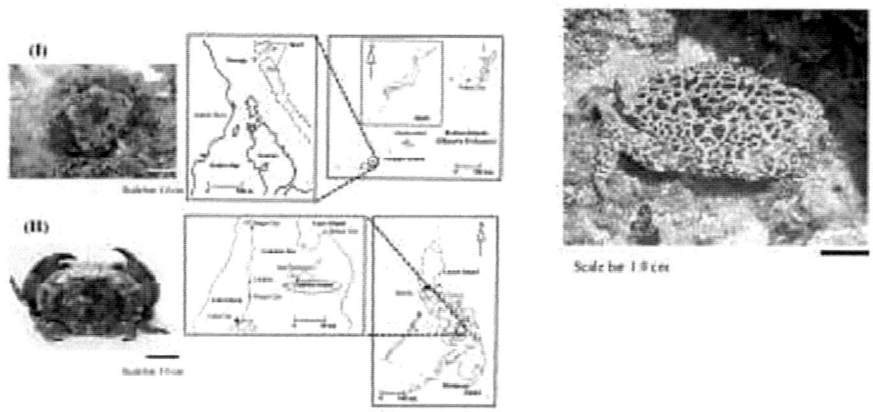

Figure 10. Highly toxic xanthid crab *Atergatis floridus* (left) from Ishigaki Is. (I), Camotes Is. (II), and *Zosimus aeneus* (right) from Ishigaki Is. Ishigaki specimen contains PSP. (A), (B): locations of sampling in Ishigaki Island. Camotes specimen contains PSP as the major components and a small amount of TTX (tetrodotoxin).

Xanthid crabs inhabiting the tropical and subtropical islands of the Pacific, possess high levels of PSP [29-33], and several cases of human poisoning and fatalities have been reported as a result of the ingesting coral reef crabs. For example, in the Southwestern Islands of Japan, at least 29 people were poisoned and at least 15 died due to the ingestion of xanthid crabs between1909 and 1988. Figure 10 shows specimens of the highly toxic xanthid crab Atergatis floridus from Ishigaki Island, Okinawa Prefecture, Japan and Camotes Island, the Philippines. The muscles of the appendages of these specimens were highly toxic, with a maximum toxicity of 4,641±972MU/g as PSP being recorded. The PSP toxicity scores in each of the tissues of Ishigaki specimens were as follows: carapace, 183 ± 47 to 807 ± 693 MU/g, viscera, 64 ± 41 to 654 ± 137 MU/g, appendages, 88 ± 40 to 1257 ± 607 MU/g and their muscle, 1,408 ± 404 to 4641 ± 972 MU/g (mean ± SD). By contrast, all of the Camotes Island specimens exhibited high toxicities in all of their tissues as follows: viscera, 105 ± 56 MU/g, appendages 221 ± 189 MU/g and muscle, 719 ± 349MU/g (mean ± SD).

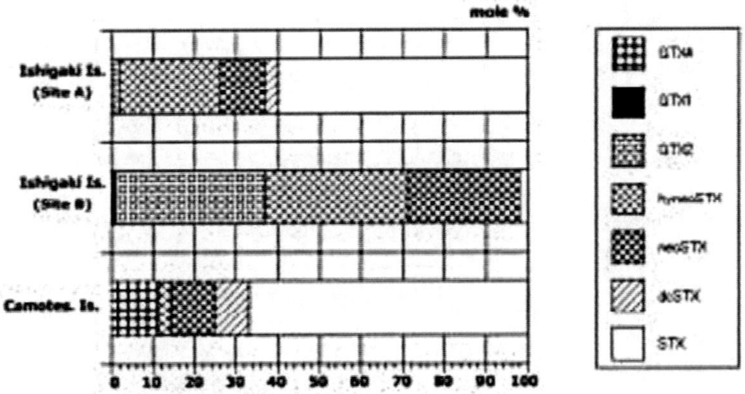

Figure 11. Comparison with the toxin profiles of *A. floridus* collected on reefs of Ishigaki Island and Camotes Island, Cebu Province, Philippines.

The toxin profiles of the viscera of A. floridus from sites A and B in June, 2007 are illustrated in Figure 11 [34]. The semi-purified toxins from the viscera of the Camotes Island samples contained rather high relative abundances (mol%) of neoSTX, dcSTX and STX were rather high (89%), along with smaller amounts of GTX4(11%) and hyneoSTX (3%). These PSP compositions were almost identical to those of the Ishigaki specimens, but the HPLC-FLD analysis also detected a trace amount of TTX (data not shown). The highly toxic PSP-bearing xanthid crab Zosimus aeneus had the highest toxicity score at 1,777 MU/g in the muscle (Figure 10). The toxicity levels of xanthid crab species are generally high, with previous reports of up to 16,500 MU/g. Since the minimum lethal dose (MLD) of PSP in humans is estimated to be 3,000 MU, as little as one gram of such a toxic crab specimen would be sufficient to kill at least five people.

4. Conclusion

The transfer of PSP to various marine organisms via the food web is shown in Figure 12. PSP is produced by toxic dinoflagellates and then transferred and accumulated throughout aquatic food webs. The occurrence of several new non-filter-feeding vectors of PSP and the secondary intoxication of edible gastropods are huge problems from a food hygiene and fisheries perspective, and so legislation should be adjusted to extend the monitoring of marine biotoxins to a wider range of species besides commercially important edible bivalves. Among non-filter feeding, non-molluscan species, the STX ngroup has been found most commonly in xanthid crabs. The origin of PSP in these toxic crabs remains unclear, as known PSP-producing toxic dinoflagellates species have not been detected in areas that are inhabited by toxic crabs. Consequently, further studies are currently in progress to elucidate the mechanisms involved.

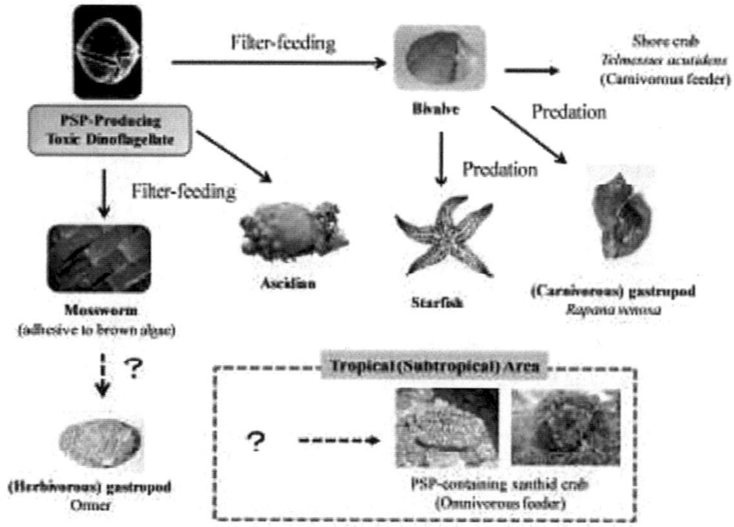

Figure 12. Transfer of PSP to various marine organisms via food web

Chapter II. Tetrodotoxin (TTX) and TTX-Bearing Organisms

1. Introduction

The pufferfish toxin, TTX is one of the most potent neurotoxins. This low molecular weight toxin (319) was first isolated by Yokoo in 1950 as a crystalline prism from a toxic pufferfish [35]. The most characteristic symptom caused by TTX is paralysis with respiratory paralysis occurring in severe cases, often resulting in the victim's death. In Japan, pufferfish have been a traditional food for many years, and so cases of TTX poisoning are frequent occurring on a regular basis not only in Japan but also in other parts of Asia, and sporadically resulting in severe poisoning or even death. Consequently, TTX poses a serious hazard to public health in several Asian countries, with more than 100 deaths per year being recorded until 1960 in Japan.

Figure 13 shows the structure of TTX and its many typical derivatives, as well as their specific toxicities, which vary greatly [36]. TTX is a heterocyclic guanide, and act as a powerful and specific sodium channel blocker. When ingested by humans, it blocks the sodium channels in nerve cells and skeletal muscles, resulting in typical symptoms such as respiratory paralysis and even death in severe cases. The lethal potency of TTX is 5,000 MU/mg, where 1 MU (mouse unit) is defined as the amount of toxin required to kill a 20g male mouse within 30 min after intraperitoneal administration, and the minimum lethal dose (MLD) for humans is estimated to be approximately 10,000 MU, which is equivalent to 2 mg of pure TTX crystals [37]. TTX was long believed to be present only in the pufferfish. Mosher et al. [38] detected TTX in California newt Taricha torosa, and since then, it has been detected in several other taxa.

IOP Conf. Series: Materials Science and Engineering **193** (2017) 012001 doi:10.1088/1757-899X/193/1/012001

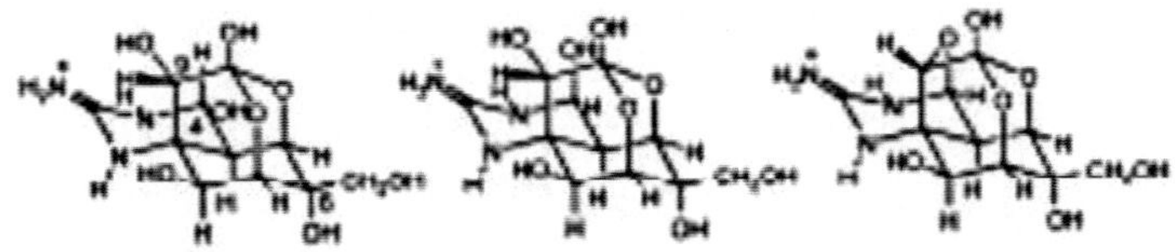

TTX [M+H]$^+$ = m/z 320 | 4-epiTTX [M+H]$^+$ = m/z 320 | 4,9-anhydroTTX [M+H]$^+$ = m/z 302

TTX	4epi-TTX	4,9-anhydroTTX
[M+H]$^+$ = 320	[M+H]$^+$ = 320	[M+H]$^+$ = 302
5000 MU/mg	710 MU/mg	92 MU/mg

Figure 13. Structure of tetrodotoxin derivatives and their specific

2. Nemertines

2.1. Toxicity of ribbon worm Cephalothrix simula

During surveillance of the toxicity of various marine fouling organisms in Hiroshima Bay, specimens of the ribbon worm Cephalothrix simula that were adhering to the cultured oyster *Crassostrea gigas* hanging onto floating culture rafts were found to contain toxins that exhibited strong paralytic action in mice [39, 40]. The i.p. injection of a 0.1% acetic acid (AcOH) extract of the worms induced restlessness, paralysis in the hind limbs, and a wobbling gait in the mice, followed by gasping, jumping, and death. Therefore, here, we examined the toxicity of these worms and attempted to identify the toxic component they contain. We carefully collected 615 ribbon worms that were adhering to cultured oysters harvested from floating rafts using tweezers (Figure 14). Based on histological studies, these specimens were identified as Cephalothrix simula by Dr.H. Kajihara, Hokkaido University [41]. We then examined the toxicity of each specimen in details.

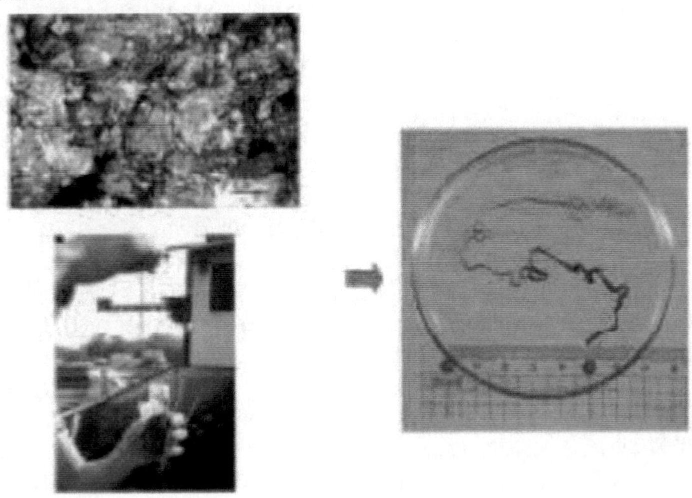

Figure 14. Ribbon worm *Cephalothrix simula* adherent to shells of oysters from Hiroshima Bay.

The toxicity and weight of each ribbon worm are shown in Figure 15. All of the specimens assayed were found to be toxic throughout the season covered, with toxicity scores ranging from 169 to 25,593 MU/g. The maximum to minimum ratio was as high as 151, but the ratio of strongly toxic (≧1,000 MU/g) to total specimens was 81%. The ratio of extremely toxic (> 2,000 MU/g) to total specimens was 53%. By contrast, the ratio of moderately toxic specimens between 100 and 999 MU/g was 100%. There was no obvious relationship between size and toxicity, as comparable sized specimens that had been collected from the same place at the same time exhibited wide variations in toxicity. The highest toxicity detected was 25,590 MU/g in a specimen collected on June 25(1999). The total toxicity of this specimen was calculated to be approximately 5,631 MU, which is equivalent to about half of the MLD of TTX in humans, 10,000 MU.

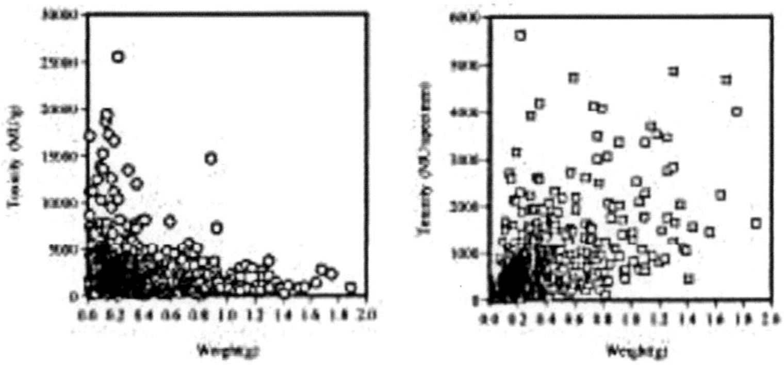

(A) (B)

Figure 15. Toxicity of ribbon worm *Cephalothrix simula* from Hiroshima Bay (1998-2005). (A) Relationship between toxicity (MU/g) and body weight (g) (B) Relationship between toxicity (MU/specimen) and body weight (g).

Although these worms were highly toxic, no paralytic toxicity was detected in the shucked meat of the oysters that were fouled with them. However, since the culturing of edible bivalves such as oysters and scallops is a flourishing industry in Japan, surveillance of the distribution of toxic ribbon worms in other areas besides Hiroshima Bay is urgently required.

2.2. Isolation of TTX as the main toxic component from Cephalothrix simula

The frozen ribbon worm specimens obtained in the above survey (390g) were semi-defrosted and their contents were extracted with three volumes of 1% AcOH in 80% methanol. The resulting supernatant, which had a toxicity of 2,897,000 MU, was then defatted by chloroform several times. The aqueous layer (2,750,000 MU) was applied to an activated charcoal column and the adsorbed toxin was eluted with 1% acetic acid in 20% ethanol. The eluate was then evaporated to dryness in vacuo, and the residue (total toxicity 2,433,000 MU; specific toxicity 99 MU/mg) was dissolved in a small amount of water and adjusted to pH5.5 with 1N NaOH. This solution was applied to a Bio-Gel P2 column (φ3.5 ×100cm). The column was washed with 3 L of water and then developed with 2 L of 0.03M AcOH. The toxicity was detected exclusively in the 0.03M AcOH fraction. Therefore, this fraction was concentrated to dryness under reduced pressure and the residue (3,300 MU/mg) was dissolved in a small volume of water. The solution was chromatographed on a Bio-Rex 70 column (H+ form, φ1.0 ×100cm) using a linear gradient of 0 to 0.03 M AcOH(Figure 16). The toxic fractions were monitored by mouse bioassay and HPLC-FLD analysis. The main toxic fractions (fr.85-100) were combined, and rechromatographed in the same manner. The toxic fractions thus obtained were then combined, freeze-dried, and then finally dissolved in 0.5 ml of 1% AcOH. Approximately 2.0 ml of methanol and 5.0 ml of diethyl ether were then added to these fractions, and the mixture was left to stand overnight in a refrigerator. During this time, plate-like crystals appeared, which were collected by decantation, and recrystallized. Bio-Gel P-2 column chromatography was found to be very effective for purifying this ribbon worm toxin, increasing the specific toxicity from 99 MU/mg to 3,300 MU/mg. Following recrystallization, the specific toxicity of this toxin was finally increased to 3,520 MU/mg. The combined homogenates had a toxicity of approximately 7,400 MU/g, from which approximately 25 mg of the stratified plate-like crystalline toxin was obtained.

A portion of the toxin was dissolved in a small amount of 1% AcOH, and electrospray ionization mass spectrometry (ESI-MS) was performed using a Hitachi M-1000 mass spectrometer using 50% methanol as the mobile phase with a flow rate of 50μl/min. The positive ion mass spectrum was then measured. The ESI mass spectrum showed an intense ion peak of (M+H)+ at *m/z* 320 and a weak ion peak of (M+H - H2O)+ at *m/z* 302. This mass spectrum agreed well with that of TTX. The molecular weight thus determined was 319, which is in accordance with the reported molecular weight of TTX. In the selected ion-monitored (SIM) mass chromatogram of the trimethylsilyl (TMS) derivatives of alkali-hydrolyzed ribbon worm toxins that were used for gas chromatography-mass spectrometry (GC MS) analysis, mass fragment ion peaks at *m/z* 376, 392 and 407, appeared at retention times of 8.33 and 8.34 min., which were almost identical to those from the TMS-C9 base derived from authentic TTX.`

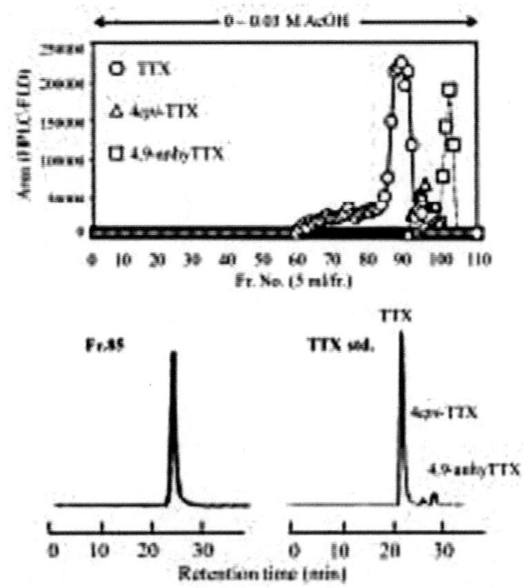

Figure 16. Elution profile of the ribbon worm *C. simula* toxin from a Bio-rex 70 (H^+) column with linear gradient from 0 to 0.03 M AcOH.

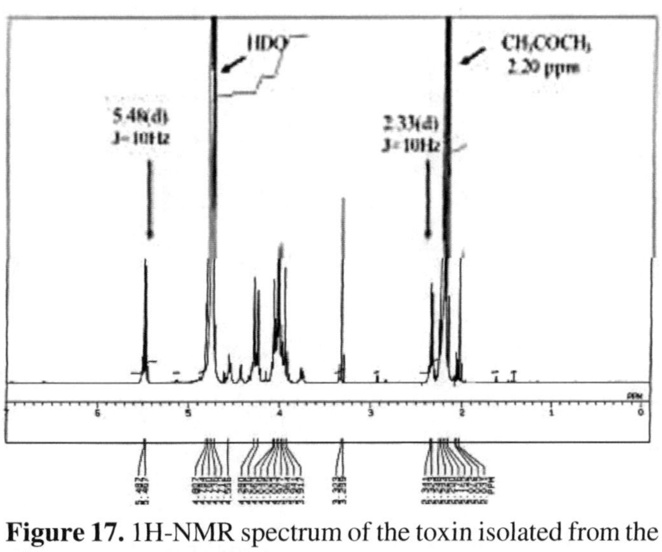

Figure 17. 1H-NMR spectrum of the toxin isolated from the ribbon worm *C. simula*

To examine the ^{1}H-NMR spectrum of the toxin isolated from ribbon worms, 5 mg of the crystals were dissolved in 0.5 ml of 1% CD3COOD in D2O, and placed in a test tube. The ^{1}H-NMR spectrum was then obtained with a 500 MHz JEOL JNM-500 spectrometer, using the methyl group proton of acetone as an internal standard. As can be seen in Figure 17, the 1H-NMR spectrum of HMT exhibited a singlet at 2.20ppm (CH3COCH3), a doublet centered at 2.33 ppm (J=10.0 Hz), a large proton peak at 4.76 ppm (HDO) and a doublet centered at 5.48 ppm (J=10.0Hz). The pair of doublets around 2.33 and 5.48 ppm, which are the hallmarks of TTX and are assigned to H-4a and H-4 respectively, were confirmed to be coupled with each other by double irradiation [42]. In comparison with 1H-NMR spectral data of the toxin isolated from ribbon worm and the structure of TTX, these results are in agreement with the corresponding data for TTX, with the signals at 4.24, 4.06, 4.28, 3.94, 4.00 and 4.02ppm are assigned to H-5, H-7, H-8, H-9 and H-11 respectively [43]. These findings demonstrate that TTX is a major component of the paralytic toxins contained in *C. simula.*

What is particularly interesting is how such a large amount of TTX could be concentrated in such a small and simple body with efficiency. Therefore, attempts to clarify this are currently in progress, results from which should be available in near future.

3. Cephalopods

Several species of octopuses secrete a venom from their posterior salivary glands that paralyzes prey organisms.

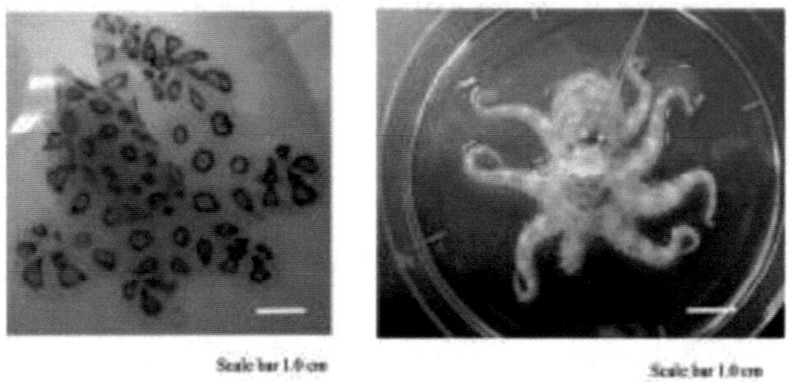

(A) (B)

Figure 18. Toxic octopus *Hapalochlaena lunulata* from Ishigaki Island, Okinawa Prefecture, Japan (2015). (A) greater blue-ringed octopus *H. lunulata* (B) posterior salivary gland

Table 2. Toxicity of greater blue-ringed octopus *Hapalochlaena lunulata* from Ishigaki Island, Okinawa Prefecture, Japan

Sample code	**N49-43-1**		**N49-43-1**	
Date Collection	Nov, 28, 2015			
Weight of Whole body (g)	9.63		5.24	
Organ	Weight (g)	Toxicity (MU/g)	Weight (g)	Toxicity (MU/g)
Posterior salivary glands	0.37	288.0	0.02	9276.0
Gonad	0.05	52.3	0.03	ND
Hepatopancreas	0.38	58.8	0.17	145.4
Body*	0.52	18.5	0.33	34.5
Arm	3.28	5.3	1.99	9.0

*Posterior salivary glands, gonad, and hepatopancreas are reserved

Bites by these octopuses cause not only pain in the area around the wound, but also occasionally fatal lesions in humans [44]. The blue-lined octopus *(Hapalochaena fasciata),* lesser blue-ringed octopus *(H. maculosa)* and greater blue-ringed octopus *(H. lunulata)* are particularly famous for their toxicity, all of which are distributed in the tropical to subtropical zone. Ishigaki Island, Okinawa

Prefecture, Japan is included in the subtropical zone, and both *H. fasciata* and *H. lunulata* are present in this region. However, information on the toxicity and toxins contained in these two species is limited. A greater blue-ringed octopus collected from Ishigaki Island in November, 2015 is shown in Figure 18. The toxicity of greater blue-ringed octopus *H. lunulata* from this region is provided in Table 2.

Selected ion-monitored liquid chromatography-mass spectrometry (LC-MS) chromatograms of toxin extracted from the posterior salivary glands in greater blue-ringed octopus showed that TTX, 4*epi*-TTX, 4,9-anhydroTTX and 6epi-TTX were present (data not shown).

4. Conclusion

Based on the findings of the present study and previous studies, the mechanism of TTX accumulation in marine animals is assumed and proposed. It is generally accepted that TTX is produced by some universal organisms such as microbes, based on the observation that it is distributed across a variety of invertebrates and vertebrates, and there are wide individual and regional variations in the content even within the same species.

TTX in toxic pufferfish is not endogenous. Therefore, it must either be derived from the food chain, or, alternatively, may be produced by symbiotic or parasitic bacteria that directly accumulated inside the pufferfish body and not obtained via the food chain (Figure 19).

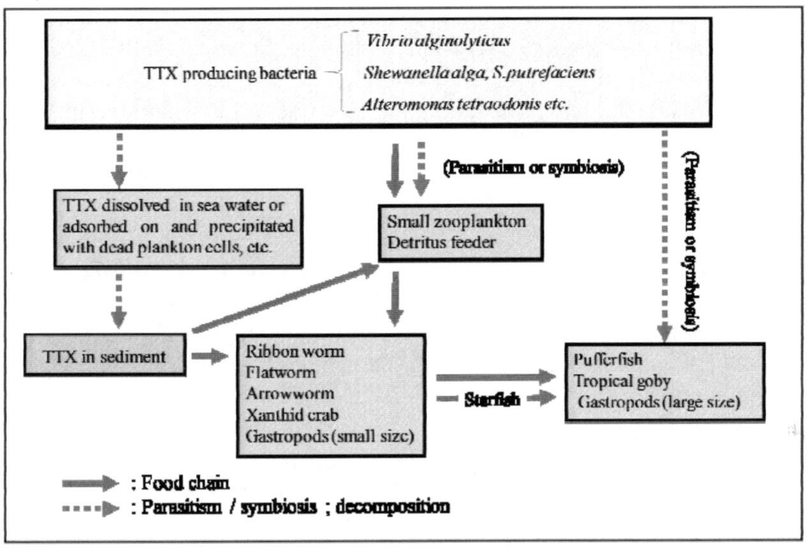

Figure 19. Proposed mechanism of TTX accumulation in marine animals.

References

[1] Asakawa M, Miyazawa K, Noguchi T 1993 *J. Food Hyg.Soc.Japan* **34** 50-54
[2] Azanza RV 1997 *Sci. Diliman* **9** (1&2) 1-6
[3] Gacutan RQ, Tabbu MY, Aujero EJ, Icatlo FJr 1985 *Mar. Biol.* **87** 223-227
[4] Oshima Y, Hasegawa M, Yasumoto T, Hallegracif G, Blackburn S 1987 *Toxicon* **25** 1105-1111
[5] Oshima Y, Blackburn SI, Hallegraeff GM 1993 *Mar. Biol.* **116** 471-476
[6] Asakawa M, Miyazawa K, Takayama H, Noguchi T 1995 *Toxicon* **33** 691-697
[7] Takatani T, Morita T, Anami A, Akaeda H, Kamijyo Y, Tsutsumi K, Noguchi T 1998 *J. Food Hyg.Soc.Japan* **39** 275-280
[8] Beppu R, Nojima K, Tsuruda S, Gomez DG, Quilantang MB, Taniyama S, Sagara T, Nishio S, Takayama H, Miyazawa K, Asakawa M 2008 *Mar. Pollu. Bull.* **56** 758-763
[9] Sagara T, Taniyama S, Yoshimatsu S, Takatani T, Hashimoto T, Nishibori N, Nishio S, Arakawa O 2010 *J.Food Hyg. Soc. Japan* **51** 170-177
[10] Corrales RA, Gomez ED 1990 *Red tide outbreaks and their management in the Philippines. Toxic Marine Phytoplankton* Graneli, E., Sundstrom. B., Edler, L., Anderson, D., eds., (New York: Elsevier)

[11] Bajarias, FFA, Relox JJr, Fukuyo Y 2006 *Coast. Mar. Sci.* **30** 104-106
[12] Noguchi T, Asakawa M, Arakawa O, Fukuyo Y, Nishio S, Tanno K, Hashimoto K 1990 *First occurrence of Alexandrium catenella in Funka Bay, Hokkaido, along with its unique toxin composition, Toxic Marine Phytoplankton,* Graneli E, Sundstrom B, Edler L, Anderson D, eds, (New York: Elsevier)
[13] Samsur M, Takatani T, Yamaguchi Y, Sagara T, Noguchi T, Arakawa O 2006 J.*Food Hyg. Soc. Japan* **48** 13-18
[14] Nishio S, Noguchi T, Onoue Y, Maruyama J, Hashimoto K, Seto H 1982 *Bull. Japan Soc. Sci. Fish* **48** 959-965
[15] Hashimoto K, Noguchi T 1989 *Pure & Appl.Chem.* **6** 7-18
[16] Asakawa M, Takayama H, Beppu R. Miyazawa K 2005 *J. Food Hyg.Soc.Japan* **46** 246-250
[17] Shimizu Y, Yoshioka M 1981 *Science* **212** 547-549
[18] Sullivan JJ, Iwaoka WT, Liston, J 1983 *Biochem. Biophys. Res. Commun.* **114** 465-472
[19] Asakawa M, Takagi M, Iida A, Oishi K 1987 Eiseikagaku **33** 50-55
[20] Oshima Y 1995 *Chemical and enzymatic transformation of paralytic shellfish toxins in marine organisms Harmful marine algal blooms* Lassus P, Arzul G, Denn EL, Gentien P, Marcaillou C eds Technique et Documentaion, (Pasir : Lavoisier)
[21] Murakami R, Yamamoto K, Noguchi T 1999 *J. Food Hyg. Soc. Japan* **40** 55-61
[22] Noguchi T, Asakawa M, Hashimoto K 1991 *Paralytic shellfish poisons in marine fouling organisms;* K Mise, JL Richard, eds. Emerging Food Safety Problem Resulting from Microbial Contamination (Tokyo: Kohken Press)
[23] Asakawa M, Nishimura F, Miyazawa K, Noguchi T 1997 *Toxicon* **35** 1081-1087
[24] Ito K, Asakawa M, Sida Y, Miyazawa K 2003 *Toxicon* **41** 291-295
[25] Ito K, Asakawa M, Reppu R, Takayama H, Miyazawa K 2004 *Mar. Poll. Bull.* **48** 1116-1121
[26] Nagashima Y, Noguchi T, Shomi K, Arakawa O, Asakawa M, Miyazawa K, Hashimoto K 1994 *J. Food Hyg.Soc.Japan* **35** 548-549
[27] Nagashima Y, Arakawa O, Shiomi K, Noguchi T 1995 *J. Food Hyg.Soc.Japan* **36** 627-631
[28] Bravo I, Reyero M.I, Cacho E, Franco JM 1999 *Toxicol* **46** 79-85
[29] Arakawa O, Noguchi T, Shida Y, Onoue Y 1994 *Toxicon* **32** 175-183
[30] Arakawa O, Nishio S, Noguchi T, Shida Y, Onoue Y 1995 *Toxicon* **33** 1577-1584
[31] Arakawa O, Noguchi T, Onoue Y 1995 *Fish Sci.* **61** 659-662
[32] Daigo K, Uzu A, Arakawa O, Noguchi T, Seto H, Hashimoto K 1985 *Bull. Japan Soc. Sci. Fish* **51** 309-313
[33] Sagara T, Taniyama S, Takatani T, Nishibori N, Nishio S, Noguchi T, Arakawa O 2009 *J.Food Hyg. Soc. Japan* **50** 237-242
[34] Asakawa M, Tsuruda S, Ishimoto Y, Shimomura M, Kishimoto K, Shida Y, Quilantang MB, Gomez-Delan G 2014 *Sci. J. Clinic. Med.* **3** 75-81
[35] Yokoo A 1950 *Nippon Kagaku Zasshi* **71** 590-592
[36] Nakamura M, Yasumoto T 1985 *Toxicon* **23** 271-276
[37] Noguchi T, Ebesu JSM 2001 *J.Toxicol.-Toxin Rev.* **20** 1-10
[38] Mosher H S, Fuhrman GJ, Fuhrman FA, Fischer HG 1965 *Science* **144** 1100-1110
[39] Asakawa M, Toyoshima T, Shida Y, Noguchi T, Miyazawa K 2000 *Toxicon* **38** 763-773
[40] Asakawa, M, Toyoshima, T, Ito, K, Bessho, K, Yamaguchi, C, Tsunetsugu, S, Shida, Y, Kajihara, H, S.F.Mawatari, Noguchi, T, Miyazawa, K 2003 *Toxicon* **41** 747-733
[41] Kajihara H, Sun SC, Chernyshev AV, Chen HX, Ito K, Asakawa M, Maslakova SA, Norenburg JL, Strand M, Subdberg Per, Iwata F 2013 *Zool. Sci.* **30** 985-997
[42] Sheumack DD, Howden MEH 1978 *Science* **199** 188-189
[43] Mahmud Y, Yamamori K, Noguchi T 1999 *J.Food Hyg. Soc. Japan* **40** 391-395
[44] Flecker H, Cotton BC 1955 *Med. J. Aust.* **2** 329-331

Identification of pork contamination in meatball using genetic marker mitochondrial DNA cytochrome b gene by duplex-PCR

E Novianty[1], L R Kartikasari[1], J H Lee[2], M Cahyadi[1*]
[1]Department of Animal Science, Faculty of Agriculture, Universitas Sebelas Maret, Surakarta, Indonesia 57126
[2]Division of Animal and Dairy Science, Chungnam National University, Deajeon 34134, Korea

E-mail: mcahyadi@uns.ac.id

Abstract. Meat based food products have a big opportunity to mix and adulterated with other meats. Muslim communities are prohibited to consume pork-containing product or other pig derivatives in food. Therefore, the high sensitivity, fast, cheap and accurate approach is needed to detect pig contamination in raw meat and meat-processed product such as meatball. The aim of this study was to identify pork contamination in meatball using genetic marker of mitochondrial DNA cytochrome b gene by duplex-PCR. Samples were prepared and designed by following the proportions 0, 1, 5, 10, 25% of pork in meatballs, respectively. The DNA genome was extracted from meatballs and polymerase chain reaction (PCR) was performed using species specific primer to isolate mt-DNA cytochrome b gene. The results showed that the DNA genome was successfully isolated from pork, beef, and contaminated meatballs. Furthermore, 2% agarose gels was able to visualize of duplex-PCR to identify pork contamination in meatballs up to very small proportion (1%). It can be concluded that duplex-PCR of mt-DNA cytochrome b gene was very sensitive to identify pork contamination in meatball with the presence of specific 398 bp DNA band.

1. Introduction

Along with the growing of meat and dairy product consumptions, food safety supervision and processed products of animal origin also must be increased. Supervision is carried out to prevent food from possible contamination of biological, chemical or other objects that can harm consumers. Food adulteration is a common problem in the food industry [1]. The production of meat-based foods have a high susceptibility to counterfeiting, especially mixing with the meat of wild animals [2]. In Bali there are cases of machines use the same production between pork and beef sausage [3]. The activity is at risk of contamination of pork products processed sausages. Many meatball traders are using pork as a raw material or mixture in their manufactures [4]. Pork is the type of meat which is often mixed into other processed meat products. Pork has a relatively cheaper price, in addition to the color and shape similar to beef, so it is often used as a compound for lowering the price of production and improving the taste [5]. Pork mixing action against the processed products would be very detrimental to the public, this is because the majority of people in Indonesia are Muslims that forbidden to eat pork [6]. The components of food containing pork in food products can be identified through the fat, protein and deoxyribose nucleutida acid (DNA). The technology polymerase chain reaction (PCR) can be used to detect any species with amplification techniques specific DNA fragments [7]. Meat species detection methods such as PCR-based DNA is able to detect DNA that has been degraded by the heating process during food processing [8]. The use of PCR technology can detect pig DNA in samples of fresh meat and processed meat [9]. Previous study reported that the gene cytochrome b using PCR technology has a high sensitivity to detect pig DNA. The use of gene cytochrome b with duplex-PCR method was able to identify the contamination of pork in fresh and cooked beef until the level of 1% [10]. Cytochrome b gene is a DNA fragment that is specific. Some researchers reported that cytochrome b gene can be used as a marker of DNA types of meat in the identification of certain types of meat [11]. Animal DNA can also be detected using DNA sequences from the mitochondrial cytochrome b gene as a biomarker. Based on these descriptions mentioned above, duplex-PCR of mitochondrial cytochrome b gene should be done to detect contamination of pork in processed meat product such as meatball.

2. Methods

2.1 Samples Preparation

Sample preparation was conducting by mixing the beef and pork that has been flavored meatballs Meatball was made by following published method [12]. The sample was designed as follows 0, 1, 5, 10, 25, 100% of pork in beef meatball, respectively. Firstly, sample was grinded and mixed with other ingredients until homogenous. Then, meatballs were boiled at 100°C for 30 minutes. Samples meatballs were cooked and then placed in a 1.5 ml tube to be analyzed in the next step, namely the isolation of DNA.

2.2 Isolation of DNA

The DNA genome was isolated from meatballs by following *Quick*-DNA™ Universal Kit (Zymo Research). A total of 25 mg meatball was weighed and transferred to a 1.5 ml microcentrifuge tube, and a solution of 95 µl aquabidest (ddH_2O), 95 µl Solid tissue buffer (blue) and 10 µl proteinase K were added. The mixture was vortexed and incubated at 55°C until tissue dissolved. Insoluble debris should be removed by centrifugation at 14.000 x g for 1 minute. Furthermore, 200 µl aqueous supernatant was transferred to a clean tube and 400 µl genomic binding buffer was added to the supernatant and then it was mixed thoroughly. The mixture was transfered to a Zymo-Spin™ IIC-XL Column in a collection tube and it was centrifuged at 14.000 x g for 1 minute and then the flow through in the collection tube was discarded. Next, 400 µl DNA pre-wash buffer was added to the column in a new collection tube and it was centrifuged for 1 minute and then the flow through in the collection tube was discarded. Furthermore, 700 µl g-DNA wash buffer was added and centrifuged for 1 minute and then the flow through in the collection tube was discarded. In addition, 200 µl g-DNA wash buffer was added and centrifuged for 1 minute, and then the flow through in the collection tube was discarded. The mixture was transferred to a clean microcentrifuge tube. To elute the DNA, 50 µl DNA elution buffer was added and incubated for 5 minutes, and then it was centrifuged for 1 minute. Then, the DNA genome was visualized by 1% electrophoresis agarose gel using gel document (Vilber Lourmat Infinity 1100126M, France) and it was ready to be used for next analysis.

2.3 Duplex-PCR

The PCR amplification was conducted in total volume of 25 µL containing 12,5 µl 2X KAPA2G Fast Multiplex Mix (Kapa Biosystems, Inc., United States), 1 µl each primer (10 µM), 1 µl DNA template, and 8,5 µl ddH_2O. Three primers which are designed by Matsunaga were used in this study to isolate DNA target [13]. Those primers were universal forward primer (5'GACCTCCCAGCTCCATCAAACATCTCATCT TGATGAAA-3') and reverse primers for bovine (5'-CTAGAAAAGTGTAAGACCCGTAATATAA G3'), and reverse primer for pig (5'-GCTGATAGTAGATTTGTGATGACCGTA-3'). The PCR reaction was carried out using GeneAmp® PCR System 9700 (Singapore) as follows: initial denaturation at 95ºC for 3 minutes, and followed by 30 cycles of denaturation at 95ºC for 15 seconds, annealing at 60ºC for 30 seconds, extension at 72ºC for 30 seconds. To make sure that PCR reaction completed, final extension was applied at 72ºC for 3 minute. The *duplex*-PCR results were viewed in the agarose gel 2% under the UV light using Gel Document (Vilber Lourmat Infinity 1100126M, France).

3. Results and Discussion

3.1 The DNA Isolation

Pork and beef samples were bought from traditional market in Surakarta, Central Java. A total of 6 samples was designed and used for DNA isolation in this study. The pork contamination in the meatballs samples was designed as much as 0, 1, 5, 10 and 25%. The result showed that DNA genome was successfully isolated from meatballs samples. The principle of genome isolation is isolated the DNA genome from other components of the cells [14]. The DNA genome still could be isolated from meat processed product such as meatball. Previous studies reported that DNA genome can be isolated from cooked meats such as sausages, cold cut and leisure meat food [15].

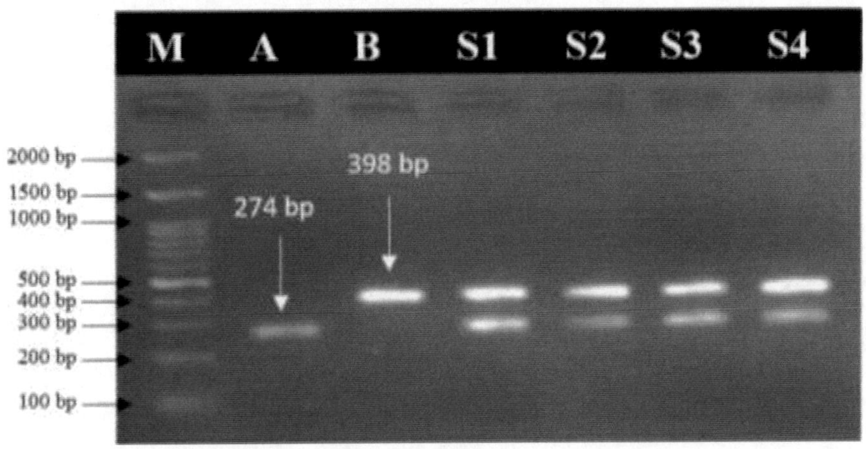

Figure 1. Agarose gel electrophoresis of PCR products amplified from beef meatball and pork DNA mixture. M is molecular marker, A is comparison between beef and pork in the mixture (100% beef : 0% pork); B is 0% beef : 100% pork; S1 is 75% beef : 25% pork; S2 is 90% beef : 10% pork; S3 is 95% beef : 5% pork; S4 is 99% beef :1% pork, respectively.

3.2 Detection of Pork Contamination in Meatballs

In figure 1. showed that M is a molecular marker, A is 100% beef meatball, B is 100% pork meatball, S1 is 75% beef meatball mixed with 25% pork, S2 is 90% beef meatball mixed with 10% pork, S3 is 95% beef meatball mixed with 5% pork and S4 is 99% beef meatball mixed with 1% pork. The existence of bovine in the sample mixtures was indicated by 274 bp of mitochondrial DNA Cytochrome b, on the other hand, the exsistence of pig species in the meat mixture was indicated by appearance of 398 bp of specific mitochondrial DNA *Cytochrome b* site for porcine. This study succesfully identified of pork contamination in meatball. It means the ingredients in meatballs sample did not influence the DNA isolation. The contamination of pork in beef products can be detected until 1% contamination of pork. Additionally, our current results proved that multiplex PCR able to identify pork and chicken in meatballs commercially marketed in Surakarta (*unpublished data*). Previous study also revealed that species contamination can be detected until 8% in the mixtures by multiplex PCR [14]. In addition, multiplex PCR assays have successfully detected existence of five species forbidden in Islamic foods (cat, dog, pig, monkey and rat meats) by designing species-specific primers of mitochondrial ND5, ATPase 6 and *Cytochrome b* genes to amplify 172, 163, 141, 129 and 108 bp DNA fragments from those species, respectively [5]. The existence of pork in beef meatballs was also detected by PCR-RFLP method in the meatball shop around Yogyakarta [16]. The result showed that *Cytochrome-b* gene using PCR method had a high sensitivity to detect pork DNA. In Moslem countries, those assays have currently been applied to verify and to authenticate whether a food product is halal or haram. Standardization and certification of halal products can use laboratory analysis such as DNA-based assays which are very sensitive to achieve appropriate assurance of halal food products and proper labelling for consumers and buyers [10].

4. Conclusion

Duplex-PCR of mt-DNA *cytochrome b* gene successfully identified the pork contamination until the level of 1% in the meatball. It may be applied to authenticate commercial meatball marketed in Indonesia.

Aknowledgment

This study was supported by the research grant "Pusat Keunggulan-Universitas Sebelas Maret (PK-UNS) project number: 632/UN27.21/LT/2016, Indonesia.

References

[1] Yuningsih R 2010 *Aspirasi* **1** 173-202
[2] Fajardo V, Gonzales I, Rojas M, Garcia T, and Martin R 2010 *Trends Food Sci. Tech.* **21** 408-421
[3] Usman 1997 *J. Halal* **17** 8-14
[4] Nurbowo 1995 *J. Halal* **6** 13-17
[5] Ali M E, Hashim U, Mustafa S, Che Man Y B and Foo K L 2012 *J. Nanomater.* **2012** 1-7
[6] Ong S B, Zuraini M I, Jurin W G, Cheah Y K, Tunung R, Chai L C, Haryani Y, Ghazali F M and Son R 2007 *ASEAN Food J.* **14** 51-59
[7] Fajardo V, Gonzalez I, Martin I, Rojas M, Hernandez P E and Garcia T 2008 *J. AOACI* **91** 103-111
[8] Aida A A, Che Man Y B, Wong C M V L, Raha A R and Son R 2005 *Meat Sci.* **69** 47-52
[9] Tanabe S, Miyauchi E, Muneshie A and Mio K 2007 *Biosci. Biotech. Biochem.* **71** 1663-1667
[10] Ni'mah A, Kartikasari Y, Pratama A D, Kartikasari L R, Hertanto B S dan Cahyadi M 2016 *J. Indonesian Trop. Anim. Agr.* **41** 7-12
[11] Hsieh H S, Tuu-Jyi C and Deng-Fwu H 2007 *Food Contr.* **18** 369-374
[12] Chakim L, Dwiloka B dan Kusrahayu 2013 *Anim. Agr. J.* **2** 97-104
[13] Nuraini H, Primasari A, Andreas E dan Sumantri C 2012 *Media Pet.* **35** 15-20
[14] Muladno, Maryatni D dan Budiarti S 1999 *Media Pet.* **23** 14-17
[15] Hou B, Meng X, Zhang L, Guo J, Li S and Jin H 2015 *Meat Sci.* **101** 90-94
[16] Erwanto Y, Abidin M Z, Muslim E Y P, Sugiyono dan Rohman A 2014 *Asian-Australas J. Anim. Sci.* **27** 1487-1492

IOP Conf. Series: Materials Science and Engineering **193** (2017) 012003 doi:10.1088/1757-899X/193/1/012003

Development of Seaweed-based Biopolymers for Edible Films and Lectins

D Praseptiangga

Department of Food Science and Technology, Sebelas Maret University (UNS), Jl. Ir. Sutami 36 A Kentingan 57126, Surakarta, Indonesia

E-mail : dpraseptiangga@yahoo.com

Abstract. Marine macroalgae (seaweeds) as one of important groups of biopolymers play an important role in human life. Biopolymers have been studied regarding their film-forming properties to produce edible films intended as food packaging and active ingredient carriers. Edible film, a thin layer or which is an integral part of food and can be eaten together with, have been used to avoid food quality deterioration due to physico-chemical changes, texture changes, or chemical reactions. Film-forming materials can be utilized individually or as mixed composite blends. Proteins and polysaccharides used for their mechanical and structural properties, and hydrophobic substances (lipids, essential oils, and emulsifiers) to provide good moisture barrier properties. In addition, bioactive substances from marine natural products, including seaweeds, have been explored for being used in the fields of medicine, food science, pharmaceutical science, biochemistry, and glycobiology. Among them, lectins or carbohydrate-binding proteins from seaweeds have recently been remarked. Lectins (hemagglutinins) are widely distributed in nature and also good candidates in such prospecting of seaweeds. They are useful as convenient tools to discriminate differences in carbohydrate structures and reveal various biological activities through binding and interacting to carbohydrates, suggesting that they are promising candidates for medicinal and clinical application.

1. Introduction

Marine macroalgae, better known as seaweeds, are an important group of marine bio-resources with multiple uses that affect the life of humans [1]. Indonesia, a tropical country with 5.8 million km square of marine waters and 95.181 km of coastlines, is one of nation with the greatest marine biodiversity in the world and one of global players in seaweed production [2-4]. Indonesia is increasingly focusing on in-country processing of the raw materials [2], since an approximately 8.6 % algae of the whole species in the world are found in Indonesia [3,4]. Indonesian ocean has a large variety of biota, including 782 algae species [5]. The production of seaweed-based biopolymers such as agar, carrageenan, and alginate showed the huge economic impact through their application in various industries, including food industries that exploit their physical properties such as gelation, water holding capacity, and their ability to emulsify [1,6]. Industrial utilization of algae is largely confined to phycocolloids and fine biochemicals including carbohydrates, proteins, lipids, minerals, and low molecular weight compounds [1,6,7]. However, in recent years algae have attracted a great deal of interest in the search for bioactive compounds to develop new drugs and healthy foods. Among them, lectins or carbohydrate-binding proteins from algae have begun to receive much attention [1,7].

Moreover, the use of renewable resources, including seaweed-based biopolymers to produce biopolymer films and coatings has gained significance with the increasing of consumers concern on high-quality and long-shelf-life-products and their awareness of environmental issues. Petrochemical-based plastics packaging put a serious problems on the environment since the material's inability to biodegrade [8-12]. Therefore, research works related to the use of biopolymers has emerged as an alternative, due to their biodegradability. In the development of edible and biodegradable materials, seaweed-based biopolymers such as carrageenans and alginates are among the most important biopolymers due to their great abundance and processability [11,12]. Thus, the utilization of seaweed-based biopolymers for edible films and lectins in a perspective of food and health applications is studied in this current article.

2. Edible Films

Biopolymers have been studied regarding their film-forming properties to produce edible films and coatings intended as food packaging [8-10]. Edibility and biodegradability are the most beneficial properties of edible films and coatings. Edibility of films and coatings could be achieved if films and coatings components including biopolymers, plasticizers, and other additives be food grade ingredients. Meanwhile, all the processes and equipment should be acceptable for food processing [13,14]. Further, the toxicity and environmental safety must be evaluated by standard analytical procedures to claim biodegradability of films and coatings [13,14].

Edible film, a thin layer made of edible material which can be placed on or between food components or which is an integral part of food and can be eaten together with [8,10,13], while an edible coating is a thin layer of edible material formed as a coating on a food product [13,15] The main difference between these two food systems is that the edible coating is applied in liquid form on the food, usually by immersing the product in the solution of edible material, and edible film is first molded as solid sheets, then applied as a wrapping for food products [13,16]. Edible films and coatings have been used to avoid food quality deterioration due to physico-chemical changes, texture changes, or chemical reactions. Moreover, the protective barrier of edible film can be formulated to prevent the transfer of moisture, gases, flavor, and therefore to maintain or improve food quality and to increase shelf life of food products [8,17]. Different biopolymers have been used in the development of edible film and coatings, such as proteins, lipids, and polysaccharides [8,10,17]. Among these biopolymers, seaweed-based biopolymers such as alginates and carrageenans have been frequently used in recent years due to their good barrier properties to oxygen, carbon dioxide, and lipids as well as their great mechanical properties (tensile strength and elongation at break). Various edible films and coatings from alginates and carrageenans have been reviewed, including productions and properties of these films [13].

Film-forming materials can be utilized individually or as mixed composite blends. Proteins and polysaccharides used for their mechanical and structural properties, and hydrophobic substances (lipids, essential oils, and emulsifiers) to provide good moisture barrier properties [8,13]. Hydrophilic or hydrophobic materials can be used in film-forming materials, but in order to maintain edibility, merely water or ethanol could be used as solvent during processing. Other components can also be added into the materials matrix to enhance its functionality, including those that can improve or modify the basic functionality of the material (plasticizers, crosslinking agents, reinforcements, and emulsifiers) and those that intend to improve the quality, stability, and safety of packaged foods (antioxidants, antimicrobial compounds, nutraceuticals, flavors, and/or color agents) [18]. Edible films and coatings showed promising systems to be employed as active ingredient carriers, however, the mechanisms of deterioration of each food and the mode of action of each package should be understood and related in order to asses a successful application of these materials as food packaging [18].

3. Lectins

Lectins are naturally occurring proteins and glycoproteins of non-immune origin which selectively bind non-covalently to carbohydrate residues and able to agglutinate cells and/or precipitates polysaccharides or glycoconjugates. They play some important roles as recognition molecules in cell-cell or cell-matrix interactions and may be regarded as valuable biochemical tools to probe protein-carbohydrate interactions since they can interpret the 'sugar code' [1,19]. They are not only useful in many research fields including glycomics, but they are promising candidates for medicinal and clinical application [1,19]. The ability to bind to carbohydrates and hence agglutinate cells is the most important property of lectins. Agglutination is due to the fact that the lectin has at least two binding sites and it is therefore able to crosslink cells through its interaction with carbohydrates on the cell membrane [20]. In general, the molecular structures and carbohydrate-binding specificity of lectins are diverse and dependent on the organisms from which they originate, although lectins are classified into several families based on amino acid sequences of carbohydrate-recognition domains, some of which are also evolutionarily conserved [1, 21-24].

Lectins occur in all classes and families of organisms, including in algae (seaweed). Recently, the presence of lectins was analyzed at about 800 algae species [1,25]. In comparison to other lectins, some characteristics patterns of algal lectins are becoming apparent. For example, many algal lectins have some common characteristics of low-molecular weight that are much smaller than those derived from land plants [26,27]. This feature may make algal lectins more suitable in drug targeting and/or drug delivery system because the smaller molecules may be expected to be less antigenic than the larger land plant lectins. Before such uses can be envisaged, however, further information is required on the precise specificity of the small glycoprotein-binding algal lectins and their primary structure [28,29]. Furthermore, lectins showing anti-HIV and anti-influenza virus activity have been recently discovered from cyanobacteria, bacteria and eukaryotic macroalgae. These lectins are promised to be a new antiviral agent because their inhibiting activities are extremely strong compared to those of lectins from other biological groups [30]. A red alga *Griffithsia sp.* [31] as well as the cyanobacteria (blue-green algae) such as *Nostoc ellipsosporum* [32], *Schytonema varium* [33] and *Microcystis viridis* [34], OAA from *Oscillatoria agardhii* NIES-240 [29,35,36], PFL from *Pseudomonas fluorescens* Pf0-1 [37], MBHA from *Myxococcus xanthus* [38], BOA from *Burkholderia oklahomensis* [39], KAA-2 from *Kappaphycus alvarezii* [40], BCA from *Boodlea coacta* [41] and ESA-2 from *Eucheuma serra* [42] are examples of this lectin family [30]. All of these lectins share binding specificity for high-mannose N-glycans and have commonly two or four tandem repeats consisting of highly conserved sequences, which may lead to subtle difference in the degree of inhibiting activities, suggesting that this lectin family may become a novel class of antiviral compounds for application [30]. Moreover, EDA from *Eucheuma denticulatum* [43] and KSA-2 from *Kappahycus striatum* [30] showing antibacterial activities, suggesting that algal lectins could be useful as antibacterial reagents and they also showed antiviral activities that might be indicate to become a candidate for a functional food ingredients that can prevent virus infection in the future. Thus, marine algae should be expected as sources of novel lectin molecules for basic research and future application in the fields of medicine, food science, pharmaceutical science, biochemistry, and glycobiology.

4. Conclusion

Edible films and coatings have been received considerable attention over the last years due to their possibility to use as edible packaging materials over synthetic ones and to reduce environmental pollution. They are promising systems to be used as active ingredient carries, since such films can be carrier of antioxidants, antimicrobials, nutraceuticals, and flavoring agents or other additives to improve the mechanical integrity, handling, and quality of food products. Among biopolymers, seaweed-based biopolymers (alginates and carrageenans) have been paid a great of interest in recent years due to their good barrier properties as well as mechanical properties. Furthermore, algal or seaweed-based industry has promising prospects to be more developed in Indonesia. Marine-derived nutraceuticals, including from marine algae (seaweeds), are alternative sources for ingredients that play an important role in human health and nutrition. Algal (seaweed) lectins, a potential algal-derivative product, could be more intensively studied, because there is no report on lectins from Indonesian marine algae (seaweeds) up to now. Hopefully, novel lectins from Indonesian marine algae with specificities distinct from other sources will be found and applied as convenient tools to contribute greatly to the era of glycomics-based biomarker investigations.

Acknowledgements

This work was supported by research project of Sebelas Maret University (PNBP UNS No. 632/UN27.21/LT/2016).

References

[1] Praseptiangga D 2015 *Squalen Bulletin of Marine and Fisheries Postharvest and Biotechnology* **10**(2) 89-98
[2] Klose J 2015 *Switzerland Global Enterprise* 1- 39
[3] Fajarningsih N D 2011 *Squalen Bulletin of Marine and Fisheries Postharvest and Biotechnology* **7**(2) 89-98
[4] Dahuri R 2003 *Keanekaragaman Hayati Laut, Aset Pembangunan Berkelanjutan Indonesia* (Jakarta: PT Gramedia Pustaka Utama)
[5] Moosa M K, Dahuri R, Hutomo M, Suwelo I S, Salim S 1996 *Indonesian Country Study on Integrated Coastal and Marine Biodiversity Management. Ministry of State for Environment Republic of Indonesia in Cooperation with Directorate for Nature Management Kingdom of Norway* (Jakarta)
[6] Khan S I, Satam S B 2003 *Aquaculture Asia* **VIII**(4) 26-29
[7] Smit A J 2004 *J. Appl. Phycol.* **16** 245-262
[8] Praseptiangga D, Fatmala N, Manuhara G J, Utami R, Khasanah L U 2016 *AIP Conf. Proc.* **1746** 020036
[9] Marsh K, Beguzu B 2007 *J. Food Sci.* **72**(3) 39-55
[10] Espitia P J P, Du W, Avena-Bustillos R J, Soares N F F, McHugh T 2014 *J. of Hydrocoll.* **35** 287-296
[11] Farhan A, Hani N M, 2017 *Food Hydrocolloids* **64** 48-58
[12] Manuhara G J, Praseptiangga D, Muhammad D R A, Maimuni B H 2016 *AIP Conf. Proc.***1710** 030043
[13] Tavassoli-Kafrani E, Shekarchizadeh H, Masoudpour-Behabadi M. 2016 *Carbohydrate Polymers* **137** 360–374
[14] Han J H, Aristippos G 2005 *Edible Films and Coatings: A Review in J H Han (Ed.), Innovations in food packaging* (San Diego: San Diego Academic Press)
[15] Kang H, Kim S, You Y, Lacroix M, Han J 2013 *LWT-Food Science and Technology* **51**(1) 393-6
[16] Falguera V, Quintero J P, Jiménez A, Mu˜noz J A, Ibarz A 2011 *Trends in Food Science and Technology* **22**(6) 292–303
[17] Krochta J M 1992 *Control of Mass Transfer in Food With Edible Coatings and Films. Advances in Food Engineering.* (Florida: CRC Press)
[18] Salgado P R, Ortiz C M, Musso Y S, Di Giorgio L, Mauri A N 2015 *Current Opinion in Food Science* **5** 86-92
[19] Sharon N, Lis H 2003 *Lectins 2nd editions* (Nertherlands: Kluwer)
[20] Hori K, Miyazawa K, Ito K 1986a *Bull. Jpn. Soc. Sci. Fish.* **52** 323-331
[21] Lis H, Sharon, N 1998 *Chem. Rev.* 637-674
[22] Dodd R G, Drickmer K 2001 *Glycobiology* **11** 71-79
[23] Praseptiangga D, Hirayama M, Hori K 2012 *Biosci. Biotechnol. Biochem.* **76**(4) 805-811

[24] Praseptiangga D 2012 *Comparative Biochemical Studies of Lectins from The Green Algal Genus Codium (Chlorophyceae, Codiaceae)* Thesis (Japan: Hiroshima University)
[25] Teixeira E H, Arruda F V S, Nascimento K S, Carneiro V A, Nagano C S, da Silva B R, Sampaio A H, Cavada B S 2012 *Carbohydrates - Comprehensive Studies on Glycobiology and Glycotechnology* 533-558
[26] Hori K, Miyazawa K, Ito K 1986 *Bot. Mar.* **19** 323-328
[27] Hori K, Miyazawa K, Ito K 1990 *Hydrobiologia* **204/205** 561-566
[28] Rogers D J, Hori K 1993 *Hydrobiologia* **260** 589-593
[29] Sato T, Hori K 2009 *Fish. Sci.* **75** 743-753
[30] Hung L D, Hirayama M, Ly B M, Hori K 2015 *Phytochemistry Letters* **14** 99–105
[31] Mori T, O'Keefe B R, Sowder R C II, Bringans S, Gardella R S, Berg S, Cochran P, Turpin J A, Buckheit R W Jr, McMahon J B, Boyd M R 2005 *J. Biol. Chem.* **28** 9345-9353
[32] Boyd M R, Gustafson K R, Mcmahon J B, Shoemaker R H, O'Keefe B R, Mori T, Gulakowski R J, Wu L, Rivera M I, Laurencot C M, Currens M J, Cardellina J H II, Buckheit R W Jr, Nara P L, Pannell L K, Sowder R C II, Henderson L E 1997 *Antimicrob. Agents Chemother.* **41** 1521–1530
[33] Bokesch H R, O'Keefe B R, McKee T C, Pannell L K, Patterson M L, Gardella R S, Sowder R C II, Turpin J, Watson K, Buckheit R W Jr 2003 *Biochemistry* **42** 2578–2584
[34] Williams D C Jr, Lee J Y, Cai M, Bewley C A, Clore G M 2005 *J. Biol. Chem.* **280** 29269–29276
[35] Sato Y, Okuyama S, Hori K 2007 *J. Biol. Chem.* **282** 11021-11029
[36] Koharudin L M I, Furey W, Gronenborn A M 2011 *J. Biol. Chem.* **286** 1588–1597
[37] Sato Y, Morimoto K, Kubo T, Yanagihara K, Seyama T 2012 *PLoS One* **7** e45922.
[38] Koharudin L M I, Kollipara S, Aiken C, Gronenborn A M 2012 *J. Biol. Chem.* **287** 33796–33811
[39] Whitley M J, Furey W, Kollipara S, Gronenborn A M *FEBS J.* **280** 2056–2067
[40] Sato Y, Morimoto K, Hirayama M, Hori K 2011a *Biochem. Biophys. Res. Commun* **405** 291–296
[41] Sato Y, Hirayama M, Morimoto K, Yamamoto N, Okuyama S, Hori K 2011b *J. Biol. Chem.* **286** 19446–19458
[42] Sato Y, Morimoto K, Kubo T, Sakaguchi T, Nishizono A, Hirayama M, Hori K 2015 *Mar. Drugs* **13** 3454–3465
[43] Hung L D, Hirayama M, Ly B M, Hori K 2015 *J. Appl. Phycol.* **27** 1657–1669

IOP Conf. Series: Materials Science and Engineering **193** (2017) 012004 doi:10.1088/1757-899X/193/1/012004

Effects of different *Pediococcus halophilus* level and fermentation time on chemical properties of fermented anchovy paste "*terasi ikan*"

S D Lestari[1*], Herpandi[1] and G R R Simamora[1]
[1] Fisheries Product Technology Study Program, Sriwijaya University, Ogan Ilir, Indonesia 30662

*E-mail: shantidwita_thi@unsri.ac.id

Abstract. The aim of this study was to investigate the *Pediococcus halophilus* addition on the chemical quality of *terasi ikan* (fermented anchovy paste) product. Two levels of bacterial concentration (10^6 CFU/mL and 10^9 CFU/ mL) were used as a single starter culture for the fermentation process. Changes in chemical characteristics were observed at day 7, 14 and 21. No differences ($p > 0.05$) in moisture and protein content were found in the analysis of variance within *terasi ikan* samples. The decrease in reducing sugar and L-lysine HCl during the fermentation was attributed to the formation of Maillard Reaction Products (MRPs) which was manifested in dark brown color of the end products. The interaction between *P. halophilus* and *terasi ikan* microbiota as well as their enzymatic activities were considered to affect vitamin B synthesis and degradation of protein into amino acids and amines. These findings facilitate further investigations on using *P. halophilus* as constituent of mixed culture, instead of as a single culture for *terasi* industry in order to produce products of well-controlled quality and safety.

1. Introduction

Terasi is a highly popular fermented seafood product among Indonesians. Its distinctive flavor and the presence of glutamic acid make it a common ingredient used in Southeast Asian dishes. *Terasi* is made of a half-dried small shrimp (*Acetes* sp.), termed as *terasi udang* or small fish /anchovies (*Stolephorus* sp.), termed as *terasi ikan*, mixed with 10-25% of solar salt, pounded, sun-dried and fermented for two days. After the first fermentation, the mixture is finely ground, sun-dried, shaped in cubes and wrapped in a banana leaves for the development of a special aroma. The second fermentation lasts about four weeks at ambient temperature (between 28°C and 30°C) leaving a soft-textured and dark-colored product. The fermentation takes place either spontaneously, or is orientated by adding a small percentage of fermented paste from the previous batch. The common practice in *terasi* making relies on spontaneous fermentation, although [1] observed the practice of using *terasi* starter which was composed of *Bacillus brevis, Bacillus pumilus, Bacillus megaterium, Bacillus coagulans, Bacillus subtilis* and *Micrococcus kristinae* in the proportion of 39.1%, 26.1 %, 8.7%, 8.7%, 8.7%, and 8.7%, respectively. The halophilic lactic acid bacteria has also been isolated from autochthonous *terasi udang* paste and identified as *Tetragenococcus halophilus* and *T. muriaticus* [2]. The former species *Pediococcus halophilus* has been raised to genus level, forming the genus Tetragenococcus [3] which its species are important in lactic fermentation of high-salt-containing food. However, little is known regarding the use of this species as a part of mix starter cultures or single starter in *terasi ikan* production.

Lactic acid bacteria have been known for their ability to produce antibacterial substances such as bacteriocins which have a wide antibacterial spectrum with potential application as food preservatives. Study on the potential bacteriocin production by *Tetragenococcus halophilus* indicated that this type of lactic acid bacteria exhibited good antibacterial activity against *B. cereus, S. aureus, L monocytogenes, P. aeruginosa, V. parahaemolyticus* and *A. hydrophila* and had antibiotic resistance against streptomycin [4].

The addition of competitive starter cultures to begin the fermentation process is an effective method of controlling the growth of spoilage organisms, food-borne pathogens and preventing the formation of undesirable end-products. In traditional *terasi* production, the fermentation process is usually unstable. The proliferation of undesirable microorganisms also raises a quality control issue. The development of starters containing lactic acid bacteria species associated with *terasi* products seems promising since in general, lactic acid bacteria offer a potential alternative as natural food-grade biocontrol agents and have the ability to improve the safety, flavor, nutritional value, and structure of the products [5]. Therefore, the present study attempted to determine the effects of *P. halophilus* concentrations, as a single starter culture, and fermentation time on some quality characteristics of *terasi ikan* (fermented anchovy paste).

2. Experimental detail

2.1. Sample preparation

Fresh anchovies (*Stolephorus* sp.) were purchased from a local fisherman in Sungsang, South Sumatera. The fish were held in ice for at least 24h before arrival at the laboratory. They were then manually beheaded, gutted, washed and drained. After sun-dried, the half-dried anchovies was mixed with solar salt (20% w/w), ground and separated into three different batches. The first batch served as control, the second and the third batch was inoculated with 10^6 CFU/mL and 10^9 CFU/mL *Pediococcus halophilus* in saline water (0.85% NaCl) respectively. The pastes were fermented over an 18-h period in a closed jar, finely ground, formed into cubes shape and sun-dried for one day. Subsequently, the paste was wrapped tightly in banana leaves and left for fermentation process at 28-30°C and 95% humidity for 21 days. Samples were removed every week (day 7, 14 and 21) from the same processing batch for analysis.

2.2. Analytical methods

Amino acids analyses were performed by using Waters Acquity UPLC H Class Amino Acid Analysis System following the method outlined in [6]. Amino acids were identified by comparing retention times with a standard mixture of 15 authentic amino acids. The content of each amino acid was calculated on the basis of the standard curve of each authentic amino acid standard.

2.3 Statistical analyses

Main effects from a factorial model of *P. halophilus* culture concentrations, fermentation periods and their interactions were analyzed by analysis of variance (ANOVA). HSD was used to determine the differences of means. P values <0.05 were regarded as significant.

3. Results and discussion

Values of the moisture, protein content, reducing sugar, total volatile basic nitrogen (TVBN) and Vitamin B_{12} (cyanocobalamin) in *terasi ikan* are presented in table 1. The levels of moisture and protein content in all samples ranged from 47.75% to 49.75%, and 12.72% to 15.48%, respectively. These results were in agreement with the previously reported data for fermented shrimp paste in Phillipines. The protein content derived from *terasi ikan* samples was similar to values reported for *bagoong alamang* which varied between 12.9–15.3 g/100g and dry matter content ranged from 38 to 53g/100g, equal to 47-62% moisture [7]. According to [8] as also stated in Indonesian National Standard (SNI-012716-1992), the typical *terasi udang* from Indonesia has 30-50% moisture.

No differences ($p > 0.05$) in moisture as well as protein content were found in the analysis of variance of main effects of *P. halophilus* culture combination or fermentation periods within *terasi ikan* samples. However, an increase in protein content was found between the 7th and the 14th day of fermentations on control and samples inoculated with 10^6 CFU/mL *P. halophilus* which was attributed to the accumulation of bacterial cells and enzyme secretion due to rapid microbial growth. After that period, the protein content decreased which was possibly due to the proteolytic activity of the microorganisms and enzymes which degraded protein into amino acids, peptides and amines. Shift in moisture as fermentation prolonged, especially in inoculated samples was likely the result of biochemical process releasing water during the fermentation.

Table 1. Chemical characteristics of *terasi ikan*

Parameters	*P. halophilus* concentration	Fermentation time		
		7 d	14 d	21 d
Moisture (%)	Control	49.00±1.41^{Aa}	47.75 ±1.06^{Aa}	49^{Aa}
	10^6 CFU/mL	48.50^{Aa}	49.00^{Aa}	49.75 ± 0.35^{Aa}
	10^9 CFU/mL	49.75± 6.01^{Aa}	49.25±3.18^{Aa}	49.75±2.47^{Aa}
Protein (%)	Control	13.00±0.18^{Aa}	15.48±0.18^{Aa}	13.42±0.18^{Aa}
	10^6 CFU/mL	13.38±0.14^{Aa}	15.58±1.40^{Aa}	13.94±1.41^{Aa}
	10^9 CFU/mL	14.89±0.03^{Aa}	13.33±1.64^{Aa}	12.72±1.70^{Aa}
Reducing sugar (%)	Control	0.64±0.01^{Aa}	0.74±0.01^{Ab}	0.73±0.03^{Ab}
	10^6 CFU/mL	0.24 ±0.06^{Aa}	0.23±0.05^{Aa}	0.25±0.06^{Aa}
	10^9 CFU/mL	0.45±0.06^{Aa}	0.48±0.01^{Aa}	0.36±0.04^{Ab}
TVB (mg N/100g)	Control	102.60 ± 0.57^{Aa}	126.4±1.70^{Aa}	199.80±11.60^{Ab}
	10^6 CFU/mL	107.80±3.11^{Aa}	82.10±0.42^{Ba}	198.40±42.25^{Ab}
	10^9 CFU/mL	95.10±42.00^{Aa}	229.90±23.90^{Cb}	212.50±30.69^{Ab}
Vitamin B12 mcg/100g	Control	3.09	3.83	3.46
	10^6 CFU/mL	3.11	3.74	3.98
	10^9 CFU/mL	2.82	3.76	3.28

Results are expressed as mean ± standard deviation; n=2.
Values within a row with different superscripts (a-c) denotes significant differences from each other, as effected by different fermentation time, $p \leq 0.05$
Values within a column with different superscripts (A-C) denotes significant differences from each other as effected by different *P. halophilus* concentration, $p \leq 0.05$

During the *terasi ikan* fermentation, organic substances such as carbohydrates, fats and protein will be transformed into simpler compounds such as monosaccharides, disaccharides, fatty acids, peptides, amino acids, and other nitrogenous compounds either by the action of endogenous enzymes or microorganisms. The interaction between monosaccharides and disaccharides which are capable of acting as reducing agent, termed as reducing sugar and amino acids through Maillard reaction will decreased their availability. The formation of Maillard reaction products (MRPs) during fishery products fermentation and their correlation with antioxidant activity has been evaluated [9]. During the Maillard reaction, in which pH and temperature also play important roles, free amino groups react with carbonyl groups to form a reversible Schiff base continued by the formation of brown polymer melanoidins at the final stage [10]. The dark brown color formation in *terasi ikan* has been attributed to the formation of MRPs from reducing sugar and lysine, as this amino acid is very sensitive to Maillard reaction. The correlations between reducing sugar and L-Lysine HCl amino acid content in *terasi ikan* are depicted in figure 1 and figure 2 respectively.

The reducing sugar and L-lysine HCl content of *terasi ikan* fluctuated at different stages of the fermentation. A sharp decrease on reducing sugar was observed from day 7 to day 14 in both control and *P. halophilus*-inoculated *terasi ikan*. A similar pattern was observed in L-lysine HCl content which could be attributed to their interaction in Maillard reaction. As the sugar and protein decomposition continued, the amount of reducing sugar and amino acid increased. The formation of dark brown color on *terasi ikan* due to melanoidin production were similar to those reported in other studies on fermented fish product, yu-lu [11] and Philippine salt-fermented shrimp paste [9]. The development of brown color formation on previous studies was assessed through absorbance measurement at 420 nm.

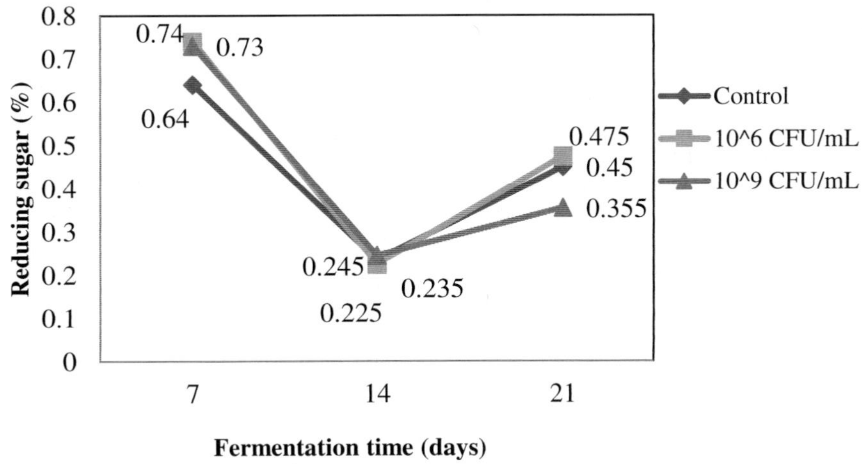

Figure 1. Changes in reducing sugar content of *terasi ikan* samples at 7, 14 and 21 days

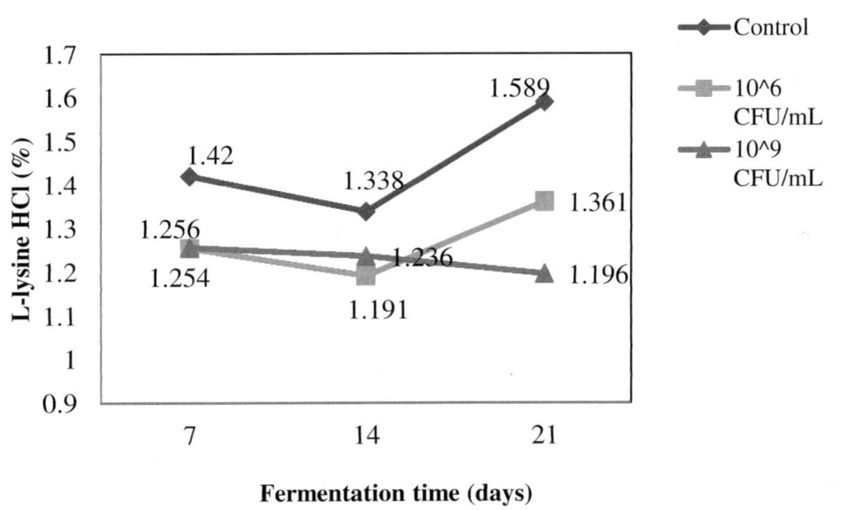

Figure 2. Changes in L-lysine HCl content of *terasi ikan* samples at 7, 14 and 21 days

During the microbial fermentation, formation of biogenic amines is considered as a result of metabolic activity in microorganisms. They are produced by the decarboxylation of free amino acids or by amination and transamination of aldehydes and ketones [12]. In case of terasi ikan fermentation, changes in l-lysine content might also be associated with its decarboxylation to produce cadaverine. Several groups of microorganisms has been known as biogenic amines producer, mainly Enterobacteriaceae, *Pseudomonas* spp., enterococci and some lactic acid bacteria [13]. Research conducted by [14] indicated the production of cadaverine by *Lactobacillus brevis* and *L. curvatus* but several strains of lactobacilli, *Leuconostoc* spp., *Weissella* spp., and pediococci showed no decarboxylase activity. The results suggested that the capability to produce amines was strain dependent. The use of *P. halophilus* as a starter and protective culture in this study and its further interaction with indigenous bacteria in *terasi ikan* as a complex food system might have contributed to chemical changes including biogenic amines formation.

The results also indicated increase on TVB during the experiment. Both *P. halophilus* concentration and fermentation time significantly affected the TVB value ($P < 0.05$) which was expressed as mg N/ 100g dry matter. The TVB value of *terasi ikan* supplemented with 10^9 CFU/mL *P. halophilus* increased by 117.4 compared to 90.6 and 97.2 of the 10^6 CFU/mL-supplementation and the control group respectively. Total Volatile Bases Nitrogen (TVB-N) is one of the most widely used methods to estimate the degree of decomposition of fish. It includes the measurement of trimethylamine *dimethylamine, ammonia* and other volatile nitrogenous compounds associated with seafood spoilage [15]. The TVB-N concentration increased as the fermentation time prolonged ($p<0.05$) despite the fluctuation over time of fermentation. When tested for *P. halophilus* culture concentration effect, there was no difference among culture concentration at day 7 and 28. The significant effect of *P. halophilus* concentration on TVB was only observed at day 14. The high TVB value on *terasi ikan* inoculated with 10^9 CFU/mL *P. halophilus* might be attributed to the rapid formation of trimethylamine which mainly originates from bacterial decomposition. Despite the high TVB values at day 21, they were still within the acceptable limits set, which are at 100-200 mg/100g for salted and dried fish products, according to FDA. A similar pattern has been reported for *terasi udang* which TVB value increased over fermentation time reaching 350 mg/100g after 28 days of fermentation [16].

The addition of microbes into the fermentation system in most cases significantly contributes to the nutritional value improvement of fermented product by taking part in the production of health-promoting compounds such as vitamins [17]. Certain LAB have capability to synthesize water-soluble vitamins specifically those included in the B-group including folic acid, riboflavin, cobalamin and biotin [18]. Vitamin B12 (cobalamin) is one of the very important vitamins and is required for the metabolism of fatty acids, amino acids, nucleic acids and carbohydrates [19]. Cobalamin cannot be synthesized by animals, plants and fungi. It is the only vitamin that is exclusively produced by bacteria and archaea, particularly by anaerobes. *P. freudenreichii, Salmonella enterica* and *Bacillus megaterium* are examples of bacteria that could perform the anaerobic biosynthesis of cobalamin, while *Pseudomonas denitrificans* is the aerobic B_{12}-producing bacterium [20]. Lactobacilli have the ability to produce vitamin B_{12}, in particular a probiotic strains of *L. reuteri* which are shown to be capable of producing cobalamin [21]. The Vitamin B_{12} values of *terasi ikan* supplemented with 10^6 CFU/mL *P. halophilus* had progressively increased over time of fermentation. It had an initial value of 3.11 and reached a level of 3.98 mcg/100g on the 21 days of fermentation. As for control and *terasi ikan* cultured with 10^9 CFU/mL *P.halophilus*, the values were fluctuated with a noticeable decrease at day 14. At the end of fermentation, the values of both samples increased but still lower than the 10^6 CFU/mL ones. The differences in cobalamine content was presumably due to the interaction between *P. halophilus* and several normal microbiota constituent of *terasi*, which can be originated from various species or strains including cobalamin producer such as *Pseudomonas* spp. and *Lactobacillus* spp..

table 2 presents the amino acid content of *terasi ikan* at various *P. halophilus* concentration and fermentation periods. The initial total of 15 amino acid content of raw anchovy sample was 14.68% with the major amino acids were L-glutamic acid, L-lysine HCl and L-aspartic acid. The research of [9] reported that the total content of free amino acid in shrimp paste changed during fermentation. It increased distinctly in the beginning, relatively stable in the mid fermentation period and then decreased as the fermentation prolonged. The same trend was observed in *terasi ikan* samples supplemented with 10^6 CFU/mL *P. halophilus,* while in control samples, the total count of amino acids still progressively increased after 21 days of fermentation. This result might indicate that the fermentation process can be terminated earlier with the use of lactic acid starter. The significant amount of *P. halophilus* may speed up the macromolecules conversion into their constituents such as amino acids, mono and disaccharides, and fatty acids. The high bacterial cell density in samples inoculated with 10^9 CFU/mL created competition between cells for oxygen and nutrients, leaving the survivors ruled the fermentation process. The decline in the amino acid content in both inoculates samples could be due to its degradation to amines, volatile acids, and other nitrogenous substances as by-products of bacterial metabolism or enzymatic decomposition [9].

Table 2. Changes in amino acid content of *terasi ikan* as influenced by different *P. halophilus* concentration and fermentation time

P. halophilus concentration	Control			10^6 CFU/mL			10^9 CFU/mL			Fresh Anchovy
Fermentation time	7d	14d	21d	7d	14d	21d	7d	14d	21d	
Amino acids (in %)										
L-Aspartic acid	0.735	0.822	1.182	0.555	0.706	0.408	0.840	0.661	0.491	1.38
L-glutamic acid	**0.832**	**0.956**	**2.076**	**0.546**	**0.504**	**0.468**	**0.865**	**0.699**	**0.552**	**2.38**
L-Serine	0.050	0.054	0.471	0.025	0.037	0.022	0.038	0.037	0.034	0.61
Glysine	0.756	1.054	1.161	0.739	0.831	0.625	0.889	0.873	0.837	1.16
L-Alanine	0.570	0.610	0.531	0.591	0.568	0.563	0.570	0.579	0.589	0.98
L-Proline	0.531	0.665	0.665	0.518	0.553	0.498	0.577	0.597	0.594	0.65
L-Tyrosine	0.264	0.316	0.422	0.318	0.422	0.346	0.322	0.338	0.415	0.55
L-Histidine*	0.160	0.205	0.301	0.174	0.216	0.177	0.193	0.198	0.209	0.41
L-Arginine*	0.516	0.704	0.823	0.557	0.680	0.565	0.546	0.650	0.654	0.97
L-Isoleucine*	0.511	0.549	0.710	0.512	0.558	0.595	0.519	0.568	0.588	0.70
L-Leucine*	0.818	0.900	1.153	0.824	0.900	0.936	0.838	0.911	0.917	1.19
L-Lysine HCl*	**1.420**	**1.338**	**1.589**	**1.254**	**1.191**	**1.361**	**1.256**	**1.236**	**1.196**	**1.51**
L-Valine*	0.558	0.562	0.747	0.551	0.589	0.615	0.567	0.605	0.593	0.74
L-Phenylalanine*	0.315	0.399	0.547	0.369	0.478	0.392	0.465	0.411	0.489	0.77
L-Threonine*	0.244	0.240	0.659	0.198	0.234	0.202	0.223	0.232	0.223	0.69
Total	8.281	9.372	13.036	7.730	8.466	7.771	8.708	8.593	8.382	14.68

*essential amino acids

As also shown in table 2, glutamic acids in control samples was prominent, whereas lower content of the same amino acids were found in inoculated samples. Glutamate plays a central role in a wide range of metabolic processes in bacterial cells including protein synthesis, glycolysis, gluconeogenesis and the citric acid cycle [21]. Under the stress condition, such as acidic stress, certain microbes are able to metabolize glutamate through the action of glutamate dehydrogenase or glutamate decarboxylase (GAD). *Lactobacillus reuteri, Lactococcus lactis* and *Lactobacillus plantarum* are examples of lactic acid bacteria which posses enzyme that catalyzes the decarboxylation of glutamate to γ-aminobutyrate [22]. The variation in amino acids composition in each samples depended on microbial interaction within fermentation system. Some microbes secreted proteins decomposing-enzymes which would increase the amount of free amino acids while others would use the amino acids in their metabolism process.

4. Conclusions

The use of *P. halophilus* as a starter and protective culture and its further interaction with indigenous bacteria in *terasi ikan* as a complex food system might have contributed to chemical changes during the product's development. Some nutritional quality parameters including the content of reducing sugar, vitamin B and also the amino acid composition were changed over time and influenced by the concentration of *P. halophilus*. Roles of many other species of bacteria during the fermentation, however, can not be neglected. These findings facilitate further investigations on using *P. halophilus* as constituent of mixed culture, instead of as a single culture for *terasi* industry.

Acknowledgments

The authors would like to thank Sriwijaya University for funding through SP DIPA Number 042.01.2.400953/2016 for Unggulan Kompetitif Research Scheme 2016. Research grant was awarded to Herpandi, Ph.D.

References

[1] Surono IS and Hosono A 1994 *Biosci. Biotech. Biochem.* **58(6)** 1167-9
[2] Kobayashi T, Kajiwara M, Wahyuni M, Kitakado T, Hamada-Sato N, Imada C and Watanabe E 2003 *J. Gen. Appl. Microbiol.* **49** 279-86.
[3] Axelsson L 2004 Lactic acid bacteria: classification and physiology *Lactic Acid Bacteria Microbiological and Functional Aspects 3rd Edition* eds S Salminen, A von Wright and C Ouwehand (New York: Marcel Dekker)
[4] Pal V, Jamuna M and Jeevaratnam K 2004 *J. Cult. Collect.* **4** 53-60
[5] Salminen S, AC Ouwehand, A von Wright and C Daly 1998 Future aspects of research and product development of lactic acid bacteria *Lactic Acid Bacteria.Microbiology and Functional Aspects* eds S Salminen and A von Wright (New York: Marcel Dekker)
[6] Aristoy MC and Toldra F 2013 Amino acids *Food analysis by HPLC 3rd Edition* eds LM Nollet and F Toldra (Boca Raton Florida: CRC Press)
[7] Pilapil AR, Neyrinck E, Deloof D, Bekaert K, Robbens J and Raes K 2016 *J. Sci. Food Agric.* **96(3)** 933-8
[8] Putro S 1993 Fish fermentation technology in Indonesia *Fish Fermentation Technology*. Eds. CH Lee, KH Steinkraus and PJ Reilly (Tokyo: United Nations University Press)
[9] Peralta EM, Hatate H, Kawabe D, Kuwahara R, Wakamatsu S, Yuki T and Murata H 2008 *Food Chem.* **111** 72-7
[10] Zhuang Y and Sun L 2011 *Afr. J. Biotechnol.* **10(35)** 6784-93
[11] Jiang JJ, Zeng QX, Zhu ZW and Zhang LY 2007 *Food Chem.* **104(4)** 1629-34
[12] Stadnik J and Dolatowski ZJ 2010 *Acta Sci. Pol., Technol. Aliment.* **9(3)** 251-263
[13] Halasz A, Barath A, Simon-Sarkadi L and Holzapfel WH 1994 *Trends Food Sci. Technol.* **5** 42-49
[14] Bover-Cid S and Holzapfel WH 1999 *Int. J. Food Microbiol.* **53** 33-41
[15] Malle P and Poumeyrol M 1989 *J. Food Prot.* **52 (6)** 419-23
[16] Soetikno N, Fitrial Y and Khairina R 2011 *Fish Scientiae* **1(1)**
[17] Steinkraus KH 1998 Bio-enrichment: production of vitamins in fermented foods *Microbiology of Fermented Foods* ed BJB Wood (London: Blackie Academic and Professional) p 603-21
[18] Capozzi V, Russo P and Duenas MT 2012 *Appl. Microbiol. Biotechnol.* **96** 1383
[19] Patel A, Shah N and Prajapati JB 2013*Croat. J. Food Sci. Technol.* **5 (2)** 85-91
[20] LeBlanc JG, Lain JE, Juarez del Valle M, Vannini V, van Sinderen D, Taranto MP, Font de Valdez G, Savoy de Giori G and Sesma F 2011 *J. Appl. Microbiol.* **111** 1297–1309
[21] Taranto MP, Vera JL, Hugenholtz J, De Valdez GF and Sesma F 2003 *J. Bacteriol.* **185** 5635-47
[22] Feehily C and Karatzas KAG 2012 *J. Appl. Microbiol.* **114(1)** 11-24

IOP Conf. Series: Materials Science and Engineering **193** (2017) 012005 doi:10.1088/1757-899X/193/1/012005

Polygalacturonase production by AR2 pectinolytic bacteria through submerged fermentation of raja nangka banana peel (*Musa paradisiaca* var. *formatypica*) with variation of carbon source and pectin

R Utami[1*], E Widowati[1], A Ivenaria[1], E Mahajoeno[2]
[1] Department of Food Science and Technology, Universitas Sebelas Maret, Jl. Ir Sutami 36A, Surakarta, Central Java, Indonesia.
[2] Department of Biology, Universitas Sebelas Maret, Jl. Ir Sutami 36A, Surakarta, Central Java, Indonesia.

*E-mail: rohulautami@staff.uns.ac.id

Abstract. Polygalacturonase (EC 3.1.2.15) catalyzes the hydrolysis of α-1,4-glycosidic bonds on galacturonic acid. Polygalacturonase can be produced from AR2 pectinolytic bacteria isolated from orange peel and vegetable waste. Commonly cost production of enzymes were high. However, with the advancement of technology, enzymes can now be manufactured at a low cost. Production of enzymes in low cost media with agro-industrial waste is interesting. Raja nangka banana peel is agro-industrial waste that is uneconomic. Therefore, this material can be used as a pectin source in polygalacturonase production. Polygalacturonase was produced by AR2 pectinolytic bacteria with the addition of various carbon sources (1% glucose, 1% galactose, 1% lactose) and variation of pectin concentrations (5%; 7.5%; 10%). This study used submerged fermentation with a cultivation temperature of 55°C and an agitation speed of 144 rpm for a 48-h incubation time. The results showed that variation of carbon sources and pectin concentrations affected the production of polygalacturonase. After 48 h fermentation, the results showed that the number of cells of samples ranged from 8.3 to 9.445 log cells/mL; the used pectin of samples ranged from 87.170-93.745%; and the polygalacturonase activity of samples ranged from 0.030 to 0.151 U/mL. The highest polygalacturonase activity was obtained by production of polygalacturonase on 1% glucose and 10% pectin medium.

1. Introduction

Polygalacturonase (EC 3.2.1.15) catalyzes hydrolysis of α-1,4-glycosidic bonds on polygalacturonic acid producing D-galacturonic [1]. Polygalacturonase has been used in foods, biofuels, and the textile industry due its thermostability; polygalacturonase remains stable at 50-60°C. Polygalacturonase is one of the most widely used enzymes in the food industry [2]. Therefore, the production of enzymes is necessary to fulfill the demand of the industry.

Polygalacturonase can be produced by microbes in the agricultural waste medium fermentation such as by *Fusarium moniliforme* on wheat bran and orange pulp substrate, *Penicillium viridicatum* on orange bagasse and wheat bran substrate, *Aspergillus awamori* on grape pomace substrate, *Candida utilis* on apple pomace substrate, and *Penicillium occitanis* on citrus pectin substrate [3]. Banana peels can also be used as a medium for fermentation in pectinase production by *Aspergillus niger* under laboratory conditions [4]. Previous research [5] showed that on the sixth day of fermentation, *Aspergillus parasiticus* produced polygalacturonase that showed enzyme activity ranging from 0.1183 to 2.4683 U/mL using banana peel as the medium fermentation substrate. Previous study also reported that banana peel is best at producing enzyme substrate polygalacturonase compared with the substrate from orange peel and pineapple skin [4]. The pectic substances of agricultural waste are good inducer of polygalacturonase production.

Carbon and nitrogen sources in medium fermentation are needed in the production of pectinase at an optimum level. However, previous study reported that the addition of 5% pectin and 0.5% ammonium sulfate in polygalacturonase production using raja nangka banana peel waste by AR2 bacteria only produced polygalacturonase with 0.134 U/mL enzyme activity [6]. This activity was lower than the activity of polygalacturonase produced by the same bacteria (AR2) using pectin citrus medium [7] that achieved 0.209 U/mL. Therefore, to increase the activity of polygalacturonase the addition of a carbon source such as sugar or pectin is needed. Some of these sugars are glucose, galactose, and lactose. Previous research mentioned that the activity of polygalacturonase produced in medium with the addition of glucose, lactose and galactose were 2.4 U/mL, 2.8 U/mL, and 31.29 U/mL, respectively [8-9]. Pectin can also be used as medium fermentation substrate for polygalacturonase production and produced enzyme activity at 41.13 U/mL by *Rhizomucor pusillus* [9]. Medium fermentation for polygalacturonase production containing sugar and pectin can be expected to produce higher enzyme activity compared with medium fermentation only from sugar.

AR2 pectinolytic bacteria have been isolated from orange peel and vegetables waste. Polygalacturonase produced by AR2 bacteria showed a juice clarification ability [7]. Polygalacturonase produced by AR2 bacteria has also already been used in the clarification of super-red dragon fruit juice [10]. Additionally, previous study also conducted research by combining enzymes polygalacturonase from AR2 bacteria with cellulase and amylase in orange juice clarification [11]. Furthermore, to increase the juice clarification ability of polygalacturonase produced by AR2 bacteria, the enzyme activity of polygalacturonase must also increase. This research investigated the effect of variations in the carbon source and pectin concentration in the fermentation medium on polygalacturonase production by AR2 bacteria. After 48 h fermentation the number of bacteria, the pectin content, and enzyme activity of polygalacturonase were determined.

2. Material and methods

2.1. Material

The raja nangka banana peels were obtained from UPKKS Bakti Kencana Karanganyar (Indonesia). AR2 pectinolytic bacteria (further identified as *Bacillus licheniformis* strain AR2 by 16S rDNA gene sequencing analysis) were isolated from orange peel and vegetables wastes from Lokal Market (Surakarta, Indonesia) and collected by the Food Microbiology and Biotechnology Laboratory of Universitas Sebelas Maret. Other materials were also prepared such as ammonium sulfate ($(NH_4)_2SO_4$) (Merck), glucose ($C_6H_{12}O_6$) (Merck), galactose ($C_6H_{12}O_6$) (Merck), and lactose ($C_{12}H_{22}O_{11}$) (Merck).

2.2 Starter cultures preparation

Subcultures were made by scraping 1 loopfull from AR2 culture media collection into pectin agar slant. The subcultures were incubated at 55°C for 24 h. Preparation starter isolates AR2 done by scraping 1 loopfull from subculture into 10 mL of liquid pectin media. Incubation was carried out at 55°C for 24 h until the cell number reached 10^8 cells/mL [6].

2.3 Fermentation medium preparation

The cleaned banana peels (100 g) were steam blanched for 6 min at 100°C to soften the tissue. Furthermore, the steamed banana peels were crushed using mortar. The banana peels pure at various concentration (5% (13.186 g); 7.5% (19.778 g) and 10% (26.371 g) were diluted with distilled water to fulfill100 mL. Furthermore, 0.5% ammonium sulfate as nitrogen source [6] and 1% carbon source such as glucose, galactose, and lactose [12] were added to the medium. The medium was sterilized for 15 min at 121°C.

2.4 Fermentation for polygalacturonase production

Each medium (100 mL) was cultured with 1 mL AR2 bacteria (10^8cells/mL). Incubation was conducted for 48 h at 55°C and 144 rpm using an incubator shaker [7]. After 48 h fermentation, the number of bacterial cells were enumerated using Hemocytometer. Determination of residual pectin content in the medium after fermentation was conducted by taking 20 mL of fermented medium and followed the previous method [13].

2.5 Enzyme partial purification

After 48 h fermentation, each medium was centrifugated (*Eppendorf 5810 R*) at 6,000 rpm and 4°C for 15 min. The supernatants were precipitated with ammonium sulfate at 50% saturation fraction with agitation, and incubated for 24 h at 4°C to separate the enzyme. After 24 h, the precipitates were separated by centrifugation (*Eppendorf 5810 R)* at 12,000 rpm at 4°C for 10 min. The resulting pellets were dissolved in 0.05 M acetate buffer [14] and pH 5.2. The ratio of pellets and buffer was 1: 1. The mixtures were dialyzed in 12 kDa cut off cellophane membrane and soaked in 600 mL of 0.05 M acetate buffer solution and pH of 5.2 with magnetic homogenization stirer for 24 h in a cool chamber, at a constant temperature of 4°C. The buffer solution was replaced and tested qualitatively using $BaCl_2$ 1% per h to ensure the ammonium sulfate particles and other impurities have been removed from the cellophane membrane [15].

2.6 Determination of polygalacturonase activity and specific enzyme activity

Polygalacturonase activity was determined by measuring the reducing sugar that released from pectin by using a 3,5-dinitrosalicyclic acid reagent assay. The reaction mixture containing 0.1 mL enzyme and 0.9 mL 0.7% citric pectin in 0.0025 M natrium acetate buffer pH 4.8 [16] was incubated at 55°C for 30 min.The absorbance of the sample was determined using a spectrophotometer UV-Vis (Shimadzu) at wavelength of 540 nm [16]. One unit of enzyme activity (U/mL) was defined as the amount of enzyme which released one μmole of galacturonic acid per min per mL of enzyme solution. The specific enzyme activity is the number of enzyme units per mL divided by the concentration of protein in mg/mL. The concentration of protein enzyme was measured using the Lowry method with absorbance value at wave length of 600 nm.

2.7. Statistical analysis

Completely Randomized Design was used to evaluate the effect of carbon sources (1% glucose, 1% galactose, 1% lactose) and pectin concentration (5%; 7.5%; 10%). All treatments were replicated twice. Data were analysed with ANOVA test (One Way ANOVA) followed with DMRT (Duncan Multiple Range Test) (α=5%) if there were the variances between samples by SPSS Statistics 16 program.

3. Result and discussion

3.1 Bacterial cells count

The result showed that the carbon source significantly affected the bacterial cells count, with the exception of the sample with 5% pectin. The bacterial cells counts of 1% glucose and 10% pectin concentration medium were significantly higher than those of 1% lactose and 1% galactose medium (Table 1). During fermentation, microbes utilize carbon sources for metabolism and cell proliferation [17]. Previous study reported that by the addition of glucose on medium, the biomass of *Aspergillus niger* was increased more than by the addition of galactose, fructose or lactose [12].

Pectin concentration (5; 7.5; 10%) on medium did not show any significant effect on bacterial cells count (Table 1). Previous research reported that pectin concentration (1-5%) on medium significantly affected the bacterial cells [6]. That reasearch indicated that bacterial cells utilized the carbon sources (glucose, galactose, lactose) more than pectin for cell growth.

The bacterial cells resulted at this research ranged from 8.310- 9.445 log cell/mL. This result was higher than the bacterial cell resulted by the fermentation using pectin as the single carbon source at 1-5% (6.000-6.738 log cell/mL) [6]. This indicated that the addition of other carbon source such as glucose, galactose or lactose on medium could increase the number of bacterial cells.

Table 1. Bacterial cells count (log cell/mL)

Carbon	Pectin		
	5%	7.5%	10%
Glucose 1%	$9.267^{a\ A}$	$9.445^{a\ B}$	$9.444^{a\ C}$
Galactose 1%	$9.281^{a\ A}$	$9.366^{a\ B}$	$9.051^{a\ B}$
Lactose 1%	$8.972^{a\ A}$	$8.481^{a\ A}$	$8.310^{a\ A}$

Lowercase different notation on the same line and the notation of different capital letters in the same column indicate significant difference at $\alpha = 5\%$

Table 2. Pectin utilization during fermentation (%)

Carbon	Pectin		
	5%	7.5%	10%
Glucose 1%	$91.185^{a\ A}$	$91.730^{a\ A}$	$93.745^{a\ A}$
Galactose 1%	$88.420^{a\ A}$	$90.293^{a\ A}$	$90.585^{a\ A}$
Lactose 1%	$87.170^{a\ A}$	$91.137^{b\ A}$	$92.333^{b\ A}$

Lowercase different notation on the same line and the notation of different capital letters in the same column indicate significant difference at $\alpha = 5\%$

3.2 Pectin utilization

The pectin content of raja nangka banana peel was 37.924%. This result was lower than the pectin content of raja nangka banana peel found in the previous research (68.47%) [6]. The pectin content of banana peels are related to the banana maturity levels [18].

The initial pectin content on medium were varied at 5; 7.5 and 10%. During fermentation, pectin was hydrolyzed into galacturonic acid by pectinase enzyme produced by microbes [19]. The result showed that the pectin utilization after fermentation ranged from 87.17-93.745 % (Table 2). Previous studies have shown that 86% pectin was hydrolyzed by AR2 during fermentation [6]. The decrease of pectin content was due to the AR2 bacteria activity to metabolize the pectin into galacturonic acid [6]. Previous study also reported that the pectin content of apple pomace was decrease during fermentation [20]. However, the carbon sources and pectin concentration variation have no significant effect on pectin utilization.

3.3 Enzyme activity

Different types of carbon source significantly affected ($\alpha<0.05$) the polygalacturonase activity. The polygalacturonase activity of enzyme produced on 1% glucose medium was significantly higher than that produced on 1% galactose or 1% lactose medium (Table 3). For most bacteria, glucose is the most common substrate and can be used directly for growth. Meanwhile, the lactose is polysaccharide which is required to breakdown into monosaccharides (glucose and galactose) in the fermentation process, which takes more time for metabolism [21]. Under the same conditions as glucose, galactose was not suitable as carbon source [22]. Previous research also reported that glucose performed as the best carbon source while lactose had no supporting role as a carbon source for pectinase production [23].

Pectin concentrations in medium also significantly affected the polygalacturonase activity (Table 3). However, higher pectin concentration did not always result in higher activity. A higher concentration of the substrate in media does not always increase the activity of produced enzymes [24]. Additionally, if the active site of enzyme has been completely filled out by the substrate, the enzyme can not degrade the other substrates [25]. So although the substrate concentration is increased, it will not increase the activity of enzymes.

Table 3. Polygalacturonase activity (U/mL)

Carbon	Pectin		
	5%	7,5%	10%
Glucose 1%	$0.129^{a\ C}$	$0.147^{a\ C}$	$0.151^{a\ C}$
Galactose 1%	$0.108^{b\ B}$	$0.087^{a\ B}$	$0.089^{a\ B}$
Lactose 1%	$0.050^{b\ A}$	$0.030^{a\ A}$	$0.054^{b\ A}$

Lowercase different notation on the same line and the notation of different capital letters in the same column indicate significant difference at $\alpha = 5\%$

Table 4. Concentration of protein enzyme (mg/mL)

Carbon	Pectin		
	5%	7,5%	10%
Glucose 1%	$0.021^{a\ A}$	$0.061^{ab\ A}$	$0.133^{b\ A}$
Galactose 1%	$0.032^{a\ A}$	$0.073^{c\ A}$	$0.059^{b\ A}$
Lactose 1%	$0.371^{a\ B}$	$0.668^{b\ B}$	$0.728^{b\ B}$

Lowercase different notation on the same line and the notation of different capital letters in the same column indicate significant difference at $\alpha = 5\%$

Table 5. Polygalacturonase Specific Activity (U/mg)

Carbon	Pectin		
	5%	7,5%	10%
Glucose 1%	$6.317^{b\ C}$	$3.207^{ab\ A}$	$1.140^{a\ B}$
Galactose 1%	$3.423^{b\ B}$	$1.195^{a\ A}$	$1.524^{a\ C}$
Lactose 1%	$0.134^{b\ A}$	$0.045^{a\ A}$	$0.074^{a\ A}$

Lowercase different notation on the same line and the notation of different capital letters in the same column indicate significant difference at $\alpha = 5\%$

The results showed that the highest polygalacturonase activity was 0.151 U/mL which found with a 1% glucose-10% pectin concentration medium while the lowest enzyme activity was 0.030 U/mL that found with a 1% lactose-7.5% pectin concentration fermentation medium. This result showed higher activity than previous report [6] which produced polygalacturonase with 0.134 U/mL enzyme activity by the fermentation using pectin as single carbon source. This indicated that the addition of other carbon source such as glucose, galactose or lactose to the fermentation medium could increase the polygalacturonase activity.

3.4 Concentration of enzyme protein

Variations in the carbon type significantly affected the concentration of the enzyme protein (Table 4). Nutrients in the fermentation medium containing sugar will provide energy for metabolic processes of microorganisms [26]. If there are better nutrients in the substrate, the cell growth can be faster and will increase the protein content [27].

Variations of pectin concentration significantly affected the concentration of the enzyme protein (Table 4). The results showed that the higher concentration of pectin the higher concentration of the enzyme protein. However, the addition of 10% pectin on galactose medium decreased the protein concentration of enzymes.

3.5 Specific activity of enzyme

Table 5 showed that variations of carbon type significantly affected the specific activity of the polygalacturonase. The specific activity of the polygalacturonase produced on glucose medium was higher than the specific activity of the polygalacturonase produced on galactose and lactose medium. This result was different from previous study [28]. The specific activity of polygalacturonase enzymes produced by *Aspergillus niger* on lactose medium were higher than that produced on glucose or galactose medium. The lowest specific activity of polygalacturonase enzymes were produced on glucose medium [28].

Meanwhile, statistically (Table 5) the variation of pectin concentration in the medium affected the specific activity of the polygalacturonase. The results showed that if greater concentration of pectin was added to the fermentation medium, the specific activity of the polygalacturonase enzyme decreased.

The highest specific activity of the polygalacturonase (6.317 U/mg) was produced on glucose medium with the addition of 5% pectin. This specific activity was higher than the specific activity of polygalacturonase produced on pectin medium (2.679 U/mg) [6]. Other study [19] also stated that the specific activity of enzyme polygalacturonase produced by *Aspergillus utus* BL5 was 4.15 U/mg. This result indicated that the variety of carbon types and concentrations of pectin affected the specific activity of the polygalacturonase.

4. Conclusion

The results showed that different of carbon sources (glucose 1%, galactose 1%, lactose 1%) and pectin concentration (5%; 7.5%; 10%) affected the production of polygalacturonase. After 48 h fermentation, the result showed that the number of cells of samples ranged from 8.3 to 9.445 log cells/mL; the used pectin of samples ranged from 87.170-93.745%; and polygalacturonase activity of samples ranged from 0.030 to 0.151 U/mL. The highest polygalacturonase activity was obtained by using 1% glucose and 10% pectin.

Acknowledgements

This work was financially supported by research project of Penelitian Unggulan Perguruan Tinggi (PUPT) No: 041/SP2H/LT/DRPM/II/2016 from Ministry of Research, Technology, and Higher Education of Republic of Indonesia.

References

[1] Pedrolli D B, Monteiro A C, Gomes E, and Carmona E C 2009 *TOBIOTJ* **3** 9-18
[2] Wang S, Lian Z, Wang L, Yang X, and Liu Y 2015 *Beijing Bioresources and Bioprocessing* **2** (33) 1-13
[3] Torres E F, Sepulveda T V, and Gonzales G F 2006 *FTB* **44** (2) 221-227
[4] Durairajan B, and Sankari P S 2014 *J. Pharm. Biosci.* **2** 50-57
[5] Kumar G P, and Suneeetha V 2014 *Int. J. Drug Dev. & Res.* **6** (3) 109-118
[6] Safitri N R D 2015 *Produksi Poligalakturonase Melalui Fermentasi Cair Limbah Kulit Pisang Raja Nangka (Musa paradisiaca) dengan Variasi Sumber Karbon dan Nitrogen Anorganik Oleh Bakteri Pektinolitik* (Surakarta : Ilmu dan Teknologi Pangan, Universitas Sebelas Maret)
[7] Kalistyatika K 2014 *Produksi dan Karakterisasi Enzim Poligalakturonase pada Klarifikasi Sari Buah Jeruk Keprok Garut (Citrus nobilis var. chrysocarpa)* (Surakarta : Ilmu dan Teknologi Pangan, Universitas Sebelas Maret)
[8] Kumar A, and Sharma R 2012 *J. phytol.* **4** (1) 01-05
[9] Siddiqui M A, Pande V, and Arif M *India Afr. J. Microbiol. Res.* **7** (3) 252-259
[10] Muthoharoh 2015 *Penggunaan Enzim Poligalakturonase dan Gelatin dalam Klarifikasi Sari Buah Naga Super Merah (Hylocereus costaricensis)* (Surakarta : Ilmu dan Teknologi Pangan, Universitas Sebelas Maret)
[11] Sulistyo A C B 2016 *Pengaruh Kombinasi Enzim Poligalakturonase, Selulase, dan Amilase pada Klarifiaksi Jus Jeruk Keprok Garut (Citrus nobilis var. chrysocarpa)* (Surakarta : Ilmu dan

Teknologi Pangan, Universitas Sebelas Maret)
[12] Mojsov K 2010 *ATI* **3** (3) 23-29
[13] Sulihono A, Tarihoran B, and Agustina T E 2012 *Jurnal Teknik Kimia* **18** (4) 1-8
[14] Barense R I, Chellegatti M A S C, Fonseca M J V, and Said S 2001 *Braz. J. Microbiol.* **32** 327-330
[15] Widowati E, Utami R, Nurhartadi E, Andriani M A M, dan Wigati A W 2014 *Jurnal Aplikasi Teknologi Pangan* **3** (1) 16-20
[16] Ordonez R G, Morlon J, Gasparian S, and Guyot J P 1998 *Folia Microbiol.* **4.3** (6) 657-660
[17] Oberman H, and Libudzisz Z 1998 *Fermented Milks, In B.J.B. Wood (Ed.), Microbiology of Fermented Foods second edition* (London: Blackie Academic & Professional)
[18] Akili M S, Ahmad U, dan Suyatma N E 2012 *Jurnal Keteknikan Pertanian* **26** (1)
[19] Yopi, Rahmani N, Andriani A, dan Dewi F 2013 *Berita Biologi* **12** (3)
[20] Villas-Boas S G, Esposito E, and de Mendonca M M 2002 *World J. Microbiol. Biotechnol.* **18** (6) 541–545
[21] Todar, K 2008 *Textbook of Bacteriology: Regulation of Bacterial Metabolism.* (Madison: University of Wilconsin)
[22] Chan V, Dreolini L F, Flintoff K A, Lloyd S J, and Mattenley A A 2002 *JEMI* **2** 130-137
[23] Jagiasi R R 2013 *National Conference Biodiversity Status and Challenges Conservation 'Favea'*
[24] Rahmani N, Andriani A, Anggraini Y S, dan Yopi 2013 *Teknologi Indonesia* **36** (3) 142-148
[25] Oktavia Y, Andhikawati A, Nurhayati T, dan Tarman K 2014 *Jurnal Ilmu dan Teknologi Kelautan Tropis* **6** (1) 209-218
[26] Buckle K A, Edwards R A, Fleet G H, and Wooton M 1985 *Ilmu Pangan* Translated by Purnomo, H. dan Adiono (Jakarta : UI Press)
[27] Fardiaz S 1992 *Mikrobiologi Pangan* (Yogyakarta: Penerbit Kanisius)
[28] Ogbonnaya N, and Chukwuemeka E I 2016 *Scientific Paper, Hemijska Industrija*
[29] Sebayang F 2005 *Jurnal Komunikasi Penelitian* **17** (3)

IOP Conf. Series: Materials Science and Engineering **193** (2017) 012006 doi:10.1088/1757-899X/193/1/012006

Amylase production potentials of bacterial isolates obtained from the gut of *Oryctes rhinoceros* larvae

P C Aryati[1], A Pangastuti[1]*, and S L A Sari[1]

[1]Department of Biology, Faculty of Mathematics and Natural Sciences, Sebelas Maret University, Ir. Sutami street No. 36A Surakarta, Central Java 57126, Indonesia

E-mail: artinipangastuti@gmail.com

Abstract. Amylase is one of the main enzymes used in industry, such as food, detergent, textile, and pharmaceutical industry. Amylase can be produced by plants, animals, and microorganisms. However, bacterial and fungal amylases have dominated application in industries. This research was aimed to determine amylolytic activity of bacteria isolated from the gut of *Oryctes rhinoceros* larvae. Based on clear zone formation, 9 from 11 isolates showed amylolytic activity. Isolates with the widest clear zone, i.e *Bacillus subtilis* GOR1, *Bacillus cereus* GOR3, and *Bacillus pumilus* GOR2, were screened for amylolytic activity based on reduction sugar production. The result showed that *Bacillus subtilis* GOR1 was the most potential as amylase producer, showed by the widest clear zone 5.224 cm^2 and highest reduction sugar production 0.0235 mg/ml. Highest amylase specific activity (0.1447 U/mg protein) was obtained at 60°C and pH 7. Amylase activity was stable for 3 hours at 60°C with residual activity respectively was 59.7%.

1. Introduction

Enzyme is protein catalyst or biocatalisator that specific against chemical reactions and importantly used in industries [1]. Amylase is one of main enzyme in industry, such as food, detergent, textile, and pharmaceutical industry [2]. Amylase can be obtained by plants, animals, and also microorganisms. But, bacterial and fungal amylases have dominated application for industries. Microbial amylase is being economical, easy to manipulate, more controllable, and relatively more stable than the enzymes derived from plants and animals [3].

Amylase is enzyme which degrade starch into glucose through a hydrolysis reaction [4]. Amylase can be divided into several types: α-amylase, β-amylase, γ-amylase, pullulanase, α-glucosidase and cyclodextrin enzyme producer [5]. Endoamylase is a well-known α-amylase which able to cleave α,1-4 glycosidic bonds present in the inner part of the amylose or amylopectin chain. Exoamylase is either exclusively cleave α,1-4 glycosidic bonds such as β-amylase or cleave both α,1-4 and α,1-6 glycosidic bonds like amyloglucosidase or glucoamylase and α-glucosidase [6].

Microbial amylases are most preferred in industry. The gut of *O. rhinoceros* larvae is potential source of amylolytic bacteria producing amylase. The industrial application of enzymes often requires to be conducted at higher temperature for enzyme reaction, because under those conditions productivity improved with less contamination. Therefore, in the present study was mainly focused to determine amylolytic activity of bacteria isolated from the gut of *O. rhinoceros* larvae and to determine the optimum conditions, such as temperature, pH and stability for amylase activity.

2. Materials and Methods

Bacterial Strain

The isolates that were used in this study were obtained from the gut of *O. rhinoceros* larvae and had been identified based on 16S rRNA protein squence. Based on the 16S rRNA sequence, 10 isolates were classified into the *Bacillus* and 1 isolate was classified into *Citrobacter* [7].

Medium

Screening for the amylolytic activity used starch agar medium which composed of 0.2 g $NaNO_3$, 0.05 g $MgSO_4$, 0.005 g K_2HPO_4, 0.001 g $FeSO_4$, 0.002 g $CaCl_2$, 0.0002 g $MnSO_4$ and was added 0.05 g yeast extract, 2% agar and 1% *soluble starch.*

Amylase production used medium contained starch which as same as screening medium, without adding agar.

Screening for the amylolytic activity

Screening for the amylolytic activity was done based on clear zone formation on starch agar plate and reduction sugar production. Eleven isolates were grown on starch agar plates and were incubated at 37°C for 48 hours. After incubation, the plates were flooded with Iodine (0.01 M I_2-KI solution) and observed for clear zone formation. 3 isolates with highest clear zone formation were screened based on reduction sugar production.

Amylase Production

Inoculum was produced by inoculating 10 ml *Luria Bertani* (LB) medium (containing 1% *soluble starch*) with 1 ml bacterial isolate and incubated at 30°C, pH 7, 150 rpm for 24 hours. 10 ml inoculum inoculated in 100 ml production medium and incubated in the same condition.

Enzyme Extraction

To obtain crude enzyme, 24-hours old cultures were transferred to centrifuges tubes and centrifuged at 6000 rpm and 5°C for 20 min. The resultant supernatant was used as crude enzyme extract.

Enzyme Assay

Amylase assay was carried out using a mixture reaction consisting of 1 ml substrate (1% *soluble starch* in 0.05 M phosphate buffer pH 7) and 1 ml crude enzyme extract. The mixture was determined with *dinitrosalicylic acid* (DNS) method. Total protein content was estimated by Bradford method using *Bovine Serum Albumin* (BSA) as the standard.

Characterization of Enzyme

2.1.1. Effect of Temperature.

To study the effect of temperature on amylase activity, 1 ml crude enzyme extract with 1 ml substrate (pH 7) was incubated at various temperature (30°, 40°, 50°, 60°, and 70°C) for 30 min.

2.1.2. Effect of pH.

To study the effect of pH on amylase activity, 1 ml crude enzyme extract and 1 ml substrate (with different pH 4, 5, 6, 7, and 8) at optimum temperature for 30 min.

Stability Enzyme

Crude enzymes extract were incubated for 1, 2, 3, 4, and 5 hours at various temperature 30°, 40°, 50°, and 60°C. Then, analyzed for amylase activity at optimum conditions.

IOP Conf. Series: Materials Science and Engineering **193** (2017) 012006 doi:10.1088/1757-899X/193/1/012006

3. Results and Discussion

Amylase Producer

The bacteria isolated from the gut of *O. rhinoceros* larvae were screened based on clear zone formation. It was found that 9 strains showed amylase activity (table 1). Then, 3 strains with the widest clear zone formation were screened based on reduction sugar production (table 2).

***Table 1**. Amylolytic activity based on clear zone formation.*

Bacteria	Clear Zone Formation (cm^2)
Bacillus subtilis GOR1	5.224*
Bacillus pumilus GOR2	4.229
Bacillus cereus GOR3	4.811
Bacillus megaterium GOR4	0.535
Bacillus thuringiensis GOR5	0.673
Bacillus aquimaris GOR6	0.239
Bacillus aryabhattai GOR7	0
Bacillus cereus GOR8	0.420
Citrobacter koseri GOR9	0
Bacillus clausii GOR10	0.946
Bacillus thuringiensis GOR11	0.808

*showed the highest amylolytic activity

***Table 2**. Amylolytic activity based on reduction sugar production.*

Bacteria	Reduction Sugar Production (mg/ml)
Bacillus subtilis GOR1	0.0235
Bacillus cereus GOR3	0.0220
Bacillus pumilus GOR2	0.0213

According the tests, it was confirmed that *Bacillus subtilis* GOR1 was the most potential bacteria as amylase producer, showed by the widest clear zone 5.224 cm^2 and highest reduction sugar production 0.0235 mg/ml. Femi-Ola and Babalola [8] were reported that *Bacillus subtilis* isolated from the gut had ability to produce amylase.

Characteristics of Amylase

3.1.1. Effect of Temperature.

The highest specific activity of amylase was at 60°C (0.1148 U/mg protein) (Figure 1). Specific activity increased from 30°C and peaked at 60°C. Above those temperature, the activity would decrease. It proved that amylase had reached critical temperature, where enzymes were denaturated due to higher temperature. Sani *et al.* [9] proved that amylase from *Bacillus subtilis* had the highest specific activity at 60°C.

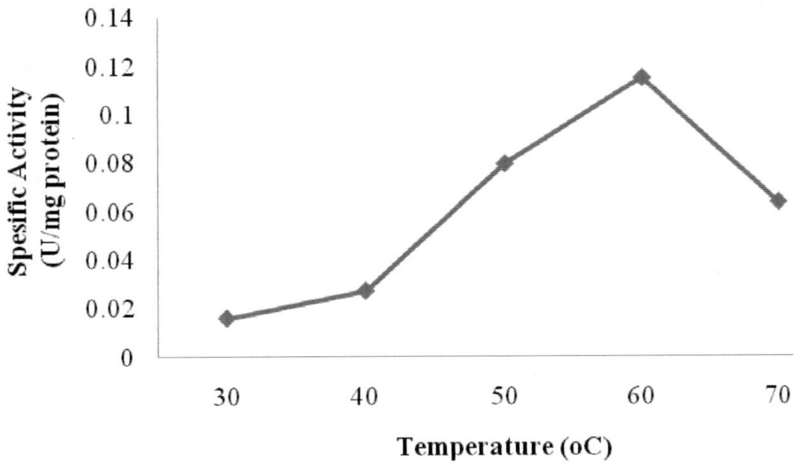

Figure 1. Effect of temperature on amylase activity *Bacillus subtilis* GOR1 at pH 7 and incubation for 30 min.

3.1.2. Effect of pH.
The effect of pH showed that the highest activity of amylase *Bacillus subtilis* GOR1 was at pH 7 (0.1447 U/mg protein) (Figure 2). The activity increased from pH 4 and peaked at pH 7. Enzymes are protein that has tertier structure and sensitive against pH condition [10]. Harris and Angal [11] reported that enzyme is very active at neutral pH. Similar results have been reported that amylase from *Bacillus subtilis* had the highest activity at pH 7 [12-13].

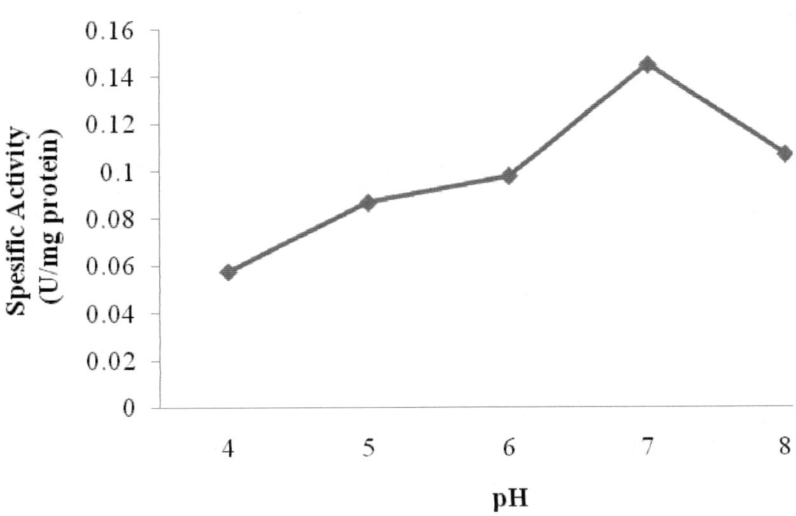

Figure 2. Effect of pH on amylase activity *Bacillus subtilis* GOR1 at optimum temperature (60°C) and incubation for 30 min.

3.1.3. Stability of Amylase.

Stability of enzymes influence by several factors, including pH and temperature. The result showed that amylase was the most stable until 3 hours incubation at 60°C with residual activity respectively was 59.7% (Figure 3). Enzyme is stable if the residual activity was more than 50% [14].

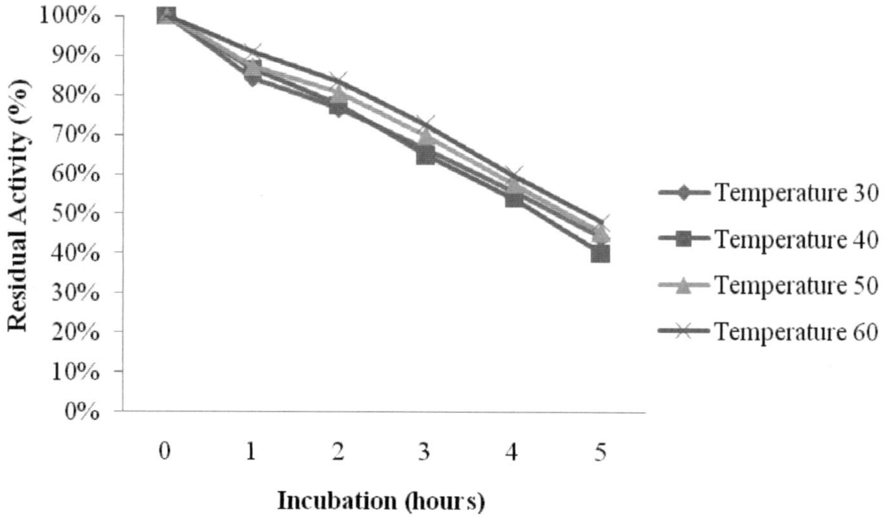

Figure 3. Effect of various temperature on stability of *Bacillus subtilis* GOR1 amylase. Activity was measured at optimum conditions (60°C, pH 7).

The results reported that the characteristics of amylase *Bacillus subtilis* GOR1 had optimum conditions at 60°C, pH 7 and most stable for 3 hours incubation at 60°C. The enzyme can be classified into thermostable enzyme that commonly used in industries.

4. Conclusion

The present study has revealed that *Bacillus subtilis* GOR1 was the most potential isolate as amylase producer, showed by the widest clear zone 5.224 cm^2 and highest reduction sugar production 0.0235 mg/ml. Highest amylase specific activity 0.1447 U/mg protein was obtained at 60°C and pH 7. Amylase activity was stable for 3 hours at 60°C with residual activity respectively was 59.7 %.

References

[1] Okoko F J and Ogbomo O 2010 *J. Micro* **4** 1-7
[2] van der Maarel M J, van der Veen B, Uitdehaag J C, Leemhuis H and Dijkhuizen L 2002 *J. Biotechnol* **94** 137-155
[3] Burhan A, Nisa C, Gokhan C, Omer A and Osman G 2003 *Proc. Chem* **38** 1397-1403
[4] Aiyer P V 2005 *African. J. Biotech* **4** 1525-29
[5] Hagihara H, Igarashi K and Hayashi Y 2001 *Appl. Environ. Microbiol* **67** 1744-50
[6] Pandey A, Nigam P, Soccol C R, Soccol V T, Singh D and Mohan R 2000 *Biotech. Appl. Biochem*
[7] Sari S L A, Pangastuti A, Susilowati A, Purwoko Tj, Mahajoeno E, Hidayat W, Mardhena I, Panuntun D F, Kurniawati D and Anitasari R 2016 *Biodiversitas* **17** 78-83
[8] Femi-Ola T O and Babalola A G 2010 *J. Insect. Sci*
[9] Sani I, Abdulhamid A, Bello F, Yahaya M and Bagudo A I 2014 *J. Microbiol. Biotech. Res* **4** 49-54

[10] Copeland A R 2000 *Enzymes: A Practical Introduction to Structur, Mechanism and Data Analysis* (New York: John Wiley-Interscience)
[11] Harris E L V and Angal S 1989 *Protein Purification Methods: A Practical Approach* (England: Oxford University Press)
[12] Gomaa E M 2013 *African. J. Microbiol. Res* **7** 3720-29
[13] Bukhari D A and Rehman A 2015 *Pakistan. J. Zool* **47** 905-911
[14] Sukmana M E, Sutrisno dan Roosdiana A 2014 *Jurnal Kimia Student* **12** 34-4

Prevalence of Antibiotic-resistance *Enterobacteriaceae* strains Isolated from Chicken Meat at Traditional Markets in Surabaya, Indonesia

R Yulistiani[1,2], D Praseptiangga[2,3], Supyani[2,4], Sudibya[2,5], D Raharjo[6], T Shirakawa[7]

[1]Department of Food Technology, Faculty of Industrial Technology, Universitas Pembangunan Nasional Veteran Jawa Timur, Jalan Rungkut Madya Gunung Anyar, Surabaya, Indonesia
[2]Graduate School Program of Sebelas Maret University (UNS), Jalan Ir. Sutami 36 A, Kentingan, 57126, Surakarta, Indonesia
[3]Department of Food Science and Technology, Faculty of Agriculture, Sebelas Maret University (UNS), Jalan Ir. Sutami 36 A, Kentingan 57126, Surakarta, Indonesia
[4]Department of Agrotechnology, Faculty of Agriculture, Sebelas Maret University (UNS), Jalan Ir. Sutami 36 A, Kentingan 57126, Surakarta, Indonesia
[5]Department of Animal Husbandry, Faculty of Agriculture, Sebelas Maret University (UNS), Jalan Ir. Sutami 36 A, Kentingan 57126, Surakarta, Indonesia
[6]Faculty of Veterinary Medicine, Airlangga University, Jalan Mulyorejo, Surabaya, Indonesia
[7]Center of Infectious Diseases, Kobe University Graduate School of Medicine, Kobe, Japan

E-mail: ratnayulistiani@yahoo.co.id

Abstract. Antibiotic resistance in bacteria from the family *Enterobacteriaceae* is an important indicator of the emergence of resistant bacterial strains in the community. This study investigated the prevalence of antibiotic-resistant *Enterobacteriaceae* isolated from chicken meat sold at traditional markets in Surabaya Indonesia. In all, 203 isolates (43 *Salmonella* spp., 53 *Escherichia coli*, 16 *Shigella spp.* , 22 *Citrobacter* spp., 13 *Klebsiella* spp, 24 *Proteus* spp., 15 *Yersinia spp.,* 7 *Enterobacter* spp., 6 *Serratia* spp., 3 *Edwardsiella* spp. were resistant to tetracycline (69.95 %), nalidixid acid (54.19 %), sulfamethoxazole/sulfamethizole (42.36 %), chloramphenicol (12.81%), cefoxitin (6.40 %), gentamicin (5.91 %). Tetracycline was the antimicrobial that showed the highest frequency of resistance among *Salmonella, E. coli, Citrobacter, Proteus* and *Erdwardsiella* isolates, and nalidixid acid was second frequency of resistance. Overall, 124 (61.08 %) out of 203 isolates demonstrated *multidrug resistance* to at least two unrelated antimicrobial agents. The high rate of antimicrobial resistance in bacterial isolates from chicken meat may have major implications for human and animal health with adverse economic implications.

1. Introduction

Antibiotic resistanceare significant health, social, and economic problem at this time. Antibiotic resistance of bacteria is biological risk, which increases morbidity and mortality of animal and human [5]. In recent years, accumulating problems with bacteria, which are resistant to antibiotics, leading to predictions that we return to the time before the discovery of antibiotics [8]. Resistant bacteria from the intestines of food animals may be transferred to retail meat products resulting from fecal contamination during various stages of the slaughter process (e.g. evisceration) and subsequent handling of animal tissue [7]. Endogenous bacterial flora may play an important role as acceptor and donor of transmissible drug resistance genes [3,13]. The *Enterobacteriaceae* family is commonly used as an indicator of fecal contamination during food microbiological analyses, and includes important zoonotic bacteria such as *Salmonella* spp., *Yersinia* spp. and *Escherichia coli*. *Enterobacteriaceae* are the significant causes of serious infection, and many of the most important members of this family are becoming increasingly resistant to currently available antimicrobials [14].

In the past decade, *multidrug-resistant Enterobacteriaceae* (MDE) has become an important challenge to disease control [15]. Gram-negative *Enterobacteriaceae* may cause severe infections and unfortunately several of the most important members of this family are becoming progressively more resistant to currently available antimicrobials [4, 6, 14]. Considering their wide spread and few options for treatment, MDE have public health importance [8].

A main goal of the present study was to investigate the prevalence of antibiotic-resistant *Enterobacteriaceae* isolates collected from chicken meat sold at traditional market in Surabaya, Indonesia. The potential repercussions of these results were in terms of microbiological safety, especially concerning the development and spread of antimicrobial resistance in the food chain.

2. Experimental

Two hundred and three (203) isolates of *Enterobacteriaceae* obtained from our preliminary study. The isolates were isolated from chicken meat sold at traditional markets (Kapas Krampung, Pacar Keling, Pucang Anom Timur, Wonokromo Baru, Soponyono, and Pabean) in Surabaya, Indonesia.

The Kirby–Bauer disk diffusion method was used to determine sensitivity or resistance to antimicrobial agents, according to guidelines developed by the Clinical Laboratory Standard Institute [2]. Antimicrobial susceptibility testing was performed for 203 isolates of *Enterobacteriaceae* from chicken meat (43 *Salmonella* spp., 53 *Escherichia coli*, 16 *Shigella spp.*, 22 *Citrobacter* spp., 13 *Klebsiella* spp, 24 *Proteus* spp., 15 *Yersinia spp.,* 7 *Enterobacter* spp., 6 *Serratia* spp., 3 *Edwardsiella* spp. and 1 *Morganella* spp.). Antibiotic discs used in this study were Tetracycline (TE 30 µg/disc, Gentamicin (CN) 10 µg/disc, Cefoxitin (FOX) 30 µg/disc, Sulfamethoxazole/ Sulfamethizole (SXT) 25 µg/disc, Nalidixid acid (NA) 30 µg/disc, Chloramphenicol (C) 30 µg/disc (Oxoid, UK).

The protocol was performed as follows: fresh cultures were inoculated into LB broth and incubated until they reached a turbidity of 0.5 using the McFarland standard. Mueller–Hinton agar plates were swabbed with these cultures, and antibiotic disks (Oxoid, UK) were placed on to inoculated plates in a sterile environment. The plates were incubated at 37 °C for 18 to 20 h. The diameters (in millimeters) of clear zones of growth inhibition around the antimicrobial agent disks were measured using a precision digital caliper (Absolute, Mitutoyo, Japan). Recommendations of the National Antimicrobial Resistance Monitoring System for *Enterobacteriaceae* isolates were utilized to define breakpoints of antibiotics and thus categorize the isolates as resistant, intermediate or sensitive classified base on the Clinical Laboratory Standard Institute [2].

3. Results and Discussion

Research findings on 203 isolates showed resistance to all antibiotics tested. The resistance levels of each antibiotic against isolates that were tested are shown in table 1. The highest resistance was to tetracycline (69.95%), followed by nalidixid acid (54.19%), sulfa-methoxazole/sulfamethizole (42.36%), chloramphenicol (12.81%), cefoxitin (6.40%) and the lowest resistance was to gentamicin (5.91 %). High resistance to tetracycline in this research can be explained by the widespread use and quite long of tetracycline on poultry farms associated with the prophylaxis, growth promotor agent, and therapeutics or their potential uses of antibiotics were not improper. Habit of improper use of antibiotics in poultry may be one trigger of antibiotic resistance particularly antibiotics are often used on the chicken farm.

Table 1. In vitro susceptibility of *Enterobacteriaceae isolates* to several antibiotics

Class and Antibiotic	Breakpoints (CLSI, 2011) S/I/R (mm)	*Enterobacteriaceae* isolates (n = 203)		
		Number of sensitive isolates (%)	Number of intermediary isolates (%)	Number of resistant isolates (%)
Tetracycline (TE)	≥ 15/12-14/ ≤ 11	47 (23.15)	14 (6.90)	142(69.95)
Aminoglycoside - *Gentamicin* (CN)	≥ 15/13-14/≤ 12	187 (92.12)	4 (1.97)	12 (5.91)
Cephalosporin - *Cefoxitin* (FOX)	≥ 18/15-17/≤ 14	185 (91.13)	5 (2.46)	13 (6.40)
Folate Pathway Inhibitors *Sulfamethoxazole/Sulfamethizole*(SXT)	≥ 16/11-15/≤ 10	112 (55.17)	5 (2.46)	86 (42.36)
Quinolones - *Nalidixid acid* (NA)	≥ 19/14-18/≤ 13	76 (37.44)	17 (8.37)	110(54.19)
Phenicols - *Chloramphenicol* (C)	≥ 18/13-17/≤ 12	159 (78.32)	18 (8.87)	26 (12.81)

Notes: R = resistant; I = Intermediate resistant, S = susceptible

The results in table 2 showed differences in the resistance level of *Enterobacteriaceae* isolates against antibiotics that were tested. *Salmonella* spp. and *E. coli* had the highest resistance to tetracycline. This is possible due to *Salmonella* spp. and *E. coli* are the two dominants species of bacteria found in chicken meat, thus the bacteria are most exposed to antibiotics and can survive that have the highest resistance.

Table 2. Prevalence of antibiotic-resistance *Enterobacteriaceae* strains isolates from chicken meat

Bacterial strain	Prevalence of antibiotic-resistance (%)					
	Tetra-cyline	Genta-micin	Cefoxitin	Sulfamethoxazole/ Sulfamethizole	Nalidixid acid	Chloram-phenicol
Salmonella spp. (n =43)	36 (83.72)	0 (0.00)	2 (4.65)	14 (32.56)	24 (55.81)	6 (13.95)
Escherichia coli (n =53)	42 (79.24)	7 (13.21)	1 (1.89)	28 (52.83)	29 (54.72)	5 (9.43)
Shigella spp. (n=16)	8 (50.00)	0 (0.00)	1 (6.25)	8 (50.00)	14 (87.50)	2 (12.50)
Citrobacter spp. (n=22)	17 (77.27)	1 (4.54)	2 (9.09)	10 (45.45)	12 (54.54)	5 (22.73)
Klebsiella spp. (n=13)	8 (61.54)	2 (15.38)	1 (7.69)	5 (38.46)	7 (53.85)	2 (15.38)
Yersinia spp. (n=15)	7 (46.67)	1 (6.67)	2 (13.33)	2 (13.33)	7 (46.67)	1 (6.67)
Proteus spp. (n=24)	17 (70.83)	1 (4.17)	1 (4.17)	12 (50.00)	12 (50.00)	3 (12.50)
Enterobacter spp. (n=7)	3 (42.86)	0 (0.00))	1 (14.28)	3 ((42.86)	2 (28.57)	1 ((14.28)
Serratia spp. (n=6)	2 (33.33)	0 (0.00)	2 (33.33)	2 (33,.33)	3 (50.00)	1 (16.67)
Erdwardsiella spp.(n=3)	2 (66.67)	0 (0.00)	0 (0.00)	2 (66.67)	1 (33.33)	0 (0.00)
Morganella spp. (n=1)	0 (0.00)	0 (0.00)	0 (0.00)	0 (0.00)	0 (0.00)	0 (0.00)

Bacteria cells become resistant to tetracycline by at least three mechanisms: enzymatic inactivation of tetracycline where an acetyl group is added to the molecule causing inactivation of the drug, efflux resistance gene encodes a membrane protein that actively pumps tetracycline out of the cell, and ribosomal protection which blocking tetracycline from binding to the ribosom [1, 10].

Table 3. Identified *Enterobacteriaceae* strains, number of isolate, percentage of antibiotic resistance and resistance profile

Bacterial strain	Number of isolate	Percentage of antibiotic resistance (%)	Resistance Profile
Salmonella spp.	43	39 (90.70)	TE, FOX, SXT, NA, C
Escherichia coli	53	46 (86.79)	TE, CN, FOX, SXT, NA, C
Shigella spp.	16	15 (93.75)	TE, FOX, SXT, NA, C
Citrobacter spp.	22	19 (86.36)	TE, CN, FOX, SXT, NA, C
Klebsiella spp.	13	10 (76.92)	TE, CN, FOX, SXT, NA, C
Yersinia spp.	15	9 (0.60)	TE, CN, FOX, SXT, NA, C
Proteus spp.	24	20 (83.33)	TE, CN, FOX, SXT, NA, C
Enterobacter spp.	7	4 (57.14)	TE, FOX, SXT, NA, C
Serratia spp.	6	5 (83.33)	TE, FOX, SXT, NA, C
Erdwardsiella spp.	3	2(66.67)	TE, SXT, NA
Morganella spp.	1	0 (0.00)	----

Resistance profile of *Salmonella spp., Escherichia coli, Shigella spp., Citrobacter spp., Klebsiella spp, Proteus spp., Yersinia spp., Enterobacter spp., Serratia spp., Edwardsiella* isolates against six types of antibiotic exhibited multidrug-resistance to tetracycline, gentamicin, cefoxitin, sulfamethoxazole / sulfamethizole, nalidixid acid, and chloramphenicol (table 3.).

Table 4. Resistance patterns in *Enterobacteriaceae* strain isolates from chicken meat

Enterobacteriaceae strain	Number of antibiotic resistance							Multi-resistant Strain (%)
	0	1	2	3	4	5	6	
Salmonella (n=43)	4 (9.30)	13 (30.23)	13 (30.23)	9 (20.94)	4 (9.30)	0 (0.00)	0 (0.00)	26 (60.46)
E. coli (n=53)	7 (13.21)	9 (16.98)	18 (33.96)	8 (15.09)	11 (20.75)	0 (0.00)	0 (0.00)	37 (69.81)
Shigella (n=16)	1 (6.25)	5 (31.25)	4 (25.00)	4 (25.00)	2 (12.50)	0 (0.00)	0 (0.00)	10 (62.50)
Citrobacter (n=22)	3 (13.64)	5 (22.73)	4 (18.18)	7 (31.82)	2 (9.09)	1 (4.54)	0 (0.00)	14 (63.64)
Klebsiella (n=13)	3 (23.08)	2 (15.38)	3 (23.07)	3 (23.07)	2 (15.38)	0 (0.00)	0 (0.00)	8 (61.54)
Yersinia (n=15)	4 (26.67)	4 (26.67)	6 (40.00)	0 (0.00)	1 (6.67)	0 (0.00)	0 (0.00)	7 (46.67)
Proteus (n=24)	4 (16.67)	6 (25.00)	6 (25.00)	5 (20.83)	3 (12.50)	0 (0.00)	0 (0.00)	14 (58.33)
Enterobacter (n=7)	3 (42.86)	1 (14.28)	1 (14.28)	1 (14.28)	1 (14.28)	0 (0.00)	0 (0.00)	3 (42.86)
Serratia (n=6)	1 (16.67)	2 (33.33)	1 (16.67)	2 (33.33)	0 (0.00)	0 (0.00)	0 (0.00)	3 (50.00)
Erdwarsiella (n= 3)	1 (33.33)	0 (0.00)	1 (33.33)	1 (33.33)	0 (0.00)	0 (0.00)	0 (0.00)	2 (66.67)
Morganella (n=1)	1 (100.00)	0 (0.00)	0 (0.00)	0 (0.00)	0 (0.00)	0 (0.00)	0 (0.00)	0 (0.00)
Total (n = 203)	32 (15.76)	47 (23.15)	57 (28.08)	40 (19.70)	26 (12.81)	0 (0.00)	0 (0.00)	124 (61.08)

Multidrug-resistance is a state of resistance to various antibiotics at once mediated by plasmid R factor (extrachromosomal DNA) [12]. Factors R as the resistance factor can be transferred and spread by way of conjugation between strains of *Enterobacteriaceae*, such as *Salmonella* spp., *Escherichia coli, Shigella* and so on. Resistance to two or more antibiotics was found in 26 isolates of *Salmonella*, followed by 37, 10, 14, 8, 7, 14, 3, 3 and 2 isolates of *Escherichia coli*, *Shigella spp.*, *Citrobacter* spp., *Klebsiella* spp, *Yersinia spp.*, *Proteus* spp., *Enterobacter* spp., *Serratia* spp., *Edwardsiella* spp., respectively, as shown in table 4. Antibiotics resistance patterns for *Enterobacteriaceae* strains isolated showed a higher level of multi-resistant bacteria. As many as 61.08 % of *Enterobacteriaceae* isolates demonstrated *multidrug resistance* to at least three unrelated antimicrobial agents. *Salmonella* spp., *Escherichia coli*, *Shigella spp.*, *Citrobacter* spp., *Klebsiella* spp, *Proteus* spp., *Yersinia spp.*, *Enterobacter* spp., *Serratia* spp. and *Edwardsiella* spp. are *Enterobacteriaceae* strains which *multidrug resistance.* The high prevalence of antimicrobial resistance identified in the present study can be explained by the widespread use of common antimicrobials as prophylactics, growth promoter agents, or in veterinary medicine.

Multidrug resistance in bacteria may be generated by one of two mechanisms. First, these bacteria may accumulate multiple genes, each coding for resistance to a single drug, within a single cell. This accumulation occurs typically on resistance (R) plasmids. Second, multidrug resistance may also occur by the increased expression of genes that code for multidrug efflux pumps, extruding a wide range of drugs [11]. Food contamination with MDR bacteria is a major problem for public health and could be transferred to bacteria of clinical significance.

4. Conclusion

These results showed that the most spread resistance in *Enterobacteriaceae* were isolated from chicken meat sold at traditional markets in Surabaya Indonesia. The antibiotic resistant against tetracycline, gentamicin, cefoxitin, sulfamethoxazole/sulfamethizole, nalidixid acid, and chloramphenicol. Tetracycline was the highest frequency of resistance among *Salmonella, E. coli, Citrobacter, Proteus* and *Erdwardsiella* isolates. Nalidixid acid showed high frequency of resistance among *Shigella, Salmonella, E. coli, Citrobacter* and *Klebsiella* isolates. High percentage (60.08%) isolates demonstrated *multidrug resistance* to at least three unrelated antimicrobial agents. These findings confirm that chicken meat is a reservoir of multidrug-resistant *Enterobacteriaceae* strains, and suggest that the farming and postharvest technology are needed to improve so that the resulting meat that is free of contaminants or bacterial contamination with a low level of resistance.

References

[1] Byarugaba D K 2010 Chapter 2 *Mechanisms of Antimicrobial Resistance* In : Sosa , A de J , Byarugaba DK, Amabile C, Hsueh PR, Kariuki S, Okeke In (Eds.) *Antimicrobial Resistance in Developing Countries,* Springer Science Business Media, http://www.springer.com/978-0-387-89369-3
[2] Clinical and Laboratory Standards Institute (CLSI) *Seventeenth informational supplement* 42- 46
[3] Davies J 1994 *Science* **264** 375-382
[4] Denton M 2007 *Int. J. of Antimicrobial Agents* **2** 9-22
[5] EFSA 2008 *Foodborne antimicrobial resistance as a biological hazard, Draft Scientific Opinion of the Panel on Biological Hazards* (Question No EFSA– Q-2007- 089)
[6] Fritsche T R, Strabala P A, Sader H S, Dowzicky M J, Jones R N 2005 *Diagn. Microbiol. Infect. Dis.* **52** 209-213
[7] Jackson T C, Marshall D L , Acuff G R , Dickson J S 2001. Meat, Poultry, and Seafood In : Doyle M P, Beuchat L R , Montville T J (Eds.) *Food Microbiology: Fundamentals and Frontiers* 2*nd edition* (Washington: ASM Press)
[8] Khan A S, Dancer S J, Humphreys H 2012 *J. Hosp. Infect.* **82** 143-151
[9] Keyser P, Elofson M, Rossel S, Wolf-Watz H 2008 *J. Intern. Med.* **264** 17-29

[10] Nguyen F, Starosta AL, Arenz S, Sohmen D, Dönhöfer A and Wilson D N 2014 *Biol. Chem* 2014 **395** 559–575
[11] Nikaido H 2009 *Annu. Rev. Biochem* 2009 **78** 119–146
[12] Pratt W B 1997 *Fundamental of Chemotherapy*. (New York: Oxford University Press)
[13] Sunde M , Fossum K , Solberg A , Sorum H 1998 *Microb. Drug Resist*. 4289-299
[14] Paterson D L 2006 *Am. J. Med.* **119** 520–528
[15] Wellington E M H, Boxall A B A, Cross P , Feil E J, Gaze W H, Hawkey P M, Johnson-Rollings A S, Jones D L, Lee N M, Otten W, Thomas C M, Williams AP 2013 *Lancet. Infect. Dis.* **13** 155-165

IOP Conf. Series: Materials Science and Engineering **193** (2017) 012008 doi:10.1088/1757-899X/193/1/012008

Effect of Incubation Time and Sucrose Addition on the Characteristics of Cheese Whey Yoghurt

E Nurhartadi, R Utami, A Nursiwi, A M Sari, E Widowati, A P Sanjaya, E A Esnadewi [1]

Department of Food Science and Technology, Faculty of Agriculture, Sebelas Maret University

E-mail: edhi.nr@gmail.com

Abstract. The effect of incubation time and concentration of sucrose addition on the characteristics of cheese whey yogurt (lactic acid content, pH, total lactic acid bacteria, antioxidant activity, viscosity) and sensory characteristics (color, odor, flavor, consistency, and overalls) were investigated. The cheese whey yogurt fermentation process was carried out for 24h and 36h with the addition of sucrose 8, 10, and 12% (w/w) of total solid, respectively. The results showed that the lactic acid content, total lactic acid bacteria, antioxidant activity, and viscosity of cheese whey yogurt were affected by the incubation time and sucrose addition. The level of pH of yogurt which was incubated at 24h and 36h were relatively in the same levels, which were 4.51 up to 4.63. Due the sensory characteristic of cheese whey yogurt the panellists gave the high score for the cheese whey yogurt which was incubated at 24h and sucrose addition 12% (w/w) of total solid. The cheese whey yogurt has 0.41% lactic acid content; pH 4.51; 7.09 log total lactic acid bacteria cells / ml; 5.78% antioxidant activity; and 5.97 cP viscosity. The best sensory and physico-chemical characteristic of cheese whey yogurt was achieved by 24h incubation time and 12% concentration of sucrose addition.

1. Introduction

Whey is a by-product from the cheese or casein manufacturing. The whey was produced from milk coagulation process in cheese making. Whey contains a high lactose content which can be degraded by lactic acid bacteria to produce lactic acid in the fermentation process. These by-products can be converted to value-added products, *e.g* yogurt. Yogurt is a fermented dairy product that produced by lactic acid bacteria. The lactic acid produced during the fermentation process coagulates milk proteins that results a gel contribute to the overall texture. During the fermentation process, starter cultures also produce flavor compounds which contribute to the flavor of yogurt [1].

The fermentation process is the most important stage of yogurt production. During this stage, the yogurt curd is formed, and its textural characteristics and specific flavor are developed. One of important factor to maintain the quality of yogurt is incubation time. Incubation time influences the flavors, texture, and appearance of yogurt [2]. During incubation, the starter culture growth results in an increase the number of bacteria. As a result, the metabolites are produced by the starter culture larger during the incubation. The concentration of sugar addition also influences the growth of the starter culture and affect to the yogurt's flavors. In this study, the incubation time and sugar addition were investigated on the cheese whey yogurt production.

IOP Conf. Series: Materials Science and Engineering **193** (2017) 012008 doi:10.1088/1757-899X/193/1/012008

2. Materials and Methods

2.1. *Materials*

Mozzarella cheese whey waste was obtained from small and medium enterprises cheese industry in Central Java, Indonesia. Sucrose "Gulaku Premium" was use as material on the sugar addition treatment. Pure culture of *Lactobacillus delbrueckii* spp. *bulgaricus* FNCC 0041 and *Streptococcus salivarius* spp. *thermophillus* FNCC 0040 were obtained from the Laboratory Processing and Process Engineering of Agricultural Products, Faculty of Agriculture, Sebelas Maret University, Surakarta, Indonesia.

2.2. *Preparation of yogurt starter*

Preparation of yogurt starter culture was described by Septiani with simple modification [3]. Pure culture was inoculated into MRS broth (Merck). Then, 0.261 g MRS broth (Merck) was dissolved in 5 ml of distilled water. The media solution was sterilized at 121°C for 15 minutes. *L. bulgaricus* and *S. thermophillus* were used as pure yogurt starter culture which then was inoculated in MRS broth (Merck) and incubated at 37°C for 24 h. The inoculated media broth was centrifuge at 1000 rpm for 10 minutes. The pellet of media broth which contained *L. bulgaricus* and *S. thermophillus* was transferred into pasteurized skim milk which was prepared by mixing 20 g of skim milk powder in 200 ml of distilled water and then pasteurized at 85°C for 30 minutes. The inoculated skim milk was incubated at 37°C for 24 h. The yogurt starter culture was ready to use when the number of bacteria was 10^6 cells / ml.

2.3. *Cheese whey yogurt production*

The cheese whey yogurt fermentation process was carried out for 24h and 36h with the addition of sucrose 8, 10 and 12% (w/w) of total solid, respectively. Each flask contained 200 ml cheese whey which had been pasteurized at 75°C for 30 min. The pasteurized cheese whey was added with 8, 10, and 12% (w/w) sucrose, respectively. Then, *L. bulgaricus* and *S. thermophillus* were inoculated as 1% (v/v) with a cell density 10^8 cells/ml into the pasteurized cheese whey which was added with 8, 10, and 12% (w/w) sucrose at 37°C. The inoculated cheese whey was incubated at 37°C with 24h and 36h of incubation time. All these experiments were performed in triplicate, and the lactic acid content, pH, total lactic acid bacteria, antioxidant activity, viscosity, and sensory characteristic of cheese whey yogurt were determined after the incubation time was 24h and 36h.

2.4. *Consumer acceptance testing*

The acceptance test was carried out with 32 consumers (aged 19–21 years), preselected according to their interest and habits of yogurt consumption. Consumer evaluation was performed according to an hedonic scoring from 1 (dislike very much) to 5 (like very much) for aspect color, odor, flavor, consistency, and overalls appreciation. Samples were presented under similar conditions (temperature and quantity) for all panelists, according to Setyaningsih [4].

2.5. *Analytical methods*

Lactic acid content was determined according to IS 1166-1973 method [5]. Total lactic acid bacteria was calculated using Petroff-Hausser Bacteria Counter. Viscosity of cheese whey yogurt was performed using Falling Ball Viscometer. All characteristic analyses were carried out in triplicate. Experimental data were subjected to one-way analysis of variance (ANOVA) using the Duncan Multiple Range Test (DMRT) test ($P < 0.05$). The IBM SPSS software 17.0 was used.

3. Results and Discussion

3.1. Chemical and Physical Properties of Cheese Whey Yogurt

The coagulation of the milk proteins on the yogurt production caused by the secretion of lactic acid by lactic acid bacteria cultures. The metabolism of lactose into lactic acid resulted in low pH level [1]. Lactic acid content indicates of the amount of lactose has been metabolized by lactic acid bacteria. Lactic acid content of yogurt with cow dairy ingredients ranged from 0.5-1.0 %. The lactic acid produced by *L. bulgaricus* is the L isomer, while that by *S. thermophillus* is the D isomer of 50-70%. Yogurt contains L-lactic acid isomer, and this lactic acid is more rapidly metabolized by the body [6].

The lactic acid content and pH profile are shown in Table 1. The highest lactic acid content was achieved by cheese whey yogurt which was added with 12% sucrose and incubated for 36 h. The lactic acid content of cheese whey yogurt which was incubated for 24h was lower than the corresponding lactic acid of incubated yogurt for 36h. The lowest lactic acid content was obtained from cheese whey yogurt which was added with 8% sucrose and incubated for 24h. Increasing sucrose addition and longer incubation time resulted in increasing lactic acid content in yogurt. The longer incubation time resulted in the increasing of total acid of yogurt [7].

Table 1 Chemical and physical properties of cheese whey yogurt

Incubation Time	Sucrose concentration	Lactic acid content (%)*	Total Lactic Acid Bacteria (log cells / ml) *	Antioxidant Activity (%)*	Viscosity (cP) *
24h	8%	0.33^a	7.255^f	3.60^c	5.994^a
	10%	0.41^b	7.221^d	7.46^f	5.808^a
	12%	0.42^b	7.093^a	5.78^e	5.975^a
36h	8%	0.49^c	7.250^e	5.14^d	6.651^b
	10%	0.49^c	7.127^c	3.47^b	5.655^a
	12%	0.53^c	7.113^b	2.06^a	9.033^c

*Notation different letters in the same column indicate significant difference at a significance level of 5%

The lactic acid content of cheese whey yogurt was 0.33% up to 0.53%. SNI No. 2981: 2009 reported that the lactic acid content of yogurt has range from 0.5% up to 2.0% [8]. Dairy Australia also reported that the standard of lactic acid content in yogurt is 0.7-1.0% [9]. The cheese whey yogurt with 12% sucrose addition and 36h time incubation has the lactic acid content which is allowed by SNI No. 2981: 2009 and Dairy Australia Standard.

The decreasing pH level during the fermentation of yogurt was caused by increasing lactic acid content. The carbohydrate addition on yogurt can be degraded by lactic acid bacteria and produce organic acid [6]. As a result, the pH of yogurt was decreased from 4.6 to 4.5. The concentration of hydrogen ions increases during the fermentation of yogurt due to a decreasing pH level [10]. The synergism effect between *S. thermophilus* and *L. bulgaricus* is based on their individual characteristics, and as a result higher lactose metabolism and lactic acid production is attained compared to each one acting individually [1]. The pH of the yogurt is not more than 4.5. Fermented milk products have standard maximum pH 4.5 [6]. The cheese whey yogurt which was incubated for 24h and 36h has the pH level 4.63-4.52 and 4.56-4.45, respectively.

During the yogurt fermentation, the symbiotic culture induces changes in the native components of the milk that are responsible for the physicochemical and sensory characteristics of yogurt [1]. In this study, *S. thermophillus* and *L. bulgaricus* were used as culture starters. *S. thermophillus* produces significant levels of lactase, which catalyzes the hydrolysis of lactose to glucose and galactose due to producing of pyruvic acid, formic acid, CO_2 and folic acid. The presence of lactose degradation product by *S. thermophillus* in the lowered oxygen tension and formic, then this condition stimulates the growth of *L. bulgaricus*. When the pH of the yogurt reaches up to 5.0, the activity of *S. thermophillus* subsides and *L. bulgaricus* gradually dominates the overall fermentation process until the fermentation process finishes [1]. *L. bulgaricus* grow rapidly after *S. thermophillus* reach stationary phase [11]. *L. bulgaricus* produces the amino acid valine, glycine and histidine required by *S. thermophillus* [12].

The highest total lactic acid bacteria were achieved by cheese whey yogurt which incubated for 24h and added 8% sucrose. The lowest total lactic acid bacteria were obtained by cheese whey yogurt with 12% sucrose addition and incubation for 24h. In this study, the increasing of sucrose addition causes the decreasing of the total lactic acid bacteria.

Milk and dairy products can be useful as an antidote to oxidation with particular mechanism. Milk and bioactive peptides are released by hydrolysis of the milk during the fermentation process can showed antioxidant activity. The antioxidant activity of whey protein caused by the presence of lactoferrin, α-lactalbumin and β-lactoglobulin contained in the milk also contributes to add the antioxidant activity of yogurt [13].

The content of DPPH on *plain* yogurt on the market is around 60 up to 80% [14]. Previous study reported that the addition of yogurt cultures increase the antioxidant activity up to 45.17% [15]. The complex reaction between peptides, amino acids and enzymes in yogurt causes the antioxidant activity. As a result, shelf life of yogurt is longer. In Table 1 showed that the antioxidant activity of cheese whey yogurt was highest in 24h incubation time and the concentration of sucrose 10%, *i.e.* 7.46%. The lowest antioxidant activity had showed by cheese whey yogurt that incubated of 36h and 12% sucrose concentration, *i.e.* 2.06%.

Viscosity is the consistency of a product that shows the resistance of a fluid to flow and stirring. During the fermentation process, sucrose and lactose is degraded by yogurt starter culture into lactic acid. The residual sucrose, lactose and organic acids which are not degraded will be counted as total dissolved solids. The higher total dissolved solids in solution will increase the viscosity level [16]. The viscosity of cheese whey yogurt with a 24h incubation time and the addition of sucrose 8, 10, and 12% were lower than that of yogurt with a 36h incubation time and the addition of sucrose 8%, 10% and 12%. The coagulation of casein that was caused by the pH of the yogurt reaches isoelectric point in an increasing of the viscosity of yogurt. The lactic acid produced by lactic acid bacteria cause an increasing the acidity of yogurt. As a result, the point of negative ion charge isoelectric occurs in milk. A part from casein and calcium binds with negative ionic charge, causing clotting [17].

3.2. Sensory Evaluation of Cheese Whey Yogurt

Cheese whey yogurt was evaluated for sensory attributes using acceptance test after production (Figure 1). The higher scores for cheese whey yogurt color value were found for cheese whey yogurt with 24h incubation time and sucrose addition 10%. The corresponding value was 3.31. The color of cheese whey yogurt is depend on the riboflavin content on whey that gives a greenish yellow color [16]. In this study, the color value of cheese whey yogurt was not affected by the incubation time and sucrose addition level. The higher scores for cheese whey yogurt odor value were found for cheese whey yogurt with 24h incubation time and sucrose addition 8%. The specific flavor of cheese whey yoghurt occurs during the fermentation process of yogurt that is produce non-volatile acid, volatile acid, carbonyl compounds and other compounds [17]. In this study, the higher concentration of sucrose addition produced the lower acceptability value of flavor characteristic.

Acceptance tests showed that no significant difference ($P > 0.05$) for the flavor and consistency of the cheese whey yogurt. In general, these results showed the good acceptance of the cheese whey yogurt. As shown in figure 1a,b, the cheese whey yogurt which was incubated for 24h had a higher number for color, flavor, and overall acceptance than that of the cheese whey yogurt with 36h incubation time. The cheese whey yogurt with 36h and 8% sucrose addition had the lowest score for overall acceptance (figure 1b). The panelists of the acceptance test give a highest overall score on cheese whey yogurt with 24h incubation time and 12% sucrose addition.

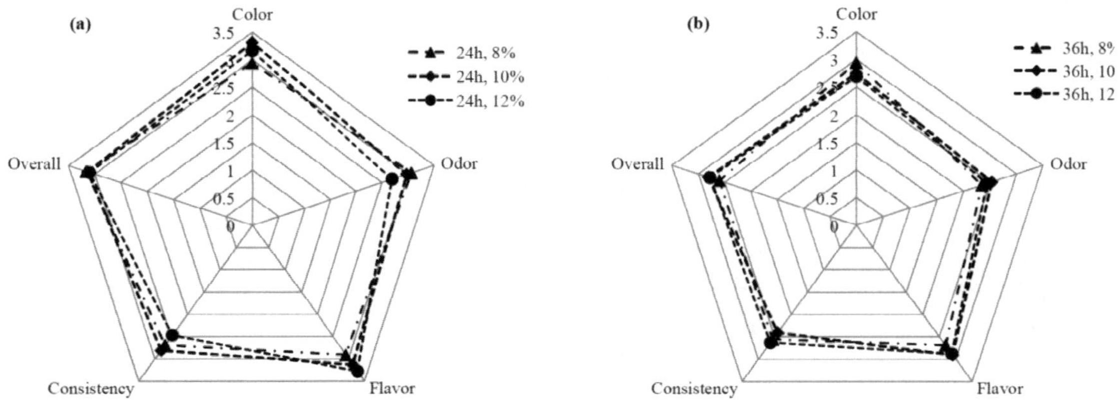

Figure 1. Graphical representation of sensory acceptance of the cheese whey yogurt with 24h (a) and 36h (b) incubation time and sucrose addition 8, 10, and 12%.

4. Conclusions

The characteristic of cheese whey yogurt, *i.e* lactic acid levels, total lactic acid bacteria, antioxidant activity, and viscosity was affected by incubation time and sucrose addition of cheese whey yogurt. Sensory acceptance test of all parameters of yogurt with 24h incubation time and sucrose addition 12% was most preferred.

Acknowledgments

This work was financially supported by research project of PNBP UNS No: 632/UN27.21/LT/2016 from Sebelas Maret University, Indonesia.

References

[1] Sfakianakis P and Tzia C 2014 *Foods* **3** 176-193
[2] Tamime A Y and Robinson R K 2007 *Yogurt: Science and Technology* 3rd ed (Cambridge: Woodhead Publishing LTD)
[3] Septiani A H, Kusrahayu and Legowo A M 2013 *Animal Agriculture Journal* **2** 225-231
[4] Setyaningsih, Dwi, Apriyantono A, dan Sari M 2010 *Analisis Sensori untuk Industri Pangan dan Agro* (Bogor: IPB Press)
[5] FSSAI 2012 *Manual of Methods of Analisis of Food: Milk and Milk Products* (India: Ministry of Health and Family Welfare)
[6] Ranjan Sharma (2013) *Sensory Quality Aspects of Yogurt* (http://www.dairyaustralia.com.au/~/media/2b52e95da7e24f54b97961f78234f3fc.pdf)
[7] Kartikasari, Dian dan Nisa F C 2014 *Jurnal Pangan Agroindustri* **2** 239-248
[8] SNI 2009 *Yogurt* National Standardization Agency ISO 2981: 2009
[9] Food Standards Australia New Zealand Act. 2014 *Fermented Milk Products* (Food Standard Code-Standard 2.5.3.)
[10] Misrianti B 2013 *Effects of Sucrose on Making Additions Buffalo Whey Fermentation against Inhibition of Bacterial Pathogens* (Thesis) (Makassar: Faculty of Animal Science University of Hasanuddin Makassar)

[11] Widodo 2003 *Biotechnology Dairy Industry* (Yogyakarta: Lacticia Press)
[12] Jannah, Alifah, Legowo A, Pramono Y, Al-Baari A dan Abduh S 2014 *J. Aplikasi Teknologi Pangan* **3** 7-11
[13] Gjoergievski, Nikola, Juliana Tomovska, Dimitrovska G, Makarijoski B and Shariati M A 2014 *JHED* **8** 88-92
[14] Gulfem Unal 2012 *Agro Food Ind. Hi Tech* **3** 39-42
[15] Sintasari, Ayu R Kusnadi J dan Ningtyas D W 2014 *Jurnal Pangan Agroindustri* **2** 65-75
[16] Hurley and Walter 2009 *Milk Composition and Synthesis: Colorful Milk* (Chicago: University of Illinois)
[17] Malcca R 2007 *Science and Dairy Processing Technology* (Makassar: Faculty of Animal Husbandry Universitas Hasanuddin)

IOP Conf. Series: Materials Science and Engineering **193** (2017) 012009 doi:10.1088/1757-899X/193/1/012009

Characteristic of Fermented Whey Beverage with Addition of Tomato Juice (*Lycopersicum esculentum*)

A Nursiwi[1], E Nurhartadi[1], R Utami[1], A M Sari[1], P W Laksono[2] and E N Aprilia[1]

[1]Department of Food Science and Technology, Faculty of Agriculture, Sebelas Maret University, Indonesia
[2]Department of Industrial Engineering, Faculty of Engineering, Sebelas Maret University, Indonesia

E-mail: edhi.nr@gmail.com

Abstract. Whey is the liquid resulting from the coagulation of milk from cheese manufacture. The availability of lactose in whey and presence of other essential nutrients for the growth of microorganisms makes it one of the potential substrate for the production of different bio-products through fermentation process. Lactic acid production through fermentation from lactic acid bacteria could be an alternative processing route for whey lactose utilization. However, a problem with such approaches is the low total solids content. Sucrose and tomato juice added to increases the total solids content. The aim of this work was to study the characteristic of fermented whey beverage with different tomato juice concentration (5, 10, 15%) using probiotic bacteria *Lactobacillus acidophilus* and *Lactobacillus plantarum*. Lactic acid content, pH, antioxidant activity, and sensory properties of fermented whey beverage samples were examined after 18hours fermentation. Fermented whey beverage with 5% tomato juice obtained the highest scores for color, aroma, flavor, texture and overall attributes. The lactic acid content and pH of fermented whey beverage ranged from 0.326 to 0.437% and from 4.13 to 4.64, respectively. The highest antioxidant activity (9.073%) was found in sample with 15% tomato juice concentration. The best formulation is the sample with 5% of tomato juice concentration.

1. Introduction

Whey is the liquid resulting from the coagulation of milk from cheese manufacture [1]. Whey treatment is indispensable for the prevention of environmental pollution, because in general the cheese whey just thrown away [2]. Sweet whey contains 6.3-7% of total solids and 4.6-5.2% of lactose [3]. The availability of lactose and presence of other essential nutrients for the growth of microorganisms make the whey one of the potential substrate for the production of different bio-products through biotechnological means [4]. The fermentation of whey by lactic acid bacteria allows the production of beverages with significantly improved characteristics [5].

However, a problem with such approaches is the low total solids content of whey, which reduces the efficiency of processing [6]. This situation causes fermented beverage products made from whey will be more watery than commercial fermented milk products, so it will affect the quality and characteristics of the product [2]. The total solid can be increased by adding of sucrose and fruit juice. Sucrose is the substrate used in the fermentation process as a carbon source and as a flavoring. Addition of 10% sucrose in fermented whey beverage produces gives good taste and have optimum level of lactic acid content and viable bacteria cells [7]. The addition of fruit juice can increase the total solid content and can disguise fresh whey flavour [8]. The addition of tomato juice on the fermented beverage substrate can stimulate and speed up the growth of *Lactobacillus acidophilus*, as well as increases the total number of bacteria. In addition, it can improve the utilization of sugar with plenty convert lactose into lactic acid, and also can lower the pH faster [9]. FAO/WHO recommended that the addition of fruit in yogurt drink ranged from 5 to 15%.

Fresh red tomato contains 4,64 g/100 g of carbohydrate, 19,1 mg/100 g of vitamin C and 4600 μg/100 g of lycopen [10]. Tomato is source of fructooligosaccharides (FOS). FOS will be fermented into lactic acid and short-chain carboxyl [11]. In addition, it has been proved that the FOS selectively stimulating the growth of *Bifidobacteria* and *Lactobacilli* [12]. Up till now, there was no study about the addition of tomato juice in fermented whey beverage. The aim of this study was to choose the best formulation of fermented whey beverage according to the sensory properties and chemical characteristic.

2. Materials and Methods

2.1. Materials

Mozzarella cheese whey waste was obtained from small and medium enterprises cheese industry in Central Java, Indonesia. Red tomato was obtained from Pasar Gede in Central Java, Indonesia and sucrose "Gulaku Premium" was use as material on the sugar addition treatment. Pure culture of *Lactobacillus acidophilus* FNCC 0051 and *Lactobacillus plantarum* FNCC 0027 were obtained from Center of Food and Nutrition Studies, Gadjah Mada University, Indonesia.

2.2. Preparation of mother culture

Each pure culture of *Lactobacillus acidophilus* FNCC 0051 and *Lactobacillus plantarum* FNCC 0027 propagated by taking 1 oose of bacterial culture then put in test tubes containing 5 ml of MRS broth sterile and then incubated at 37°C for 24 hours. The result of bacterial culture in MRS broth was then centrifuged at 1000 rpm for 10 minutes. 3 oose of obtained natan then inoculated in 5 ml of 10% skim milk solution which has passed the pasteurization process at 85°C for 30 minutes. Then incubated at 37°C for 24 hours.

2.3. Preparation of starter culture

100 ml of 10% skim milk pasteurized at 85°C for 30 minutes and then cooled to 37°C. After that, inoculated with 2% mother culture, then incubated at 37°C for 24 hours.

2.4. Preparation of tomato juice

Fresh red tomatoes weighed, cut in small parts and put in a blender for mashed, then filtered to obtain tomato juice.

IOP Conf. Series: Materials Science and Engineering **193** (2017) 012009 doi:10.1088/1757-899X/193/1/012009

2.5. Fermented whey beverage production

Cheese whey (100, 95, 90, 85%), tomato juice (0, 5, 10, 15%), and sucrose (10% w/v of the total volume of cheese whey and tomato juice) mixed and pasteurized at 75°C for 30 minutes, then add 0.7% high methoxyl pectin (w/v of the total volume of cheese whey and tomato juice). Subsequently cooled to 37°C and inoculated using 1% of starter culture *Lactobacillus acidophilus* FNCC 0051 and *Lactobacillus plantarum* FNCC 0027 with ratio of 1:1 (v/v of the total volume of cheese whey and tomato juice) with the number of bacteria 10^8 CFU, then mixed until homogeneous. Furthermore incubated at 37°C for 18 hours.

Each formulation of fermented whey beverage has 3 repetitions of samples and 2 replications analysis. The pH values of the samples were determined using a pH meter (pH Testr 20), the lactic acid content were determined with NaOH titration, the antioxidant activity were determined using DPPH method. Twenty five semi trained panelists were selected for sensory evaluation using 5-points hedonic scale. The results were analyzed statistically by Analysis of Variance (ANOVA) using completely randomized design with significance at $p<0.05$. The analysis was performed using SPSS 16.0.

3. Results and Discussion

3.1. Sensory Characteristic of Fermented Whey Beverage with Addition of Tomato Juice

Table 1. Sensory characteristic of fermented whey beverage

Formulation	Parameter				
	Color	Aroma	Taste	Consistency	Overall
0%	2,52±0,82^{a}	2,88±0,93^{ab}	3,28±1,02^{ab}	2,88±0,88^{ab}	3,12±0,78^{a}
5%	3,64±0,76^{b}	3,32±0,75^{b}	3,60±0,82^{b}	3,20±0,91^{b}	3,68±0,75^{b}
10%	2,84±1,18^{a}	2,72±0,79^{a}	2,68±1,31^{a}	2,64±0,70^{a}	2,76±1,05^{a}
15%	2,72±0,61^{a}	2,72±0,74^{a}	2,68±1,07^{a}	2,56±0,87^{a}	2,64±0,76^{a}

*Notation different letters in the same column indicate significant difference at a significance level of 5%

Score 1: extremely dislike, 2: dislike, 3: neutral, 4: like, 5: extremely like

3.1.1. Color

Color parameter for sensory assessment showed that the variation of tomato juice concentration provides the only significant difference in the concentration of 5% tomato juice only. The addition of tomato juice provides color red tinge on the sample. It can be seen that the panelists preferred the samples with the increased variety of tomato juice with a tinge of red compared to the sample without the addition of tomato juice with greenish yellow color.

3.1.2. Aroma

Aroma parameter showed that the variation of tomato juice concentration provides significant difference. The addition of 5% tomato juice on the fermented whey beverage can cover the undesirearoma of whey. However, the addition of tomato juice with higher concentration (10 and 15%) less preferred because it produces too sour scent.

3.1.3. Taste

Taste parameter indicated that the variation of tomato juice concentration provides significant difference in the 5% tomato juice. The addition of tomato juice on fermented whey beverage formulations are preferred by the panelist than samples without the addition of tomato juice, because the addition of tomato juice can mask the unfavorable taste of fresh whey. However, the addition of tomato juice with higher concentration (10 and 15%) less preferred because it produces too sour taste.

3.1.4. Consistency

Consistency parameter showed that the variation of tomato juice concentration provides significant difference in the 5% tomato juice. This is probably due to the influence of the other parameters so that the panelists preferred the 5% tomato juice formulation as the sample with the most preferred consistency.

3.1.5. Overall

Overall parameter showed that the variation of tomato juice concentration provides significant difference in the sample with 5% concentration of tomato juice, it is similar to the other parameters that placed the sample with the addition of 5% tomato juice as the most preferred formulations. Data presented in Table 1 indicate that all parameters had the impact on the overall acceptance of the fermented whey beverage. Disappearance of the cooked flavour and taste of fresh whey in the examined new beverages could probably be related to the metabolising of n-hexanal and pentanal by lactic acid bacteria as well as due to the masking effect of the new pleasant attributes, emerging after fermentation and fruit supplementation [13]. Nonvolatile compounds may have contributed significantly to explaining flavor variation, especially in differentiation of tomato juice concentrations. These compounds should therefore be considered and analyzed further as some of them may be relevant to the flavor of processed whey beverages [6].

3.2. Chemical Characteristic of Fermented Whey Beverage

Table 2. Chemical characteristic of fermented whey beverage

Formulation	pH	Lactic acid content (%)	Antioxidant Activity (%)
0%	4,64±0,03[d]	0,326±0,00[a]	8,703±0,15[a]
5%	4,51±0,03[c]	0,356±0,01[b]	8,732±0,33[a]
10%	4,37±0,02[b]	0,390±0,01[c]	8,987±0,14[b]
15%	4,13±0,07[a]	0,437±0,01[d]	9,073±0,13[b]

*Notation different letters in the same column indicate significant difference at a significance level of 5%

3.2.1. pH

The result indicate there is statistically significant difference between all formulations of fermented whey beverage samples. During the 18 hours of the fermentation process, the pH decreased from the initial pH of cheese whey that is in the range of 5.6 to 6.12 to less than 4.7. Decreasing pH during the fermentation process occurs because of the breakdown of lactose by β-galactosidase into lactic acid thus increasing acid levels in line with the fermentation time. Increased lactic acid will be followed by increase of H^+, which means decrease in pH [14]. The addition of tomato juice causes the decreasing of pH because of the characteristics of tomato juice itself is acidic, so the more tomato juice added, the pH will be lower.

3.2.2. Lactic acid content

The result indicate there is statistically significant difference in the levels of lactic acid content in all formulations of fermented whey beverage samples. The more nutritional components in the material, then the metabolic activity of the lactic acid bacteria will be higher, so the pH level decreased and lactic acid content increased throughout the fermentation process [15].

3.2.3. Antioxidant activity

The result indicate there is statistically significant difference between the antioxidant activity. From the analysis it was found that the higher the concentration of tomato juice, the higher the antioxidant activity of fermented whey beverage. This is because in red tomatoes contained high antioxidant lycopen [16]. The antioxidant activity of fruit associated with the fenolic content, and will increase during fermentation by lactic acid bacteria. In addition, processing tomatoes will optimize antioxidant activity [10].

4. Conclusion

Variations of the tomato juice concentration had effect on the characteristics of the fermented whey beverage. The higher tomato juice concentration added, the higher the levels of lactic acid content and antioxidant activity. The higher tomato juice concentration added, the smaller pH value of fermented whey beverage. The best formulation of fermented whey beverage according to the sensory properties and chemical characteristic is the sample with 5% of tomato juice concentration.

Acknowledgments

This work was financially supported by research project of PNBP UNS No: 632/UN27.21/LT/2016 from Sebelas Maret University, Indonesia.

References

[1] Onwulata C I and Peter J H 2008 *Whey Processing, Functionality and Health Benefits* (State Avenue: IFT Press)
[2] Fatma, Soeparno, Nurliyani, Hidayat C and Taufik M 2012 *J. Agritech* **32** 352-61
[3] Jeličić I, Božanić R and Tratnik L 2008 *J. Mljekarstvo* **58** 257-74
[4] Panesar P S, Kennedy J F, Knill C J and Kosseva M 2010 *Braz. arch. biol. technol.* **53** 219-26
[5] Bulatović M Lj, Rakin M B, Mojović L V, Nikolić S B, Sekulić M S V and Vuković A P Đ 2014 *Chem. Ind. Chem. Eng. Q* **20** 1-8
[6] Gallardo-Escamilla F J, Kelly A L and Delahunty C M 2005 *J. Dairy Sci.* **88** 2689–99
[7] Kar T and Misra A K 1999 *Revista de Microbiologia* **30** 163-9
[8] Djurić M, Carić M, Milanović S, Tekić M and Panić M 2004 *Eur Food Res Technol* **219** 321-8
[9] Babu V, Mital B K and Garg S K 1992 *Int J Food Microbiol* **17** 67-70
[10] Kailaku S I, Dewandari K T and Sunarmani 2007 *Buletin Tek Pascapanen Pertanian* **3** 50-8
[11] Dominguez A L, Rodrigues L R, Lima N M and Teixeira J A 2013 *Food and Bioprocess Technol* **6**
[12] Young Ng S, Chia L W, Padam B S and Chye F Y 2014 *J. Pharm. Nutr. Sci.4* 92-9
[13] Dalev D, Bielecka M, Troszynska A, Ziajka S and Lamparski G 2006 *Pol. J. Food Nutr. Sci.* **15** 71-7
[14] Nisa A K dan Agustin K W 2016 *J. Pangan dan Agroindustri* **4** 367-76
[15] Cho Y H, Shin I S, Hong S M and Kim C H 2015 *Korean J. Food Sci. An.* **35**189-96
[16] Thompson K A, Marshall M R, Sims C A, Wei C I, Sargent S A and Scott J W 2000 *J. Food Sci.* **65** 791-5

Detection of chicken contamination in beef meatball using duplex-PCR Cyt b gene

E P Sari[1], L R Kartikasari[1] and M Cahyadi[1]
[1]Department of Animal Science, Faculty of Agriculture, Universitas Sebelas Maret, Surakarta, Indonesia 57126

E-mail: mcahyadi@uns.ac.id

Abstract. Beef is one of expensive animal protein sources compared to other meats, on the other hand, chicken is cheap animal protein source. Mixing of chicken into beef meatball is possibly performed to decrease production cost. The aim of this study was to detect chicken contamination in beef meatball using Cytochrome b (Cyt b) gene by duplex-PCR. Sample was designed and prepared as follows, 100% of chicken meatball, 100% of beef meatball and serial level of chicken contaminations in beef meatball (1, 5, 10 and 25%, respectively). Isolation of DNA genome from meatball was according to the guideline of gSYNC^{TM} DNA Extraction Kit for animal tissue. The PCR reaction was carried out using KAPA2G Fast Multiplex Mix. This study found that the DNA genome was succesfully extracted. Moreover, chicken contamination in beef meatball was indicated by the presence of 227 bp DNA band on 2% of agarose gels. Current study revealed that duplex-PCR using Cyt b gene as a genetic marker was able to detect chicken contamination in beef meatball until 1% of chicken meat in the sample. It can be effectively used to identify contamination and also authenticate species origin in animal products to protect consumer from undesirable contents in the food.

1. Introduction

Meat is one of animal protein resources which can be processed into various meat products. One of the meat products favored by Indonesian people is meatball [1]. The main raw material from meatball is beef which prices are more expensive than other meats. These conditions trigger meatball producers to mix with other meats from different species. Mixing of chicken into beef meatball is possibly performed to decrease production cost [2], on the other hand, chicken is easily to be found in the market [24]. There are many cases of mixing chicken into beef meatball at different cities in Indonesia to decrease production cost [3] [4] [5]. Therefore, species detection method is required to authenticate species origin in animal products to protect consumer from undesirable contents in the food. Polymerase chain reaction (PCR) is an enzymatic method to multiply exponentially a certain nucleotide sequence specific by in vitro. The PCR method is highly sensitive, so it can be used to amplify the DNA molecule [6]. PCR method has been used for identification of chicken adulteration in different meat products and raw meat [7] [8]. However, substitution of high-cost meats with low-cost meat species is considered fraudulent practices if it is not indicated on the label. Adulteration has been violated consumer protection laws. This regulation creates a consumer protection system containing elements of legal certainty and transparency of information and access to information [9]. Analysis of PCR from specific sequence of mitochondrial DNA is a method that is easy, fast and can be identified for several species at the same time [10], cooking process did not affect to the DNA identification [11]. One of the genes in the DNA of mitochondria is Cyt b [12] which has a sequence variation that is used as a marker to distinguish material originating from different animal species [13]. Cyt b have been used to characterized various vertebrates [14], it can be used to identify vertebrates species which contain specific species information [15]. Here, we performed duplex-PCR to amplify two target fragments from beef, chicken and serial level of chicken contaminations in beef meatball sample. Therefore, the aim of this study was to detect chicken contamination in beef meatball using Cyt b gene by duplex-PCR.

2. Methods

2.1 Sample Collection

Beef and chicken were obtain from Pasar Gede traditional market in Surakarta, Central Java. Each sample was labeled to avoid undesireable contamination.

2.2 Meatball Sample Preparation

A total of 30 meatball was designed and prepared. Meatballs were made by grinding the meat with different grinder in order to avoid undesireable contamination. Beef and chicken were mixed with the following serial level of chicken contaminations, 100% of chicken meatball, 100% of beef meatball and serial level of chicken contaminations in beef meatball (1, 5, 10 and 25%, respectively). Contaminated meat were prepared and mixed until homogeneous with flour, salt, pepper and garlic [16] then they were cooked in a boiled water for 5-10 minutes until the meatball rised to the surface.

2.3 Isolation of DNA Genome

Isolation of DNA genome from meatball sample was according to the guideline of gSYNC™ DNA Extraction Kit for animal tissue. 25 mg of each meatballs was weighed and transfered into a 1.5 ml microtube, and then 200 µl of GST buffer and 20 µl of proteinase K was added. The mixture was mixed thoroughly using the vortex and then incubated at 60°C overnight until the sample lysate becomes clear. During incubation, 1200 µl elution buffer (200 µl for each sample) were transfered to a 1.5 ml microtube and heated 60°C. The supernatant was transfered to a new microtube, then 200 µl of GSB buffer was added and vortex thoroughly for 10 seconds until sample and GSB buffer were mixed thoroughly. A total 200 µl of absolut ethanol was added to the sample lysate and mixed immadiately by vortex thoroughly for 10 seconds. Furthermore, GS column was placed in a 2 ml collection tube, all the mixture were transfered to the GS column then centrifuge at 14,000-16,000 × g for 1 minute. The 2 ml of collection tube containing the flow-through was discarded then the GS column was transfered to a new 2 ml collection tube. For the next step, 400 µl of W1 buffer was added to the GS column and it was centrifuged at 14,000-16,000 × g for 30 seconds then the flow-through was discarded. The GS column was placeced back into 2 ml collection tube. 600 µl of wash buffer was added to the GS column, centrifuged at 14,000-16,000 × g then the flow-through was discarded. The GS column was placed back into 2 ml collection tube, centrifuged again for 3 minutes at 14,000-16,000 × g to dry the column matrix. The dried GS column was transfered to a clean 1.5 microtube, and then 100 µl of pre-heated elution buffer was added into center of the column matrix for 3 minutes to make sure the elution buffer was completely absorbed. After that it was centrifuged at 14,000-16,000 × g for 30 seconds to elute purified DNA. The purified DNA was transfered into a clean 1.5 microtube. Finally, purified DNA were stored in refrigerator until used.

2.4 Polymerase Chain Reaction (PCR)

The PCR reaction was conducted using PCR machine (GeneAmp PCR System 9700, Singapore). The total volume of 25 µl containing of 12.5 µl of KAPA2G Fast Multiplex Mix PCR Kit (Kapa Biosystems, Inc., USA), 1 µl of forward primer, 1 µl of reverse primer for bovine and chicken respectively, 1 µl of DNA template and 8.5 µl aquabidest was carried out. Oligonucleotide primers were forward primer (5′-GACCTC-CCAGCTCCATCAAACATCTCATCTTGATGAAA-3′), reverse primer for chicken (5′-AAGATACAGATGAAGAAG-AATGAGGCG-3′) and reverse primer for bovine (5′-CTAGAAAAGTGTAAGACC-CGTAATATAAG-3′) [17]. Duplex-PCR was carried out in phases: initial denaturation at 95°C for 3 minutes, followed by 30 cycles of denaturation at 95°C for 15 seconds, annealing at 60°C for 30 seconds, extention at 72°C for 30 seconds and followed by a final extension at 72°C for 3 minutes. Duplex-PCR result was visualized using gel documentation on 2% of agarose gels.

3. Result and Discussion

3.1 Isolation of DNA Genome

The samples were designed and prepared for DNA isolation. Six DNA samples were extracted from meatballs with serial level of chicken contamination. Each DNA samples was extracted from three different meatballs in composites. Specific site for various species was amplified by the primer of Cyt b gene [18]. Mitochondrial DNA has many advantages over nucleus DNA for authenticate species origin in animal product studies. Mitochondrial DNA has several folds of copy number and relatively more abundant in total nucleid acid preparation than nucleus DNA [19]. This study performed that the DNA genome was succesfully isolated from all of meatballs. Isolation of DNA genome did not affect by cooking process. Previous studies reported that DNA genome can be isolated from cooked meats [17]. The presence of DNA was visualized using gel documentation on 1% agarose gel.

3.2 Polymerase Chain Reaction (PCR)

Duplex-PCR was performed to detect chicken contamination in beef meatball. Mitochondrial DNA Cyt b gene from different species were amplified by PCR. Double stranded DNA denaturated to single stranded DNA then take up a secondary structure that is depend on its sequence [20]. Chicken contamination in beef meatball was indicated by the presence of 227 bp DNA band specific for chicken, on the other hand, the existance of bovine species was indicated by the presence of 274 bp of specific mitochondrial DNA Cyt b site for bovine (Figure 1). The presence of DNA was visualized using gel documentation on 2% agarose gel. This study was succesfully detected chicken contamination in beef meatball. The chicken contamination in beef meatball can be detected until 1% of chicken contamination level. The cooking process did not affect the identification of chicken in beef

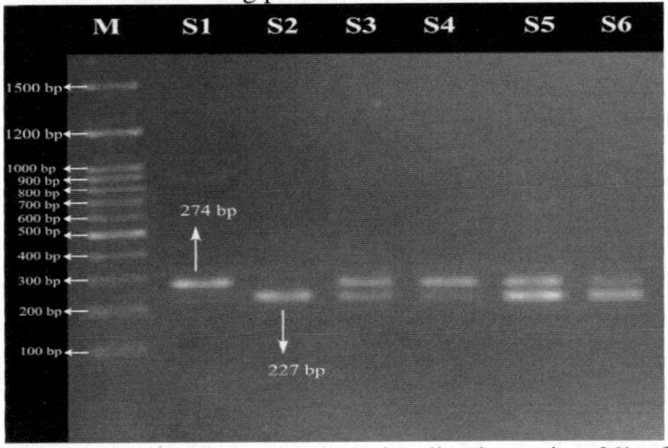

Figure 1. Gel Documentation visualization using 2% of agarose gel. M was 100 bp marker, S1 was 100% of beef meatball, S2 was 100% of chicken meatball and S3 to S6 was serial level of chicken contamination in beef meatball (1, 5, 10 and 25%, respectively).

meatballs. Previous study reported that detection of the other species in processed meat products with PCR [21] was succeeded in detecting the presence of a mixture of pork in beef meatballs. Furthermore, previous study successfully identified the species of meat in processed products using the PCR method including cows, goats, sheep, pigs and horse [17]. Duplex-PCR of Cyt b gene was able to identify pork in fresh and cooked beef until the level of 1% contamination [22]. In addition,

PCR methods have been used for identification of meat and meat product [23]. This current study revealed that PCR method may be applied to authenticate species origin in animal products to protect consumer from undesirable contents in meat-based food.

4. Conclusion

Duplex-PCR of mitochondrial DNA Cyt b gene was able to detect chicken contamination in beef meatball in very small proportion in the samples (1%). It may be very useful tools to authenticate species origin in animal products to protect consumer from undesirable contents in the animal products.

Acknowledgement

This study was supported by the research grant "Pusat Keunggulan-Universitas Sebelas Maret (PK-UNS)" Project Number: 632/UN27.21/LT/2016.

References

[1] Suradi K 2007 *J. Ilmu Ternak* **7** 52-57
[2] Fajardo V, Gonzalez I, Martin I, Rojas M, Hernandez P E and Garcia T 2008 *J. AOACI* **91** 103-111
[3] Agustinus M (2017, January 16) Begini cara pedagang bakso siasati mahalnya daging sapi https://finance.detik.com/ekonomibisnis/2998727/begini%20cara%20pedagang%20bakso%20siasati%20mahalnya%20daging%20sapi
[4] Purnama A (2017, January 16) Bakso dicampur daging ayam http://edisicetak.joglosemar.co/berita/bakso-dicampur-daging-ayam-108366.html
[5] Hapsari E 2013 (2017, January 16) Daging mahal, begini aksi pedagang bakso http://www.republika.co.id/berita/nasional/umum/13/02/08/mhw717-daging-mahal-begini-aksi-pedagang-bakso
[6] Yuwono T 2006 *Theory and Applications of Polymerase Chain Reaction: PCR Experiments Guide for Solving Current Biology Problem* (Yogyakarta: Andi)
[7] Ghovatti S, Nassiri M R, Mirhoseini SZ, Moussavi A H and Javadmanesh A 2009 *Food Contam.* **20** 696-699
[8] Dalmasso A, Fontanella E, Piatti P, Civera T, Rosati S and Bottero M T 2004 *Mol. Cell. Probes* **18** 81-87
[9] Undang-undang Nomor 8 Tahun 1999 tentang Perlindungan Konsumen Pasal 3D ayat (1)
[10] Erwanto Y, Abidin M Z, Muslim E Y P, Sugiyono and Rohman A 2014 *Asian Australas J. Anim. Sci.* **27** 1487-92
[11] Rastogi G, Dharne M, Bharde A, Pandav S, Ghumatkar S V, Krishnamurty R, Patole M S and Shouche Y S 2004 *Current Sci.* **87** 1278-81
[12] Meyer A 1994 *Trend Eco. Evo.* **9** 278-280
[13] Primasari A 2011 *Proc. Bogor Agricultural University* (Bogor : Bogor Agricultural University)
[14] Ong S B, Zuraini M I, Jurin W G, Cheah Y K, Tunung R, Chai L C, Haryani Y, Ghazali F M and Son R 2007 *ASEAN Food J.* **14** 51-59
[15] Parson W, Pegoraro K, Niederstatter H and Foger M 2000 *Int. J. Legal Med.* **144** 23-28
[16] Chakim L, Dwiloka B and Kusrahayu 2013 *Anim. Agr. J.* **2** 97-104
[17] Matsunaga T, Chikuni K, Tanabe R, Muroya S, Shibata K, Yamada J and Shinmura Y 1999 *Meat Sci.* **51** 143-148
[18] Kocher T D, Thomas W K, Meyer A, Edwards S V, Pääbo S, Villablanca F X and Wilson A C 1989 *Proc. Nat. Acad. Sci U.S.A.* **86** 6196-6200
[19] Alberts B, Johnson A, Lewis J, Raff M, Roberts K and Walter P 2003 *Molecular biology of the cell (4th ed.) Ann. Bot.* **91** 401
[20] Lockley A K and Bardsley R D 2000 *Trends in Food Sci. Tec.* **11** 67-77
[21] Fibriana F, Widianti T, Retnoningsih A dan Susanti 2012 *Biosaintifika* **4** 10-7

[22] Ni'mah A, Kartikasari Y, Pratama A D, Kartikasari L R, Hertanto B S and Cahyadi M 2016 *J. Indonesian Trop. Anim. Agr.* **41** 7-12
[23] Chikuni K, Tabata T, Kosugiyama M, Monma M and Saito M 1994 *Meat Sci.* **37** 337-45
[24] Ilham K 2009 *Analisis Kebijakan Pertanian* **7** 211-21

IOP Conf. Series: Materials Science and Engineering **193** (2017) 012011 doi:10.1088/1757-899X/193/1/012011

Production of Mozzarella Cheese Using Rennin Enzyme from *Mucor miehei* Grown at Rice Bran Molasses Medium

I H Rusdan[1] and J Kusnadi[2]
[1] Departement of Nutrition Science, Faculty of Medicine, Universitas Brawijaya, Malang, Indonesia.
[2] Department of Agricultural Product Technology, Faculty of Agricultural Technology UB

Email: ilzamha@ub.ac.id

Abstract. The research aimed to study the characteristic and yield of Mozzarella cheese produced by using rennin enzyme from *Mucor miehei* which is grown at rice bran and molasses medium. The popularity of Mozzarella cheese in Indonesia is increased caused by the spreading of western foods in Indonesia such as pizza and spaghetti that use Mozzarella cheese for ingredient. In Italy, Mozzarella and pizza cheeses are dominating 78% of the total Italian Cheese products. In producing Mozzarella cheese, rennin enzyme is always used as milk coagulant. Even now, Indonesia has not produced the rennin enzyme yet. The rennin enzyme from *Mucor miehei* growing at rice bran and molases medium which have the availability can be managed purposively within short period of time. The completly randomized design methode used to get the best crude extracts of *Mucor miehei* rennin enzyme, then is employed to produce mozzarella cheese. The result of Mozzarella cheese has various characteristics such as the yield's weight is 9.1%, which consists of 50% moisture content, 36.64% peotein levels, 0.1 melting ability and 82.72% stretch ability or 0.79/N. With that characteristic it is concluded that rennin enzyme from *Mucor miehei* grown at rice bran molasses medium has the potential to alternatively subtitute calf rennin to produce Mozzarella cheese, and the characteristics fulfill the standart.

1. Introduction

Mozzarella cheese is originally manufactured from high-fat bufallo milk. It is a soft, unripened cheese variety of the Pasta –filata family whose origin is from the Battipaglia region of Italy [1]. In the other European countries and the U.S., Mozzarella cheese is also made from cow milk with certain modification [2].

The recent increasing in global cheese production and the scarcity oncalf rennet which is usually obtained from the fourth stomach of suckling calves, become the driving force to look for some new rennet substitutes. In the recent decades, due to a shortage of calf rennet in world markets, alternative milk-clotting enzymes from different origins have been investigated [3, 4] . *Mucor miehei* is preferred by the industry to produce milk-coagulating enzyme because of its high *R* factor (milk-clotting activity/proteolytic activity), especially for certain peptide bonds in casein, good cheese quality, and yield [5,6] .

Solid State Fermentation (SSF) technique is one of the recent techniques particulary suitable for milk-clotting enzyme production [7, 8], while rice bran and molasses have the needed nutritions [9,10] to grow*Mucor miehei*. This work is conducted to study the characteristic and yield of Mozzarella cheese produced by using rennin enzyme from *Mucor miehei* which is grown at rice bran and molasses medium.

2. Material and Method

2.1 Material

Fresh cow milk was obtained from the village unit cooperative Junrejo, Malang. *Mucor miehei* was obtained from the chemical laboratory of Universitas Brawijaya. The fungal culture was periodically subcultured and maintained in potato dextrose agar medium. Molasses was obtained from Tasikmadu's sugar cane factory, Karanganyar, Central Java. Rice bran was obtained from the local market Dinoyo, Malang, East Java.

2.2 Method

This research used completly randomized design methode, there are 2 factors and repeted three times. The factors are additional molasses concentration, and incubation time. The data analysed by ANOVA (Analysis of Variance) methode and continued with posthoc test used least significance different (LSD) test 5%.

2.2.1 Growing Medium and Condition

This work employed SSF (Solid State Fermentation) system as the growing medium, which is 5 ml spore suspention of *Mucor miehei* was inoculated in petridish containing 1:1 of sterilized rice bran and molasses medium. Molasses was diluted by using 10% of aquadest, and added by 2% of casein. And incubated at 37°C for 96 hours.

2.2.2 Extraction of Rennin Enzyme Mucor miehei

After the incubation period, solid state fermentation medium was extracted by using destilled water (1:10 w/v) and 5 ml of tween 80, in low speed blender for 5 minutes. The obtained filtrate was filtered by using filter papper, and centrifuged at 4000 rpm for 20 minutes at 4°C. The supernatan was filtered again with whatman 42 papper and used as crude enzyme source to produce Mozzarella cheese . The obtained enzyme has 21,946 unit/ml/minute of proteolytic activity and has 5,637.251094 unit/mg protein/minute of milk-clotting activity.

2.2.3 Production of Mozzarella Cheese [11]

Firstly, cow fresh milk was pasteurized at 72°C for 5 minute, then cooled down until the milk's temperature was 40°C. Rennin enzyme from *Mucor miehei* with 0.55% (v/v) and 0.5% (v/v) concentration of citric acid was added to the pasteurized milk, and incubated at 43°C for 5 minutes (or until the curd can be cut). It was then cut and left undisturbed for 15 minutes at 43°C until the curd was separated from the whey, and the whey was separated from the curd. After that, the curd was collected in the bowl. Hot water (75° C) was added to the curd in a bowl, then the curd was mixed and kneaded by hand until homogenous paste was obtained. The curd was soaked into ice water for 1 hour, then soaked into saturated brine for 15-20 minutes. Finally the Mozzarella cheese was made.

2.2.4 Cheese Evaluation

After the Mozzarella cheese was obtained, the amount of yield is examined by measuring the cheese's weight. The moisture content was determined by AOAC [12]. The total protein content was determined by measuring the total nitrogen by using the Kjeldahl method [13]. The meltability was determining by using Tunick et al method [14], and the stretchability was determined by using Kuo and Gunasekaran method [15].

3. Result and Discussion

Rennin enzyme from *Mucor miehei* obtained from this research does not belong to a good milk-clothing activity yet. However, when *Mucor miehei's* enzyme is supported by even a little bit of commercial enzymes it could produce a better yield of cheese than a pure commercial rennin. There is a possibility of synergy between commercial rennin and rennin enzyme from *Mucor mihei*. The commercial renet used in this research to support the clotting activity is Marzyme Supreme, the double microbial-strain rennin manufactured by a USA company called Danisco. The equation used in producing *Mozzarella cheese* is 0.5% of renin enzyme from *Mucor miehei* and supported by 0.01% commercial rennin. The Mozzarella cheese which is employed for comparison (the control cheese) uses 0.25% of the commercial enzyme, according to the company rules.

3.1 Yield Percentage of Mozzarella Cheese

The comparison between the yield percentage of the Mozzarella cheese sample which is repeated three times (R1, R2 and R3) and the control cheese are presented in figure 1. The cheese yield is calculated from 100 gram of cheese per liter of milk times 100%. It is obtained that the average yield percentage of Mozzarella cheese produced by *Mucor miehei's* rennin (9.1%) is higher than Mozzarella cheese produced by commersial rennin (5%).

3.2 Moisture Percentage in Mozzarella Cheese

The result in figure 2 shows the moisture presentage in Mozzarella cheese. The average value of Mozzarella cheese produced by *Mucor miehei's* rennin (50.04%) is not significantly different from Mozzarella cheese produced by commersial rennin (49.93%).

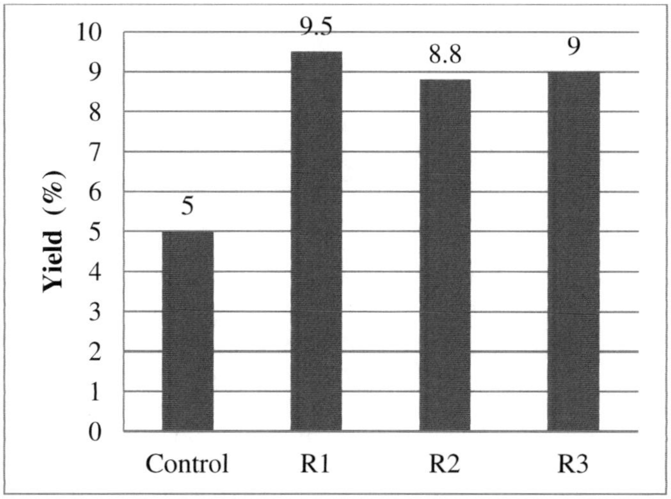

Figure 1. Mozzarella cheese yield

IOP Conf. Series: Materials Science and Engineering **193** (2017) 012011 doi:10.1088/1757-899X/193/1/012011

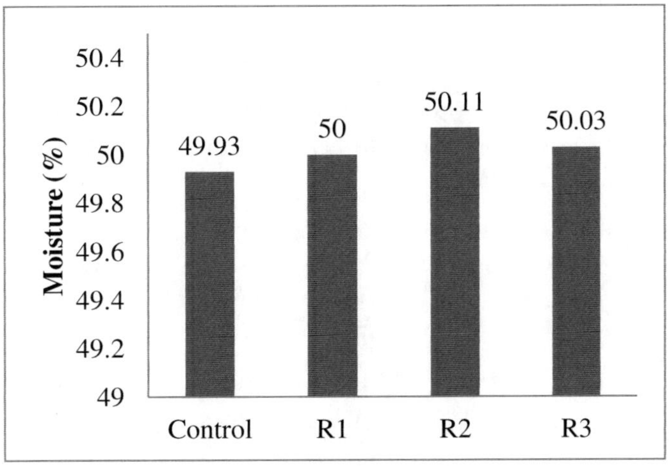

Figure 2. Moisture content of Mozzarella cheese

The percentage of moisture level from both of the products still fulfill the Mozzarella cheese moisture standart according to USDA specification for Mozarella cheeses (2012) which mentions that the moisture content in Mozzarella cheese should be more than 45% but not more than 52%.

3.3 Protein Percentage in Mozzarella Cheese

The average of protein content from Mozzarella cheese produced by *Mucor miehei's* rennin (36.64%) is lower than the protein percentage of the Mozzarella cheese produced by commersial rennin (44.7%). The data is showed at Figure 3.

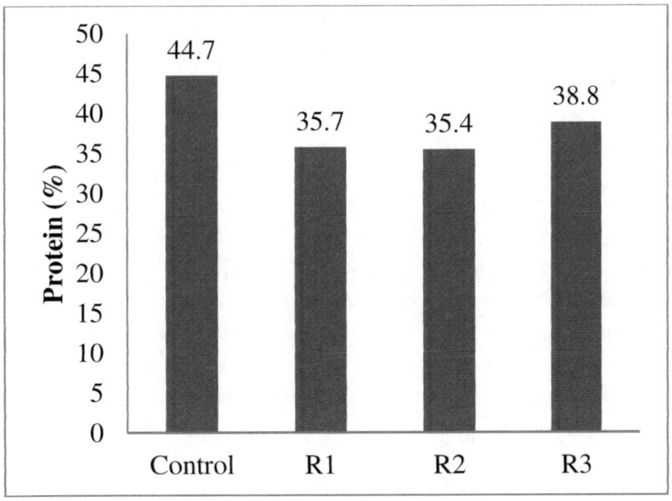

Figure 3. Percentage of protein content in Mozzarella cheese

The lower value of protein content of Mozzarella cheese produced by *Mucor miehei*'s enzyme probably because of high proteolytic activity in rennin enzyme from *Mucor miehei* grown at rice bran and molasses medium.

IOP Conf. Series: Materials Science and Engineering **193** (2017) 012011 doi:10.1088/1757-899X/193/1/012011

3.4 Meltability in Mozzarella cheese

Meltability in Mozarella cheese is one of its important functional features. The comparison of meltability between Mozzarella cheese sample which is repeated three times and the control cheese are presented in figure 4. The data shows that the meltability of the Mozzarella cheese produced by *Mucor miehei's* rennin is not significantly different from the control cheese. All of them have meltability no more than 10 mm, or have the meltability value of 1.0

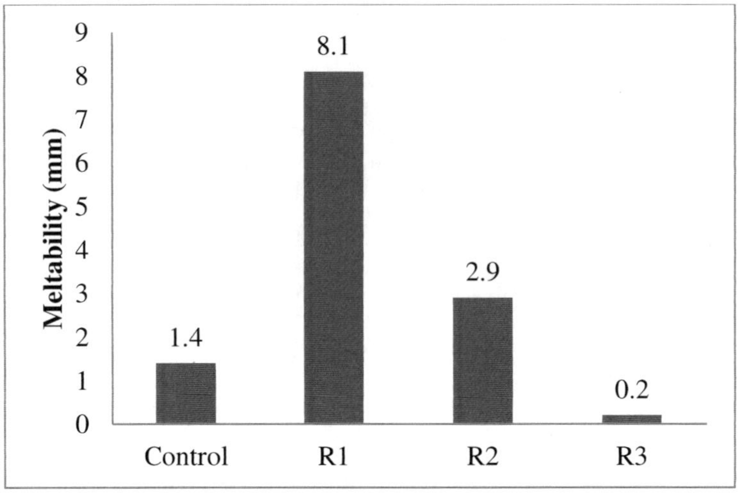

Figure 4. Meltability of Mozzarella cheese

The meltability of cheese is defined as the ease and extent of cheese to melt and spread upon heating [16]. It is found that the melting area is determined by the fat and moisture content [17]. The low meltability in this research might be caused by the low fat content in both of Mozzarella. According to Alexandra et al [18], solid state fermentation of *Mucor miehei* has the highest lipolytic activity compared to the other Mucoromycotina strain. It makes this rennin enzyme have the capability to divide lipid in cheese and decrease the lipid content significantly.

3.5 Stretchability in Mozzarella cheese

The comparison between the stretch ability of the cheese sample and the control cheese are presented in figure 5. Stretch ability percentage of Mozzarella cheese produced by *Mucor Miehe's* rennin is higher than Mozzarella cheese produced by commercial rennin or control. This is based on Kuo and Gunasekaran [15] who report that stretch ability for pasta filata ranges between 0.5- 1.2 1/N, and the average of 1/N value obtained from this research is 0.79/N or 82.72%.

The functional quality of Mozzarella cheese is defined by its ability to melt and stretch, so the value of melt ability and stretch ability are important.

4. Conclusion

In short, Mozzarella cheese produced by rennin enzyme from *Mucor miehei* has better yield percentage (9.1%), protein percentage (36.64%), and stretchability (82.72% or 0.79/N), but the value in the moisture content (50%) and meltability (0.1) are not significantly different. It is concluded that rennin enzyme from *Mucor miehei* grown at rice bran molasses medium has the potential to alternatively subtitute calf rennin to produce Mozzarella cheese, and the characteristics fulfill the standart. The low fat content of the product may even be considered as a benefit for those who prefer low fat diet.

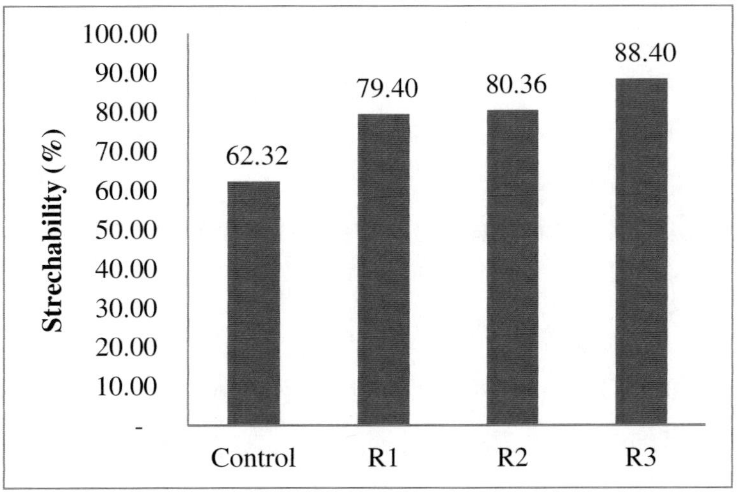

Figure 5. Strechability of Mozzarella cheese

Acknowledgement
The author wishes to thank Dr. Joni Kusnadi, Universitas Brawijaya, and Indofood Research Nugraha programe for the assistance and financial support in this research.

References

[1] Citro V 1981 *Scienza-e Tecnica lattiero-asearia* **32** 263-27
[2] Ghosh, BC and S Singh 1992 *J. Dairy Sci* **45** 199-202
[3] Escobar J and Barnett S M 1993 *Enzyme Microb. Technol.* **15** 1009–1013
[4] Thakur M S , Karanth N G and Nand K 1990 *Appl.Microbiol. Biotechnol.* **32** 409–413
[5] Leasure LL 1980 *Mycol.* **72** 83–86
[6] Levadoux W Johnson M and Andr G 1989 *J. Dairy Sci.* **72** 627–634
[7] Pandey A C R Soccol and D Mitchell 2000 *Process Biochemistry.* **35** 1153-1169
[8] Silveira G G and Contiero J 2001 *European J. Pharmaceutical Sci,* **13** S38
[9] Faria SADSC,Bassinello PZ and Penteado MDPC 2012 *Brazilian J. Pharmaceutical Sci.* **48**
[10] Curtin, Leo 1983. *Molasses general consideration.Molasses in Animal Nutrition* (Iowa: National Feed Ingredients Association West Des Moines)
[11] Carroll, R 2002. *Home Cheese Making : Recipes for 75 Homemade Cheeses.* 3th Edition. (New England)
[12] Cunniff P 1999 *Official method of analysis of AOAC* (Maryland: AOAC International)
[13] AOAC 1990 *Official methods of analysis*, 15th edition (Washington: Association of Official Analytical Chemists)
[14] Tunick M H E L, Malin P W, Smith J J, Shieh BC, Sullivan K. L, Mackey, and V H Holsinger 1993 *J. Dairy Sci.* **76**: 3621- 3628
[15] Kuo, M. I. and S. Gunasekaran 2003. *J.Dairy Sci.* **86** 1108- 17
[16] Gunasekaran S & Ak M M 2003 Cheese rheology and texture (CRC Press)
[17] Gunasekaran S, Kuo M I & Wang Y C 1998 *Australian J. Dairy Technol.* **53** 111
[18] Alexadra K, Brigitta N, Csba V, Tamas P, and Miklos Tako 2014 *Foodtechnol. Biotechnol.* **52** 73-82

IOP Conf. Series: Materials Science and Engineering **193** (2017) 012012 doi:10.1088/1757-899X/193/1/012012

Physicochemical Properties of Dietary Fibers from *Artocarpus camansi* Fruit

V Suryanti*, T Kusumaningsih and Y S Rumingtyas
Department of Chemistry, Faculty of Mathematics and Natural Sciences, Sebelas Maret University, Jl. Ir. Sutami 36A Surakarta, Central Java 57126 Indonesia

*Email: venty@mipa.uns.ac.id

Abstract. The objective of this work was to investigate the dietary fiber (DF) contents of *Artocorpus camansi* (breadnut) fruit and examine their physicochemical properties, such as water-holding capacity (WAC), oil-holding capacity (OHC) and water absorption capacity (WAC). This fruit flour contained of both water soluble fibers (SDF), such as pectin (1.95%) and gum (0.4%), and water insoluble fibers (IDF) (89.25%). The IDF content of this fruit was significantly high in respect to other DF sources. The WHC, OHC and WAC of IDF were 4.10, 2.60 and 4.0%, respectively. Moreover, the WHC, OHC and WHC of total dietary fibers (TDF) were 4.2, 4.3 and 4.6%, respectively. The results showed that the DF of fruit flour had good physicochemical properties. The findings suggested that there is a potential application of *A. camansi* of fruit as functional ingredients in the food industry.

1. Introduction

Dietary fibre (DF) is composed of a mixture of plant carbohydrate. It is categorized into two major groups, which are water soluble fibers (SDF), such as pectin and gum, oligosaccharides and some hemicelluloses, and water insoluble materials (IDF), such as cellulose, hemicellulose and lignin [1-4]. DF is found naturally in cereals, nuts, vegetables and fruits. DF has several interesting physical properties such as water and oil holding capacities, swelling capacity, increasing viscosity and/or gel formation. The capability of the DF to absorb the water is the important physiological effect of DF on human health. This swelling properties occur due to the presence of carbohydrates with free polar groups, interaction with hydrophilic links or retention within the matrix [5].

Each group of DF has different physiological effects on human. The IDF is correlated to both water absorption and intestinal regulation, whereas the SDF is related with the reduction of cholesterol in blood and the decrease of glucose absorption by the small intestine [6,7]. Generally, the SDF and IDF is useful in the prevention of cardiovascular disease and colon cancer, respectively. Many studies have reported that high-fiber consumption protects and cures against obesity, cardiovascular diseases, diabetes and some types of cancer [4]. Therefore, growing interest in searching for new sources of dietary fiber content with good physicochemical properties.

Artocorpus camansi (breadnut) is found throughout the tropics region which is native to New Guinea, Indonesia and Philippines. It belongs to Moraceae family. Trees grow to heights of 10–15 m. The spiny fruits is dull green to green yellow when ripe and have little pulp. Breadnut is usually grown for its nutritious seeds. The seeds from the mature fruits are high content in protein and relatively low in fat. The seeds are also a good source of minerals and niacin. They are usually boiled or roasted. The fruits are usually consumed as staple food. However, the information on the nutritious value of the fruits is still limited. In Indonesia, the immature fruits, including the seeds, are thinly sliced and cooked as a vegetable. This work reports the dietary fiber contents and physicochemical properties of the *A. camansi* immature fruits in order to find new sources of dietary fiber.

2. Experimental

2.1. Materials and Instruments

The mature fruits were obtained from local market. The fruit was peeled and then spread in tray and dried in oven at 50-60°C for overnight. The dried fruit was powdered in a mechanical grinder and passed through a 150 mesh sieve. The dried sample powder then were stored in hermetic bags at room temperature (25°C). High Performance Liquid Chromatograph (HPLC-UV Vis) LC-10AD Shimadzu was used.

2.2. Free Fat Sample

Dried sample (10 g) was macerated with a mixture of 15 mL methanol and 45 mL chloroform for 2 h and then filtered. The solvent was removed using a rotary evaporator. This free fat sample was used to obtained pectin and IDF.

2.3. Pectin Isolation

The free fat sample (2 g) in 100 mL of 0.1 M buffer phosphate pH 7.5 was added 85.7 mg protease (0.7 unit/mg) and incubated overnight at 37°C. It was mixed thoroughly during this period. The mixture was then filtered. Distilled water (200 mL) was added to the residues and the mixture was acidified with HCl to pH 1.5-2. The mixture was placed in water bath at 80°C for 4 h and then filtered. The resulted filtrate was mixed with 95% ethanol in the ratio filtrate:ethanol = 1:4 and left for 1 h. The mixture was then centrifuged at 15,000 rpm for 22 mins. The supernatant was discarded and the pellet was then analysed by HPLC.

2.4. Gum Isolation

The dried sample (100 g) was dispersed in 2 L distilled water. The NaOH was added to the mixture to reach pH 9.4 and was heated to 45°C for 1 h. It was mixed thoroughly during this period. The mixture was then centrifuged at 5,000 rpm for 15 mins. The supernatant was acidified with HCl to pH 4.6. The mixture was then left at cold room for overnight and then centrifuged at 10,000 rpm for 15 mins. The NaOH was then added to the supernatant to pH 8.3 and the mixture was heated to boil until the volume of the mixture was reduced by one-half. The mixture was added with 96% ethanol until the volume of ethanol reached 40% of the mixture. The mixture was mixed thoroughly and left overnight. The mixture was centrifuged at 10,000 rpm for 10 mins. The supernatant was discarded and the pellet was then analysed by HPLC.

2.5. IDF Isolation

The free fat sample (2 g) in 100 mL of 0.1 M buffer phosphate pH 7.5 was added 85.7 mg protease (0.7 unit/mg) and incubated for 2 h at 37°C. The mixture was then placed at 25°C and added EDTA 2 g. The mixture was left for 1 h and then filtered. The residue was then dispersed in 0.1 M buffer acetate pH 4 and 15.4 mg pectinase (3 unit/mg) was added to the mixture. The mixture was then placed in water bath at 25°C for 2 h. It was mixed thoroughly during this period. The mixture was then filtered and the filtrate was the insoluble dietary fibers.

2.6. Characterization of DF

Water holding capacity (WHC) and oil holding capacity (OHC) were determined following published procedures with some modification [8-9]. About 0.02 g of each sample was placed in a centrifuge tube, followed by the addition of 0.1 mL of oil or distilled water. The tube was left for 1 min at room temperature (25°C) with agitation. Subsequently, the mixture was centrifuged at 3.000 rpm for 10 mins and the oil supernatant was decanted. The wet sample was weighed, dried overnight and weight again to determine the water content. The OHC was expressed as g of oil held per g of sample, and WHC was quantified as g of water held per g of fibers. The water absorption capacity (WAC) was determined by comparing the mass of the wet sample and the dried sample.

3. Results and Discussion

3.1. Dietary Fiber Contents

In the preparation of dried samples, 25 of fruits with approximately 200 g of each fruit were used. The dried sample was obtained as 500 g of light brown powder. The SDF (such as pectin and gum) and IDF content of *A. camansi* fruits are shown in Table 1. Pectin are structural components of plant cell walls, whereas gums are found in secretory plant cells. They are highly water-soluble and have gelling ability. The gelling ability of pectin can reduce the rate of gastric emptying and alter small intestinal transit time [1,10].

The pectin and gum values of *A. camansi* fruit were 1.955 and 0.40%, respectively. The pectin value was relatively lower than that of pectin content of coconut residue which is 3.60-4.32% [11]. The IDF of some cereal derivatives, fruits and vegetables have been reported, such as for wheat bran (41.6%), barley bagasse (41.4%), oat bran (20.2%), pear (22.0%), orange (24.3%), peach (26.1), asparagus (38.6%) and artichoke (44.5) [8]. Gorinstein (2001) have reported the IDF content of apples (0.37-0.46%) and persimmons (0.66-0.87%) [12]. It is noted that the IDF content of *A. camansi* fruit, which was 89.25%, was significantly higher than the reported DF sources.

Although the SDF content for pectin and gum were relatively small, the IDF content of *A. camansi* fruit was high. The relatively high IDF content in *A. camansi* fruit recommends possible applications in dietetic–physiological products. Consumption of this fiber feasibly enhanced functioning of the digestive system and preventing disorders such as constipation and colon cancer.

Table 1. Dietary fibers content of *A. camansi* fruit

Dietary Fibers	Weight (%)
SDF	
- Pectin	1.95
- Gum	0.40
IDF	89.25

3.2. Characterization of DF

Table 2 shows the physicochemical properties of DF content of *A. camansi* fruit. The WHC represents the ability of the materials to retain water. Number of hydroxyl groups in the fiber structure allow water interaction through hydrogen bonding which caused water absorption. Water has important role on food changes during baking such as on gelatinization, denaturation, yeast and enzyme inactivation, flavor and color formation [12, 13]. The WHC is related to physiological role in intestinal function and blood sugar level control [8,14].

The WHC of the SDF from pectin and gum were lower than that of the IDF and TDF. The WHC of the IDF and TDF value were similar. The WHC values of IDF and TDF from *A. camansi* fruit were higher than that of for maize hull and (2.32 g/g) dan wheat hulls (2.48 g/g). However, they are lower than that of the soybean fiber (4.9 g/g), wheat bran (6.1 g/g) and shoyu mash residue (10.85 g/g). The results indicated that the WHC of IDF and TDF from the *A. camansi* fruit had comparable WHC of the reported DF sources. Vegetable fibers with high WHC have been added as meat product ingredients [8,15]. This finding suggested that the potential application of DF from *A. camansi* fruit in food products such as in functional bakery and meat products ingredients, in order to increase the dietary fiber intake.

The OHC of IDF and TDF from A. camansi fruit was 2.60 and 4.30 g/g, respectively. The OHC of the several DF sources have reported such as chia FRF. (2.02 g/g), jack bean (2.3 g/g) and barley (2 g/g) and the OHC of IDF from orange peel was 3.36 g oil/g sample. Particle size and the method used to isolate the fiber affect the OHC values [16]. In this case, the smaller the particles the more contact surface, as consequences, it could hold more oil [5]. It is possible to increase the OHC value of IDF from A. camansi fruit by reduction of the particle size and isolation of DF with alcohol to expose the lipophilic components. Due to its relatively low in OHC value, the A. camansi fruit is a potential ingredient in fried products since it would create a non-greasy sensation.

Table 2. Physicochemical properties of DF from *A. camansi* fruit.

No	Physicochemical Properties	SDF		IDF	TDF
		Pectin (g/g)	Gum (g/g)	(g/g)	(g/g)
1.	Water-holding capacity	1.80±0.19	1.20±0.31	4.10±0.38	4.20±0.35
2.	Oil-holding capacity	2.10±0.59	2.40±0.40	2.60±0.42	4.30±0.48
3.	Water-absorption capacity	1.80±0.08	2.30±0.16	4.30±0.05	4.60±0.10

The WAC express the fiber structure ability to spontaneously absorb water when soaked in water or placed in contact with a constantly moist surface. Water absorption initially occurs at surface, however, it can lead to swell and event solubilize [17]. The WHC of IDF and TDF from *A. camansi* fruit were higher than that of wheat hulls (2.91 g/g) carrot (0.82 g water/g sample), beet (1.58 g water/g sample) and soybean fiber (1.42 g/g). However, it was lower than that of chia FRF (11.73 g/g). The relatively high value of WAC probably due to the protein of the *A. camansi* fruit may be denaturalized by the solvent and heat treatment during defatting. A large number of exposed hydrophilic sites occurred and then interacted with water and increased the WAC. However, denaturalization would also expose hydrophobic sites and thus favor protein–protein interaction that would reduce WAC [8, 18]. In this case, the balanced of hydrophilic and hydrophobic sites might be exposed which gave relatively high value in WAC.

4. Conclusion

Overall, it could be concluded that *A. camansi* fruit had high IDF content. The WHC, OHC and WAC were relatively high and comparable with other known DF sources. This work approves that DF obtained from *A. camansi* fruit have great potential in food application, especially in development of functional foods.

Acknowledgments

The authors acknowledge their thanks to Sebelas Maret University for the financial support.

References

[1] Dhingra D, Michael M, Rajput H and Patil R T 2012 *J. Food Sci. Technol.* **49**(3) 255-266
[2] Borchani C, Besbes S, Masmoudi M, Blecker C, Paquot M and Attia, H 2011 *Food Chem.* **125** 1194-1201.
[3] Vergara-Valencia N, Granados-Pérez E, Agama-Acevedo E, Tovar J, Ruales J and Bello-Pérez L A 2007 *LWT Food Sci. Technol.* **40** 722-729
[4] Figuerola F, Hurtado M L, Estévez A M, Chiffelle I and Asenjo F 2005 *Food Chem.* **91** 395-401
[5] López G, Ros G, Rincón F, Periago M J, Martínez M C and Ortuño J 1996 *J. Agric. Food Chem.* **44** 2773-2778
[6] Grigelmo-Miguel N, Gorinstein S and Belloso O M 1999 *Food Chem.* **65** 175-181
[7] Yangilar F 2013 *J. Food Nutr. Res.* **1**(3) 13-23
[8] Alfredo V-O, Gabriel R-R, Luis C-G and David B-A 2009 *Food Sci. Technol.* **42** 168-173
[9] Chau C, Cheung K and Wong Y 1997 *J. Agric. Food Chem.* **45** 2500-2503
[10] Jenkins D J A, Wolever T M S, Leeds A R, Gassull M A, Haisman P, Dilawari J, Goff D V, Meta G L and Albert K G M M 1978 *Br Med J* **1** 1392-1394
[11] Ng S P, Tan C P, Lai O M, Long K and Mirhosseini H 2010 *J. Food Agric. Environt.* **8**(2) 172-1777

[12] Gorinstein S, Zachwieja Z, Folta M, Barton H, Piotrowicz J, Zemser M, Weisz M, Trakhtenberg S and Belloso O M 2001 *J. Agric. Food Chem.* **49** 952-957
[13] Schneeman, B. O. 1987 *Food Technol.* **41**(2) 81-82
[14] Wolever T 1990 *Am. J. Clin. Nutr.* **51** 72–75
[15] Dalgetty D and Baik B 2003 *Cereal Chem.* **80**(3) 310-315
[16] Chau C F and Huang Y L 2003 *J Agric Food Chem* **51** 2615–2618
[17] Hassana F A, Ismaila A, Hamidd A A, Azlana A, Al-sherajia S H 2011 *Food Chem.* 126 283-288
[18] Yeh H, Su N and Lee M 2005 *J. Agric. Food Chem.* **53** 4361-4366

IOP Conf. Series: Materials Science and Engineering **193** (2017) 012013 doi:10.1088/1757-899X/193/1/012013

Effect of Fermentation on Compositional Changes of *Cinnamomum osmophloeum* Kaneh Leaves

AD Kurniawati[1], TC Huang[2], J Kusnadi[3]
[1]Nutrition Department, Faculty of Medicine, Universitas Brawijaya, Malang, Indonesia 65145
[2]Department of Biological Science and Technology, College of Agriculture, National Pingtung University of Science and Technology, Pingtung, Taiwan 912
[3]Department of Agricultural Product Technology, Faculty of Agricultural Technology, Universitas Brawijaya, Malang, Indonesia 65145

E-mail: adel.kurniawati@ub.ac.id

Abstract. *Cinnamomum osmophloeum* Kaneh is known as "indigenous cinnamon" with the chemical constituents of its leave's essential oil are similar to the famous *C. cassia* inner bark oil. Its oil has long been used as a medicinal plant. Fermentation is one of the processes in tea production, which could change the compound's composition. This research aims to study the compositional changes of *C. osmophloeum* leaves during fermentation compared to unfermented leaves and commercial tea leaves. The main bioactive secondary metabolites in *C. osmophloeum* leaves extract are two flavonol glycosides. Both of this glycosides changed into aglycone during fermentation. By using HPLC and LC-MS analysis the major components and their derivative were identified. The retention time of kaempferol aglycone was 35.21 minute and the concentration showed increased from 0.46 to 46.8 μg. mL^{-1} after fermentation. There are 3 major groups lactic acid bacteria isolated from fermented *C. osmophloeum* Kaneh leaves, *Bacillus coagulans, Lactobacillus plantarum,* and *Pediococcus pentosaceus,* which it plays the key role during the compositional changes of glycosides into aglycone.

1. Introduction

Cinnamomum osmophloeum Kaneh is native to Taiwan. It is commonly known as "indigenous cinnamon" with Taiwan producing cinnamon tea leaves. Cinnamomum species or usually known as cinnamon, belonging to the *Lauraceae* family. Its widely distribute in South and Southeast Asia as spices and traditional medicines. Previous studies reported that *C. osmophloeum* has a bioactivity, such as decreasing the risk of cardiovascular diseases [1], cancer chemoprevention activity [2] and cholesterol lowering effects [3]. Taiwan produced many famous teas, due to the land conditions and the climate, which it very suitable for tea production. However it has been difficult due to the market competition. Since early 2000, Taiwan produced cinnamon tea leaves as a new product variation. Fermentation is process which complex organic compounds, such as glucose, are broken down by the action of enzymes into simpler compounds without the use of oxygen. Fermentation plays an important key to tea quality in terms of appearance, aroma, liquor and infusion [4]. Lactic acid bacteria usually produces secondary metabolites, such as enzymes during fermentation. The β-glucosidase is one of enzyme produced by the lactic acid bacteria in the fermented leaves hydrolyzes the flavonol glycosides to aglycone [5]. This experiment focused on the effects of fermentation on the compositional changes of *C. osmophloeum* leaves and identified the microorganism involved during fermentation.

2. Experiment

2.1 Plant materials and Fermentation

The leaves of *C. osmophloeum* Kaneh were collected from Yuli, Hualien County in Eastern Taiwan, dried overnight in room temperature and cut. Fresh leaves were ensilaged in an airtight mason jar as spontaneous fermentation for 4-12 weeks of storage at 37°C, and extracted every 2 weeks for analysis. Fresh leaves and fermented, extracted using 100% methanol (1:3), then shaken at 200 rpm for 24 h at 37°C. The solvent was removed then kept at -4°C for short storage.

2.2 HPLC Analysis

The analytical HPLC system consisted of a Hitachi (Tokyo, Japan) L-7100 pump, and a Model L-7400 diode array detector (DAD). Separation was carried out with a C18 column (4.6mm×100 mm, 3 μm particle size). The mobile phase consisted of acetonitrile (solvent A) and water containing 0.1% trifluoroacetic acid (solvent B). A linear gradient program was used as follows: 10% A in the first 0 min, linearly gradient to 30% A over 20 min, then hold for 10 min. The mobile phase flow rate was 0.8 mL/ min, and the detector was monitored at 265 nm. All the chromatographic operations were carried out at ambient temperature, 25 °C [2].

2.3 LC-MS Analysis

The mass spectrometer was run in electrospray ionization mode (ESI). Samples were loaded on to LC symmetry C18 column bio basic with diameter 150 x 2.1 mm, particle size 5 μm. The mobile phase consisted of 5% acetonitrile + 0.1 formic acid in deionized water (solvent A) and 95% acetonitrile + 0.1% formic acid (solvent B). The mass spectrophotometer operated in positive ion mode using ion trap analyzer and mass range scan of 150 – 1000 m/z with a flow rate of 200 μm/min. The sheath gas flow rate was 50 arb, spray voltage applied for full mass scan was 4 kV and the capillary voltage was 30v with capillary temperature of 300°C. Mass spectral data was acquired using Thermo-Xcalibur™ data acquisition (Thermo Scientific).

A linear gradient program was used as follows for the HPLC gradient: 10% A in the first 0 min, linearly gradient to 30% A over 20 min, then hold for 10 min. The LCQ Fleet ion trap mass spectrometer was operated in positive electrospray mode.

2.4 Bacterial Isolation

One gram of *C. osmophloeum* leaves, fresh and fermented, were cut, then diluted with 90 ml sterile saline solution (0.85% sodium chloride), then a serial dilution was conducted. For the last 3 serial dilution and original samples then plated onto de Man-Rogosa-Sharpe (MRS) agar supplemented with $CaCO_3$, and incubated at 37°C, 48 hours. The lactic acid bacteria produced a clear zone.

The DNA cells were extracted using KIT DNA extraction (FavorPrep Tissue Genomic DNA Extraction Mini Kit #FATGK001, Favorgen Biotech Corp., Taiwan), and confirmed by mean of specific species by PCR. The PCR amplification was then sent to the Genomics Bioscience and Technology Co. Ltd, Taiwan for DNA sequencing.

The PCR analysis was conducted using two primers (partial 16S rRNA gene amplification). 5'-AGAGTTTGATCCTGGCTCAG-3' as forward primer (P1b 16-F #12041224)) and 5'-GGCTGCTG GCACGTAGTTAG-3'as reverse primer (P1b 16-R #12041225), were purchased from Genomic Biosd & Tech. The reaction mixture (30 μL) contained 10 μL PCR master mix (Thermo Starr), 6 μL DNA template, and 2 μL for each primer. After an initial denaturation of 4 minutes at 96°C, 30 cycles of 30 sec at 96°C, 30 sec at 56°C, 45 sec at 72°C, and final extension at 72°C for 4 min were performed.

2.5 Measurement of kaempferol aglycone

The quantification of kaempferol aglycone in fermented leaves of C. osmophloeum was performed using calibration with standard solution of kaempferol (Sigma) in the range from 5, 10, 20, 40, 80, and 100 $\mu g.mL^{-1}$. The peak area were correlated with the concentrations according to standard. Standard stock solution were made by diluted kaempferol with methanol 100 $\mu g.\ mL^{-1}$. All data are expressed as mean standard deviation of triplicate independent experiment (n=3).

3. Results and Discussion

3.1 Physical Changes during Fermentation

Fresh leaves were ensilaged in an airtight mason jar. A spontaneous fermentation started to occur after 4 weeks of storage at 37°C until 8 weeks, and extracted every 2 weeks for analysis. This experiment focused on fermentation of *C. osmophloeum* Kaneh, fresh and fermented, then extracted using 100% methanol, and incubated in a shaker for 24 hours. During tea fermentation, tea flavor and color are complex problems since there a number of volatile and nonvolatile compounds formed during this oxidation process on fine granules of tea.

(A) (B) (C)

Figure 1. The color change of *C. osmophloeum* Kaneh leaves before extraction using 100% methanol from three different samples, (A) fresh leaves (without fermentation), (B) fermented leaves after 8 weeks fermentation, and (C) commercial cinnamon tea (produced by Wenberli Biotechnology Co)

The fresh leaves were green, while the fermented leaves, which have a similarities with the commercial cinnamon tea leaves, were brown. The quality of made tea can be determined by visual inspection of the optimum color during the process as shown at figure 1. The flavor of the fermented leaves also increased significantly compared to the fresh leaves, and it has similar flavor as the commercial one.

The situation is further complicated by the act that the flavor is unstable and its odor is changing with time as and when chemical composition of the tea changes. These changes are contribute towards the color and flavor of made-tea [6], that a series of chemical reactions take place during this process due to the severe damage of the leaf cells, which is mostly influenced by oxidized flavonol formed during fermentation. Proteins get degraded, the chlorophyll are transformed into pheophytins and some volatile compounds are generated due to transformations of certain aroma precursors present in leaves. Sharma and Bhattacharya (2013) reported that at the time of fermentation grassy smell continues change to the floral smell due to some complex chain of biochemical reaction inside the tea leaf and greenish color changes into the coppery brown.

3.2 Chemical Changes during Fermentation

Every samples from fermentation time of *C. osmophloeum* leaves were extracted, following the procedure above. The volume of 20 µL of the filtered solution was injected directly and separated under the optimum condition mentioned earlier. Each sample was determined in triplicate. Peaks in the chromatograms were identified by comparing the retention times and on-line UV spectra with those of the standards. Retention time for compounds 1 and 2 were 23.4 and 25.1 minute, and compound 3 were identified as kaempferol aglycone has retention time around 35.21 minute, respectively (see figure 2).

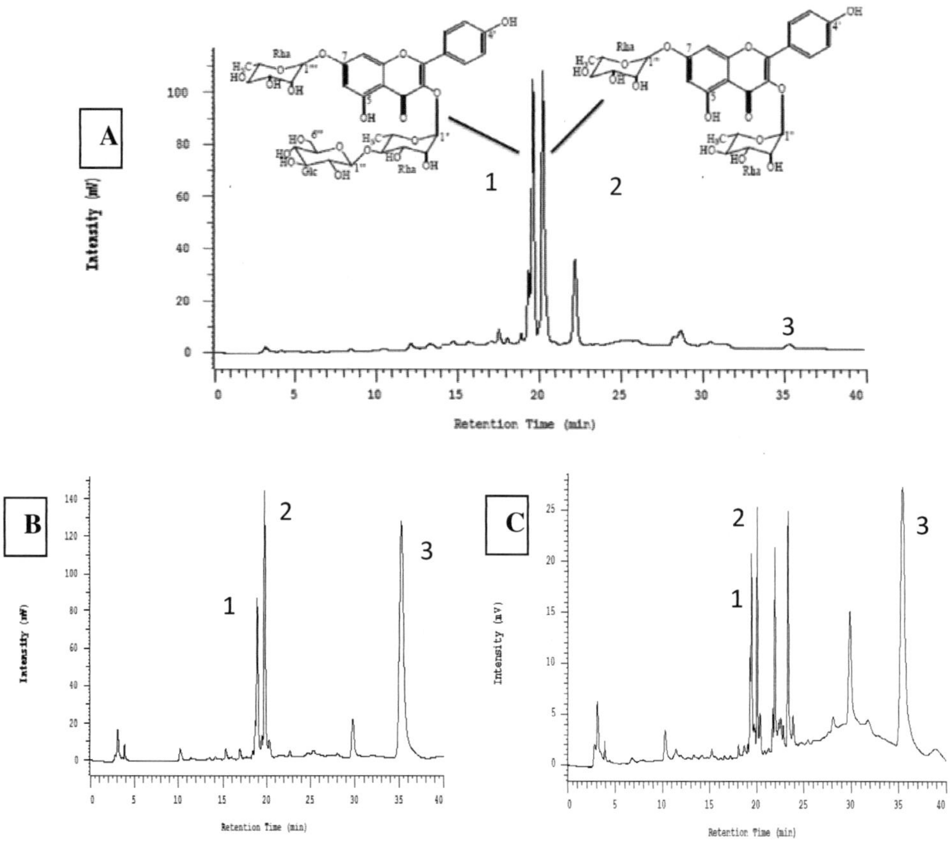

Figure 2. Representative HPLC profiles of methanol extract of *C. osmophloeum* Kaneh leaves. These figures compared among the extraction of 3 different samples. A) Fresh leaves, B) 8 weeks fermented leaves, C) commercial tea. Compound 1 identified as kaempferol 3-O-β-D-glucopyranosyl-(1→4)-α-L-rhamnopyranosyl-7-O-α-L-rhamnopyranoside, and compound 2 is kaempferitrin (in accordance with previous research [2]), compound 3 is kaempferol (aglycone) which identified by compared with kaempferol standard (Sigma) then continued with LC-MS identification.

Fresh *C. osmophloeum* Kaneh leaves showed two major glycosides after analysis using high performance liquid chromatographic (HPLC) as shown in figure 2. Compound 1 and 2 were identified as kaempferol 3-O-β-D-glucopyranosyl-(1→4)- α-L-rhamnopyranosyl-7-O-α-L-rhamnopyranoside and kaempferitrin as identified in the previous research, by using HPLC and NMR analysis [2]. Both compounds are glycosides.

These results are in accordance with the previous research [2] about determination of kaempferol glycosides in *C. osmophloeum* leaves by HPLC. Glycosides are water soluble and less reactive and volatile than aglycone, possible explaining why plant (fresh leaves) store a great number of compounds in glycosides form [7]. The determination were initially by isolation and characterization of compounds with explained method [2].

Figure 2. Shows the similar pattern, peaks 1 and 2 were found in three different samples, but peak 3 only showed in fermented leaves and commercial tea. Commercial tea is also already fermented, which means that compound 3 is produced after fermentation. The methanol extract of fermented *C. osmophloeum* leaves was then analyzed using LC-MS, as shown in figure 3, selection ion chromatogram in m/z 287. Kaempferol has molecular weight 286 and the *m/z* 287 for positive ionization [8]. Kaempferol found in the retention time of 8.07 minutes. This means that kaempferol was present in the methanol extract of fermented *C. osmophloeum* Kaneh leaves.

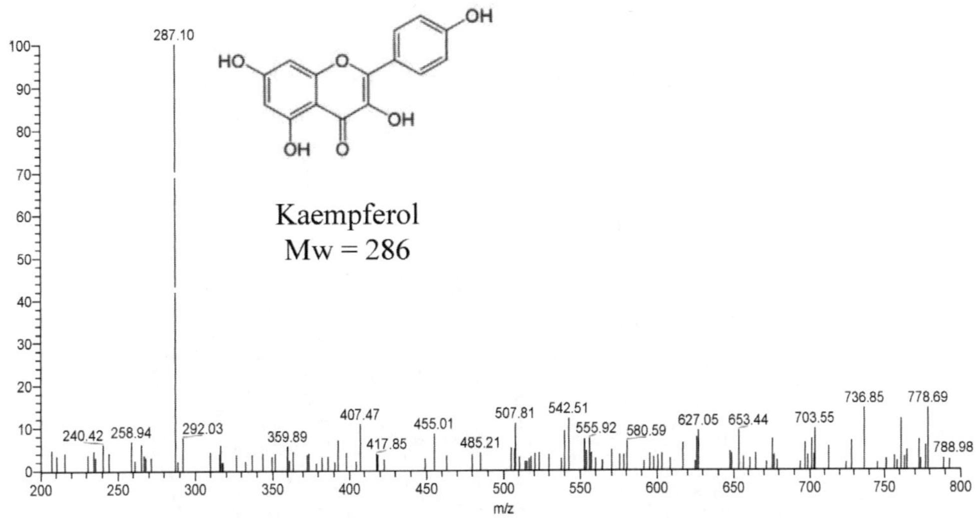

Figure 3. Mass spectrum (LC-MS) of kaempferol from crude extract of fermented leaves of *C. osmophloeum* Kaneh found in the retention time of 8.07 minutes, with selection ion chromatogram in m/z 287.

With longer fermentation the amount of kaempferol aglycone in the leaves increased significantly. Measurement was done by comparing the peak area from HPLC analysis, which by week 8 of fermentation the amount of kaempferol aglycone increased significantly compared with 0 week fermentation, from 0.46 to 46.81 μg.mL^{-1}. It is increased about more than 100 fold.

Figure 4 showed that length of fermentation effects the composition of the *C. osmophloeum* Kaneh leaves. These results explained the similar HPLC pattern between fermented leaves and commercial tea leaves, that the activity of β-glucosidase produced by lactic acid bacteria during fermentation of *C. osmophloeum* Kaneh leaves converted two kaempferol glycosides stored at a fresh leaves become kaempferol aglycone. Yang, *et al.* reported that the concentration of aglycone in meju was increased to reach to approximately 40% of total isoflavones in 90 day meju fermentation. Some amino acids increased and others remained almost constant with increased fermentation time [9].

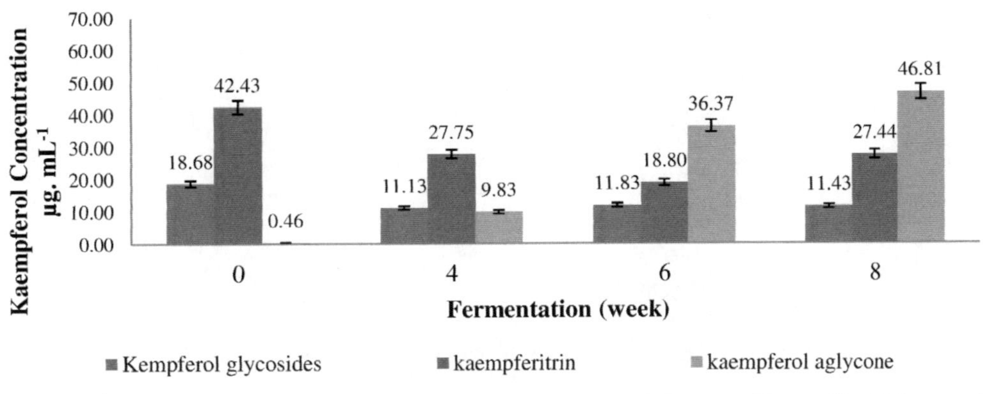

Figure 4. The effects of fermentation on the compositional changes of *C. osmophloeum* Kaneh leaves shows that the longer fermentation done the kaempferol aglycone (green graphic) increased significantly. Both glycosides (blue and red graphic) were decreased after fermentation.

3.3 Microbiological Analysis

Lactic acid bacteria plays an important role in the fermentation process, producing lactic acid, ethyl alcohol, carbon dioxide and aromatic compounds, which contributed to changes of color, flavor, and odor during fermentation. Ten strains were isolated from 4 weeks fermented *C. osmophloeum* Kaneh leaves. The DNA of each strain extracted using Tissue Genomic DNA Extraction Mini KIT (FavorPrep, FATGK001), then continued by PCR analysis.

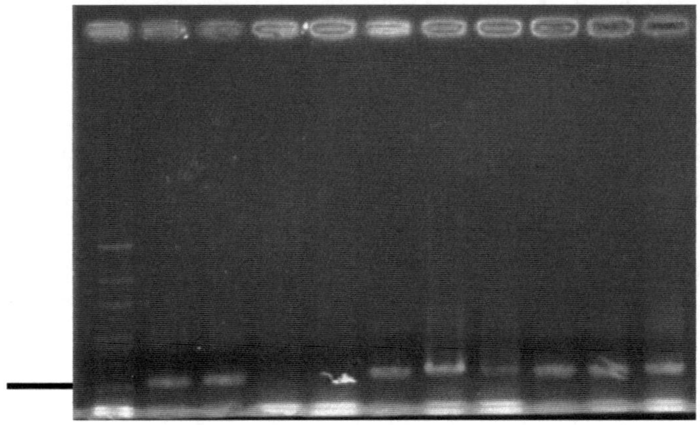

Figure 5. Lane A-J shown the results of PCR amplification products (500 bp) of LAB isolated from fermented tea leaves. Lane C and D didn't showed an any bend in 500 bp

In this study, PCR amplification products was 500 base pair fragment (figure 5). After PCR amplified, continued to sequences the 16S rRNA gene from bacteria, then use BLAST (Basic Local Alignment search Tools, Program BLASTN 2.2.28+) on the complete NCBI genomic database to identify the species that most closely matches their sequence results for the genes. This experiment showed that there are three groups of lactic acid isolated from the fermented leaves, *B. coagulans*, *L. plantarum*, and *P. pentosaceus* (table 1). The activity of β-glucosidase produced by those lactic acid bacteria during fermentation of *C. osmophloeum* Kaneh leaves converted two kaempferol glycosides stored at a fresh leaves become aglycone form, kaempferol.

Table 1. Sequencing DNA then compared to the world gen bank data by BLAST (Basic Local Alignment search Tools, Program BLASTN 2.2.28+)

Strain Code	Species Identified*	Max. Identified (%)
A	*Bacillus coagulans*	99
B	*Bacillus coagulans*	99
E	*Lactobacillus plantarum*	98
F	*Pediococcus pentosaceus*	97
G	*Pediococcus pentosaceus*	97
H	*Lactobacillus plantarum*	97
I	*Lactobacillus plantarum*	98
J	*Pediococcus pentosaceus*	97

Naturally, the lactic acid bacteria involved in food fermentation process produced enzymes as secondary metabolites. β-glucosidase activity was found in wine making by *L. plantarum* [10]. β-glucosidase also produced by lactic acid bacteria from local fermented food of Indonesia [11]. Its enzymes generated isoflavone aglycones during fermentation by enzymatic hydrolysis of glycosides [5]. Similar results also reported by [12] that *Bacillus* strain enhanced the aglycone production during soybean fermentation.

The molecule of aglycone is smaller than glycosides which made aglycone easily incorporated into target cells through simple diffusion mechanism because of its high hydrophobicity. The isoflavone aglycones were absorbed faster and greater amounts than their glycosides in human, which means it may be more effective than glycoside products in preventing chronic disease [13]. These results demonstrated that consumption of fermented tea leaves of *C. osmophloeum* Kaneh are way better than the fresh leaves.

4. Conclusion

Fermentation could change the phsyco-chemical and organoleptic properties of *C. osmophloeum* Kaneh leaves by 3 major groups of lactic acid bacteria activity, *B. coagulans, L. plantarum* and *P. pentosaceus*. The HPLC analysis showed that the chemical composition changed with the length of fermentation, compared to fresh leaves. The presence of kaempferol aglycone produced during fermentation was confirmed by LC-MS. However, its need further research to understand the stability of kaempferol aglycone during fermentation and the possibility that it will use by other bacteria.

Acknowledgment

This experiment was supported by National Pingtung University of Science and Technology, Taiwan and Universitas Brawijaya, Indonesia from Beasiswa Unggulan DIKTI for Double Degree Program.

References

[1] Woodman OL and Chan EC 2004 *Clin. Exp. Pharmacol. Physiol.* **31** 786–90
[2] Rao YK, Geethangili M, Chan HS, Wu WS and Tzeng YM 2009 *Int. J. Appl. Sci. Eng. Technol.* **7** 1–9
[3] Kurowska EM and Manthey JA 2004 *J. Agric. Food Chem.* **52**(10) 2879–86
[4] Sharma M, Ghosh D and Bhattacharya N 2013 *International Journal of Engineering Science Invention* **2**(3) 56–60
[5] Richelle M, Pridmore-Merten S, Bodenstab S, Enslen M and Offord EA 2002 *J. Nutr.* **132** 2587–92
[6] Kim JM, Lee EK, Kim DH, Yu BP and Chung HY 2010 *Springer* **32** 197–208
[7] Matthews A, Grimaldi A, Walker M, Bartowsky E, Grbin P and Jiranek V 2004 *Appl. Environ. Microbiol.* **70** 5715–31
[8] He M, Wang P, Xiang Y, Qi Y, Sun H and Phillips J 2007 Profiling and characterization of polyphenol polymers from cinnamon using an ion trap mass spectrometer *Thermo Scientific* Application Note: 392
[9] Yang HJ, Kwon DY, Kim MJ, Kang S and Park S 2012 *Nutr. Metab.* **9** 1-12
[10] Sestelo ABF, Poza M and Villa TG 2004 *World J. Microbiol. Biotechnol.* **20** 633–7
[11] Sumarna 2009 *Malays J. Microbiol.* **6** 30-40
[12] Dajanta K, Chukeatirote E, Apichartsrangkoon A and Frazier RA 2009 *Acta Biologica Szegediensis* **53** 93-8
[13] Izumi T, Piskula MK, Osawa S, Tobe K, Saito M, Kataoka S, Kubota Y and Kikuchi M 2000 *J. Nutr.* **130** 1695-9

IOP Conf. Series: Materials Science and Engineering **193** (2017) 012014 doi:10.1088/1757-899X/193/1/012014

The effect of nisin from *Lactococcus lactis subsp. lacti*s on refrigerated patin fillet quality

S N Adilla[1*], R Utami[1], A Nursiwi[1], and E Nurhartadi[1]

[1] Departmenet of Food Science and Technology, Universitas Sebelas Maret, Ir Sutami Street No. 36A, Surakarta, Central Java, Indonesia
*E-Mail: sarah.nur.a@gmail.com

Abstract. The effect of nisin from *Lactococcus lactis subsp. lactis* with spraying method application on quality of patin fillet during refrigerated storage (4±1°C) was investigated. The quality of patin fillet based on total plate count (TPC), pH, TVB-N, and TBA values during 16 days at 4±1°C. Completely Randomized Design (CDR) was used in one factor (nisin activity) at 0 IU/ml, 500 IU/ml, 1000 IU/ml, and 2000 IU/ml. The observation was done at 0, 4th, 8th, 12th, and 16th days of storage. The result showed that variation of nisin activity significantly affected the quality of fillet according to TPC, pH, and TVB-N values, however no significant difference on the obtained of TBA value. Nisin in 500 IU/ml, 1000 IU/ml, and 2000 IU/ml could extend the shelf-life of fillet until 4th, 8th, and 12th days respectively based on standard in all parameters.

1. Introduction

Patin (Pangasius hypophthalmus) are freshwater fish that are farmed in Indonesia. Patin fillet has become an export commodity, and it is also in high demand in the global market, especially the United States and Europe [1]. Fillet has some advantages such as being practical and easy to process; however, fillet can be damaged rapidly because of autolysis, microbial growth, and the oxidation process. Microbial growth is the major factor that leads to spoilage, and fillet then cannot be consumed [2]. Low temperature storage is a common method to decrease the spoilage rate of fillet; however, this method is unable to inhibit growth of psychrophilic bacteria such *Acinetobacter, Pseudomonas spp,* and *Brochothrix thermosphacta* species. In this case, the use of an antimicrobial substance is recommended to retain the quality of fillet during cold storage.

Bacteriocins are ribosomally synthesized peptides from bacteria, especially lactic acid bacteria (LAB) with bactericidal or bacteriostatic effects. Bacteriocins produced by LAB have several properties that make them suitable for food preservation: (i) they are generally recognized as safe (GRAS) substances, (ii) they are inactive and nontoxic on eukaryotic cells, (iii) they become inactivated by digestive proteases, having low effect on the gut microbiota, (iv) they are usually pH and heat-tolerant, (v) they have a relatively broad antimicrobial spectrum against many food-borne pathogenic and spoilage bacteria, (vi) they show a bactericidal mode of action, usually acting on the bacterial cytoplasm membrane, and (vii) their genetic determinants are usually plasmid-encoded, facilitating genetic manipulation [3]. Nisin is a bacteriocin that has been used as food preservative to inhibit Gram positive bacteria, *Eschericia coli*, and sporing bacteria. Nisin is composed of 34 amino acids with 3354 Da molecular weight produced by *Lactococcus lactis subsp. lactis* [4]. In 1988, nisin was approved by FDA for specific uses and as a GRAS (Generally recognized as safe) substance [5]. It is also approved by Ministry of Health, Republic Indonesia [6] and in over 40 countries as a food preservative [7].

The application of nisin as food preservative has examined for many products such as cheese, fermented beverage, and canned food to inhibit the growth of Gram positive bacteria and the spores [5,7]. For fishery products, nisin alone or combined with another substance has been used against Lister*ia monocytogenes, Salmonella sp., Vibrio spp., Clostridium botulinum*, and other pathogenic bacteria that usually contaminate the product [4,8]. In this study, the effect of nisin from *Lactococcus lactis subsp. lactis* with spraying method on the quality of patin fillet were investigated. The patin fillet quality observed based on total plate count (TPC), pH, TVB-N and TBA values during 16 days in refrigerated storage (4±1°C).

2. Material and method

2.1. Sample preparation

Patin siam fish used in the experiments were obtained from Lembah Hijau Multifarm Sukoharjo, Central Java (Indonesia). The fish were immediately filleted and washed with tap water and stored on ice during preparation. The nisin solution was made by dissolving an amount of 1000 mg of nisin powder (activity: 10^6 IU/g from Sigma-Aldrich Pte Ltd, product number: N5674) in 100 ml HCl (0,02 N) to obtain an initial concentration of 10^4 IU/ml. About 1.25 ml, 2.5 ml, and 5 ml of this solution were diluted with HCl 0,02 N up to 25 ml to obtain the final concentrations 500 IU/ml, 1000 IU/ml, and 2000 IU/ml respectively [9]. The fillets were randomly assigned into four group consisting of control (0 IU/ml nisin), and three groups sprayed with nisin solution in 500 IU/ml, 1000 IU/ml, and 2000 IU/ml until the whole sample became wet. All samples were packed in plastic plates covered with plastic wrapping, then stored under refrigeration (4±1°C) for 16 days. Total plate count (TPC), pH, total volatil base (TVB), and thiobarbituric acid tests were observed on the the 0, 4th, 8th, 12th, and 16th days of storage days of storage.

2.2. Microbial analysis

Determination of microbial growth in the sample was done using the pour plate method according to SNI 2332.3 [10]. 10 g of fillet were aseptically homogenized with 90 ml sterilized NaCl dilution water (0.85% NaCl w/v) using a mechanical blender (Philips). From this dilution, other decimal dilutions were prepared and plated in Plate Count Agar (PCA) media (Merck) in duplo. Enumeration of the total plate count was determined after two days incubation at 36°C. The data were transferred into logarithms of the number of colony-forming unit per gram of fillet (Log CFU/g).

2.3. Chemical analyses

The pH value was measured using digital pH meter (Hanna Instrument) at room temperature [11]. The TVB-N value was determined using the Conway micro diffusion method according to SNI 2354.8 [12]. The data of TVB-N value was shown in mg N per 100 gram fillet. To measure the TBA value, 10 g of fillet was homogenized with 50 ml distilled water and transferred into a distillation flask. Then, 47.5 ml distilled water and 2.5 ml HCl (4M) were added to the flask and the mixture was distillated in high temperature. 5 ml of distillate was pipetted in a reaction tub and 5 ml of TBA reagent was added. The mixture was heated in boiling water for 35 minutes then cooled. The absorbance (A) was measured using a spektrofotometer (528 nm) [13].

2.4. Statistical analysis

Completely Randomized Design (CRD) was used in one variable was nisin activity in 0 IU/ml, 500 IU/ml, 1000 IU/ml, and 2000 IU/ml in two replication for each sample. Data were analyzed with an ANOVA test (One Way ANOVA) followed with DMRT (Duncan Multiple Range Test) (α=5%) if there were the variances between samples.

IOP Conf. Series: Materials Science and Engineering **193** (2017) 012014 doi:10.1088/1757-899X/193/1/012014

3. Result and discussion

3.1. Microbial analysis

The deterioration process of fish results from three mechanisms: enzymatic autolysis, lipid oxidation, and microbial growth. Microbial growth is the major factor that leads to spoilage, and the product cannot be consumed because of the production of new compounds that are responsible for changes in odor and flavor [2]. Some species of bacteria such as *Pseudomonas spp, Shewanella, Micrococcus, Alcaligenes, Photobacterium phosphoreum,* and *Brochothrix thermosphacta* are spoilage microorganisms that are usually present in fishery products, producing amine compounds such as trimethilamine (TMA), dimethylamine (DMA), and other biogenic amines [14]. During storage, the microorganisms have different abilities to tolerate the preservation conditions. In this study, changes of microbial count in patin fillets throughout storage are shown in Fig. 1.

The initial microbial count of control and samples treated ranged from4.76-5.38 log CFU/g, which showed all the samples were in good condition according to the SNI standard of fresh fish [15]. The TPC value of all the samples increased gradually with storage; however, the sample with nisin had lower counts than the control throughout storage. The control sample exceeded the microbiological acceptability limit of 7 log CFU/g [16] in the 4th day of storage and increased faster than the treated samples. The sample with 500 IU/ml and 1000 IU/ml nisin exceeded the limit after the 8th and 12th days respectively, while the sample with 2000 IU/ml nisin did not reach 7 log CFU/g until the end of storage. The TPC values in the control sample and samples treated with 500 IU/ml, 1000 IU/ml, and 2000 IU/ml nisin at the end of storage were 9.58 log CFU/g, 9.25 log CFU/g, 7.37 log CFU/g, and 6.50 log CFU/g respectively.. These results indicated that nisin had an inhibitory effect on microbial growth in refrigerated patin fillet and extended its shelf life, in approache with previous reports that informed the self life of refrigerated vacuum packaged rainbow trout fillet that was sprayed with nisin (100 ppm) could reach 16 days of storage [17]. These result also accordance with other report that informed the used of 5% nisin solution with dipping method could inhibit microbial growth in pompano fillet about 1.22 log CFU/g lower than control in 12 days [18].

The effect of nisin on the quality of patin fillet is related to its ability to be an antimicrobial substance. Nisin is active against Gram positive bacteria and its spores. The inhibition effect of nisin against Gram positive bacteria is based on pore formation in the cytoplasmic membrane of the target bacteria. It leads to loss of these substances and depletes the proton motive force, which interferes with cellular biosynthesis and causes cellular death [5]. It is also been reported that nisin interferes with cell wall biosynthesis by binding with a peptidoglycan precursor (lipid II) [8]. Whereas the inhibition effect of spores is nisin interacts with membrane sulfhydryl that destrubes spore germination [4].

3.2. Chemical analyses

3.2.1. pH

pH had correlations with microbial counts and total volatile base nitrogen (TVB-N) values of the sample. Fish tissue generally has neutral pH and decreases into slight acidity in early storage; however, the pH gradually increased because of microbial activity that produced alkaline compounds such trimethylamine, dimethylamine, and biogenic amines [2,19]. In this study, changes of pH in patin fillets throughout storage are shown in Fig. 2.

The initial pH of patin fillets ranged from 6.35-6.44 and gradually increased for all samples until the end of storage. There were no significant differences (α=0,05) between the control and sample treated with 500 IU/ml nisin; however, the sample treated with nisin 1000 IU/ml and 2000 IU/ml showed significant differences (α=0,05) with the control. The sample with 2000 IU/ml nisin showed the lowest increase of pH among other treatments without any significant differences (α=0.05) until the 12th day of storage. At the end of storage, the pH of each sample reached 8.39; 8.49; 7.48; and 7.25 for the control and samples treated with 500 IU/ml nisin, 1000 IU/ml, and 2000 IU/ml respectively. The results indicated that 1000 IU/ml nisin and 2000 IU/ml activities could keep the pH of the fillets under control during storage.

In the previous study informed that the used of 100 ppm nisin sprayed on rainbow trout fillet that was kept under vacuum showed no significant difference ($p \geq 0,05$) with a control until the end of storage. Another report also reported that the used of 5% nisin with dipping method on pompano fillet showed no significant difference ($p \geq 0,05$) with a control during storage [18]. These differences might be because the species and the conditions used in the experiments were different [20].

3.2.2. Total Volatil Base Nitrogen (TVB-N)

TVB-N is widely used as an indicator of fish spoilage. This measurement based on total amount of ammonia (NH_3) produced by the deamination of amino acid of spoilage bacteria, dimethylamine (DMA) produced by autolytic enzymes during storage, and trimethylamine (TMA) produced by the reduction of trimethylamine oxide (TMAO) of spoilage bacteria that are formed in non-fermented food products during storage [21, 22]. In this study, the changes of TVB-N in patin fillets during storage are shown in Fig. 3.

The initial TVB-N of patin fillet was variated at 11.39 mg N/100 g, 15.84 mg N/100 g, 13.36 mg N/100 g, and 16.42 mg N/100 g for the control, sample treated with nisin 500 IU/ml, 1000 IU/ml, and 2000 IU/ml, respectively. The TVB-N of all samples gradually increased until the end of storage. Samples with 1000 IU/ml nisin and 2000 IU/ml had much lower TVB-N values than the control; however, the sample with 500 IU/ml nisin had the highest values of the other treatments. This might be caused by the low effectivity of nisin at 500 IU/ml, meaning it could not inhibit spoilage bacteria in the samples. However, in the higher activities (1000 IU/ml and 2000 IU/ml) nisin could maintain the TVB-N values below those of the control until the end of storage. Previous study showed that the used of nisin at the lowest concentration (0.1%) with the dipping method in grass carp fillet gave a higher TVB-N value than control in the end of storage, whereas at the higher concentration (0.2%) nisin could maintain the TVB-N value below that of the control [23]. In this study, the control and samples treated with 500 IU/ml nisin exceeded the acceptability limit of 35 mg N/100 g fillet [21] on the 4th day of storage, the sample with 1000 IU/ml nisin and 2000 IU/ml exceeded the limit on the16th day of storage. The TVB-N values inthe control sample, and the samples treated with 500 IU/ml nisin, 1000 IU/ml, and 2000 IU/ml at the end of storage were 307.54 mg N/100 g, 424.25 mg N/100 g, 79.47 mg N/100 g, and 51.90 mg N/100 g,respectively.

3.2.3. Thiobarbituric Acid (TBA)

The TBA is one of the indicator related to lipid oxidation occuring in the fishery product during storage, it is also closely related to off-odour and off-flavor attributes [24]. The maximum level of TBA value indicating the good quality of fishery product (frozen, chilled, or stored with ice) is 5 mg malonaldehyde/kg, while the fish may be consumed up to the level of 8 mg malonaldehyde/kg [25]. In this study, the changes of TBA value in patin fillets during storage are shown in Fig. 4. All the samples had very low initial TBA values at 0.05-0.06 mg malonaldehyde/kg and gradually increased during storage. However, no significant diferences were found the any samples at the end of storage.The similar results also reported in the previous study that examined the effect of nisin on the TBA values of pompano fillet during refrigerated storage [18]. The results showed that during the entire storage period, TBA values of sample treated with nisin were similar to the control, which indicated that nisin had no effect on reducting the lipid oxidation of fish. The addition of 100 ppm nisin alone in the buffalo meat sausage also did not alter the TBARS (thiobarbituric acid reactive substances) level from that of the control [26].

At the end of storage, the TBA values of all the samples were still below the maximal acceptable limit of 8 mg malonaldehyde/kg. The control sample and sample treated with nisin 500 IU/ml, 1000 IU/ml, and 2000 IU/ml reached 0.37 mg malonaldehyde/kg, 0.31 mg malonaldehyde/kg, 0.33 mg malonaldehyde/kg, and 0.34 mg malonaldehyde/kg, respectively.

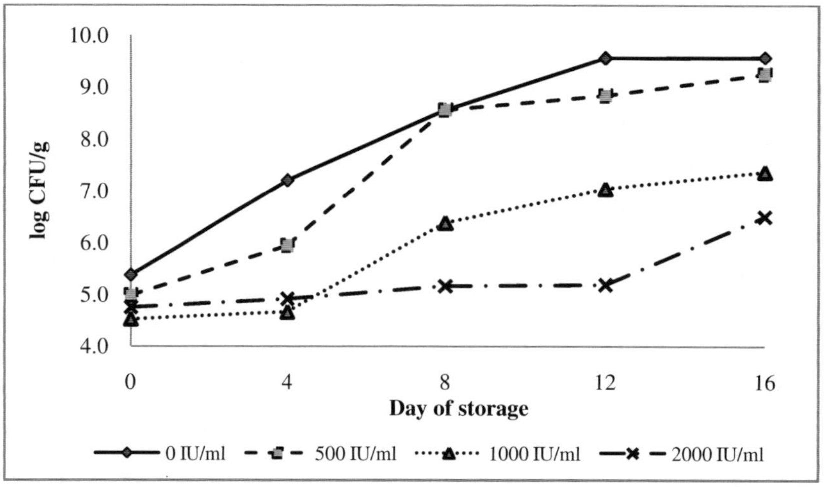

Figure 1. Changes of microbial count in patin fillets during storage

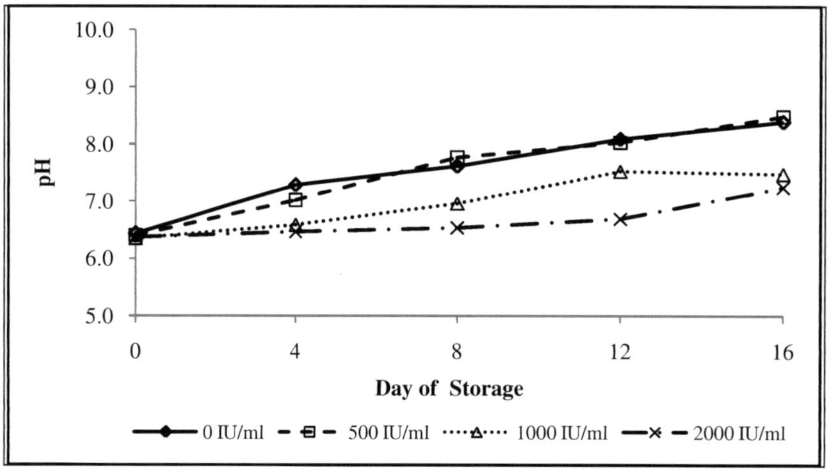

Figure 2. Changes of pH in patin fillets during storage

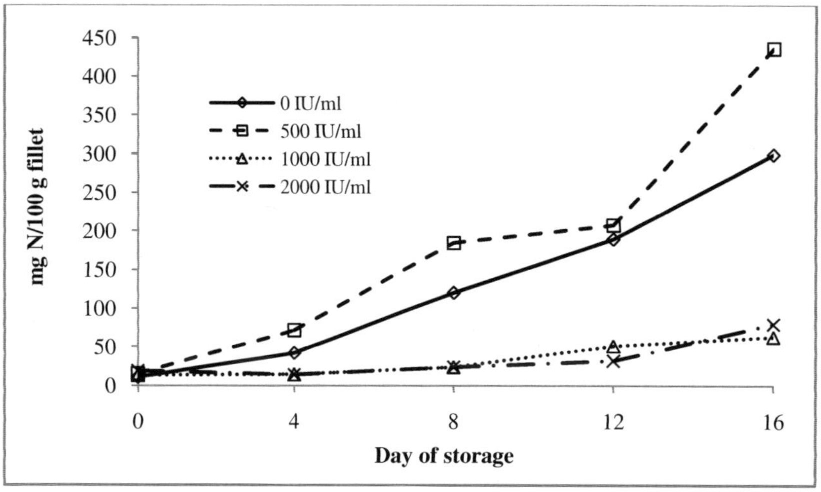

Figure 3. Changes of TVB-N in patin fillets during storage

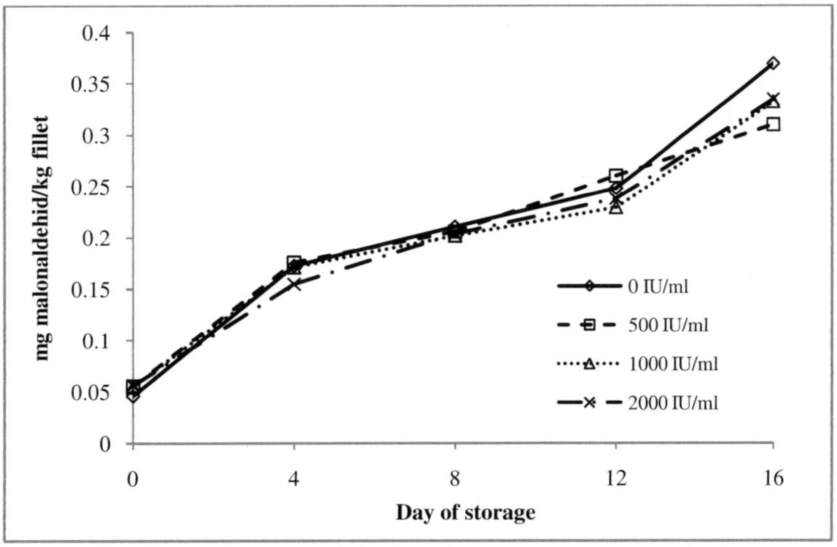

Figure 4. Changes of TBA in patin fillets during storage

4. Conclusion

The microbial and chemical analyses of the patin fillet during 16 days cold storage (4±1°C) have examined to explain the effect of nisin as food preservative on the quality of patin fillet. According to TPC, pH, and TVB-N results, patin fillets quality could be maintained by the use of nisin at 1000 IU/ml and 2000 IU/ml. Due to all standards, patin fillets sprayed with nisin activity 1000 IU/ml and 2000 IU/ml could extend the shelf life until 8th and 12th days of storage respectively, while control and sample with nisin actvity 500 IU/ml have been unable to consume in the 4th day of storage.

References.

[1] Ministry of Trade Republic of Indonesia 2013 Ikan *Patin Hasil Alam Bernilai Ekonomi dan Berpotensi Ekspor Tinggi* Warta Ekspor, October 2013.
[2] Ghaly A E, D Dave, S Budge and M S Brooks 2010 *American J. Appl. Sci* **7** (7) pp 859-877.
[3] Hwanhlem N, Jean-Marc C and Aran H K 2014 *Food Control* **41** pp 202-211.
[4] Jeevaratnam K, M Jamuna, and A S Bawa 2005 *Indian J. of Bioechnology* **4** pp 446-454.
[5] Cleveland J, Thomas J M, Ingolf F N and Michael L C 2001 *Int. J. Food Microbiology* **71** pp 1-20.
[6] Ministry of Health Republic of Indonesia. 2012 *Bahan Tambahan Pangan* Permenkes RI No. 033 (2012).
[7] White S B 2011 *Antibacterial efficacy of phosvitin, carvacrol, or nisin alone or combined against foodborne human enteric pathogens* (Iowa: Graduate Theses and Dissertations Iowa State University)
[8] Suganthi V, Selvarajan, C Subathradevi, V Mohanasrinivasan 2012 *Int Research J. Of Pharm* **3** pp 13-19.
[9] Millette M, C Le Tien, W Smoragiewicz and Monique L 2007 *Food Control* **18** pp 878–884.
[10] National Standardization Agency of Indonesia 2015 *Cara uji mikrobiologi bagian 3: Penentuan Angka Lempeng Total (ALT) pada produk perikanan*. (Jakarta: National Standardization Agency of Indonesia)
[11] AOAC 1995 *Official Methods of Analysis* 16th Ed. (Washington DC)
[12] National Standardization Agency of Indonesia 2009 *Cara uji kimia bagian 8: Penentuan kadar Total Volatile Base Nitrogen (TVB-N) dan Trimetil Amin Nitrogen (TMA-N) pada produk perikanan* (Jakarta: National Standardization Agency of Indonesia).
[13] Apriyantono A, Dedi F, Ni Luh P, Sedarnawati and Slamet B 1989 *Petunjuk Laboratorium*

Analisa Pangan (Bogor: IPB-Press).

[14] Ray B 2003 *Handbook of Fundamental food microbiology* 3rd Ed. (New York: CRC Press) chapter 19 pp 275-276.

[15] National Standardization Agency of Indonesia 2009 *Batas maksimum cemaran mikroba dalam pangan* (Jakarta: National Standardization Agency of Indonesia).

[16] International Commission on Microbiological Specifications for Foods (ICMSF) 1986 *Sampling plans for fish and shellfish, microorganisms in foods 2, sampling for microbiological analysis: principles and specific applications* 2nd ed. (Toronto: University of Toronto Press).

[17] Behnam S, Mohammad A, Masoud R, Siyavash S and Reza S 2013 *J. Food Sci and Tech.*

[18] Gao M, Lifang F, Tianjia J, Junli Zhu, Linglin Fu, Dongxia Yuan and Jianrong Li 2014 *J. Food Control* **37** pp 1-8.

[19] Abbas K A, A Mohamed, B Jamila and M Ebrahimian 2008 *Am. J. Biochem. & Biotech.* **4** (4) pp 416-421

[20] Liviawati E and Afrianto E 2010 *Penanganan Ikan Segar* (Bandung: Widya Padjadjaran).

[21] Jinadasa B K K K 2014 *Nature and Science* **12** (5) pp 106-111.

[22] Li T, Jianrong Li, Wenzhong Hu, Uguang Zhang, Xuepeng Li, Jin Zhao 2012 *Food Chemistry* **135** pp 140–145.

[23] Langroudi H F, Mehdi S, Abolghasem K, Mohammad R G, Seyed E H, Soottawat B and Zoheir H 2011 *Afr. J. Biotechnol.* **10**(42) pp 8484-8490.

[24] Shakhtour F G M and Babji A S 2014 *Int. J. ChemTech Research* **6**(2) pp 1193-1203.

[25] Sallam K I 2007 *Food control* **18**(5) pp 566-575.

[26] Sureshkumar S, Kalaikannan A, Dushyanthan K, Venkataramanujam V 2010 *J. Food Sci. Tech* **47**(3) pp 358–363.

IOP Conf. Series: Materials Science and Engineering **193** (2017) 012015 doi:10.1088/1757-899X/193/1/012015

Isolation and Screening of Pectinolytic Fungi from Orange (*Citrus nobilis* Tan.) and Banana (*Musa acuminata* L.) Fruit Peel

K R Amilia[1], S L A Sari[1], R Setyaningsih[1]

[1]Department of Biology, Faculty of Mathematics and Natural Sciences, Sebelas Maret University, Ir. Sutami street No. 36A Surakarta, Central Java 57126, Indonesia

E-mail: arumprasojo@gmail.com

Abstract. Pectinase is the one of most important enzyme which is used in food industry such as fruit and vegetable juice extraction, oil extraction and fermentation of coffee, cocoa and tea. Pectinase can be produced by microorganism such as bacteria and fungi. Fungi are known as potent producer of pectinase. This research was conducted to isolate and screen of the pectinolytic fungi from rotten orange and banana fruit peels. This research succeeded to isolate 10 fungal isolates from rotten orange peels and 5 fungal isolates from rotten banana peels. These isolates were screened in pectinolytic activities based on clear zone formation on pectic medium which is stained by cetyl trimethyl ammonium bromide. The screening result showed that fungal isolates which showed pectinolytic activity were O2, O3, O4, O7, O8, O10, B3, and B5. Based on morphological characters, pectinolytic fungi were identified as *Fusarium* O4 and O10, *Penicillium* O2, *Aspergillus* O3, O7, B3 and B5 and *Trichoderma* O8. The highest pectinolytic activity was showed by *Penicillium* O2 which was isolated from orange peel.

1. Introduction

Pectinase is a group of enzymes that catalyze a degradation of pectin through depolymeration (hydrolases and lyases) and de-esterification (esterases) reactions [1]. Based on their mode of action, the pectinase is classified into three types: pectin esterase, hydrolases and lyases. Pectin esterase catalyze the de-esterification of the methoxyl group of pectins, forming pectic acid. Hydrolases (polygalacturonases and polymethylgalacturonases) catalyze the hydrolytic cleavage of α-1,4-glycosidic linkage in pectic acid and pectin. Lyases (polygalacturonate lyase and polymethylgalacturonate lyase) catalyze the cleavage of α -1,4-glycosidic linkage in pectic acid and pectin, forming unsaturated galacturonates and methyl galacturonates [2].

Pectinases share about 25% in global sales of food enzymes [3]. Pectinases are widely used in food industries, such as fruit processing industry for extraction and clarification of fruit juice, preparation of fruit cordials [4], enhance process efficiency of wine making and wine quality [5], oil extraction and fermentation of coffee, cocoa and tea [3]. Since pectinases are widely used enzyme for different industrial application, it is necessary to improve pectinase production, including selection for potential source of pectinase. Many experiments have attempted to increase the performance of enzyme production [6]. The new microbes with high extracellular pectinase activity, stability over wide range of temperature and pH for a longer period of time, with their cost-effective production have been the focus of recent research [2].

Nowadays, commercial pectinase preparations are obtained from fungi [5]. Fungal pectinases are commonly used in fruit industry because fungi are potent producers of pectic enzymes and the optimal pH of fungal enzymes are very close to the pH of many fruit juices, in the pH range of 3 to 6 [7]. This research was conducted to isolate and screen of pectinolytic fungi from orange and banana fruit peels.

2. Methodology

2.1. Medium

This research used modified pectic agar medium to isolate and screen a pectinolytic fungi [8]. The composition of modified pectin agar medium (g/l, wt/vol.) were: Pectin 5.00, K_2HPO_4 0.50, $MgSO_4.7H_2O$ 0.10, NaCl 0.20, $CaCl_2.2H_2O$ 0.20, $FeCl_3.6H_2O$ 0.01, Yeast extract 1.00, Agar 20.00.

2.2. Isolation and Screening of Pectinolytic Fungi

The fungi were isolated from rotten orange fruit peels collected in Jember East Java and rotten banana fruit peels collected in Lumajang East Java. A slice of rotten peels was put on pectic agar medium and had been incubated for 2-3 days. Whereas, the different fungal colonies were inoculated on potato dextrose agar (PDA) for purification. The spores of pure culture were maintained in PDA medium in test tubes sealed with parafilm and stored at 4 °C for further use.

The screening of the pectinolytic activity was conducted based on a clear zone formation on pectic agar medium after it was stained by Cetyl trimethyl ammonium bromide. Fungal isolates were inoculated on pectic agar medium and had been incubated at a room temperature for 48 hours.

2.3. Identification of Pectinolytic Fungi

Pectinolytic fungi were identified based on macroscopic and microscopic morphological characters. These characters were compared to Pictorial Atlas of Soil and Seed Fungi, Second Edition [9] until genus level.

3. Result and Discussion

3.1. Isolation and Screening of Pectinolytic Fungi

This research succeeded to isolate 10 fungal isolates from rotten orange peels named as isolate O1 until O10, 5 fungal isolates from rotten banana peels named as B1 until B5. These isolates were screened in pectinolytic activities based on clear zone formation on pectic agar medium which is stained by cetyl trimethyl ammonium bromide also known as cetrimide or CTAB. The screening result showed that fungal isolates which showed pectinolytic activity were O2, O3, O4, O7, O8, O10, B3, and B5. The diameter of clearing zone of each pectinolytic isolates are shown in table 1. Among them, isolate O2 showed the highest pectinase activity.

Table 1. Pectinolytic activity of fungal isolate showed by clear zone formation on pectin medium after incubation at room temperature for 48 hours.

Code of Isolate	Clear Zone Diametre (mm)
O2	3.44
O3	1.14
O4	1.42
O7	1.00
O8	2.30
O10	1.46
B3	0.88
B5	1.05

This research showed that the rotten orange peels was a more potential source of pectinolytic fungi than the rotten banana peels. Citrus peels especially orange peels are a rich source of pectin since it contains the most microbes such as bacteria and fungi. Orange peels have been reported to contain 18.3% pectin and is utilized for commercial extraction of pectin [10].

Pectinolytic isolates that depolymerized pectin produced colourless hydrolytic zones around colonies on pectic agar medium after colored by CTAB (Figure 1). The cetrimide or CTAB is an insoluble acidic precipitant of polysaccharide. Enzymatic hydrolysis of the pectic substrates inhibited their precipitation by using CTAB, which then lead to the appearance of cleared zones in front of the pectin hydrolases and lyases [11].

Figure 1. Isolate O2 showing clear zone on pectic agar medium. Clear zone formation indicating a pectinolytic activity.

3.2. Identification of Pectinolytic Fungi

Based on morphological characters, pectinolytic fungi was identified as belonging to 4 genera, i.e. *Penicillium* (O2), *Aspergillus* (O3, O7, B3, B5), *Fusarium* (O4, O10) and *Trichoderma* (O8) (table 2).

Table 2. Identification result of pectinolytic fungi isolated from Orange and Banana peels based on morphologycal characters.

Code of Isolate	Genera
O2	*Penicillium*
O3, O7, B3, B5	*Aspergillus*
O4, O10	*Fusarium*
O8	*Trichoderma*

3.2.1. Penicillium

Isolate O2 which had highest pectinolytic activity was identified as belonging to the genera of *Penicillium*. Morphologycal characters of isolate O2 is showed on figure 2. This isolate produced white mycelium with dark green conidia. Whereas, a branched conidiophore showed a brush-like apearance and condiophore which was end with bottle shaped phialides. This microscopic character is a great taxonomic importance of *Penicillium* [12]. *Penicillium* have been known as pectinase producer, i.e. *Penicillium chrysogenum* [13], *Penicillium citrinum* [8].

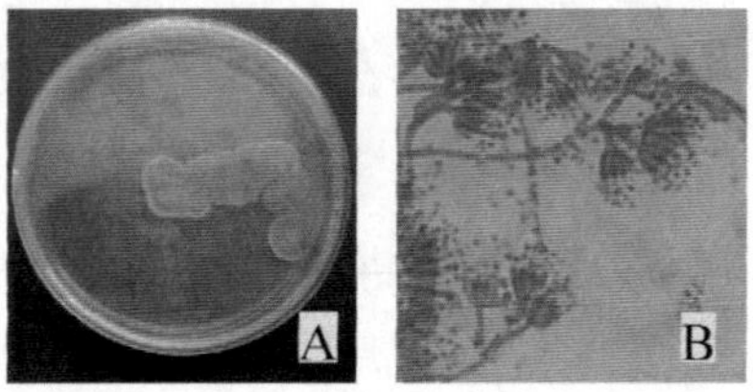

Figure 2. *Penicillium* O2 on PDA, 6 days old (A) and brush-like conidiophores (B).

3.2.2. Aspergillus

Isolates O3, O7, B3 and B5 were identified as belonging to the genera of *Aspergillus*. Morphological characters of these isolates are showed in figure 3. These isolates showed the same microscopic characters, conidiophores upright, simple, terminating in vesicles bearing philiades. These characters were great important taxonomic characteristics for *Aspergillus* [9]. Isolate O3 and B3 had the same macroscopic characters, formed white mycelium with radial zonation, black conidia and yellow pigmentation on the center of reverse site. Isolate O7 and B5 formed white yellowish mycellium with yellow greenish conidia. *Aspergillus* is one of pectinolytic fungi that often produce acid pectinases used in food processing industry [1]. Pectinolytic thermophilic fungi were isolated from soil, vegetables and fruit waste. The most potential isolate as pectinase producer was reported as *Aspergillus fumigatus* which produce thermophilic pectinases that active in low pH [14].

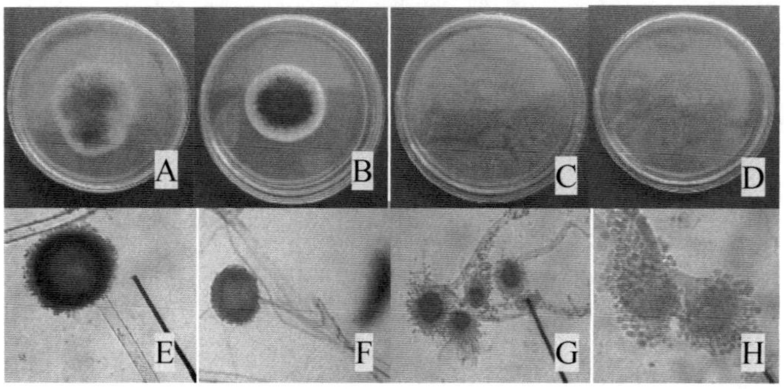

Figure 3. Colonies of *Aspergillus* strains on PDA plate, 6 days old. *Aspergillus* O3 (A) and B3 (B) showed black conidia. *Aspergillus* O7 (C) and B5 (D) showed yellow greenish conidia. These isolates formed conidiophores terminating in vesicles (E,F,G,H).

3.2.3. Fusarium

Isolates O4 and O10 were identified as *Fusarium.* Morphological characters of these isolates are showed in figure 4. *Fusarium* O4 produced white mycelia with yellow pigmentation on the center of the colony. *Fusarium* O10 produced mycelium with pink pigmentation (dorsal and reverse). *Fusarium* O4 and O10 produced lunar shape macroconidia with septation. These characteristics were regarded as taxonomically useful characteristics for *Fusarium* [9]. Shorter macroconidia was shown by *Fusarium* O10. *Fusarium* is known as pectinase producer. Strong activity of pectinase was detected from F. *oxysporum* and F. *proliferatum* [15].

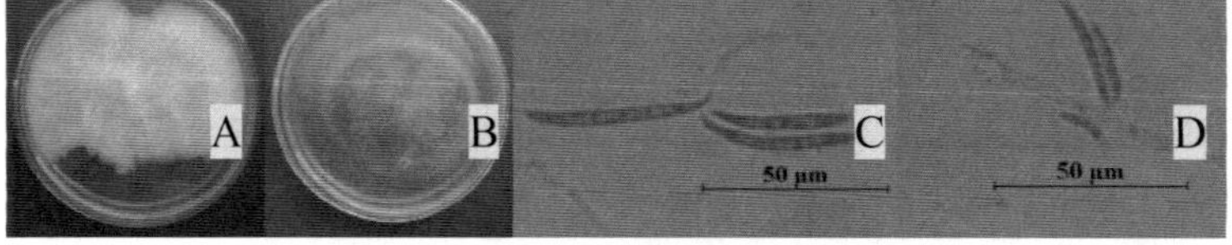

Figure 4. Colonies of *Fusarium* strains on PDA plate, 6 days old (top view), *Fusarium* O4 (A) and O10 (B) and macroconidia produced by *Fusarium* O4 (C) and O10 (D).

3.2.4. Trichoderma.

Isolate O8 was identified as *Trichoderma.* Morphological characters of isolate O8 was showed in figure 5. This isolate formed white mycelium with grey conidia, branched conidiophores and elipsoidal conidia. The colour of conidia was green when it puts under the light microscopes. These characters were regarded as taxonomically useful characteristics of *Trichoderma* [9]. The production of notable, safe and highly active pectinase of *Trichoderma* was reported [16]. This enzyme was used in fruit juices clarification with remarkable stability.

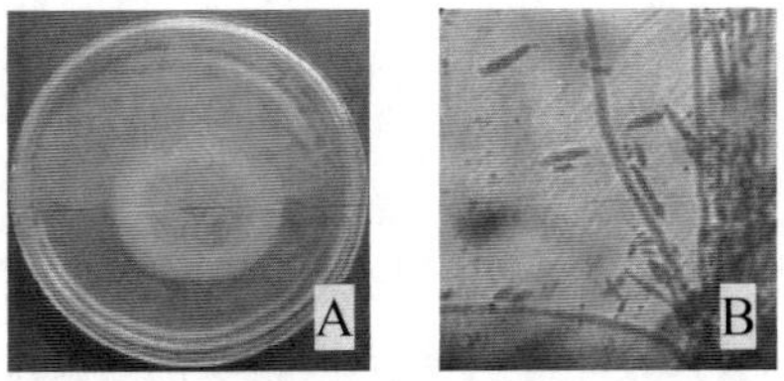

Figure 5. Colony of *Trichoderma* O8 on PDA plate, 6 days old (A) and micro morphological view (B).

4. Conclusion

The Total of 15 fungal isolates were obtained in this research consist of 10 isolates from Orange peel and 5 isolates from Banana peel. Fungal isolates which showed pectinolytic activity were O2, O3, O4, O7, O8, O10, B3, and B5. Based on morphological characters, pectinolitic fungi were identified as *Penicillium* O2, *Aspergillus* O3, O7, B3 and B5, *Fusarium* O4 and O10, and *Trichoderma* O8. The highest pectinolytic activity was showed by *Penicillium* O2 which was isolated from rotten orange peel.

References

[1] Pedrolli D B, Monteiro A C, Gomez E, and Carmona E C 2009 *The Open Biotech J.* **3** 9-13
[2] Garg G, Singh A, Kaur A, Singh R, Kaur J, and Mahajan R 2016 *Biotech* **6** 47
[3] Sieiro C, Gracia-Fraga B, Lopez-Seijaz J, da Silva A F, and Villa T G 2012 *Food Industrial Processes-Metodes and Equipment www.intechopen.com*
[4] Tapre A R, and Jain R K 2014 *Int Fd Res J.* **21(2)** 447-453
[5] Merín G M, Martín M C, Rantsiou K, Cocolin L, and Morata de Ambrosini V I 2015 *Braz J. Microbiol* **46(3)** 815-823
[6] Molina S M W, Pelissari F A, Vitorello C B M 2001 *Braz J. Microbiol* **32(4)** 320-326.
[7] Jayani R S, Shukla S K, and Gupta R 2010 *Enzime Res* **31** 306785
[8] Sandhya R, and Kurup G 2013 *Int Res J. Biol Sci* **2(9)** 34-3
[9] Watanabe T 2002 *Pictorial atlas of soil and seed fungi: morphologies of cultured fungi and key to species*
[10] Jabeen A, Hanif Q A, Hussain M, Munawar A, Farooq N, and Bano S 2015 Sci Lett **3(2)** 42-45
[11] Hadj-Taieb N, Tounsi H, Chabchoub A, Abid N and Gargouri A 2011. *Appl Biochem Biotechnol* **165**:1652–1660
[12] Visagie C M, Houbraken J, Frisvad J C, Hong S.-B, Klaassen C H W, Perrone G, Seifert K A, Varga J, Yaguchi T, and Samson R A 2014 *Studies in Mycology* **78** 343–371
[13] Banu A R, Devi M K, Gnanaprabhal G R, Pradeep B V, and Palaniswamy M. 2010 *Indian J. Sci. Technol.* **3(4)** 377-381
[14] Phutela U, Vikram D, Shobhna S, and Chadha B S 2005 *Braz J. Microbiol* **36** 63-69
[15] Kwon H W, Yoon J H, Kim S H, Hong S B, Cheon Y, and Ko S J 2007 *Mycobiology* **35(3)** 162–165
[16] Ismail A-M S, Abo-Elmagd H I, and Housseuny M M 2016 *J.Genet Eng Biotechnol* **14** 153-159

IOP Conf. Series: Materials Science and Engineering **193** (2017) 012016 doi:10.1088/1757-899X/193/1/012016

Preliminary Characterization of Crude Lectin Fraction of the Red Alga, *Acrocystis nana* from Wediombo Beach of the Southern Coast of Java Island, Gunung Kidul, Yogyakarta, Indonesia

C Anam[1,3], D Praseptiangga[1], M A Nugraheni[1], T Nurhayati[1], N D Fajarningsih[2], D S Zilda[2], E Chasanah[2], A Yunus[3]

[1] Department of Food Science and Technology, Sebelas Maret University (UNS), Jl. Ir. Sutami 36 A, Kentingan 57126, Surakarta, Indonesia

[2]Research and Development Center for Marine and Fisheries Product Competitiveness and Biotechnology, Ministry of Marine Affairs and Fisheries, Republic of Indonesia, Jl. KS Tubun Petamburan VI, Slipi, Jakarta, Indonesia

[3] Graduate School Program of Sebelas Maret University (UNS), Jl. Ir. Sutami 36 A, Kentingan, 57126, Surakarta, Indonesia

E-mail: Dikchoir@yahoo.com

Abstract. Lectins, bioactive compounds that found in algae, are bind to sugars or glycoproteins reversibly with high specificity, but are devoid of catalytic activity, and in contrast to antibodies, are not products of an immune response. *Acrocystis nana* is a species of red alga that is collected from the Wediombo beach, coastal region of Gunungkidul, Yogyakarta. Hemaglutination activity of a crude lectin fraction from *Acrocystis nana* was examined using trypsin-treated rabbit erythrocytes and its chemical properties, including its protein content, stability on pH, temperature, and divalent cations were determined as a preliminary characterization phase. Crude lectin of *Acrocystis nana* showed a titer of 2^{12} on hemagglutination activity assay and the protein content was 6225.44 μg/ml. Hemagglutination activity of this crude lectin was stable after treatment at various pH from 3 to 10 and its activity was lost by heating at 50°C until 100°C. Moreover, the hemagglutination activity was slightly affected by divalent cations treatment, indicating that the presence of divalent cations may require for its activity, however, further studies are still needed for a more comprehensive understanding about its properties.

1. Introduction

Indonesia is a maritime country with a coastline of 81,000 km [1] and marine area of 5.8 million km^2 [2]. Wediombo Beach is a beach in the southern coast of Java Island, which locates at 8° 08' 42.7" S and 110° 35' 58.1" E. One of the potential of marine resources is algae. Based on the Van Bosse expedition from 1899 to 1900, Indonesia has 555 out of 8642 species of algae in the world (6.42% of all species of algae in the world). The number of red algae species is more than most of other types of algae. *Acrocystis nana* is one of the species of red algae found in Wediombo Beach. It has a shape like human fingers, wrinkled, and clustered like a glove. Its color is yellowish green and the biggest size has a height of 30 cm and a width of 4 cm [2].

One of the bioactive compounds in algae are lectins. Lectins derived from Greek "Legere" which means to pick up or to take. Lectins are proteins that bind to sugar or glycoprotein, non-immune, can agglomerate cells and/or precipitate glycoconjugates. Lectins have at least two binding sites [3]. Lectins specifically recognize and bind some carbohydrates, so it can be a useful tool in some research fields such as immunology, cell biology, membrane structures, and cancer [4]. Lectins are also able to act as anti-viral, anti-cancer, and anti-tumor [5].

With respect to lectins from algae, it has been reported on Praseptiangga [6] that algal lectins surveys have been conducted from Puerto Rican [7], English [8-10], Japanese [11-13], Spanish [14,15], United States[16,17], Brazilian [4,18,19], and Vietnamese marine algae [20]. Most investigators have detected lectins by agglutination experiments using human and animal erythrocytes and/or other cells [6]. The improvement in the methodologies of both, extraction, and hemagglutination activity assays could increase the number of positive species. In fact, there appears to be coincidence that the rabbit erythrocytes treated with enzyme (trypsin and/or papain) are most suited for the detection of hemagglutination activity in marine macroalgae [13, 18, 21]. Even with the increase in the publications related to algal lectins, biochemical and structural information on algal lectins is still limited [6]. There have been merely 31 lectins from Rhodophyceae and 17 lectins from Chlorophyceae were isolated and characterized up to now [21]. *Acrocystis nana* has been investigated on its history, nutrition, and benefits to the human body [22]. However, lectin from *Acrocystis nana* has not been studied yet. Thus, this paper deals with preliminary studies on the chemical properties of crude lectin from the red alga, *Acrocystis nana* .

2. Experimental

The red alga, *Acrocystis nana* was collected from Wediombo Beach of the southern coast of Java Island, Gunung Kidul, Yogyakarta, in December 2015. It was packed in plastic bags, and immediately transferred to the laboratory on ice. It was then washed with reverse osmotic water, and kept at -20°C until use. All chemicals used in this study were of an analytical grade.

Crude lectin (salting-out fraction) was prepared by procedures as follow: a hundred grams sample was thawed, cut into small pieces and ground in a mortar, then 2 volumes of 20 mM phosphate buffer pH 7.0 containing 0.85% NaCl (PBS) was added. The powdered alga was then stirred and kept at 4°C overnight. The mixture was centrifuged (10000 rpm, 30 min, 4°C) and to the supernatant, solid $(NH_4)_2SO_4$ was added to attain a final concentration of 75%-saturation. The mixture was kept overnight at 4°C and centrifuged (10000 rpm, 30 min, 4°C). The precipitates were dissolved in a small volume of PBS and thoroughly dialyzed against the same buffer at 4°C for 8 h with buffer replacement every 2 h [23]. The inner fraction was further centrifuged (10000 rpm, 30 min, 4°C) and the supernatant was recovered as a crude lectin (salting-out fraction).

Hemagglutination activity was determined on a 96-well microtiter V-plate using a 2% suspension (v/v) of trypsin-treated rabbit erythrocytes (TRBC) [12,24]. TRBC was prepared as follow [24]. Rabbit blood was obtained from Bioassay Laboratory of Research and Development Center for Marine and Fisheries Product Competitiveness and Biotechnology (Ministry of Marine Affairs and Fisheries, Jakarta, Indonesia). The blood cells were washed three times with 50 volume of 0.85% NaCl (saline) and suspended in saline to give a 2% (v/v) suspension of native blood cells. A tenth volume of 0.5% (w/v) trypsin in saline was added to a 2% suspension of native blood cells and the mixture was incubated at 37°C for 60 min. After incubation, rabbit erythrocytes were washed four times with saline and a 2% suspension (v/v) of TRBC was prepared in saline [24].

Determination of protein content using the BCA protein assay reagent kit by Pierce Biotechnology (2002). The standards used in this test is a bovine serum albumin (BSA). A total of 25 mL of the lectin put into 96-wells microtter flat-plate, added 200 mL of working reagent (reagent A: B 50: 1), incubated (37°C, 30 min), the absorbance at λ 562 nm.

Effects of pH, temperature, and divalent cations on hemagglutination activity were examined as follows [24]. To examine the effect of pH, 500 µl each of a test solution was dialyzed against 100 mL of a 50 mM buffer of various pH values between 3 and 10 at 4°C for overnight, and then dialyzed against PBS pH 7.0 with buffer replacement every 2 h. Fraction after dialyzed with PBS was assayed for hemaglutination activity. To determine the effect of temperature, 500 µl each of a test solution was incubated at various temperatures from 30°C to 100°C for 30 min and subjected to hemagglutination activity assay after cooling. To examine the effect of divalent cations, 500 µl of a test solution was dialyzed against 50 mM EDTA in PBS at 4°C overnight and the inner fraction was measured for hemagglutination activity. Additionally, an equal volume of 20 mM $CaCl_2$ or 20 mM $MgCl_2$ in saline was added to the inner fraction, kept for 2 h at room temperature, and then measured for hemagglutination activity [24].

3. Results and Discussion

Crude lectin (salting-out fraction) from red alga *Acrocystis nana* show a titer of 2^{12} on hemaglutination activity assay, while the crude extract's titer was 2^9 (table 1.). The volume of crude extract and salting-out fraction were 201 ml and 11.5 ml, respectively, therefore, values of total hemagglutination activity (THA) of crude extract and salting-out fraction were 102,912 and 47,104, respectively (table 1.). Protein content of crude extract was 513.28 µg/ml, while salting-out fraction's protein content was 6,225.44 µg/ml (table 1.).

Hemagglutination activity of this crude lectin was stable after treatment at various pH between 3 and 10 (Fig. 1a.) and the activity of crude lectin was not changed at 30°C, but decreased at 40°C, and it was inactivated as incubation temperature exceeded 50°C (Fig.1b.). Hemaglutination activity of salting-out fraction was unchanged after dialyzed againts EDTA in PBS, however the presence of divalent cations, such as Ca^{2+} and Mg^{2+} slightly change the activity.

Table 1. Hemagglutination activity and protein content of crude lectin from *Acrocystis nana*

Acrocystis nana	Volume (ml)	HA[a]	THA[a]	Protein Content (µg/ml)
Crude extract	201	2^9	102,912	513.28
Salting-out fraction	11.5	2^{12}	47,104	6,225.44

[a]HA, hemagglutination titer, the reciprocal of the highest two-fold dilution exhibiting positive hemagglutination; THA, total hemagglutination titer (volume x HA).

Figure 1 (a) shows that effect of pH on hemaglutination activity of crude lectin from red alga *Acrocystis nana* and titer of 2^3 was used for preliminary characterization of crude lectin *Acrocystis nana* . Hemagglutination activity of this crude lectin was slightly decreased (2^2) on pH 3.0 to 5.0 and pH 8.0 to 10.0, however, titer was stable (2^3) at pH 6.0 and 7.0. It may suggested that the hemagglutination activity of crude lectin from the red alga, *Acrocystis nana* was stable against the influence of pH, as reported for some algal lectins [25].

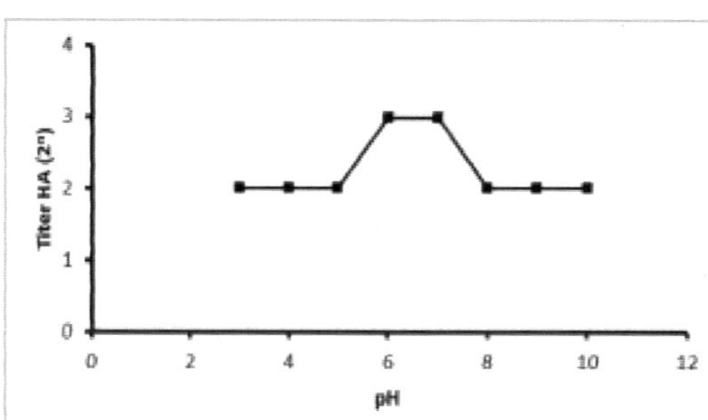

Figure 1. Effect of pH on hemaglutination activity of crude lectin from *Acrocystis nana*

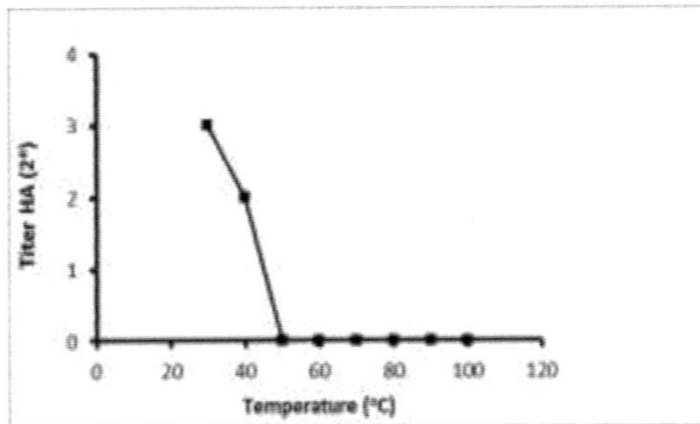

Figure 2. Effect of Temperature on hemaglutination activity of crude lectin from *Acrocystis nana*

Effect of temperature on hemaglutination activity is shown in Fig. 1b. Titer of 2^3 was stable at a temperature of 30°C, while at 40°C temperature treatment, the crude lectin's hemaglutination titer has decreased insignificantly, therefore it might be considered that hemagglutination activity at a temperature of 30°C to 40°C was still stable. However, the activity of this crude lectin was lost at temperature treatment of 50°C to100°C, suggesting this crude lectin was not thermostable. In several lectin studies from purified red macroalgal lectins reported that some of them are thermostable [12, 25]. Meanwhile, according to [26] the activity of lectins that isolated from the marine sponges were stable up to a temperature of 60°C. Moreover, *Pterocladiella capillaceas* is one example of red algal lectins that was unstable on high temperature treatment, as its activity was lost at 60°C [15]. Further, effect of divalent cations on hemaglutination activity of crude lectin from *Acrocystis nana* was also determined in this study. Titer of 2^3 was also used for this analysis (data not shown). Dialysis against EDTA and addition of divalent cations, such as Ca^{2+} and Mg^{2+}, slightly affect the hemagglutination activity, indicating that the activity of crude lectin from the red alga *Acrocystis nana* may required divalent cations for their hemagglutination.

4. Conclusion

Hemagglutination activity, total hemaggutination activity, and protein content of crude lectin from the red alga, *Acrocystis nana* was 2^{12}, 47,104 and 6,225.44 µg/ml, respectively. The activity of this crude lectin was stable after treatment at various pH, and the crude lectin lost its activity when heating temperature was exceeded 50°C. Moreover, the hemagglutination activity was slightly affected by divalent cations treatment, indicating that the activity may depend on the presence of divalent cations. However, further studies are still needed for a more comprehensive understanding about its properties.

Acknowledgements

This work was supported by research project of Sebelas Maret University (PNBP UNS) No. 501/UN27.11/PN/2014.

References

[1] Ditjen PEN 2013 (Jakarta: Kementerian Perdagangan Republik Indonesia)
[2] Starzomki B and Fretweli K (2015, October 29) http://www.centralcoastbiodiversity.org
[3] Goldstein I J, Hughes R C, Monsigny M, Osawa T, Sharon N 1980 *Nature* **285**
[4] Ainouz I L, Sampaio A H 1991 *J. of Bot. Mar.* **34** 211-213
[5] Smith A J 2004 A Review *J. of Applied Phycology* **16** 245-256
[6] Praseptiangga D 2015 *Squalen Bull. of Mar. & Fish. Postharvest & Biotech.* **10**(2) 89-98
[7] Boyd W C, Almodovar L R, Boyd L G 1966 *J. of Transfudion* **6** 82-83
[8] Blunden G, Rogers DJ, Farnham WF 1975 *Lloydia* **38** 162-168
[9] Blunden G, Rogers DJ, Farnham WF 1978 Hemagglutinins in British marine algae and their possible taxonomic value. In: Irvine, D.E.G., Price, J.H. (eds), *Modern approaches to the taxonomy of red and brown algae* (London: Academic)

[10] Rogers DJ, Blunden G, Topliss JA, Guiry MD 1980 *Bot. Mar.* **23** 569–577
[11] Hori K, Miyazawa K, Ito K 1981 *Bull. Jpn. Soc. Sci Fish.* **47** 796-797
[12] Hori K, Miyazawa K, Ito K 1986 *Bull. Jpn. Soc. Sci. Fish.* **52** 323-331
[13] Hori K, Oiwa C, Miyazawa K, Ito K 1988 *Bot. Mar.* **31** 133-138
[14] Fabregas J, Llovo J, Muñoz A 1985 *Bot. Mar.* **28** 517-520
[15] Fabregas J, Lopez A, Llovo J, Munoz A 1992 *Comp. Biochem. Physiol. A.* **103** 307-313
[16] Chiles TC, Bird KT 1989 *Comp. Biochem. Physiol. B.* **94** 107-111
[17] Bird KT, Chiles TC, Longley RE, Kendrick AF, Kinkema MD 1993 *J. Appl. Phycol.* **5** 212–213
[18] Ainouz IL, Sampaio AH, Benevides NMB, Freitas ALP, Costa FHF, Carvalho MC, Pinheiro-Joventino F 1992 *Bot. Mar.* **35** 475–479
[19] Freitas ALP, Teixeira DIA, Costa FHF, Farias WRL, Lobato ASC, Sampaio AH, Benevides NMB 1997 *J. Appl. Phycol.* **9** 495–501
[20] Hung LD, Hori K, Nang Q 2009 *J. Appl. Phycol.* **21** 89-97
[21] Teixeira EH, Arruda FVS, Nascimento KS, Carneiro VA, Nagano CS, da Silva BR, Sampaio, AH, Cavada BS 2012 *Carbohydrates - Comprehensive Studies on Glycobiology and Glycotechnology* **533-558**
[22] Rahmadi P, Ratih P, Susanto AB 2010 *J. of Mar. Bio. and Biotech.* **4**(1) 15-23
[23] Praseptiangga D 2013 *J. Teknologi Hasil Pertanian* **VI**(1) 1-6
[24] Praseptiangga D, Hirayama M, Hori K 2012 *Biosci Biotech. Biochem.***76**(4) 805-811
[25] Hung LD, Hori K, Quang NH 2009 *J. Appl. Phycol.* **21** 93-96
[26] Moura R M, Queiroz A F S, Fook J M S L L, Dias A S F, Monteiro N K V, Robeiro J K C, Moura D E D D, Macedo L L P, Santos E A, Sales M P 2006 *J. of Comparative. Biochem. and Phys.* **145** 517

IOP Conf. Series: Materials Science and Engineering **193** (2017) 012017 doi:10.1088/1757-899X/193/1/012017

Cytotoxicity of Crude Lectins from Red Macroalgae from the Southern Coast of Java Island, Gunung Kidul Regency, Yogyakarta, Indonesia

C Anam[1, 3] E Chasanah[2] B P Perdhana[1] ND Fajarningsih[2] Yusro N F[2] A M Sari[1] A Nursiwi[1] D Praseptiangga[1] A Yunus[3]

[1] Department of Food Science and Technology, Sebelas Maret University (UNS), Jl. Ir. Sutami 36 A, Kentingan 57126, Surakarta, Indonesia
[2] Research and Development Center for Marine and Fisheries Product Competitiveness and Biotechnology, Ministry of Marine Affairs and Fisheries, Republic of Indonesia, Jl. KS Tubun Petamburan VI, Slipi, Jakarta, Indonesia
[3] Graduate School Program of Sebelas Maret University (UNS), Jl. Ir. Sutami 36 A, Kentingan, 57126, Surakarta, Indonesia

E-mail: ekowatichasanah@gmail.com

Abstract. Lectins or carbohydrate-binding proteins, are widely distributed in nature, including in marine algae. It may have been considered that binding specificity of lectins to some carbohydrates provokes to produce many unique biological activities, including cell agglutination, mitogenic activity, and antitumor activity. The aim of this study was to determine the cytotoxicity of crude lectins from red macroalgae collected from the southern coast of Java Island, Gunung Kidul Regency, Yogyakarta against MCF-7 and HeLa cancer cells. *In vitro* MTT assay was used in this study. The results showed that less than 50% of MCF-7 and HeLa cancer cells growth were inhibited by the crude lectins from five species of red macro algae used in this study. The highest inhibition ability shown in the red alga *A. nana* was able to kill 47.68% of HeLa cervical cancer cells.

1. Introduction

Lectins are carbohydrate-binding proteins of non-immune origin that agglutinates cells and/or precipitates polysaccharides or glycoconjugates [1]. Lectins were one of many bioactive compounds founded in macroalgae [2]. Lectins algae have several advantages compared to most terrestrial plant lectins, i.e, they had a small molecular weight, had no affinity for monosaccharides or oligosaccharides but can be inhibited by glycoproteins [3]. Specific binding of lectins lead to a variety of unique biological activities such as cell agglutination, precipitation of polysaccharides, mitogenic activity, antitumor activity and toxicity [2]. Algal lectin research was first started by Boyd *et al.* [4]. Algal lectins have been widely reported from various areas, such as Japan [2], Spain [5,6], Brazil [7,8], China [9], Antarctica [10,11], and Vietnam [12]. More than 250 species of algae were reported contain lectin [13]. In this study, algal lectin from Gunung Kidul regency, Yogyakarta, Indonesia will be reported. The aim of this research was to investigate the cytotoxicity of crude lectins from red macro algae from the southern coast of Java Island, Gunung Kidul regency, Yogyakarta, Indonesia.

2. Experimental

The main materials used in this research were red macro algae from the southern coast of Java Island, Gunung Kidul, Yogyakarta, Indonesia namely *Nitophylium punctatum, Acanthophora spicifera, Acrocystis nana, Helminthora divaricata* and *Gloiocladia repens*. All red macroalga were cleaned from debris and sand, and were rinsed with cleaned water. Red algae were then transported in the laboratorium per species by putting in the cool box with ice outside the plastic bag of the macroalgal sample. All chemicals used in all analysis were in analytical grade.

Crude lectins fractions were prepared by procedures as follow: A hundred grams of frozen samples that have been identified were thawed and grounded to a powder using a mortar with addition of liquid nitrogen. The extraction process was conducted by adding a 20 mM phosphate buffer solution containing 0.85% NaCl (PBS, pH 7.0) to the sample with a ratio of 1: 2 (w /v) [9]. Then the mixtures stirred overnight (at least 8 hours) at 4^0C and centrifuged at 10,000 rpm for 30 min at 4^0C. Supernatant was taken and to be precipitated by the addition of ammonium sulfate slowly to obtain a final concentration of 75% saturation. Samples were stored overnight at 4^0C and centrifuged again (10,000 rpm) at 4^0C. The precipitates dissolved by the addition of a minimum buffer and dialysed against PBS for 8 hours with replacement of PBS 3 times every 2 hours. Fraction (inner fraction) obtained was referred to the crude lectins fraction (salting-out fraction) which was then stored in a freezer at -200C until used for further analysis [3.14].

Hemagglutination activity was determined on a 96-well microtiter V-plate using a 2% suspension (v/v) of trypsin-treated rabbit erythrocytes (TRBC) [15]. It was a routine assay to determine the hemagglutination activity on crude lectins and was expressed as a titer, the reciprocal of the highest two-fold dilution exhibiting positive hemagglutination.

Cell culture was prepared by seeding the cell in flask culture with complete medium. After cell culture 70-80% confluent (about one week), medium was removed and cell culture was washed by 5 ml PBS. Flask culture was adding by trypsin then incubated in $37^0C/5\%$ CO_2 for 3 minutes to make the cells rounding up. The cells were then ready to be counted and used.

Cytotoxicity assay was determined by using 3-(4,5-dimethylthiazol-2-yl)-2,5-diphenyltetrazolium bromide (MTT) [16]. The assay was started with the preparation of $1x10^4$ cells/ml cancer cells and add 100 µl in 96-well micro titer plate then incubated for 12 hours at 37^0C at 5% CO_2 condition. After the media was removed, add 100 µl crude lectin 100 µg/ml and incubated at 5% CO_2 incubator for 24 hours at 37^0C. Then the culture medium was removed and 100 µl of MTT 5 mg/ml was added and incubated for 4 hours at 37^0C at 5% CO_2. After 4 hours, 100 mL of SDS was added and immediately incubated in the dark at the ambient temperature for 12 hours. The calculation was conducted by measuring absorbance at λ=570 and 690 nm. Using data obtained from measurements of absorbance, determination of the percentage of cells inhibition was conducted by using the following formula:

$$\% \text{ Mortality} = \frac{\text{Control cells absorbance} - \text{treatment cells absorbance}}{\text{Control cell absorbance}} \text{ X } 100\% \quad (1)$$

3. Results and Discussion

Based on preliminary research, five (5) species of red algae (table 1) were considered as red algae with the highest hemagglutination activity. [17] Samples then were extracted to obtain crude lectins and their hemagglutination activity were tested (table 1.) [14].

Lectins activity can be ascertained from its ability to cause agglutination of erythrocytes. The ability of lectins in causing agglutination of erythrocytes is caused by cross-linking by lectins in the surface of erythrocytes, and therefore it causes precipitation [18]. Hemagglutination activity assay used erythrocytes of rabbit blood since rabbit erythrocytes have higher sensitivity compared to other animal erythrocytes [19]. Red macroalgae lectins crude fraction which was examined showed various hemagglutination activity. The result of hemagglutination activity testing of *N. punctatum* showed that the hemagglutination activity was 2^6. Meanwhile, other macroalgae showed hemagglutination titer of 2^{12}, 2^6, 2^{24},2^{13} for *A. nana, H. divaricata, G. repens, A. spicifera* respectively.

Table 1. Hemagglutination Activity of Crude Lectins from Five Species of Red Macroalgae

No.	Species	Volume (ml)	Hemagglutination Activity (HA)*	Total Hemagglutination Activity (THA)*
1.	*N. punctatum*	20	2^6	1280
2.	*A. nana*	11.5	2^{14}	188416
3	*H. divaricata*	6	2^6	384
4.	*G. repens*	10	2^{24}	167772160
5.	*A. spicifera*	16	2^{13}	131072

*HA, hemagglutination titer, the reciprocal of the highest two-fold dilution exhibiting positive hemagglutination; THA, total hemagglutination titer (volume x HA).

Cytotoxicity assay of the crude lectins was analysed using MTT method. The principle of the MTT method was the reduction of yellow tetrazolium salt MTT (3-(4,5-dimetiltiazol-2-yl) 2,5-difeniltetrazolium bromide) by reductase system. Succinate tetrazolium of which included in the respiratory chain in the mitochondria of living cells would form insoluble purple formazan crystals. Dead cells were unable to metabolize the various kinds of tetrazolium salts [20]. *In vitro* testing using MTT was used because it was more specific, faster and required minimal test materials when compared with *in vivo* testing and could also limit the use of experimental animals [21].

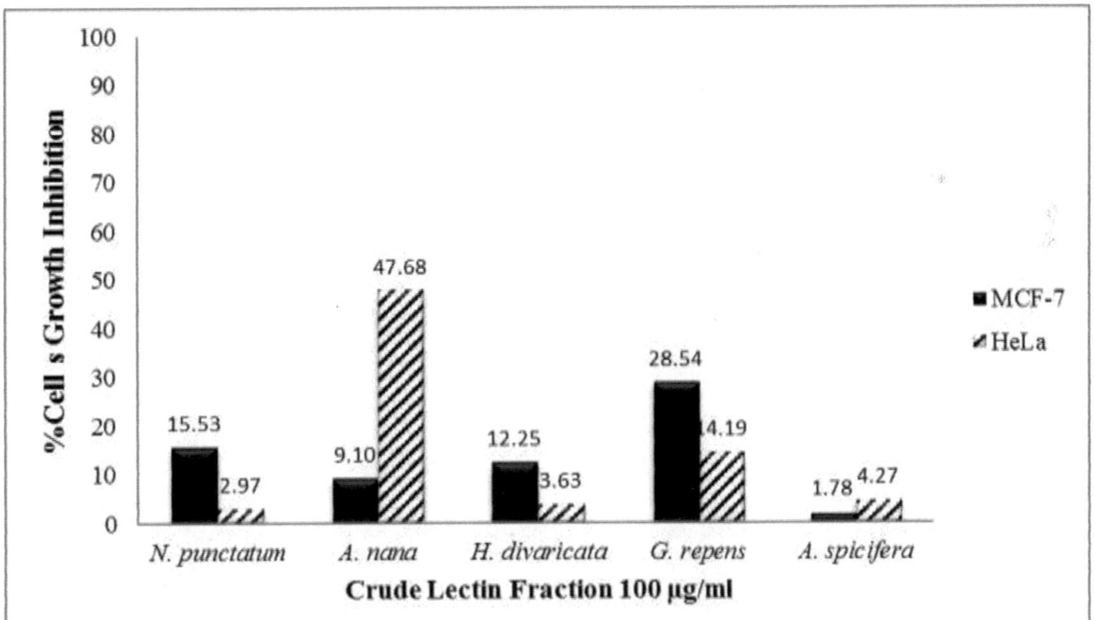

Figure 1. Result of Cytotoxicity Assay of Crude Lectins from Five Species of Red Macroalgae

The cytotoxicity assay was conducted using HeLa cervical cancer cells and breast cancer cells MCF-7. HeLa cells were first taken from the cervix Henrietta Lacks, who died in 1951 because of a malignant cancer of the cervix [22]. Breast cancer cells MCF-7 is a good *in vitro* model to study the mechanism of tumor response or complex linkages between bond and biological reaction of hormones [23].

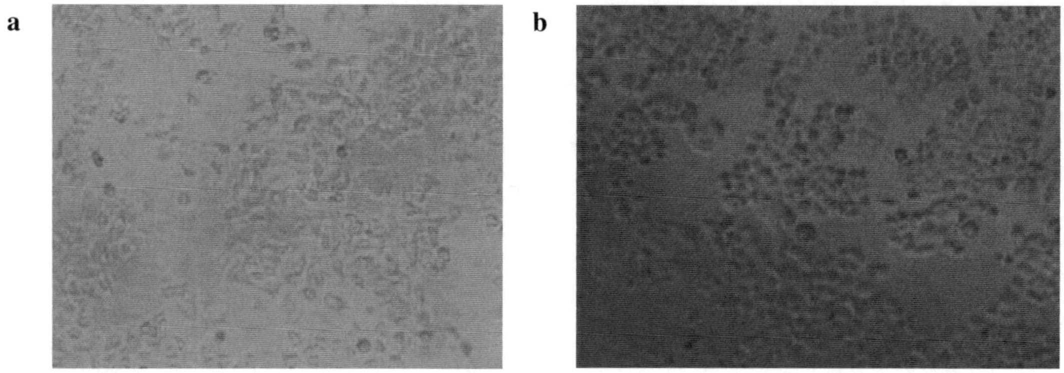

Figure 2. a) MCF-7 and b) HeLa Cancer Cells Morphology

Tests were performed at a concentration of 100 µg/ml for all samples against MCF-7 and HeLa cancer cells. The results showed that none of the crude lectins reached 50% inhibitory activities. National Cancer Institute (NCI) explained that the crude extract of natural materials is claimed to be active as anticancer agents if performing IC_{50} <30 µg/ml, moderate active when concentrations 30 µg/ml $\leq IC_{50}$ <100 µg/ml, whereas for the nature fractional with a concentration $\geq$ 100 µg/ml is considered as inactive [24,25]. Referring to the NCI, it can be concluded that all samples that tested were not performed potent cytotoxic effects on MCF-7 and HeLa cells. It may indicate that the crude fraction still contain impurities compounds. Thus, the crude lectins need to be further purified by a combination of purification procedures include conventional chromatography, such as gel-filtration chromatography, ion exchange chromatography, hydrophobic interaction chromatography, and/or affinity chromatography [26].

Of 5 samples tested, the highest activity was from sample *A. nana* which had an average inhibition of 47.68% against HeLa cells but giving lower inhibition against MCF-7 cells. It means that crude lectin from *A. nana* cell was more selective to HeLa, on the contrary, crude lectin extracted from *G. repens* was more selective to MCF-7. From this research, it may conclude that *A. nana* was potentially used as an anticancer compound, however purification steps is needed to obtain higher activity. Previous research reported that lectin from red algae *Eucheuma serra* was able to cause cell death when tested on some cancer cells. The mechanism of cell death was caused by lectins by apoptosis mechanisms [27]. Most lectins follow a general pathway involving carbohydrate recognition of receptors. Lectin-induced apoptosis occurs through binding of lectin molecules to a specific receptor, internalization into the cell via endocytosis, and further pathway cascades leading to apoptosis [26].

4. Conclusion

Based on this study it can be concluded that crude lectin from red algae studied showed selectitivity. The 5 crude lectins extracted were considered not potential as anticancer shown by less than 50% inhibition of breast cancer cells MCF-7 and HeLa cervical cancer cells. the highest inhibitory activity was detected by the crude lectins of red algae *A. nana*, showing its ability to inhibit HeLa cancer cell growth up to 47.68%. From this result, *A. nana* is needed to be purified to obtain higher activity of the lectin.

Acknowledgement

This work was supported by research project of Sebelas Maret University (PNBP UNS No. 632/UN27.21/LT/2016).

IOP Conf. Series: Materials Science and Engineering **193** (2017) 012017 doi:10.1088/1757-899X/193/1/012017

References

[1] Praseptiangga D 2015 *Squalen Bull. of Mar. & Fish. Postharvest & Biotech.* 10(2) 89-98
[2] Hori K, Miyazawa K, Ito K, 1981 *Bulletin of the Japanese Society Of Scientific Fisheries,* 47(6) 793-798
[3] Hori, Kanji, Keisuke Miyazawa and Keiji Ito 1990 *Hydrobiologia* 204/205 561-566
[4] Boyd WC, Luis R, Almodovar, Lyle G, Boyd, 1966 *Journal of Transfudion* **6** 82-83
[5] Fabregas J, Munoz A, Llovo J 1986 *J. Exp. Mar. Biol. Ecol.* 97 213-219
[6] Fabregas J, Llovo J, Munoz A 1985 *J. of Bot. Mar.* XXVIII 517-520
[7] Ainouz I, Lima, Sampaio AH 1991 *J. of Bot. Mar.* 34 211-214
[8] Ainouz I, Lima, Sampaio AH, Freitas LP 1995 *J. of R. Bras. Fisiol. Veg* 7(1) 16-19
[9] Yi Z, Hai-sheng L 2002 *Chinese J. Oceanol. Limnol.* 20 (3) 256-260
[10] Souza B, Inacio DA, Teixeira FBA, Andrade FK, Melo MRS, Munoz AM, Freitas ALP, 2007 **11**(1) 122-130
[11] Souza B, Inacio DA, Teixeira FBA, Mansila A, Freitas ALP 2010 *J. of Polar Biol.* 33 1311-1318
[12] Hung LD, Sato T, Shibata H, Hori K 2009 *Fish Sci.* 75 723–730
[13] Hung LD, Ly BM, Trang VTD, Ngoe NTD, Hoa LT, Trinh PTH. 2012 *J Appl. Phycol.* 24 227-235
[14] Praseptiangga D 2013 *Jurnal Teknologi Hasil Pertanian* VI (1) 1-6
[15] Praseptiangga D, Hirayama M, Hori K 2012 *Biosci. Biotechnol, Biochem,* 76 (4) 805-811
[16] Wahyuni F S, Sutma S, Aldi Y 2011 *Jurnal Sains and Teknologi Farmasi*, 16 (2) 209-215
[17] Nurhayati T 2015 *Penapisan and Karakterisasi Awal Senyawa Bioaktif Lektin Alga Merah dari Pesisir Pantai Gunung Kidul, Yogyakarta* (Skripsi) (Surakarta:Program Sarjana, Universitas Sebelas Maret)
[18] Lis H, Sharon N. 1998. *Chem. Rev.* 98 637-674
[19] Hori, Kanji, Miyazawa K and Keiji I 1981 *Bul.l Jpn. Soc. Sci. Fish* 47(6) 793-798
[20] Rode, H J 2008 *Apoptosis, Cytotoxicity and Cell Proliferation 4th Edition* (Germany: Roche Applied Science)
[21] Wikanta T, Gusmita D, Rahayu L, Marraskuranto E 2012 *JPB Perikanan* **7** (1) 1–10
[22] Lucey, Brendan P, Walter A, Nelson-Ress, Grover M. Hutchins 2009 *Arch Pathol Lab Med* 133
[23] Sethilraja P, Khatiresan K 2015 *Journal of Applied Pharmaceutical Science* 5 (03) 080-084
[24] Inthe MG, Tarman K, Safithri M 2014 *J. Teknol. and Industri pangan* 25 (1)
[25] Scheuer P J 1987 *Bioorganic Marine Chemistry Volume 1.* (USA: Springer)
[26] Yau T, Xiuli, Cheuk C, Ng W, Ng TB 2015 *Molecules* 20 3791-3810
[27] Suhagara T, Ohama Y, Fukuda A, Hayashi M, Kawabuko A, Kato K 2001 *Cytotechnology* 36 93–99

IOP Conf. Series: Materials Science and Engineering **193** (2017) 012018 doi:10.1088/1757-899X/193/1/012018

Effect of temperature and pH on polygalacturonase production by pectinolytic bacteria *Bacillus licheniformis* strain GD2a in submerged medium from Raja Nangka (*Musa paradisiaca* var. *formatypica*) banana peel waste

E Widowati [1], R Utami [1], E Mahadjoeno[2] and G P Saputro [1]

[1] Department of Food Science and Technology, Faculty of Agriculture, Universitas Sebelas Maret
Jl. Ir. Sutami 36A Kentingan Jebres Surakarta 57126 Central Java Indonesia
[2]Department of Biology, Faculty of Mathematic and Natural Science, Universitas Sebelas Maret
Jl. Ir. Sutami 36A Kentingan Jebres Surakarta 57126 Central Java Indonesia

E-mail: esti_widowati@yahoo.com

Abstract. The aim of this research were to determine the effect of temperature (45°C, 55°C, 65°C) and pH (5.0; 6.0; 7.0) on the increase of total cell count and polygalacturonase enzyme activity produced from raja nangka banana (*Musa paradisiaca* var. *formatypica*) peel waste by pectinolytic bacterial *Bacillus licheniformis* strain GD2a. This research applied two sample repetition and one analysis repetition. The result showed temperature and pH affect total cell count. The total cell count on 45°C and pH 7 recorded the highest number at 9.469 log cell/ml. Temperature and pH also affected pectin concentration at the end of fermentation. The lowest pectin concentration recorded at 45°C and pH 7 was 0.425 %. The highest enzyme activity recorded at 65°C and pH 7 was 0.204 U/ml. The highest enzyme protein concentration was recorded at 65°C and resulted as 0.310 mg/ml on pH 6. The highest specific activity was 19.527 U/mg at 65°C and pH 7. By this result, could be concluded that optimum condition process on polygalacturonase production was at 65°C and pH 7 because it gave highest enzyme activity result (0,204 U/ml).

1. Introduction

Polygalacturonase (EC 3.2.1.67) was an enzyme that catalyzed hydrolysis of α-1,4-glycosidic bond on polygalacturonic acid chain into monomers. This enzyme could be produced by fermentation on raja nangka banana peel waste medium with 0.134 U/ml as the highest enzyme activity [23].

Two factors influenced enzyme activity on fermentation process were temperature and pH. Optimum condition on polygalacturonase production by *Bacillus sp.* was on 40°C of temperature and pH 9 [13], 50°C and pH 7 [26], 50°C and pH 9 [16] and 37°C and pH 6.5 [5]. The rising of temperature would affect the solubility of protein, so that enzyme activity increased. If temperature was too high, it would denature the enzymes thus decreased activity [4; 28]. Incompatible pH could inhibit the growth of microorganism, so that inhibited the enzymes production and decreased enzyme activity [12].

Bacillus licheniformis strain GD2a pectinolytic bacterial had been isolated from traditional market vegetable waste. It was chosen in polygalacturonase production since capability to produce enzyme with the highest activity of 0.209 U/ml and it was enzyme gave the best ability to clarify Garut orange juice. The morphology of these isolates were rod-shaped, chain form, Gram positive and 0.5 μm in size [9].

Production of polygalacturonase by *Bacillus licheniformis* strain GD2a isolate [23] were incubated at 55°C at 48 hours without any initial pH adjustment. Based on this study, further research was needed to determine the effect of temperature and pH during fermentation to enzyme activity. Various temperature used as 45°C, 55°C and 65°C and variation of pH were 5.0; 6.0 and 7.0. From this research, the best temperature and pH to produce polygalacturonase would be selected by the highest activity of enzyme.

2. Methods

2.1. Material Preparation

Raja nangka banana peel waste as the main material was collected from UPKKS Bhakti Kencana Karanganyar, Indonesia and *Bacillus licheniformis* strain GD2a bacterial isolates from food microbiology and biotechnology laboratorium Department of food science and technology collection Universitas Sebelas Maret.

2.2. Banana Peel Pectin and Protein Total Assay

Raja nangka banana peel waste were collected and sorted. Sample was washed, dried and grinded into powder. Method by [25] was used to determine pectin and Kjeldahl method [2] was used to determine total protein on banana peel sample.

2.3. Subcultures and Starter Preparation

Subculture made by one loopful of *Bacillus licheniformis* strain GD2a culture collection isolate was striked on pectin agar media. It was incubated (SELECTA) at 55°C for 24 hours. After that, starter was prepared as one loopful of *Bacillus licheniformis* strain GD2a subculture dispersed into the pectin broth media then incubated at 55°C for 24 hours until it was 10^8 cell/ml amount [23].

2.4. Medium Preparation

Banana peel was washed, blanched and crushed. It was taken 13.18 g of slurry and diluted with distilled water into 100 mL (5% media pectin based on pectin assay result). This media added with 0,5 % of ammonium sulfate [23] and pH was adjusted on 5.0; 6.0; and 7.0 by using hidrogen cloride (HCl) 2 M and natrium hydroxide (NaOH) 3M. Media were sterilized by autoclaving (SELECTA) for 15 minutes at 121°C.

2.5. Enzyme Production

One mL of *Bacillus licheniformis* strain GD2a starter was added into 100 mL of production media. Incubation was carried out at 45°C [13], 55°C [17] and 65°C [26] in *incubator shaker* at 144 rpm of speed [29] for 48 hours [13].

2.6. After Fermentation Assay

Haemocytometer (NEUBAEUR ASSISTANT) bacterial count method [8] was used to count cell amount on media. Final pectin on fermentation media was assayed by modified [25] method. It was taken 20 g of fermentation media and analysed same as [25] method [23].

2.7. Enzyme Isolation and Partial Purification

75 mL of production media was centrifuged (HETTICH) at 6000 rpm for 15 minutes at 4°C [29] to separate supernatant and pellet. Supernatant precipitated with ammonium sulfate at 50% saturation [9] for 24 hours at 4°C. This precipitate was centrifuged at 12.000 rpm at 4°C for 10 minutes [29] to get the pellet. This pellet was dissolved in 0.05 M acetate buffer pH 5.2 (1: 1). Dialysis membrane MWCO 12 KDa was used to dialysis the pellet. It was soaked in 300 mL 0.05 M

acetate buffer solution pH 5.2 in a 600 mL of glass beaker. Magnetic stirrer was used to homogenize this process for 24 hours in a cool chamber, at 4°C [29]. Buffer solution was replaced every six hours.

2.8. Partial Purificated Enzyme essay
Partial purificated enzyme activity was measured with DNS methods [15; 3]. Protein enzyme concentration was measured using the method of Lowry [24]. From the results of enzyme activity and protein concentration could be determined the specific activity of the enzyme to determine the purity of the enzyme [30].

2.9. Data Analysis
Data was analyzed by One Way Analysis of Variance (ANOVA) by using SPSS 16.0 software. If there was a difference in response between the treatment then continued by Duncan Multiple Range Test (DMRT) at $\alpha = 5\%$.

3. Result and discussion
3.1. Banana Peel Pectin and Protein Total Assay
The analysis showed that average pectin content of raja nangka banana peel samples were 37.924 ± 1.250%. This result was lower than the content of pectin in raja nangka banana peel samples on [23], it was 68.47%. This different of pectin content caused of banana ripening level [1]. While the total protein content of raja nangka banana peel samples was 4.742 ± 0.196%. These results were higher than [6] was 2.15% on a kepok banana peel sample. The difference was caused of different types of banana [14].

Analysis of pectin and total protein content of banana peels were important to determine. Pectin as a carbon source and protein as source of nitrogen which would be utilized during the production polygalacturonase. Both of these compounds were the main energy source of the pectinolytic bacteria during the fermentation process.

3.2. After Fermentation Total Cell Assay

Table 1. Effect of Temperature and pH Variation to Total Cell Count

End Cell Count (log cells / ml)			
Temperature (°C)	**pH**		
	5	**6**	**7**
45	8.763 ± 0.03 ^{aB}	9.005 ± 0.10 ^{bB}	9.469 ± 0.03 ^{CB}
55	8.351 ± 0.10 ^{aA}	8.802 ± 0.03 ^{bA}	8.975 ± 0.16 ^{bA}
65	8.665 ± 0.02 ^{aB}	8.878 ± 0.03 ^{bAB}	9.111 ± 0.01 ^{cA}

Description: Different notations in the same line and different capital letters notation in the same column indicate significant differences at $\alpha = 5\%$.

Table 1 showed data after fermentation cell count (log cells/ml) after fermentation process. Statistically, the difference in pH variations gave a significant effect ($\alpha \leq 5\%$) against the after fermentation cell count on the variation of temperature of 45°C, 55°C and 65°C. Increasing in pH was followed increase the number of cells at all temperatures. The highest cell count at pH 7.

Differences in temperature variations gave a significant impact ($\alpha \leq 5\%$) of the after fermentation cell count at various pH 5, pH 6 and pH 7. Data showed that the highest number of cells count in general at a temperature of 45°C in each variation of pH. These results were supported by [22] which explained that there was a positive relationship between enzyme activity and cell growth during fermentation.

3.3. After Fermentation Pectin Content Assay
Determination of pectin of the after fermentation process was intended to determine the yield of the pectin in the fermentation media. Initial Pectin content of the fermentation medium was 5%. By

knowing the pectin contained in the media, could be attributed the microbial activity to degrade pectin media.

Table 2 showed data of the final pectin content (%) after fermentation. Statistically, difference in pH variations gave a significant impact (α≤5%) against after fermentation pectin content at a temperature of 65°C. While at a temperature 45 ° C and 55°C, pH variation did not give effect to pectin content. The highest level of pectin was pH 5. While the temperature variations gave a significant effect (α≤5%) against after fermentation pectin content at pH 5 and pH 7. Variations in temperature of 45°C, 55°C and 65°C did not give effect to final pectin content at pH 6. The highest pectin content in general at a temperature of 55°C.

Table 2. Effect of Variations in Temperature and pH to Pectin Content (%)

Temperature (°C)	After Fermentation Pectin Content (%) pH 5	6	7
45	0.565 ± 0.0 1 ^{aA}	0.472 ± 0.1 3 ^{aA}	0.425 ± 0.01 ^{aA}
55	0.632 ± 0.00 ^{aB}	0.615 ± 0.04 ^{aA}	0.605 ± 0.01 ^{aB}
65	0.570 ± 0.0 2 ^{bA}	0.428 ± 0.01 ^{aA}	0.540 ± 0.04 ^{bB}

Description: Different notations in the same line and different capital letters notation in the same column indicate significant differences at α = 5%.

From initial pectin content 5% and after fermentation pectin content data, could be calculated the pectin that degradated during fermentation process. Hydrolyzed pectin number on this study was about 4.368% to 4.575%. It was not much different from the research by [23], where the levels of pectin degraded in treatment variations pectin 5% with 0.5% ammonium sulfate was 4.790%.

3.4. Partial Purificated Enzyme Assay

Table 3. Effect of Temperature and pH Variation to Partial Purified Enzyme Activity (U / ml)

Temperature (°C)	Enzyme activity (U / ml) pH 5	6	7
45	0.129 ± 0.00 ^{aC}	0.084 ± 0.01 ^{aB}	0.150 ± 0.0 5 ^{aA}
55	0.066 ± 0.00 ^{bB}	0.050 ± 0.00 ^{aA}	0.082 ± 0.00 ^{cA}
65	0.029 ± 0.00 ^{aA}	0.026 ± 0.01 ^{aA}	0.204 ± 0.01 ^{bB}

Description: Different notations in the same line and different capital letters notation in the same column indicate significant differences at α = 5%.

Table 3 showed that variation of different pH gave a significant impact (α≤5%) at the production temperature 55°C and 65°C. However, at 45°C, pH variation did not give effect to polygalacturonase activity. The highest enzyme activity was at pH 7. The lowest enzyme activity was at pH 6.

The results of this study supported by [11], [7] and [18] which stated that the optimum pH conditions to produce polygalacturonase by *Bacillus sp.* was at pH 7. The highest enzyme activity at pH 7 with regard to the after fermentation cell count (table 1.) which also showed the highest cell count at pH 7. [22] revealed that the high activity of the enzyme related with number of microbes existing and further described by [19] that microbes had an optimum pH for growth. If the pH is not appropriate, microbial metabolism would be disturbed and its growth would be stopped. But in this study, after fermentation pH measurements needed to know to determine the pH change media that might be affected to enzyme activity.

Table 3 also showed significant effect (α≤5%) variations in production temperature 45°C, 55°C and 65°C to the activity of an enzyme produced in pH 5, pH 6 and pH 7. On temperature 45°C produced the highest activity and temperature of 65°C resulted in the lowest activity at pH 5 and pH 6 during production. While at pH 7, incubation temperature 65°C produced the enzyme with the highest activity and temperature of 55°C produced an enzyme with lowest activity.

Temperature was a factor that very influential in the production of enzymes. [18], explained if the temperature was too low, the cell metabolism would run slower, but if the temperature was too high, it would damage the enzymes cell growth, which killed the cell itself. The result of this study were generally at temperature 45°C produced the highest activity, in accordance with the number of cells in table 1, which stated that the highest number of cells also recorded at 45°C.

However at 65°C activity of the enzyme gave low result generally, except at pH 7. This phenomenon occurred due environmental factors related to the bacterial enzyme. pH environments and extreme temperatures damaged the enzyme secreted microbial cells. [22] revealed that the production of the enzyme pectinase influenced by the environment.

[20] explained that incompatible temperature would be caused denaturation of enzymes. Denaturation of the enzyme was due to the release of covalent and hydrogen bond in the structure of the enzyme protein. The liberation of these bonds results in the disintegration of the enzyme protein folds that caused enzyme inactivation. While the influence of pH, described by [27] related with pH balance of ionization on active site enzyme. Ion balance affected the bonding of enzyme molecules. If there was no balance of ions, the 3D-shape of active site on enzyme structure would be disturbed so that the substrate could not be bounded in to the enzyme. Such conditions would lead to a decrease in the activity of enzymes.

3.5. Enzyme Protein Concentration Assay

Table 4 showed the enzyme protein concentration data produced at various pH and temperature. pH variation did not give effect to the protein concentration of the enzyme at each temperature of incubation except at 45°C. In general, the highest enzyme concentration was at pH 6. For a given temperature variation, there was significant effect ($\alpha \leq 5\%$) for enzyme protein concentration at each pH production. The highest protein concentration at temperature of 55°C produced at all pH conditions given.

Table 4. Effect of Temperature and pH variation to Enzyme Protein Concentration (mg / ml)

	Partial Purificated Enzyme Protein Concentration (mg / ml)		
Temperature (^{O} C)	**pH**		
	5	**6**	**7**
45	0.014 ± 0.00 ^{aA}	$0.140 \pm 0.0 4$ ^{bA}	$0.075 \pm 0,00$ ^{aB}
55	0.279 ± 0.07 ^{aB}	0.310 ± 0.07 ^{aB}	0.227 ± 0.01 ^{aC}
65	$0.021 \pm 0.0 1$ ^{aA}	$0.022 \pm 0.0 1$ ^{aA}	$0.013 \pm 0.0 1$ ^{aA}

Description: Different notations in the same line and different capital letters notation in the same column indicate significant differences at $\alpha = 5\%$.

Concentration of the enzyme protein associated with the enzyme purity. [10] conducted a purification of enzymes decreased concentration of soluble protein. The decreasing concentration of this protein occurred during purification process, the protein would be cleaned.

On this study, there was also a relationship between amount of protein and enzyme activity, which on table 4, the production conditions of temperature 65°C and pH 7 resulted in a lower protein concentration. However, on table 3 under the same conditions gave high enzyme activity. This indicated that production on 65°C and pH 7, polygalacturonase purity level was high.

Table 5. Specific Enzyme Activity (U / mg)

	Specific Enzyme Activity (U / mg)		
Temperature (^{O} C)	**pH**		
	5	**6**	**7**
45	9.360 ± 2.44 ^{bB}	0.636 ± 0.26 ^{aA}	2.018 ± 0.69 ^{aA}
55	1.561 ± 0.39 ^{abA}	1.062 ± 0.20 ^{aA}	2.333 ± 0.22 ^{bA}
65	1.453 ± 0.37 ^{aA}	1.154 ± 0.14 ^{aA}	19.527 ± 13.19 ^{aA}

Description: Different notations in the same line and different capital letters notation in the same column indicate significant differences at $\alpha = 5\%$.

From table 5, it could be seen that pH variation gave significant effect ($\alpha \leq 5\%$) to the specific activity of the enzyme in the production temperature 45°C and 55°C. However, the pH did not give effect to the specific activity of the enzyme at 65°C. pH 5 resulted in the highest specific activity of enzyme at 45°C and pH 7 recorded highest specific enzyme activity value at 55°C. While the temperature variation gave significant effect ($\alpha \leq 5\%$) to the specific activity enzyme only on pH 5, and the highest was at 45°C. Meanwhile at pH 6 and pH 7, temperature variations did not give effect to the specific activity of the enzyme.

The highest specific activity of polygalacturonase was at 65°C and pH 7 in the amount of 19.527 U/mg. This result was smaller compared to [21] which produced polygalacturonase with specific activity of 1.015 U/mg. This difference caused by bacteria activity and media used in fermentation. [21] used *Bacillus licheniformis* KIBGE IB-21 with a fermentation medium of 1% apple pectin.

Selection of the optimum conditions for enzyme production polygalacturonase treatment based on temperature and pH which gave a significant impact on the parameters of the final number of cells, the levels of pectin, enzyme activity and protein enzyme and specific activity of enzyme. But the main point in this study was activity polygalacturonase enzyme. So that the optimum conditions for produced polygalacturonase was at a temperature of 65°C and at pH 7 because it gave highest enzyme activity (0.204 U / ml).

4. Conclusion

Temperature and pH gave an effect to the production of polygalacturonase, cell counts, pectin content, enzyme activity, enzyme protein concentration and specific activity of the enzyme polygalacturonase. There was a relationship that in conditions of high cell counts, enzyme activity polygalacturonase was high too. Selected Condition fermentation process for producing polygalacturonase was at 65°C and pH 7, due this condition produced enzyme polygalacturonase with the highest activity at 0.204 U / ml.

Acknowledgment

This research was granted by Direktorat Jenderal Pendidikan Tinggi from Hibah Penelitian Unggulan Perguruan Tinggi (PUPT) 2016.

References

[1] Akili M S, Ahmad U and Suyatma N E 2012 *Jurnal Keteknikan Pertanian* **26** 1

[2] AOAC 2005 *Official Method of Analysis of the Association of Official Analitycal Chemists* 18th ed. (Maryland: AOAC International. William Harwitz (ed). United States of America).

[3] Apriyantono A, Fardiaz D, Puspitasari N L, Sedamawati and Budiyanto S 1989 *Analisis Pangan* (Bogor : PAU Pangan dan Gizi. IPB Press)

[4] Castilho L R, Medronho R A and Alves T L M 2000 *Bioresource Technol.* **71** 45-50

[5] Darah I, Nisha M, and Lim S H 2013 *Adv. Stud. Biol.* **5** 5 173 – 189

[6] Dewati R 2008 *Limbah Kulit Pisang Kepok Sebagai Bahan Baku Pembuatan Ethanol.* (Surabaya : UPN Veteran).

[7] Dey A, Karmakar M and Ray R R 2011 *Der Pharmacia Lettre*, **3** 2 358-367

[8] Hadioetomo R S 1993 *Mikrobiologi Dasar dalam Praktek* (Jakarta : Gramedia)

[9] Kalistyatika K 2014 *Produksi dan Karakterisasi Enzim Poligalakturonase Pada Klarifikasi Sari Buah Jeruk Keprok Garut (Citrus nobilis* var. *chrysocarpa)* Skripsi Ilmu dan Teknologi Pangan. Universitas Sebelas Maret Surakarta.

[10] Kant S, Vohra A and Gupta R 2013 *Protein Expr. Purif.* **87** 11–16

[11] Kumari B L, Sudhakar P, Hemmalini K, Sree N S, and Vijetha P 2014 *R. RJPBCS* 5 6 330-339

[12] Kuru G 2013 *Investigations on The Effects of Carbon Sources and pH on Exo-Polygalacturonase Production by Aspergillus sojae*. Thesis. Izmir Institute of Technology

[13]Madu, Osuigwe J, Torimiro, Nkem, Emuebie O R, Isreal J and Kayode A F 2014 *Nature and Science* 12 8 : 110-116

[14] Nagarajaiah S B and Prakash J 2011 *Asian J. Food and Agro-Industry*, Vol. **4** 01 : 31-46
[15] Ordonez R G, Morlon J, Gasparian S, and Guyut J P 1998 *Folia Microbiol.* **4.3** 6 657-660
[16] Patil R C, Murugkari T P and Shaikh S A 2012 *Intl J .Pharm. and Bio Sciences.* **3** 1 261-266
[17] Rahayu A 2014 *Screening dan Karakterisasi Pektinesterase Sebagai Enzim Potensial Dalam Klarifikasi Sari Buah Jeruk Keprok Garut (Citrus nobilis var. chrysocarpa)* Skripsi Jurusan Ilmu dan Teknologi Pangan Universitas Sebelas Maret Surakarta
[18] Raju E V N and Divakar G 2013 *Intl J.Pharm. Sci. And Research* **4** 7 2615-2622
[19] Ray, A K, Bairagi A, Ghosh K S, and Sen S K. 2007 *Acta Ichthyologica Et Piscatoria* **37** 1 47–53
[20] Reece J B, Urry L A, Cain M L, Wasserman S A, Minorsky P V, and Jackson R B 2011 *Campbell Biology 9th ed.* (New York : Benjamin Cummings Pearson Education)
[21] Rehman H U, Qader S A U and Aman A 2012 *Carbohydr. Polym.* **90** 387– 391
[22] Roosdiana A, Prasetyawan S, Mahdi C and Sutrisno 2013 *J. Pure App. Chem. Res* **2** 1 35-41
[23] Safitri N R D 2015 *Produksi Poligalakturonase Melalui Fermentasi Cair Limbah Kulit Pisang Raja Nangka (Musa paradisiaca) Dengan Variasi Sumber Karbon dan Nitrogen Anorganik oleh Bakteri Pektinolitik.* Skripsi Jurusan Ilmu dan Teknologi Pangan Universitas Sebelas Maret Surakarta
[24] Sudarmadji S, Bambang H, and Suhardi 1997 *Prosedur Analisa Untuk Bahan Makanan dan Pertanian* Edisi Keempat (Yogyakarta :Liberty)
[25] Sulihono A, Tarihoran B, and Agustina T E 2012 *Jurnal Teknik Kimia* **18** 4 1-8
[26] Swain M R, Shaktimay K, and Ray R C 2009 *Braz. J. of Microbiol.* **40** 636-648
[27] Teal A R and Wymer P E O 1994 *Enzymes and Their Role In Biotechnology* (London : Biochemical Society)
[28] Uzuner S and Deniz C 2015 *J. Mol. Catalysis B: Enzymatic* **113** 62–67
[29] Widowati E, Utami R, Nurhartadi E, Andriani M A M, and Wigati A W 2014 *J. Aplikasi Teknologi Pangan* **3** 1 16-20
[30] Winarno F G 1983 *Enzim Pangan* (Jakarta : Gramedia Pustaka Utama)

IOP Conf. Series: Materials Science and Engineering **193** (2017) 012019 doi:10.1088/1757-899X/193/1/012019

Antibacterial Activity of Ethyl Acetate the Extract of Noni Fruit (*Morinda citrifolia* L.) Against Bacterial Spoilage in Fish

E R Nugraheni[1], G R Adriani [2], H Munawaroh[3]

[1]Department of Pharmacy, Faculty of Mathematic and Natural Sciences, Sebelas Maret University, Jl. Ir. Sutami 36A Surakarta 57126 INDONESIA
[2]Departement of Biology, Faculty of Mathematic and Natural Sciences, Sebelas Maret University, Jl. Ir. Sutami 36A Surakarta 57126 INDONESIA
[3]Inorganic Materials Research Group, Faculty of Mathematics and Natural Sciences, Sebelas Maret University, Jl. Ir. Sutami No. 36A Surakarta, INDONESIA

E-mail: retnaningtyas_n@yahoo.co.id

Abstract. Noni fruit (*Morinda citrifolia L.*) contains compounds that have potential as antibacterial agent. Antibacterial compounds produced noni fruit (*M. citrifolia L.*) can inhibit bacterial growth. This study was conducted to test the antibacterial activity of ethyl acetate extract of noni fruit (*M. citrifol*ia L.) against spoilage bacterial in fish. *Pseudomonas aeruginosa, Bacillus cereus, Escherichia coli, Klebsiella oxytoca,* and *Enterobacter aerogenes* isolates and examine antibacterial phytochemical profile. Extraction of noni compounds was done by maceration, followed by partition with ethyl acetate to obtain the soluble and insoluble ethyl acetate fraction. Previews result show that the ethyl acetate extract had very strong activity. Extraction process continued by separation and isolation used preparative thin layer chromatography method, so that obtained five isolates and mark them as A, B, C, D and E. Antibacterial activity assay performed on isolates A, B, C, D, and E with 20 and 30% concentration. The test results showed that isolates A could not be inhibit the growth of bacteria, isolates B, C, D, and E has antibacterial activity with weak to strong inhibition. Isolate B had the greatest inhibition activity against the *B. cereus*, whereas isolates E had the greatest inhibition activity against *P. aeroginosa*. MIC (Minimum Inhibitor Concentration) and MBC (Minimum Bactericidal Concentration) test result showed that MIC and MBC values could not be determined. Analysis of compounds by TLC showed that isolate B suspected contains coumarin or flavonoids compounds that have antibacterial activity.

1. Introduction

Indonesia is a tropical country with high humidity. It can be live to most bacteria, fungi and other microorganisms. Foodstuffs are intermediate or substrate for the development of microorganisms. One of wet semi food ingredients susceptible to decay (perishable food) due to bacterial contamination is fish [1, 2]. The water content was contained in the fish as the main factor causing the damage. Reduction of water content can be done with the compounds which can bind water preservative, as well as limit and kill microbial activity destroyer foodstuffs [3].

IOP Conf. Series: Materials Science and Engineering **193** (2017) 012019 doi:10.1088/1757-899X/193/1/012019

Foodstuffs preservation with synthetic preservatives such as benzoic acid, potassium nitrate, BHA (Butylated hydroxyanisole), BHT (Butylated hydroxytoluene) and TBHQ (Tertiary Butylated hydroxyanisole) will cause toxic if prolonged usage [4]. The solution needs to be sought natural preservative that could potentially antibacterial which the opportunity as an alternative to synthetic preservatives.

Noni fruit (*Morinda citrifolia L.*) is a plant that has antibacterial compounds, such as alkaloids, flavonoids, anthraquinone, scopoletin, glycosides, and glucuronic acid [5]. The study reported that ethanolic extract of noni fruit (*M. citrifolia L.*) has antibacterial activity with optimum inhibition at 70% concentration [6]. Therefore, it is necessary to separate the isolation of compounds from the dissolved ethyl acetate which have higher activity than the insoluble one. This study focus in the antibacterial activity of noni fruit isolate (*M. citrifolia L.*) on *P. aeruginosa, B. cereus, E. coli, K. oxytoca* and *E. aerogenes*. The final test was conducted to determine the value of MIC and MBC of isolates with the most active concentration.

2. Experimental

2.1 Isolation of Section Ethyl Acetate with Preparative Thin Layer Chromatography (TLC)

Ethyl acetate fraction was isolated using TLC. GF254 silica gel plate was made by 7 grams of powdered silica and 14 mL of distilled water (1: 2), TLC preparation in a glass plate activated by oven on 50°C ± 1 hour. Samples then spotted on a TLC 3 times. TLC elution were done by using (chloroform: ethyl acetate 3: 1) as mobile phase. The formed spot scraped to obtain isolates. Silica gel and isolate then separated using methanol: chloroform with the ratio 1: 1 then filtered to obtain a filtrate. The filtrate is dried into powder and weighed.

2.2 Thin Layer Chromatography and Determination Compound of Isolation Results

Isolate analysis was determined by TLC. Silica gel GF254 used as the stationary phase while chloroform: ethyl acetate (3: 1) used as the mobile phase. TLC analysis then identified under UV (254 nm and 366 nm). Furthermore, the TLC plate was sprayed with cerium (IV) sulphate to detect the presence of general compounds with $FeCl_3$ and $AlCl_3$ and then calculated the Rf (retention factor).

2.3 Test of Antibacterial of Isolates Noni Fruit (M. citrifolia L.)

2.3.1 Well Diffusion Method

Sterile media of MHA 10 mL and inoculum poured in the petri dish where inoculated with the bacteria of *P. aeruginosa*, *B. cereus*, *E. coli*, *K. oxytoca*, and *E. aerognes*. Then made sinks with perforator. 25 microliters sample with a concentration of 20 and 30% put on pitting, each concentration of 2 replicates. Furthermore, incubated at 37 °C for 24 hours.

2.3.2 Test of MIC and MBC

Isolates with optimum concentration 40 mg/ mL (2x of the optimum concentration). 50 mL of isolates and 50 mL inoculum were mixed in eppendorf. Then, incubated for 24 hours at 37°C. After 24 hours, the result of incubation displacement in a microplate which contains media of MHA, 25 mL for each repetition, then leveled with glass spreader. Then, incubated for 24 hours at 37°C. If showed bacterial growth were expressed as the Minimum Inhibitory Concentration (MIC). While the absence of bacterial growth was expressed as the Minimum Bactericidal Concentration (MBC) [7].

3. Result and Discussion

3.1 Separation and Isolation Compound

1.7 g isolate from ethyl acetate fraction separated and isolated using TLC method. The result isolation of ethyl acetate soluble 70% on UV_{254} showed that 5 isolates could be separated as isolates A, B, C, D, and E. Isolate B has the greatest weight which indicated contains the major compound in noni fruit (*M. citrifolia L.*).

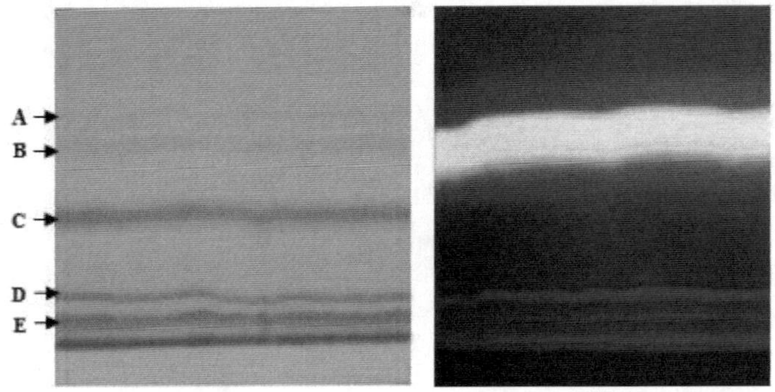

Figure 1. Profile chromatogram of the isolates A, B, C, D and E with (a) UV254 (b) UV_{366}

The result of isolates then identified 5 bacteria. The bacteria is *B. cereus* as gram-positive bacteria and *P. aeruginosa*, *E. coli*, *K. oxytoca*, and *E. aerogenes* as gram-negative bacteria.

3.2 Test of Antibacterial

3.2.1 Well Diffusion Method

Isolates with the effective concentration of partition 20 and 30% were identified in five bacteria. The bacteria are *P. aeruginosa*, *B. cereus*, *E. coli*, *K. oxytoca*, and *E. aerogenes*. Test of the plate was incubated at 37 °C for 24 hours used amoxicillin controls 0.25%, CMC 1% and 70% ethyl acetate.

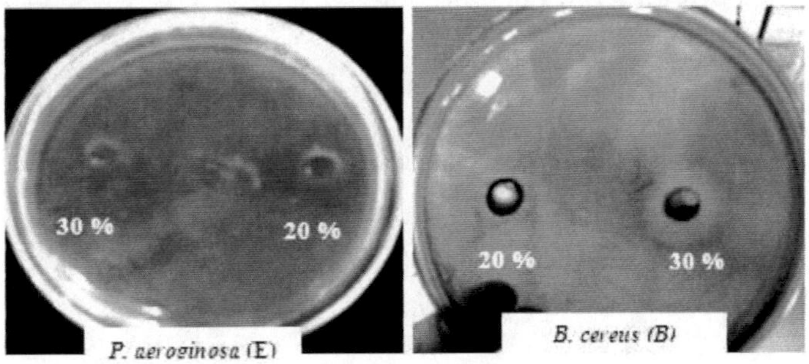

Figure 2. The antibacterial activity test result with the largest inhibition zone.

The inhibitory activity shown isolates B at a concentration of 30% against the bacteria *B. cereus* by 10 mm were considered to have moderate activity and isolates E at a concentration of 20% against the bacteria *P. aeroginosa* equal to 12 mm is considered to have strong activity. In general, B isolates thought to have compounds that can inhibit gram-positive bacteria and especially *B. cereus*. *B. cereus* is a gram-positive bacterium that is composed of peptidoglycan cell wall more (90%) than the lipid [3, 8]. The cell wall of gram-positive bacterial contain polysaccharides which are water-soluble polymers, serves as a positive ion transport. The nature of this water soluble indicating that the cell wall of gram-positive bacteria is polar.

Table 1. The results of measurement of inhibitory zone of isolates noni fruit A, B, C, D, E

Bacteria / Isolates	Diameter of inhibitory zone (mm)									
	E. coli		*B. cereus*		*K. oxytoca*		*E. aerogenes*		*P. aeroginosa*	
	20%	**30%**	**20%**	**30%**	**20%**	**30%**	**20%**	**30%**	**20%**	**30%**
A	-	-	-	-	-	-	-	-	-	-
B	3	5	6	10*	4	5	1	4	4	4
C	4	4	5	5	5	4	6	7	3	8*
D	1	2	2	3	2	3	2	5	2	6*
E	4	5	6	5	2	3	5	5	12*	5
Amoxicillin 2.5%	36		29		24		16		28	
CMC 1%	0		0		0		0		0	
ethyl acetate 70%	0		0		0		0		0	

Description: * The size of inhibition zone of each isolate

P. aeroginosa bacteria is gram-negative bacteria which is composed of layers of peptidoglycan in the cell wall is thinner than the lipid [3, 8]. The peptidoglycan layer builds structures and defenses of bacteria, which can affect the mechanism of inhibition of antibacterial compounds. The mechanism of inhibition was common in the antibacterial compound that inhibits cell wall synthesis. All cells, including bacteria, have a cell membrane that allows the movement of substances in and out of the cell with a controlled way. Antibacterial compounds can affect the integrity of cell membranes, cause damage or even death of the bacterial cell [9]. Inhibitory activity was shown the results of isolates ranges from weak to strong, but overall isolates antibacterial B, C, D, and E have a broad spectrum of activity. It can inhibit the growth of bacteria positive and negative.

3.2.2 Test of MIC and MBC

The test was performed on the isolates had bacterial growth inhibitory activity on antibacterial agar diffusion test. From the results of the concentration of 20% and 30% showed no significant. Then, the test of MIC / MBC used a concentration of 20%. The value of MIC was determined by the absence of bacterial growth media test after incubation of 24 hours at a temperature of 37 °C which it showed the presence of growth after being transferred to a new medium. And the value of MBC was determined by the absence of growth after incubation of 24 hours at a temperature of 37 °C or after being transferred to new media with certain concentration.

The test results of MIC/ MBC show all the bacteria grow at a concentration of 20% after 24 h incubation at a temperature of 37 °C. This means isolates B, C, D, and E with a concentration of 20% can not be bacteriostatic or MIC. According to Liana [10], fraction 6 (F6) of methanol extract of the "Senggani" leaves are bacteriostatic at concentrations above 300 mg / mL against *Staphylococcuc aureus* bacteria. It means that the characteristic of bacteriostatic or bactericidal affected by concentrations of antibacterial and conditions the bacteria. Antibacterial bacteriostatic works to inhibit the growth of bacteria, while bactericidal works to kill bacteria [11].

IOP Conf. Series: Materials Science and Engineering **193** (2017) 012019 doi:10.1088/1757-899X/193/1/012019

3.3 Detection compound of Isolates with Reagent Spray and UV

Detection of compounds with spray reagent $FeCl_3$, $AlCl_3$ and UV rays performed on all isolates to suspect compounds in each isolate. Results of detection isolate A, B, C, D, E shown in table 2. And the result chromatogram of spray reagent shown in figure 3.

Table 2. Results of detection of isolates A, B, C, D, E with reagent spray

No	Isolate	Reagent Spray			
		$FeCl_3$	$AlCl_3$	$AlCl_3$ and UV_{366}	Hypothesis
1	A	Light yellow	-	Reddish yellow	Phenol
2	B	Black brown	Yellow	blue phosphorescent	Flavonoid or coumarin
3	C	-	-	-	-
4	D	-	-	Blue	Flavonoid or coumarin
5	E	-	-	Blue	Flavonoid or coumarin

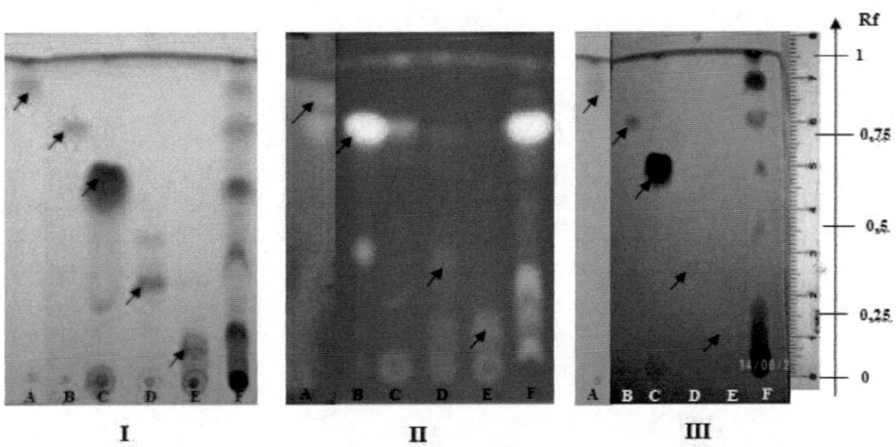

Figure 3. Profile chromatogram compound detection results. (I) Detection of isolates A, B, C, D, E with UV_{254} (II) Detection of isolates A, B, C, D, E with UV_{254} (III) Detection of isolates A, B, C, D, E with Cerium (IV) sulfate. Stationary phase: Silica gel 60 GF_{254} Mobile phase: chloroform: ethyl acetate 3: 1 (v / v)

The results of detection compound with reagent spray $FeCl_3$ against isolates of A, B, showed positive results (+) phenol compounds, but to isolate C, D, and E showed a negative result (-). Detection of compounds with reagent spray $AlCl_3$ followed UV_{365} showed positive results (+) to isolate B (table 2), so that suspected isolates B contains flavonoids. According to Wagner [12] showed that the compounds detected by UV_{365} blue fluorescence or blue-green indicate these compounds including flavonoids or coumarin in a class of phenolic. Isolates E also showed blue luminescence under UV_{365} that can be suspected to contain coumarin compounds or flavonoids as well. But the predictions need to be further analyzed to determine the type of compounds.

Flavonoids and coumarin synthesized by two main pathways for biosynthesis of the aromatic ring, namely the shikimate pathway and acetate malonate. Coumarin is phenol compound derived from the shikimate pathway while flavonoid derived combination [13]. Coumarin showed a broad spectrum of biological activities such as antibacterial activity, antifungal, anticoagulant, anti-inflammatory, antitumor and anti-HIV [14]. The other study revealed that a compound coumarin and xanthones have the ability to inhibit the growth of bacteria *Staphylococus aureus* but not for bacteria *Escherichia coli, Acinobacter spp,* and *Crypto Cocus neofumas* [15, 16].

The flavonoids reported activity inhibit pathogenic agents in humans and is an antibiotic which can inhibit gram-positive and negative bacteria [17]. The other studies showed flavonoids contained in extracts of *P. macrocarpa* which have antimicrobial activity at concentrations of 0.3 mg/disk to 8 of the gram-negative bacteria and one of gram-positive bacteria such as *E. coli* [18]. It indicated that isolates was contained flavonoids or coumarin compounds are an antibacterial candidate, especially isolates of B. Detection of compounds must be strengthened with the identification using either UV-Vis spectrophotometry, IR, or NMR to obtain a pure chemical structure of the compound antibacterial candidates.

This study could be the basic for further research in the purification of isolates. It could known the specific compound that activity as an antibacterial to reduce the problem of decay fish or fish preservation. Fish is source of high protein, fat, vitamins, and minerals which are very good and prospective. According to Siswono [19], the main advantage of fish protein compared to other products is the completeness of the amino acid composition and simplicity to digest. The role of nutrition for healthy fish should be a good fish processing system with natural ingredients that are able to inhibit the activity of microbes or enzymes that improve the quality of fish as food.

4. Conclusion

The antibacterial activity of isolates (B, C, D, E) of the 70% ethyl acetate soluble portion had inhibitory activity. The inhibitory activity showed isolates B in *B. cereus* and isolates E in *P. aeruginosa* bacteria. The concentration of 20% ethyl acetate isolates soluble part based antibacterial activity test was enough optimum inhibit growth bacteria of *P. aeruginosa*, *B. cereus*, *E. coli*, *K. oxytoca*, and *E. aerogenes*. Test MIC and MBC of isolates B, C, D, and E in *B.cereus* and *P. aeruginosa*bacteria with concentration 20 % could not be determined.

IOP Conf. Series: Materials Science and Engineering **193** (2017) 012019 doi:10.1088/1757-899X/193/1/012019

References

[1] Wikandari P R, Suparmo, Y Marsono, dan E S Rahayu 2012 *Jurnal Natur Indonesia* **14(2)**: 120–125.
[2] Fardiaz S 1992 *Mikrobiologi Pengelolaan Pangan* Departemen Pendidikan dan Kebudayaan Direktorat Jendral Pendidikan Tinggi Pusat Antar Universitas Pangan dan Gizi Bogor: Institut Pertanian Bogor.
[3] Wulandari S, Sayuti, dan I Asmaini 2005 *Jurnal Biogenesis* **2(1)**: 30–35.
[4] Barus P 2009 *Pemanfaatan bahan pengawet dan antioksidan alami pada industri bahan makanan* Medan: Universitas Sumatera Utara.
[5] Collins C H, Patricia M Lyne, J M Grage 1989 *Microbiological Methods* 6th Edition London: Butterworth.
[6] Retnaningtyas E, E Purwani, dan T Purwoko 2009 *Pemanfaatan Ekstrak Buah Mengkudu (Morinda citrifolia L) dan daun Pandan (Pandanus amaryllifolius Roxb.) sebagai Pengawet Alami Daging dan Ikan Segar* LPPM UNS: Surakarta.
[7] Zakaria Z A, A M Desa, K Ramasamy, N Ahmat, A S Mohamad, D A Israf and M R Sulaiman 2010 *Afr. J. Microbiol. Res.* **4(1)**: 071-075.
[8] Jawetz E, J L Melnick, and E A Adelberg 2001 *Mikrobiologi untuk Profesi Kesehatan (Review of Medical Mikrobiology)* translated by H Tomang Jakarta: Penerbit EGC.
[9] Vinodh R, L K Elumalai, D Sangeetha, and R Rengasamy 2009 *J. Biosci. Tech* **1(1)**: 45-51.
[10] Liana I 2010 *Aktivitas Antimikroba Fraksi dari Ekstrak Metanol Daun Senggani (Melastoma candidum D. Don) terhadap Staphylococcus aureus dan Salmonella typhimurium serta Profil Kromatografi Lapis Tipis Fraksi Teraktif* Jurusan Biologi Fakultas Matematika dan Ilmu Pengetahuan Alam Universitas Sebelas Maret: Surakarta.
[11] Sunny A J 2011 *Potensi Antibakteri Propolis Lebah Trigona spp Asal Bogor* Departemen Biokimia Fakultas Matematika dan Ilmu Pengetahuan Alam IPB: Bogor.
[12] Wagner H 1983 *Plant Drug Analysis* Translation of the German Edition
[13] Kristanti A N, N S Aminah, M Tanjung, dan B Kurniadi 2008 *Fitokimia* Jurusan Kimia FMIPA UNS: Surakarta.
[14] Dekic B R, N S Radulovic, V S Dekic , R D Vukicevic, and R M Palic 2010 *Molecules* **15**: 2246-2256.
[15] Pattalung P N, Thongtheeraparpn W, Wiriyachitra P, and Taylor W C 1994 *Planta Med.* **60**: 365–368.
[16] Verotta L, Lovaglio E, Vidari G, Finzi P V, Neri M G, Raimondi A, Parapini S, Tara melli D, Riva A, and Bombardelli E 2004 *Phytochem.* 2867-2879.
[17] Bylka W, I Matlawska, and N N Pilewski 2004 *Review Article JANA* **7** **(2)**: 24-31.
[18] Hendra R, S Ahmad, M Y Shukor, and E Oskoueian 2011 *Int. J. Mol. Sci.* **12**: 3422-3431.
[19] Siswono 2003 Ikan Air tawar kaya Protein dan Vitamin (http : gizi.net.com)

IOP Conf. Series: Materials Science and Engineering **193** (2017) 012020 doi:10.1088/1757-899X/193/1/012020

Food safety management and risk assessment in the fresh produce supply chain

L Jacxsens[1], M Uyttendaele[1], P Luning[2], A Allende[3]
[1] Laboratory of Food Microbiology and Food Preservation, Department of Food Safety and Food Quality, Ghent University
[2] Product Design and Quality Management Group, Department of Agrotechnology and Food Sciences, Wageningen University, P.O. Box 8129, NL-6700 EV Wageningen, The Netherlands.
[3] Research Group on Quality, Safety and Bioactivity of Plant Foods, Department of Food Science and Technology, CEBAS-CSIC, P. O. box 164, Espinardo, Murcia, E-30100, Spain

E-mail: liesbeth.jacxsens@ugent.be

Abstract. This paper is the output of several years of scientific research coordinated by Laboratory of Food Preservation and Food Microbiology at UGent, within the EU FP7 Research project Veg-i-trade (www.vegitrade.org), in collaboration with among other partners, Wageningen University and Cebas-CSIC. Fresh produce and derived products are globally traded and subjected to an inherent sensitive towards enteric pathogens as Salmonella and pathogenic E. coli due to their cultivation practices. As fruits and vegetables are increasingly being consumed raw, a potential health risk towards consumers is present. In the Veg-i-Trade project the extend of presence of pathogens in leafy greens and strawberry fruit and their cultivation environment (as water, soil, manured soil, etc.) was analysed. Insight in the food safety management system enlighted the need for further fostering and guidance towards farmers in good practices in order to reduce the potential pressure of the presence of the pathogens both in EU and non EU countries. Exposure assessment calculations demonstrated the usefulness of mathematic modelling to gain more insight in fragmented microbiological analysis and information of cultivation practices, as such the impact of contamination of irrigation water and the impact of a flooding event. Veg-i-Trade was a challenging project both in scientific and management perspective as 23 partners collaborated.

1. Introduction

Fruits and vegetables are increasingly being consumed and also raw consumption became important [1]. The current quality assurance and control tools and methods to prevent and/or to control microbiological risks associated with fresh produce are challenged due to the following pressures upon the food supply chain, i.e. changing consumption patterns, globalization and climate change. It demonstrates the need for scientific research and development of new and/or improved tools, techniques and practices to adapt the current risk management systems. A conceptual research approach is presented to analyse the complexity of the climate change and globalization challenge on the fresh produce supply chain taken as a case study. The factors which affect the vulnerability of the fresh produce chain demand a multidisciplinary research approach. The proposed knowledge-based modelling system is believed to be a most appropriate way to identify problems and to offer solutions to monitor and prevent microbiological food safety risks during all phases of food production and supply. To explore the potential impact of climate change and globalization, baseline information can be obtained by surveillance and performance measurement of implemented food safety management systems. Simulation of climate change scenarios and the logistic chain of fresh produce, along with mathematical models to optimize packaging technology to maintain quality and safety of fresh produce are tools to provide insights in the complex dynamic ecosystem. They are the basis for elaboration of risk assessment studies to scientifically support management options and decisions to new microbiological threats related to globalization and climate change in the fresh produce supply chain. This research concept as such will contribute to develop strategies in order to guarantee the (microbiological) food safety of fresh produce on the long term [2].

2. Fresh produce and derived food products and microbiological safety

Fresh produce and derived food products are food from non-animal origin and are more and more linked with biological food safety issues, such as presence of parasites, pathogenic bacteria as *Salmonella* or pathogenic *E. coli* and foodborne viruses as Norovirus. In particular, as on a world level still quantities of fresh fruit and vegetables traded represent a limited part of what is produced, the concept of safe food for all is a key issue. The main objective of the present study is to investigate whether and which factors could be identified as universal risk factors for pathogen contamination of fresh produce across farms in various countries with variable climate and agro-technical management practices. For this purpose leafy greens, strawberries and their primary production environment (soil, water, contact surfaces) were analyzed for the presence of *Salmonella*, STEC, *Campylobacter* and the amount of generic *E. coli* using a similar sampling plan at a variety of farms in Belgium, Brazil, Egypt, Norway and Spain within the framework of the European Veg-i-Trade project, executing research on the topic of microbiological (and chemical) safety of fresh produce in a global context.Water samples were more prone to containing pathogens (54 positives out of 950 analyses) than soil (16/1186) and produce on the field (18/977 for leafy greens and 5/402 for strawberries). The prevalence of pathogens also varied markedly according to the sampling region. Flooding of fields increased the risk considerably, with odds ratio (OR) 10.9 for *Salmonella* and 7.0 for STEC. A significant association between elevated numbers of generic *E. coli* and detection of pathogens (OR of 2.3 for STEC and 2.7 for *Salmonella*) was established. Generic *E. coli* was found to be a suitable index organism for *Salmonella* and STEC, but to a lesser extent for *Campylobacter*. Guidelines on frequency of sampling and threshold values for *E. coli* in irrigation water may differ from region to region [3].

3. Food safety management systems (FSMS)

The fresh produce chains are nowadays increasingly challenged to also respond to demands on environmental and social accountability. It is clear that climate and climate change should not be ignored as a factor, and will in particularly affect the introduction of contaminants at the pre-harvest stage. Fresh produce supply chains natural evolved to obtain effective food safety management which is also embedded in the striving to bring sufficient and nutritious good quality fresh produce to the consumer. Three self-assessment tools were developed which were shown useful as an internal audit system to track the current status of "best practices" or maturity of the "management systems" implemented at primary production, processing or trade. They consist of 64 to 69 indicators and address core control and assurance activities in the prevention and reduction of microbiological, mycotoxin and pesticide residue contamination. Each indicator uses grids with supporting information for self-assigning to situation 1 (basic level), 2 (average level) or 3 (advanced level). Selected indicators and grids as such provide a 'bird's eye view' of the performance level of the current systems in place, the risk level of the context wherein the actor has to operate, and an indication of the system's output. Primary production companies (n = 118), located in the EU and in international cooperation partner countries exporting to the EU, were assessed by using a diagnostic tool. The results from the study indicated that several factors have a dominating effect on the status of FSMSs in the global fresh produce chain. International export supply chains promote capacity building within companies in the chain, to answer the stringent requirements of private brand standards. This was shown to be an important factor in emerging and developing countries, where local institutional environments often fail to support companies in setting and implementing their FSMSs. Moreover, the legislative framework in these countries still requires improvements in the establishment and enforcement. All this has negative consequences for the FSMSs in companies supplying the local markets. In companies located in the EU, sector and other produce organisations facilitate the sampling for pesticide residues and collaboration in the sector. Overall, farmers showed less knowledge and overall awareness regarding microbiological hazards, which is related to the less attention paid to these in the current legislation and standards. Furthermore, standards are an important tool to trigger the maturation of the systems as companies that were lacking any pressure to comply to standards operated at a very basic level - with only few activities implemented. The insights from this study indicate the need of stratified measures and policies to support companies in the fresh produce chain in designing and operating their FSMSs according to the institutional environment in which they operate [4].

4. Exposure and Risk assessment

A quantitative exposure model for generic ***E. coli*** was developed to estimate the occurrence and levels of ***E. coli*** under defined circumstances of agricultural production of leafy greens. It was noted that apart from the regional variable baseline distribution of ***E. coli*** numbers in the irrigation water or the soil, the cultivation season and interlinked with this the length of the crop cycle, rainfall and extent of solar radiation influences the ***E. coli*** levels substantially in case of open field production. (based on Castro Ibanez et al., submitted for publication). A probabilistic quantitative exposure assessment model was developed to compare the distribution of contaminated portions of basil, strawberry and butterhead lettuce to which Belgian and Spanish consumers are exposed per year. The highest number of contaminated portions was found for basil compared to butterhead lettuce and strawberries which can be explained by the observed higher prevalence of enteric pathogens, their better survival during storage and less consumers washing the basil before consumption. A daily consumption of basil leads to exposure of five to nine contaminated portions a year, both in the Belgian and Spanish situation. The lowest exposure to the pathogens was observed for strawberries in the Spanish situation due to lower consumption frequency in combination with higher die-off of the enteric pathogens on strawberries. Scenario analysis revealed that the main driving force in exposure is prevalence of pathogens, a moderate effect of consumption frequency and to a minor extent consumer behavior. This quantitative exposure ranking demonstrates that also niche products such as basil can have an potential impact on public health equal or higher than lettuce, resulting in a priority in monitoring and surveillance (based on Jacxsens et al., submitted for publication).

Finally, a gap analysis was conducted via a systematic review of current published information on the impact of contaminated water towards the potential risk of pathogens on fresh produce. Both papers from water science groups and food research groups are included [5]. Risk assessment related to use of water and safety of fresh produce orient from both water and food microbiology studies. Although the set-up and methodology of risk assessment in these two disciplines may differ, analysis of the current literature reveals some common outcomes. Most of these studies from water perspective focus on enteric virus risks, largely because of their anticipated high concentrations in untreated wastewater and their recalcitrance to common wastewater treatments. Risk assessment studies from the food perspective rather focus on bacterial pathogens such as Salmonella and pathogenic E. coli. Few sitespecific data points were available for most of these microbial risk assessments, meaning that many assumptions were necessary which are retaken in many studies. Specific parameters lacking hard data included rates of pathogen transfer from irrigation water to crops, pathogen penetration, and survival in or on food crops. Data on these factors have been investigated over the last decade and this should improve the reliability of future microbial risk estimates. However, the sheer number of different foodstuffs and pathogens, combined with water sources and irrigation practices, means that developing risk models that can span the breadth of fresh produce safety will be a considerable challenge. The new approach using microbial risk assessment is objective and evidence-based and leads to more flexibility and enables more tailored risk management practices and guidelines. Drawbacks are however capacity and knowledge to perform the microbial risk assessment and the need for data and preferably data of the specific region [5].

5. References

[1] Jacxsens L, Castro Ibanez I, Gomez-Lopez V, Fernandes J, Allende A, Uyttendaele M, I Huybrechts 2015 *J. Food Prot.* **78** 784-95.

[2] Jacxsens L, Luning P, Van de Vorst J, Devlieghere F, Leemans R, Uyttendaele M 2010 *Food Res. Int.* **43** 1925-35

[3] Ceuppens S, Johannessen GS, Allende A, Tondo EC, El-Tahan F, Sampers I, Jacxsens L, Uyttendaele M 2015 *J. Environ. Res. Public Health* **12** 9809-31.

[4] Kirezieva K, Luning P, Jacxsens L, Allende A, Johannessen G, Tondo E, Rajkovic A, Uyttendaele M, van Boekel M, 2015 *Food Control* **52**, 85-97.

[5] De Keuckelaere A, Jacxsens L, Amoah Ph, Medema GJ, McClure P, Jaykus LA, Uyttendaele M 2015 *Compr. Rev. Food Sci. Food Saf.* **14**, 387-410

Optimization and Characterization of Cinnamon Leaves (*Cinnamomum burmannii*) Oleoresin

L.U. Khasanah, Kawiji, P. Prasetyawan, R. Utami, W. Atmaka, G.J. Manuhara, A.P. Sanjaya
Food Science and Technology, Sebelas Maret University, Surakarta

E-mail : liaumikhasanah@yahoo.co.id

Abstract. This research aimed to determine the optimum yield condition on cinnamon leaves oleoresin production at various temperature and contact time during maceration and to find out the characteristics of cinnamon leaves oleoresin such as active compound, cinnamon leaves oil content, and solvent residue levels at optimum yield. This research used the variations of extraction temperature (70, 75 and 80°C) and extraction time (4, 5 and 6 h). Based on Response Surface Methodology (RSM), the equation of cinnamon leaves oleoresin sample optimization as follow: Y = 13 – 1.0167X1- 0.2833X2- 0.6833X12- 0.5833X22- 0.3250 X1X2. The optimum yield of cinnamon leaves oleoresin (13.3790%) was obtained at 77.754°C for 4.9185 h. The characteristics of cinnamon leaves oleoresin that showed the optimum yield were 59.56% eugenol level, 9.50% cinnamon leaves oil content and 22700 ppm solvent residue level.

1. Introduction

Cinnamon is an important export commodity of Indonesia. The product is exported to the United States of America (46 %), the Netherlands (11 %), Germany (4 %), and Singapore (4 %). In 2015, the most exported cinnamon product (95%) was cinnamon bark quill and the other was cinnamon powder [1]. Indonesian cinnamon production reached 45 % of the world's total production, and the volume of Indonesian cinnamon export was the largest in the world (26%) [2]. The main product of cinnamon tree is cinnamon bark. The by-products of cinnamon tree are branches and leaves. However, all parts of the cinnamon tree contain essential oil and oleoresin that mainly found in the bark, stems, and leaves, and only a small portion is found in the wood part [3].

Besides from the cinnamon bark, oil may be distilled from the cinnamon leaves. The oil of C. *zeylanicum* leaf contains eugenol as the main component (80 - 90%), whereas the main content of the oil of C. *burmanii* leaf and C. *cassia* leaf is cinnamaldehyde that same with the content of the bark oil [4].

The main chemical components of the essential oil of cinnamon leaves (*Cinnamomum burmannii*) are cinnamicaldehyde (63.61 %) and eucalyptol (17.27 %) [5]. *Cinnamomum zeylanicum* oleoresin contains an active compound of eugenol, as much as 87.2% [6]. *Cinnamomum tamalla* oleoresin contains an active compound of eugenol, as much as 17.3 % [7].

Oleoresin is a mixture of oil and resin or gum, which is obtained from the extraction, evaporation, and standardization of essential oil (essential oil and non-volatile components of the herbs and spices). The oleoresin characteristics usually are viscous liquid, paste, or solid. Many researchers have studied various kinds of oleoresin such as *Curcuma zanthorrhiza* oleoresin, galangal

IOP Conf. Series: Materials Science and Engineering **193** (2017) 012021 doi:10.1088/1757-899X/193/1/012021

oleoresin, and cinnamon oleoresin. Oleoresin has many advantages. It is more popular and hygienic because its aroma and flavor are stronger than its raw material; it is stable during storage; and it needs less space for storage [8]. Oleoresin has the following characteristics: its tends to have antimicrobial characteristics, it is more hygienic, and it contains natural anti-oxidants, it is enzyme-free, it has a longer shelf life, it allows more economical storage, it has a lighter weight for transport, and it is free from the danger of mildew on spices [9].

Extraction is an important phase of oleoresin production. One-phase extraction (direct extraction) is a direct process of oleoresin production by adding solvent which produces pure oleoresin with the comparison of essential oil and resin as genuinely found in the raw materials. A two-phase extraction is a process of oleoresin-production through distillation to obtain the essential oil, and follow by oleoresin extraction which produces oleoresin that contain essential oil and resin in a certain ratio [8].The previous research stated that at the pilot plant scale, oleoresin produced by one-phase extraction with the ethanol solvent was 35% while oleoresin produced by two-phase extraction was 31.25% [8].

The extraction process is affected by several factors such as size of the raw material, solvent, duration of the extraction process, and extraction temperature [10]. In this research, optimization of extraction temperature and extraction time was performed to investigate the optimum conditions that resulting the highest oleoresin yield.

2. Experimental

The *Cinnamomum burmannii* leaves were chopped (1 cm) and extracted through maceration method. The extraction varied at several temperaturs (70, 75, and 80°C) and contact time (4, 5, and 6h). The solvent was ethanol (70%). The ratio between the material and the solvent was 1:6. Filtering process was performed to separate the dregs and filtrates. The filtrates were then separated with a *rotary vacuum evaporator* to obtain the *Cinnamomum burmannii* leaf oleoresin. The data of oleoresin yields obtained from the treatment were analyzed using the computer software of Matlab Version 7.0 to determine the optimum conditions. The oleoresin of the optimum yields was analyzed to determine the characteristics such as eugenol content using GCMS (SHIMADZU, column: Rastek RX-i-1MS, inert gas: Helium), essential oil content with the distillation method [11], and residual solvent content using GC.

3. Results and Discussion

3.1. Cinnamomum burmannii leaf Oleoresin Yield Optimization

Table 1 showed the *Cinnamomum burmannii* leaf oleoresin yields. The yields were analyzed by using the *Response Surface Methodology* (RSM) and Matlab 7.0 to investigate the optimum yields. The RSM is a statistical technique to find the optimum conditions of a response affected by complex variables. The original concept of this method was introduced by Box and Wilson in 1951 [12].

In this research, the RSM was employed to determine the optimum conditions that would result in the higher *Cinnamomum burmannii* leaf oleoresin yields (% b/b) with two factors which are extraction temperatures (X_1) of 70, 75, and 80 °C, and extraction time (X_2) of 4, 5, and 6h. Polynomial quadratic regression model was made to predict the optimum Y variable (oleoresin yields in the percentage (%) unit). The mathematical model of the formed response is as follows: Ŷ = b0 + b1x1 + ... + bnxn + b11x12 + ... + bnnxn2 +b12x1x2 + ... + bn-1.nxn-1xn

The result of the *Cinnamomum burmannii* leaf oleoresin yield data optimization with the computer software and Mathlab 7.0 and the factorial experiment design resulted the following response function: Y = 13 – 1.0167X1- 0.2833X2- 0.6833X12- 0.5833X22- 0.3250 X1X2.

Table 1. The *Cinnamomum burmannii* Leaf Oleoresin Yields

Run	X_1 (Extraction Temperature)	X_2 (Extraction Time)	Yield (%)
1	70 (-1)	4 (-1)	11.5
2	70 (-1)	5 (0)	13.3
3	70 (-1)	6 (1)	12.8
4	75 (0)	4 (-1)	13.3
5	75 (0)	5 (0)	12.8
6	75 (0)	6 (1)	11.5
7	80 (1)	4 (-1)	10.9
8	80 (1)	5 (0)	10.6
9	80 (1)	6 (1)	10.4

Figure 1 showed the optimum response of the effects of extraction temperatures (X1) and time (X2) on the *Cinnamomum burmannii* leaf oleoresin yields at the stationary points of -0.7349 and 0.0381. Of these points, the extraction temperature of 77.754°C and the extraction time of 4.9185 h gave the optimum *Cinnamomum burmannii* leaf oleoresin yield (13.3790%).

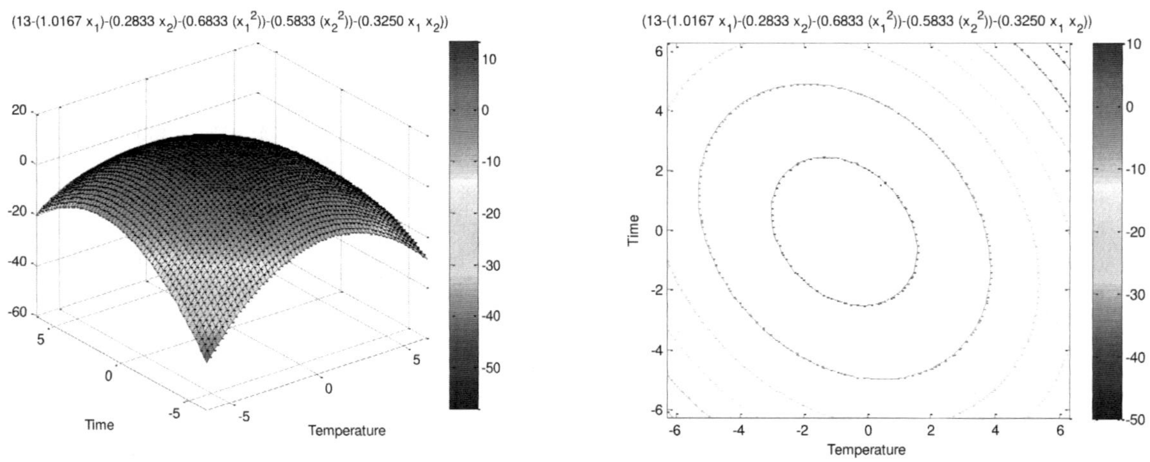

A B

Figure 1. Response Surface Methodology (A) Response surface (B) Response contour

The increase of extraction temperature caused the molecular motion of ethanol as a solvent to be faster but random. In addition, the increase of temperature caused the pores of solids to expand so that the ethanol diffused and entered the materials through the pores and then dissolved the oleoresin. As a result, the oleoresin was interacting greatly and this caused the solute mass transfer from the feed solids to the solvents to be greater [13].

Figure 1 showed that when the temperature was higher and the time was longer, the extraction yield increased. However, when the temperature and time exceeded the optimum points (77.754°C and 4.9185h), the extraction process became ineffective. The temperature above the solvent's boiling point and the long time of extraction caused the solvents to be wasted. In addition, the volatile components of oleoresin were brought by the evaporated solvents so that the yields decreased. The extraction could be promptly performed at a high temperature. However, in the oleoresin extraction, such a treatment could cause damage to several components found in the spices [14]. The longer the extraction time, the higher the yields obtained until the six-hour limit as the contact time between the materials and the solvents was greater. Above the six-hour contact time, the yields decreased as the solution had already reached the saturation point [15].

3.2. Cinnamomum burmannii leaf Oleoresin Quality Characteristics at Optimum Condition

The optimum *Cinnamomum burmannii* leaf oleoresin yield was obtained at the temperature of 77.754°C and extraction time of 4.9185 h. This treatment became the basis for the production of *Cinnamomum burmannii* leaf oleoresin whose quality characteristics would be tested in terms of yield, active compound content, essential oil content, and residual solvent content. The optimum *Cinnamomum burmannii* leaf oleoresin was then verified. The extraction was performed at 78°C for 4h and 55 mins. The *Cinnamomum burmannii* leaf oleoresin obtained from the verified condition showed the characteristics displayed in Table 2.

Table 2. Quality Characteristics of *Cinnamomum burmannii* leaf Oleoresins

Parameter	Research Result	References
Yield	12.6 %	6.9%[a] – 23.02%[b]
Active compound content	Eugenol (59,56%)	Eugenol (17.3%[c] - 87.2%[a])
Essential oil content	9.5 %	15%[d]
Residual solvent content	22700 ppm	5000 ppm[e]

Notes: [a] [6]; [b] [16]; [c] [7]; [d] [17]; [e] [18]

3.2.1. Yield

The yield of *Cinnamomum burmannii* leaf oleoresin that was extracted at 78°C for 4h and 55 mins was 12.6%. This result was greater than that of *Cinnamomum zeylanicum* oleoresin of 6.9% [6]. The *Cinnamomum zeylanicum* leaf powder was extracted with soxhlet for 2h using the acetone solvent [6]. The *Cinnamomum burmanii* leaf oleoresin yield of this research was smaller than the *Cinnamomum cassia* yield that extracted by using supercritical fluid extraction (23.02%) [16]. The difference between the yields was due to the difference in chemical compositions of different plant materials or the soil characteristics and agro-climatic conditions, including other factors such as the effectiveness of the solvent in dissolving the endogen compounds. There were several factors that affected the yield and quality of oleoresin, such as variety, condition and size of the spice powder, solvent selection, extraction conditions, and the solvent evaporation process [19].

3.2.2. Active Compound Content

Figure 2 showed the result of GCMS analysis. Table 3 showed the chemical components detected in the *Cinnamomum burmannii* leaf oleoresin such as eugenol with the relative content of 59.56%, trans-caryophyllene of 3.96%, 1,4,8-Cycloundecatriene of 1.72%, Glycerol triacetate of 2.02%, dodecane, 4,6-dimethyl of 1.42%, Benzyl benzoate of 7.78%, trans-2-tridicenal of 1.51%, tetradecanamide of 2.41%, and oelic acid amide 19.58%.

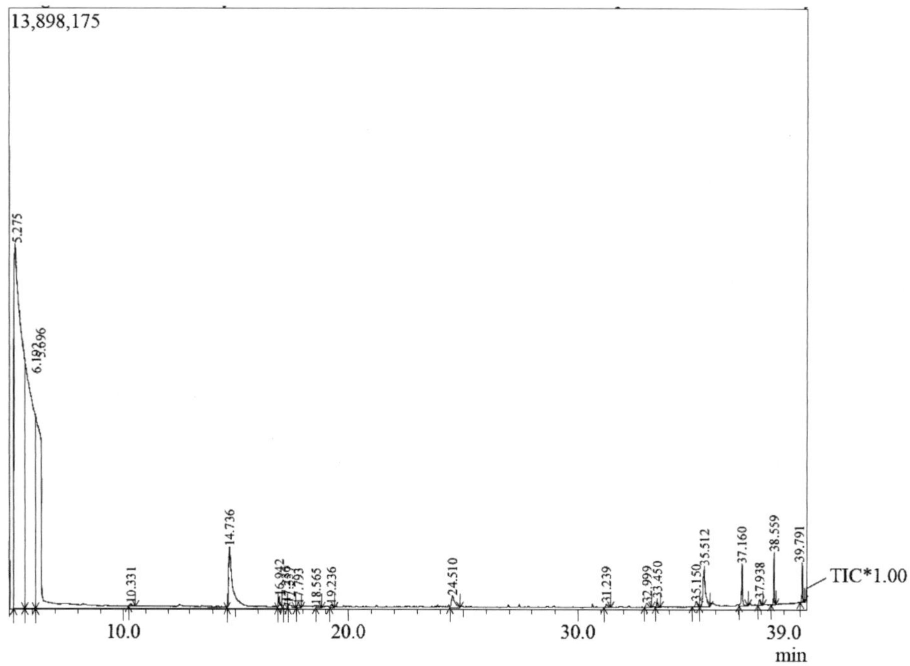

Figure 2. GCMS Spectra

Table 3. The Chemical Components in The *Cinnamomum Burmannii* Leaf Oleoresin

Peak	R.Time (min)	Area	Area (%)	Compound Name
1	14.736	20451277	53.99	Eugenol
2	16.942	1501091	3.96	trans-Caryophyllene
3	17.236	1005380	2.65	Eugenol
4	17.357	1104457	2.91	Eugenol
5	17.793	651937	1.72	1,4,8-Cycloundecatriene
6	18.565	765686	2.02	Glycerol triacetate
7	19.236	540143	1.42	Dodecane, 4,6-dimethyl- (CAS)
8	24.510	2949177	7.78	Benzyl benzoate
9	31.239	574605	1.51	Trans-2-Tridecenal
10	32.999	914525	2.41	Tetradecanamide
11	35.512	7420131	19.58	Oelic Acid Amide

The dominating active compound of the *Cinnamomum burmannii* leaf oleoresin was eugenol with the relative content of 59.56 %. Previous research state that eugenol content of the *Cinnamomum zeylanicum* leaf oleoresin was 87.2% [6] while other reported that the eugenol content of *Cinnamomum tamala* leaf oleoresin was 17.3% [7]. The eugenol content of *Cinnamomum burnanii* leaf oleoresin was higher than that of *Cinnamomum tamala* leaf oleoresin but smaller than that of *Cinnamomum zeylanicum* leaf oleoresin. The extraction of *Cinnamomum zeylanicum* leaf oleoresin used acetone solvent by using a soxhlet extractor for 2h, while the extraction of *Cinnamomum tamala* leaf oleoresin used the CCl_4 solvent for 3h. The different solvents and extraction methods affected the active compound content of oleoresin. In addition, the eugenol content of the materials was also influenced by many factors such as rainfall, altitude, harvest time, drying method, and harvest-produce storing [20].

Eugenol is slightly soluble in water but easily soluble in organic solvents. In addition to its use as an aroma enhancer, eugenol is also antiseptic, and therefore it is used in soaps, detergents, toothpastes, perfumes, and pharmaceutical products. Eugenol also has antimicrobial properties so it can be used as a natural preservative.

3.2.3. Essential Oil Content of Cinnamomum burmannii Leaf Oleoresin

Essential oil content is one of the most frequently tested qualities of oleoresin of spices because it is used particularly due to its volatile oil that very much determines its flavor. The greater the essential content found in the oleoresin, the better its quality is. In terms of quality standard, the essential oil content applicable in international trade must be greater than 15% [21]. If compared with the distillation-generated essential oil, the essential oil of oleoresin has weaker aroma and smell, but it stays longer and diffuses better [19].

The essential oil content of *Cinnamomum burmannii* leaf oleoresin obtained in this research was 9.50%, which was less than 15% so it did not fulfill the standard of international trade. The low content was probably due to the evaporation process which used the temperature of 80°C, which may have caused the essential oil to evaporate together with the solvent and enter the distillation round bottomed three-neck flask. Evaporation at a high temperature and over a long period could damage the essential oil components found in the oleoresin. In addition, many components of the oleoresin easily evaporated and were brought by the evaporated solvent [19].

Previous research reported that the essential oil content of the *Cinnamomum burmanii* extracted with ethanol solvent ranged from 7.28 - 7.72% [22]. This yield did not fulfill the international trade standard. The other research showed that the essential oil content of the cinnamon bark oleoresin ranged from 1.09% to 2.79% [23]. The longer the extraction time and the higher the temperature, the greater the probability that the essential oil will evaporate.

3.2.4. Residual solvent content

The solvent is an important factor in oleoresin extraction. The solvent used in the extraction process must easily evaporate so that during the evaporation it can evaporate thoroughly or partly. The volume of residual solvent in oleoresin can influence its quality and flavor.

Analysis of residual solvent in oleoresin is useful for its further application in industry and pharmacy, which must be in compliance with the terms and conditions of the FDA on organic residual solvent. If the residual solvent exceeds the determined threshold, the evaporation process must be optimized as it will influence its application in food and pharmacy [22]. Ethanol has bad impacts and effects on human health. It may disrupt liver functions, damages the hearts increases breast cancer risk, and its continued use can eventually be fatal.

The standard used in determining the residual solvent refers to the standards enacted by the FDA (*Food and Drugs Administration*). Based on the result of testing with GC, the residual solvent of *Cinnamomum burmannii* leaf oleoresin reached 2.27% (22700 ppm). According to the standard of the FDA, the threshold of residual solvent in food and drug products is 5000 ppm.

Thus, the result of research showed that the residual solvent of *Cinnamomum burmannii* leaf oleoresin was far above the threshold determined by the FDA. This was because the evaporation process did not work perfectly, and some solvent was trapped in the oleoresin. In addition, the imperfect performance of vacuum during the evaporation process caused the amount of residual solvent in the *Cinnamomum burmannii* leaf oleoresin to be large. The result of research on oil and oleoresin extraction from the cinnamon bark (*Cinnamomum burmanii*) showed a high percentage of ethanol residual solvent (3.18%) [24]. The residual solvent of cassiavera (*Cinnamomum burmanii* BL) oleoresin extracted with ethanol solvent ranged from 5.66% to 11.98% [22]. The residual was fairly high due to the distillation conditions and evaporation of the solvent in the rotary vacuum evaporator. The column distillation and column packing distillation can be used to as methods to reduce residual solvent of *Cinnamomum burmannii* oleoresin [25].

4. Conclusion

The optimum yield of *Cinnamomum burmannii* leaf oleoresins was 13.3790%, which was obtained at an extraction temperature of 77.754°C and with an extraction time of 4.9185 h. The quality characteristics of *Cinnamomum burmannii* leaf oleoresins at the optimum yield condition were 59.56% eugenol level, 9.50% cinnamon leaves oil content and 22700 ppm solvent residue level.

Acknowledgments

This work was financially supported by research project of PNBP UNS No: 632/UN27.21/LT/2016 from Sebelas Maret University, Indonesia.

References

[1] Towaha J dan Indriati G 2008 *Warta Penelitian dan Pengembangan Tanaman Industri* **14** 14-16
[2] Zaki I 2011 *Tiga Klon Unggulan Kayu Manis dan Teknik Perbanyakan Secara Vegetatif* (Surabaya: Balai Besar Perbenihan dan Proteksi Tanaman Perkebunan)
[3] Zarfis 2008 *Buletin BIPD.* **16**
[4] Leung A 1980 *Encyclopedia of Natural Ingredients* (New York: John Wiley & Sons)
[5] Tampubolon D F 2011 *Isolasi dan Analisi Komponen Minyak Atsiri Daun Kayu Manis (Cinnamomum burmanii) dengan GCMS* (Medan)
[6] Singh G 2007 *Food and Chem. Toxicol.* **45** 1650-61
[7] Kapoor I P S 2009 *Nat. Product Radiance* **8** 106-116
[8] Sari E 2010 *Jurnal Ekotrans* **10** 108-117
[9] Pruthi J S 1980 *Spices and Condiments, Chemistry, Microbiology, Technology* (New York: Academic Press)
[10] Anam C 2010 *Jurnal Pertanian MAPETA* **12** 71-144
[11] Sundari E 2002 *Pengambilan Minyak Atsiri dan Oleoresin dari Kulit Kayu Manis* (Bandung: Thesis)
[12] Shieh C J 2000 *J. Agric. Food Chem.* **48** 1124-28
[13] Treyball R E 1981 *Mass-Transfer Operations* (New York: Mc Graw-Hill)
[14] Sujarwadi E T 1996 *Kajian Jumlah Pelarut dan Lama Ekstraksi Rimpang Kencur terhadap Rendemen dan Mutu Oleoresin Kencur* (Bogor: Thesis)
[15] Suryandari S 1981 *Pengambilan Oleoresin Jahe dengan cara Solvent extraction* (Bogor: Balai Besar Industri Hasil Perkebunan)
[16] Hong Yang C 2012 *Molecules* **17** 7294-7304
[17] Nurdjanah N 2007 *Teknologi Pengolahan Pala* (Bogor: Balai Besar Penelitian Dan Pengembangan Pascapanen Pertanian)
[18] Food and Drug Administration 2012 *Guidance for Industry Q3C-Tables and List* (US: Department of Health and Human Services)
[19] Lestari W E W 2006 *Pengaruh Nisbah Rimpang dengan Pelarut dan Lama Ekstraksi Terhadap Mutu Oleoresin Jahe Merah (Zingiber officinale var.* rubrum) (Bogor: Thesis)
[20] Nurdjanah N 1988 Buletin Penelitian Tanaman Rempah dan Obat **3** 72-75
[21] Sutianik 1999 *Pengaruh Suhu Pengeringan dan Ukuran Bahan Terhadap Rendemen Dan Mutu Oleoresin Jahe (Zingiber officinale*, Roscoe) (Bogor: Thesis)
[22] Yusmeiarti S and Syarif R 2007 Buletin BIPD **15**
[23] Sulaswaty A 2001 *Kajian Awal Hasil Ekstraksi Minyak dan Oleoresin Dari Kulit Kayu Manis (Cinnamomum burmanii* blume) (Yogyakarta: Proceeding)
[24] Solehudin M 2001 *Ekstraksi Minyak dan Oleoresin dari Kulit Kayu Manis (Cinnamomum burmanii)* (Bogor: Thesis)
[25] Khasanah L U, Anandhito B K, Utami R, Kusumaningrum P, Waningsih, M 2014 *Jurnal Teknologi Industri Pertanian* **24(1)** 53-60

IOP Conf. Series: Materials Science and Engineering **193** (2017) 012022 doi:10.1088/1757-899X/193/1/012022

The performance of Bali Cattle in transmigration area, south Kalimantan: a financial analysis

Sumanto[1]
[1]Indonesian Research Institute for Animal Production, PO Box 221 Bogor

E-mail: sumanto_d@ymail.com

Abstract. This research was conducted in the sub-district of Takisung, district of Tanah Laut, South Kalimantan Province in 2012. This paper was aimed to know the farming performance of the Bali Cattle rearing system in the transmigration area reared by the transmigrant groups of Harapan Makmur, Sumber Rejeki, Tunas Muda and Darmarwulan, that were selected using purposive sampling as the primary data. The agroecosystem that supported the area consisted of palm oil, rubber and agricultural land. Secondary data was obtained from province agriculture officer and district livestock officer. The results of the study showed that the benefits of calf crop beef cattle rearing system (traditional breeding system) in the group of Harapan Makmur was around IDR 5,295,200/year with B/C ratio of 1.28, whereas the value of the benefits of the group of Tunas Muda was around IDR 4,105,800/year with B/C ratio of 1.22. The value of the benefit of feedlot cattle rearing system in the groups of Sumber Rejeki was around IDR 2,110,000/3 months with B/C ratio of 1.07 and the value of the benefits of the group of Damarwulan was around IDR 5,760,000/3 months with B/C ratio of 1.09. Land resource and source of manpower strongly supported the development of cattle farming with the ownership of 2-4 head/family, both the cows and the bull are very economical, effective and efficient in those business so that would benefit to the farmers group.

1. Introduction

The population of cattle in Indonesia recorded was around 15.494.300 [2], scattered throughout the provinces such as South Kalimantan Province, in this province the population of cattle was 150.900 (0.99%). The beef cattle business is part of the most important components in the transmigrants farming in South Kalimantan, because the beef cattle is kept by most (80%) farmers, kept in small-scale and semi-intensive farming system. This kind of business can help the economy for the transmigrant farmers group by using the natural resources of forages as animal feed which are available abundantly in the surroundings.

Biologically, beef cattle is well adaptive in the environmental agroecosystem conditions in South Kalimantan [1]. The development of cattle in the Island of Java, generally directed at the area of transmigration for additional income, considering the amount of natural resources which are very potential, besides that its development will hopefully be faster [11]. The challenges often faced in the business development of beef cattle, is that there is no assurance of the availability of good quality feed in the form of forage/ roughage and waste product of food crops [5].

The availability of agricultural land, of empty land plantation, dryland, rice-fields, are potential land to provide greenish (livestock green feed), wether good grass or various agricultural waste, which can be used for development of beef cattle farming [7]. Business on the development of beef cattle especially Bali Cattle seem to be very suitable in the condition of agricultural land because Bali Cattle is known easily to adapt to various conditions of agro-ecosystem in the rural environment and is a complementary business to another business of agriculture [10,12].

The farmers group of beef cattle farming in the transmigration area in South Kalimantan was integrated with palm oil plantation and other plantation in dry field as the source of livestock woof, such as elephant grass, brachiaria grass, gliricidea, leucaena, caliandra, fields grass, king grass, racquets grass, *jampang* grass, and the remaining of agricultural waste such as the waste of corn crops, soy beans, sweet potato and cassava. Based on those potentials mentioned above, a research needs to be conducted to assess the revenue of the beef cattle farming, in addition to the major business of food crops or others. The purpose of this paper is to see the performance of the farmers group of beef cattle farming especially the financial analysis in the transmigration area in South Kalimantan

2. Methodology

The research was carried out in the sub-district of Takisung of Tanah Laut Regency, South Kalimantan Province, in 2012. The location of the research was in the area of palm oil plantation, coconut, rubber and food crops farming, that is good enough for the development of Bali and PO cattle. The research conducted, using survey method. Interviewed against transmigrant farmers groups of beef cattle, that are Harapan Makmur, Sumber Rejeki, Tunas Muda and Darmarwulan, that were selected by purposive sampling. The determination of the location of the research was recommended by the Education Department of Agriculture and Animal Husbandry, South Kalimantan, farmers must have Bali or PO cattle. The information obtained among other things are maintenance/management pattern of cattle, the number of ownership of cattle, mating system, calving interval, livestock prices, labor wages and selling cattle.

The productivity of labor (IDR/day) shows the greatness of the reward obtained by farmers who poured out for 8 hours (oneday work). The volume of labor allocation can be achieved from the calculation of the total acceptance/revenue of cash and non-cash in one year divided by the total labor used in one year period. In fact, the farmers never calculate the labor costs [8]. The secondary data were obtained from the local of Animal Husbandry Department and Statistic Agents. The economic value (IDR/farmer /year) is calculated based on the difference between the acceptance and the expenditure (Cost and benefit analysis) [9], [6]. The primary data and secondary data collected were processed and analyzed economic financial and then displayed in the form of tables and descriptively.

3. Results and Discussion

3.1. Group profile of beef cattle farmers selected

The population of South Kalimantan in 2012 was around 3.62 million with the population density of around 97 people/km2, the spread of the density in the district is still in the medium density of around 50-300 people/km2, in Banjarmasin city and Banjarbaru are categorize and classified as a densely populated area, that were more than 500 people/km2, meanwhile the people who work in the field of agriculture almost 60 percent and the remainder are working as traders, employees and others. Transmigration is one form of government (central and local government) concern, to reduce the number of the inhabitants of the crowded area, so directed to the area that is still rarely populated.

The government policy in order to offset the population density, balance to a form as system which is very good and seem to be able to improve the welfare of the community, it was expected that the labor and natural resources can be used more optimally. Hopefully the spread of the population will be more balanced and equal, so that the life quality of the people will be increased. This can be seen from the purpose of the program on the development of animal husbandry in the area of transmigration, among others are to help farmers in the transmigration area, the provision of the means of production of agriculture and increase the income of the farmers, create jobs, and the provision of the beef cattle stock. Palm oil plantations is the largest section in the region of transmigration area of South Kalimantan, following the vineyard mixture and rice fields that are also integrated with the Bali and PO cattle.

South Kalimantan consists of 9 districts with a population of ruminants of 229.418,51 AU in 2012. The spread of the population of the beef cattle is 173.304,16 AU (75,5%), then followed by buffalo 40.958,97 AU (17,85%) and the rest are the sheep, goats and horses. The number of three largest population of beef cattle are in the Tanah Laut district (61.872,73 AU), then in Tanah Bumbu (23.966,07 AU) and Tapin (12.271,57 AU). Group profile of beef cattle farmers selected is shown in table 1.

Table.1. The Profile of transmigration farmers of beef cattle in South Kalimantan

No	Item	Name of farmer's group			
		Harapan Makmur	**Sumber Rejeki**	**Tunas Muda**	**Damarwulan**
1	Location (village, sub district)	Telaga, Karang Taruna	Banuang Tengah, Takisung	Banuang Tengah, Takisung	Rantau Badan, Rantau Badan
2	Number of farmers	31	20	20	10
3	Total Number of cattle (head)	75	40	60	39
4	Cattle ownership (head/ family)	2-4	2	2-5	2-4
5	Breed of cattle	Bali	Simental, PO	Bali	Simental, PO
6	Management pattern	Semi-intensive	Intensive	Semi-intensive	Intensive
7	Feed	Grass/agriculture waste	Grass/agriculture waste	Grass/agriculture waste	Grass/agriculture waste & cassava meal
8	Type of business	Cow-calfsystem	Fattening system	Cow-calfsystem	fattening system

Table 1, shows that the family labor which is poured out in the beef cattle rearing in transmigration farmers group does not increase the number of cattle kept, that is still around 2-4 head/farmer. The labor that is used by the transmigration group of farmers of beef cattle rearing does not interfere with the business activities of agriculture, trade and other business, farmers feel happy in an effort to keep the beef cattle with a reason that it is easy to find livestock, easy to get forage, easy way sales of livestock, easy to manage and easy to restore the manure as fertilizer in their own farm that will benefit to the fertility of agricultural crops of the farmers them selves.

Each location of the research has a source of green feed enough for the needs of the cattle. There are several superior grass that are already familiar and known by the group of beef cattle farmers in the transmigration area in South Kalimantan such as elephant grass, brachiaria grass, gliricidea, leucaena, caliandra, fields grass, king grass, racquets grass, *jampang* grass and the remaining are agricultural waste.

[3] report, that forage/roughage is the major feed for both large and small ruminants for the provision of livestock feed throughout the year and supposed to be sufficient for sustainable livestock feed. In the last few years. The scale of the beef cattle maintained by the farmers groups in the transmigration area is around 2-4 head/farmer. Even though the business is very real and obtain attention from the local government, some farmers keep their cattle in the stalls and some keep and leave them in the pastures, in palm oil plantations or in the dry land surrounding.

3.2. Financial analysis of Cow calf and Feedlot rearing system in the transmigration area in South Kalimantan

Business profile of cow-calf and feedlot farming in four groups of farmers is shown in table 2. Each group of farmers attempting to do either 'breeding' and fattening/feedlotting farm traditionally, the farmers kept their cattle in the traditional system. The farmers that do the 'traditional breeding' or cow-calf rearing system earn their income from selling the calves, young bull, adult bull (steers) and cows. While the source of revenue of the feedlotter comes from the sale of feedlots bull (steers) over a certain time.

Tabel 2. Financial Analysis of cow-calf ('traditional breeding') and Feedlot rearing system

Item	Farmers group			
	Harapan Makmur	Sumber Rejeki	Tunas Muda	Damar Wulan
Average number of cattle (head)	3	2	3	3
Period of rearing (month)	12	3	12	3
A. Investment/Fix cost&depression				
Cows	8,250,000	-	8,500,000	-
Feeder bull	-	29,500,000		59,000,000
Heifer	5,175,000	-	5,000,000	-
Calf	1,750,000	-	1,500,000	-
Depression of animal shade	501,500	500,000	500,000	600,000
Depression of an.shade equipment	215,500	230,000	225,000	250,000
B. Variabel cost				
Labour	3,232,800	740,000	3,139,200	760,000
Medicine	30,000	30,000	30,000	50,000
Cassava meal	-	200,000	-	500,000
Forage/roughage	-	650,000	-	1,000,000
Sum of A+B (IDR)	**19,154,800**	**31,850,000**	**18,894,200**	**62,160,000**
C. Revenue (cattle sale)				
Cow (IDR)	10,150,000	-	11,050,000	-
Adult female/cow (IDR)	7,200,000	-	6,850,000	-
Young bull/steer (IDR)	7,100,000	-	5,100,000	-
Bull (IDR)	-	33,960,000	-	67,920,000
Total revenue (IDR)	24,450,000	33,960,000	23,000,000	67,920,000
D. Benefit (IDR) /period	5,295,200	2,110,000	4,105,800	5,760,000
E. Benefit (IDR) /month	441,267	703,333	342,150	1,920,000
F. Benefit (IDR) /month/head	147,089	351,667	114,050	640,000
G. B/C Ratio	1.28	1.07	1.22	1.09

Table 2. shows that the cow-calf ('traditional breeding') and feedlot/fattening business in four groups of farmers is still profitable because the value of the B/C ratio is >1. The largest value of the benefits/month/head occurs in the business of feedlots in the group of Damarwulan, as much as IDR 640,000-/month/head. The value of the B/C ratio is used to determine the feasibility of beef cattle business on each of the groups of farmers.

There are several factors of production that can be pressed the business financing such as the cost of purchasing the stock of beef cattle, the labor and the cost of feedstuff so that the acceptance can be optimalized [4]. Forage and other roughage is still a problem for the farmers in doing business in beef cattle, when facing the dry season, where the availability greenish feedstuff decreases. But usually the farmers group will deal with it by using the any forage (feedstuff) that grows at any place, including post harvesting paddy field, palm oil plantations and waste of food crops. Cash receipts only concentrated on the sale of beef cattle yield per year and not allocated sales of natural fertilizer, because all manure is used as fertilizer for their own farms.

4. Conslusion

Four research locations in the district of Tanah Laut have agroecosystem that is good enough for the development of Bali and PO cattle. Both cow-calf (traditional breeding) rearing system and feedlot of beef cattle rearing system that still reared traditionally provide benefits.

Benefits of business gain in cow-calf rearing sytem (traditional breeding) to the farmers group of Harapan Makmur is around IDR 5,295,200/year with B/C ratio of 1.28, then the value of the benefits to the farmers group of Tunas Muda is around IDR 4,105,800/year with B/C ratio of around 1.22. The value of business advantage in the feedlot cattle of the groups of Sumber Rejeki is around IDR 2,110,000/3 months with B/C ratio of around 1.07 and the value of the benefits to the farmers group of Damarwulan is around IDR 5,760,000/3 month with B/C ratio of around 1.09.

References

[1] Ahmad S N, Siswansyah D D, dan Swastika O K S 2004 *Jurnal Pengkajian dan Pengembangan Teknologi Pertanian* **7** 155–170
[2] BPS 2016 *Badan Pusat Statistik Indonesia*
[3] Dewi F dan Mairika L 2008 *Jurnal Peternakan* **5** 28-37
[4] Hadi PU dan Ilham N 2002 *Jurnal Litbang* **21** 148-157
[5] Hoddi HA, Rombe M B, dan Fahrul 2011 *Jurnal Agribisnis Unhas* **10** 98-109
[6] Hermawan S 2012 *Jurnal Pengkajian dan Pengembangan Teknologi Pertanian Bogor* 140-148
[7] Kuswandi 2007 *Wartazoa Buletin Ilmu peternakan Indonesia* **17** 176-184
[8] Rusdiana S, Wibowo B dan Praharani L 2010 *Prosiding Seminar Nasional Teknologi Peternakan dan Veteriner* 20-29
[9] Soekartawi, Soehardjo K L, Dillon dan Hardaker J B 1995 *Ilmu Usahatani dan Penelitian untuk Pengembangan Petani Kecil* (Jakarta: UI Press)
[10] Syamsu J A, Lily A, Sofyan, K Mudikdjo dan E Gumbira S 2003 *Jurnal Wartazoa* **13** 30-37
[11] Suryana 2009 *Jurnal Litbang Pertanian*, **28** 29-39
[12] Winarso B R, Sajuti dan Muslim C 2005 *Jurnal Forum Penelitian Agro-Ekonomi* **23** 61–71

IOP Conf. Series: Materials Science and Engineering **193** (2017) 012023 doi:10.1088/1757-899X/193/1/012023

The role of grazing land on the buffalo population dynamics in Brebes regency

Sumanto[1]
[1]Indonesian Research Institute for Animal Production, PO Box 221 Bogor

E-mail: sumanto_d@ymail.com

Abstract. Brebes District is one of the centre of grazing buffalo in Indonesia that involve thousands of rearers usually kept as a family savings. This paper highlighted the availability of land and the role of the grazing land for the durability of the maintenance of buffalo in Brebes Regency. The information obtained is from interviewed the livestock facilitators in the sub-district (primary data) and from statistic of agriculture in Brebes Regency 2014 (secondary data). Generally the buffalo kept semi-intensively and commonly the buffaloes graze in the fields that are not used from morning until evening and during nights buffaloes are placed in the stalls. Rearers chose the semi-intensive system in rearing the buffalo because it is considered easy to manage and they do not need to provide fresh money to prepare the roughage for feed, because commonly the grazing buffalo are shepherd by herdman that will receive buffalo as their payment. The population density is very high (1.056 heads/km^2), the buffalo ownership is between 2-4 head/households; generally the location of the grazing land is in the forest, rice fields fallow, and sleeping land, and estimated that greenfeed stock is still available abunandtly, on the other hand the urban land is less capacity of feed. The spread of buffalo is only in 125 villages from 297 villages in Brebes. The acceptance of buffalo business is around IDR 3.5 million to IDR 7.5 million/family/year. The availability of grazing land strongly influence the maintaining of buffalo farming by rearers.

1. Introduction

Buffalo population in Indonesia in 2014 reached more than 1.3 million head spread in all provinces, but the number of population varies. There are 9 provinces that the Buffalo population is fairly high (above 50 thousand), the province of East Nusa Tenggara is the most highes (150,038 head), followed by Nangroe Aceh Darussalam (131,494 head), West Java (130,157 head), Banten (123,143 head), North Sumatera (114,289 head), West Nusa Tenggara (105,038 head), West Sumatera (100,310 head), South Sulawesi (96,515 head) and Central Java (75,674 head) [10]. The trend of Buffalo population growth is likely continuously to decrease in the last ten years, buffalo population in 2004 was around 2.5 million head and in 2014 decreased up to 1,32 million head [3,13]. Even so, buffalo still contribute to the production of national meat as much as 2.11% [9]. The type of buffalo in Indonesia is mostly swamp buffalo (95%) and the remaining 5 percent is river buffalo that formed in North Sumatera.

In Central Java, Brebes Regency is the most populated buffalo following Pemalang Regency [8] and the growth of the population also tend to decrease from 13,504 head in 2005, become 7,713 head in the year 2014 [5,7]. According to [12], the general obstacles in the development of buffalo is caused by internal factors such as : slow in sex maturity, ten months of gestation period, late weaning time, silent heat so that the signs is difficult to detect, the calving interval often takes long time, the quality of bull is not good and external factors such as : management sytem, low quality of livestock, less capital, mating system is uncertain and unknown, knowledge of ranchers/farmers is low especially on the buffalo reproduction and social culture of the community.

Eventhough in that such condition, thousands of ranchers still maintain buffaloes.The major role of buffalo for the rural is as a savings account and plowing the rice fields to help the economic development of agriculture in rural area of Brebes regency. The buffalo is generally maintain semi intensively (shepherd) and the ownership farmer is generally also small as producer. Buffaloes also contribute to help farmers in the cultivation as drought animal power, milk for butter or 'dunke', dali and leather [2]. Since 2013, Brebes Regency is one of ten Districts or cities in Indonesia that were chosen to be the source of the buffalo feeder origin by Director General of Livestock and Animal Health Ministry of Agriculture, due to its population that is still high [14].

Buffalo can grow well in the range of a very broad agroecosystem conditions of the region that is very wet to the extreemly dry condition, buffalo is well developed even in swamp land in South Kalimantan [11,16]. Brebes region with 166,296 ha wide, about 80.4 percent is in the form of the rice fields, dry land, forests and other plantations that produce roughage as the source of green stuff and plants waste that used as ruminants feed. Further more those land can be used as grazing land for ruminants, including buffalo. Studies on natural roughage and waste plants that support its capacity for ruminants feed in Brebes indicates that the supporting capacity index is about 5.1 (safe limits =3), which means that Brebes is still safe for the development of the ruminants [15].The availability of land for ruminant's grazing is still wide enough in Brebes, this mean that there is a chance to maintain the existence of buffalo.

2. Methodology

The primary data on grazing land for buffalo per sub-districts in Brebes is obtained through the list of contents questioner, covering the role of buffalo, the average ownership of buffalo, the grazing area, rearing patterns and income per year earned from buffalo farming. Primary data collect from 17 sub-districts were conducted by the officers of animal husbandry of Brebes regency for 2 months in the year of 2014.

While the secondary data that were collected as supporting data, obtained from the BPS and the Livestock office in Brebes, the supporting capacity of the region in Brebes [15] and information related to the study of the buffalo rearing was also collected. The results of the data collection was processed, tabulated, and analysed descriptively.

3. Results and Discussion

3.1. The Profile of The Brebes Regency

Brebes Regency is one of the autonomous region in the province of Central Java, with an area of around 1,662.69 km^2, Brebes Regency is administratively divided into 17 sub-districts, which consists of 292 villages and 5 *kelurahan*, Brebes is a second largest district in Central Java Province. Most of the area consist of lowlands. The southwestern part is a high land (with the peak of *Pojoktiga* mountain and the *Gunung Kumbang*), while the southeast part are mountains which is part of *Gunung Slamet*. The condition makes the area of the great potential for the development of agricultural products such as rice, horticulture, plantation, fishery and livestock. Land Use data in the region Brebes shown in Table 1.

The total population in the year of 2014 is about 1,773,379, the average of the population growth is around 0.22 % per year, whereas the population density is 1066 people/km2. Brebes is a district with the most highest population in Central Java and the most widest regency in Central Java after Cilacap Regency. Of 297 villages/subdistricts spread in Brebes Regency, only 4 villages that have low population density, 11

IOP Conf. Series: Materials Science and Engineering **193** (2017) 012023 doi:10.1088/1757-899X/193/1/012023

villages in the status of medium, 58 villages were high density status and 224 were villages in the status of the very high density [1].

Tabel 1. The land use in Brebes Regency 2014

Land Use	Width of area	
	Ha	%
Wetland/ rice field	62.703	37,7
Yard and Building	19.250	11,6
Dry Land/plantation	17.499	10,5
Unused land	279	0,2
Pond/swamp	9.001	5,4
Foresty	52.265	31,4
Plantation	1.252	0,8
Others	4.047	2,4
Total	166.296	100

Source: [7]

3.2. The Population and The Development of The Buffalo in Brebes Regency

Besides the agriculture and plantations sector, Brebes Regency has the potential source of greenfeed and roughage abunandtly for livestock, that spread over almost in every sub-district [1]. The condition makes this district become the source of the development of the ruminants, including buffalo. Since 2013, Brebes Regency was appointed by the Directorate General of PKH as the centre for the development of buffalo [14]. In fact up to now, the farmer keeps the buffalo as saving for the family (55%), and also for the purpose of plowing the land (45%). By the increasing use of the tractor to plow the land, the benefits of buffalo decreases because it is not used as the drought animal power so that the buffalo much more often to be sold and slaughtered rather than to be maintained. The population of buffalo and its development from the year of 2008-to 2014 shown in Table 2.

Tabel 2. The development of the buffalo population in Brebes Regency 2008-2014

Year	Population (head)	Development	
		head	%
2008	8,373	-	
2009	10,949	2,576	30.8
2010	11,582	636	5.8
2011	8,774	-2,808	-24.2
2012	8,867	93	1.1
2013	7,885	-982	-11.1
2014	7,713	-172	- 2.2
Average	9,163	-109.5	- 1.2

Source : [5,6]

The Table 2, shows that the number of the population of buffalo in Brebes was fluctuated from 2008 until 2014, but there is a tendency to continue to decrease in the year of 2008 buffalo population was 8,373 head and in the year 2014 has decreased up to 7,713. In the period of 2008-2014 the buffalo population decreased 1.2% /year in average.

3.3. Ownership, farm management and farmers' income earn from buffalo rearing

The spread of the population, the role and management system of buffalo per sub-districts in Brebes is shown in Table 3. Buffalo is not found in some villages in Brebes Regency. Of 297 villages in Brebes regency, the spread of the buffalo population is only in 125 villages (42.1 percent from 297 villages) with the level of the scale of the ownership of buffalo is 2-4 head/household. The pattern of buffalo management generally is semi intensive, where in the day time, the buffalo shepherd in the community land such as in the forest, rice fields, river bank, shepherd or grazing land and during the afternoon until the next morning of the buffalo kept in the stall/animal shed. The quality of livestock feed is generally not good, but according to [2] the production and the reproduction of the buffalo are still better than the cattle at the same conditions. In addition in certain conditions the buffalo is also kept for several weeks in place of shepherd area in the edge of the forest, because the source of livestock feed (forage/roughage) still plenty enough and usually the herd are guarded by 2-3 ranchers interchangeably, as happened in the group "Sidamulyo" in the village of Pamulihan, sub-districts of Larangan, Brebes. The percentage of land, where the buffalo were herded on village-based shown in figure 1.

The Figure 1. shows that the buffalo management pattern is mostly semi-intensive and still depend on the availability of livestock green feed (forage/roughage) that located within the region which is not owned by farmers and ranchers not issued for the cost of the daily woof. For the night buffalo also not given additional feeds. Forest land and/or paddy fields (>100 villages) region is often used by ranchers to herd and keep the buffalo. The rice fields can only be used when after harvest or at the time of the rice fields in the state of the fallow (not used for plants). Shepherd land (grazing land) is only available in 18 villages. Through the semi intensive of buffalo management, the ranchers obtain income varies in the range of IDR 3.5 to 7.5 million/year.

Tabel 3 : The buffalo population , the role and the management system in Brebes Regency

Sub district	**Total Number of village**	**Number of village found buf.**	**Av. of owner ship (head/hh)**	**Buf. role (*)**	**Shepherd location (**)**	**Manag. system**	**Income from buf (IDR mil./ year)**
Salem	21	19	2	1 and 2	1 and 2	semi intensive	2.5-3
Bantar kawung	18	12	1	2	1 and 2	semi intensive	5 -10
Bumiayu	15	14	2	1 and 2	2 and 1	semi intensive	3
Paguyangan	12	9	2	1 and 2	3	semi intensive	3
Sirampog	13	9	2	1 and 2	3	semi intensive	2,5
Tonjong	14	14	3	1 and 2	1 and 2	semi intensive	7
Larangan	11	3	2	2	1	semi intensive	5 -10
Ketanggu-ngan	21	5	4	2	1	semi intensive	5 - 8
Banjarharjo	25	7	4	2	1	semi intensive	5 -15

IOP Conf. Series: Materials Science and Engineering **193** (2017) 012023 doi:10.1088/1757-899X/193/1/012023

Losari	22	1	4	2	4	semi intensive	7
Tanjung	18	0	0	0	0	semi intensive	0
Kersana	13	1	6	2	2	semi intensive	10
Bulakamba	19	2	4	2	2	semi intensive	8
Wanasari	20	4	4	2	2 and 4	semi intensive	5 -10
Songgom	10	4	4	2	2 and 4	semi intensive	5 -10
Jatibarang	22	10	2	2	1	semi intensive	5 -10
Brebes	23	6	4	2	1	semi intensive	5 -10
Total	297	125	2 -4	1 and 2	1 and 2	semi intensive	3,5-7,5
%		42.1					

Remark : * = 1.cultivation, 2. saving;** = 1.forest; 2= rice field, dry land ; 3. Shepherd land ; 4. River side

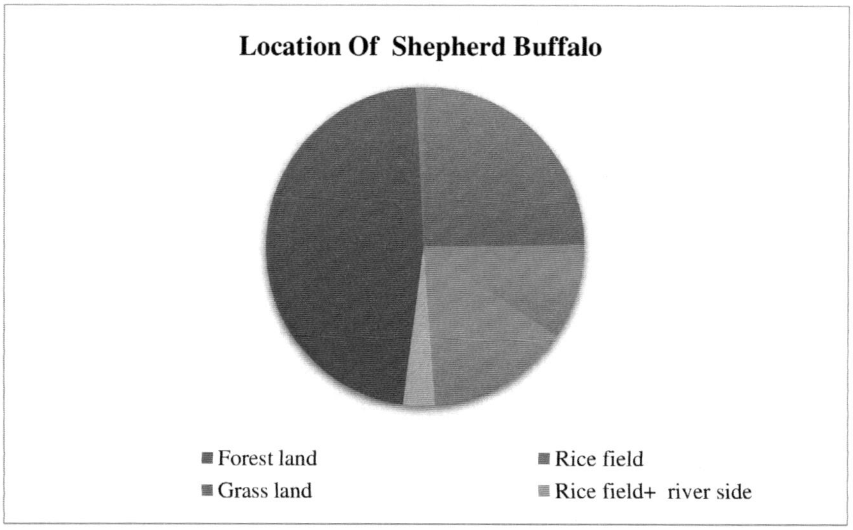

Figure 1. Shepherd Location for Buffalo on the Village Base in Brebes Regency

4. Conslusion

The dominant pattern of buffalo rearing in Brebes is semi-intensive, considered easy and do not need to provide fresh money, where buffalo herded by others with the payment with buffalo also. The spread of cattle buffalo only in 125 villages from 297 villages in Brebes.

Buffalo ownership range between 2-4 head/households and generally the location of the land of the shepherd is in the forest and rice field (>100 villages), available area of 133.064 ha estimated still able to provide livestock feed (forage/roughage) sufficiently.

The acceptance of buffalo business is around IDR 3.5 million to IDR 7.5 million/HH/year. Availability of land shepherds strongly influence the maintaining of the ownership of buffalo by farmers.

References

[1] Balitnak 2014 *Dinas Peternakan Kabupaten Brebes dengan Balai Penelitian Ternak Ciawi*
[2] Bamualim A M dan Zulbardi Muhammad 2007 *Seminar dan Lokakarya National Usaha Ternak Kerbau* 32-39
[3] BPS 2015 *Badan Pusat Statistik Indonesia*
[4] BPS 2005 *BPS Kabupaten Brebes*
[5] BPS 2008 *BPS Kabupaten Brebes*
[6] BPS 2012 *BPS Kabupaten Brebes*
[7] BPS 2015 *BPS Kabupaten Brebes*
[8] BPS Jateng 2015 *BPS Jawa Tengah*
[9] Ditjen Peternakan 2006 *Direktorat Jenderal Peternakan* (Jakarta: Deptan)
[10] Ditjen PKH 2015 *Ditjen PKH* (Jakarta: Kementan)
[11] Hamand A, Rohaeni E S dan Subhan A 2006 *Prosiding Lokakarya Nasional Usaha Ternak Kerbau Mendukung Program Kecukupan Daging Sapi.* (Sumbawa: Pusat Penelitian and Pengembangan Peternakan)
[12] Muhlisin M (2016, August 21) http://muhammadmuhlisinfapet.blogspot.co.id
[13] Pusdatin 2015 *Kementrian Pertanian*
[14] Nugroho F E 2015 *Laporan Tribun Jateng*
[15] Sumanto, Wibowo B dan Suratman 2014 *Seminar di UNDIP*
[16] Suryana 2007 *Jurnal Litbang Pertanian*

IOP Conf. Series: Materials Science and Engineering **193** (2017) 012024 doi:10.1088/1757-899X/193/1/012024

Effect of drying temperatures on physical characteristics of sorghum flour modified with lactic acid

G J Manuhara[1], B S Amanto[1], T A Astuti[1]
[1] Department of Food Science and Technology, Faculty of Agriculture, Sebelas Maret University, Jl. Ir. Sutami no.36A, Surakarta, Central Java, Indonesia

E-mail: godrasjati@yahoo.com

Abstract. Different drying temperatures produce different starch structure, while the remains of lactic acid in sorghum grain might continue starch modification during drying. The purpose of this study was to determine the effect of drying temperatures (40, 50, 60° C) on physical characteristics of sorghum flour modified with lactic acid. Higher drying temperatures produced sorghum flour, with higher viscosity and swelling power, but lower whiteness and solubility. Modified sorghum flour showed higher value than the unmodified flour at all characteristics. Infrared spectroscopy showed that modification shifted absorption peak from 1,334 cm^{-1} to 1,373 cm^{-1}, but there was no difference between samples dried at different temperatures.

1. Introduction

Sorghum is the most important cereal in the world after wheat, rice, corn and barley. In fact, sorghum plant is adaptable to dry land, high temperature, and floods as compared to other cereals, and is harvested after 100-110 days cultivated so the production cost is relatively low. Sorghum grains contain 10.4% protein.

Although in certain countries, sorghum grains are utilized for bioethanol and industrial raw materials, processing sorghum grain into an acceptable food is still limited. Moreover, in the developing countries, sorghum grain are only supplied for cattle. The sensory quality of food produced from sorghum is also not good as the product of other types of cereals, causing its application in food production is not attractive. For example, the texture of the bread and biscuits produced from a mixture of wheat flour and sorghum flour tend to be drier, sintered and harder, whereas sorghum flour is known to perform high gelatinization temperature and low water holding capacity [1]. The higher gelatinization temperature, the more cooking time or more energy required.

The modification of starch or fluor using acid hydrolysis has been widely applied to cleave the glycosidic bonds of polysaccharides chains, lower gelatinization temperature, reduce cooking time, increase water holding capacity or swelling power [2, 3]. The similar modification using lactic acid produced sorghum fluor with higher swelling power, whiteness, protein and sugar content than using acetic acid.

It is important to note that acid was also absorbed into the starch granules during the modification, so it was still possible to continue the hydrolytic activity during the subsequent processing step (drying). Higher drying temperature might affect the reaction kinetic, so physical characteristic of flour sorghum might also be affected. This study aimed to determine the effect of different drying temperatures in the modification process using lactic acid to whiteness, solubility, viscosity, and the swelling power of sorghum flour. In addition, the FTIR analysis was also performed to detect any changes in functional groups of the starch structure in the flour as the result of the modification.

2. Materials and Methods

2.1. Materials

Sorghum grains (Numbu variety) were obtained from the Perseroan Terbatas Perkebunan Nusantara (PTPN) XII Surabaya, Indonesia, while the lactic acid p.a was supplied by MERCK.

2.2. Methods

Lactic acid solution 1% was prepared in 600 ml of distilled water. Sorghum grains (200 grams) were soaked in the lactic acid solution, whereas the unmodified sorghum grains were soaked in 600 ml of distilled water. The grains in the lactic acid solution were stirred periodically while heated in a water bath at 45°C for 60 minutes [4]. After being soaked, the grains were dried with cabinet dryer at 40, 50, and 60°C until the equilibrium moisture content was reached finally. The dried grains were then sized reduced by using miller and sieved into 80 mesh size.

Sorghum flour was then analyzed to obtain its moisture content, whiteness, viscosity, swelling power, solubility and infrared spectrum. Factorial randomized design consisting of two factors: the concentration of lactic acid and drying temperature, was used. Each treatment was performed in three repetitions and then analyzed in two repetitions. The data was analyzed by the analysis of variance (ANOVA). Duncans Multiple Range Test (α = 0.05). was then performed after the ANOVA demonstrated any significant effect of the drying treatments.

3. Result and Discussion

Table 1. Physical properties of sorghum flour

Drying temperature	Moisture (%)	Whiteness (%)	Viscosity (cP)	Swelling power	Solubility (%)
Unmodified flour					
40°C	10.38 ± 0.21	81.10 ± 0.07	821 ± 177.9	3.14 ± 0.18	2.74 ± 0.59
50°C	9.06 ± 0.31	80.92 ± 0.59	1,424 ± 31.8	3.27 ± 0.13	2.53 ± 0.67
60°C	8.14 ± 0.32	80.74 ±0.64	1,470 ± 95.0	3.30 ± 0.19	1.92 ± 0.49
Modified flour					
40°C	9.73 ± 0.28	81.25 ± 0.35	1,888 ± 232.3	3.65 ± 0.06	4.45 ± 0.14
50°C	8.66 ± 0.39	81.21 ± 0.29	2,409 ± 166.5	3.80 ± 0.11	3.87 ± 0.83
60°C	8.11 ± 0.33	81.19 ± 0.19	2,721 ± 607.1	3.99 ± 0.31	2.14 ± 0.45

3.1. Moisture content

The moisture content of the modified sorghum flour was significantly lower than the unmodified sorghum flour (table 1). The starch modification by acid probably produce starch granules with the lighter structure that led to the rapid evaporation of water molecules during drying.

As the higher drying temperature, the moisture content of all flour decreased (table 1). At higher drying temperature, the thermal energy received by materials is also greater, so that water molecules bound with starches is more easily released and evaporated.

The moisture content of modified or unmodified flour ranged between 8.11% - 10.38%. It has already met the Codex Standards which specify the requirements is 15% [5]. Lower moisture content prevent the growth spoilage microorganisms and slow down the rate of degradation reaction associated with water in food stuff so that product shelf life is prolonged. The results demonstrated that there was significant difference ($p<0.05$) between the modified and unmodified flour. The significant differences were also noted between all different drying temperatures (table 1).

3.2. Whiteness

Modified sorghum flour showed higher whiteness (table 1). The increase in whiteness was possibly, due to the pigment degradation and the inhibition of enzymatic browning reaction in sorghum flour associated with the decrease in pH.

Sorghum grains contain anthocyanin pigments (0-2800 mg/g) and proanthocyanidin (0-68000 mg/g). Both are water-soluble pigment which its degradation rate increased in acidic conditions. Sorghum fluor modifications in lactic acid solution lowered the pH, result in the inactivation of phenolase [6]. Phenolase enzyme serves as a catalyst in the enzymatic browning reaction. Thus, the browning reaction in modified sorghum flour was inhibited, and resulted in the whiter flour. Kawali variety sorghum fluor contain phenolic compounds to 975.61 mg/100 g dry weight [7], which is contained in the testa, a layer of pigment in sorghum grain.

The lower whiteness due to the longer drying causing discoloration into darker appearance. The decrease in whiteness was also probably caused by the Maillard reaction whereas interaction between reducing sugars with amino acids, peptides or proteins was involved that result in a very dark products called with melanoidin. The rate of the Maillard reaction is tend to greater when the higher drying temperature was employed, so that sorghum flour produced from drying at 60°C performed the lowest whiteness.

The whiteness of sorghum flour ranged from 80.74 to 81.25%, whereas the previous study showed that Numbu variety performed 83.41% whiteness. Based on ANOVA test, there was no significant difference at various drying temperatures ($p>0.05$), but the significant difference was noted between modified and unmodified flour.

3.3. Viscosity

The viscosity of modified sorghum flour was higher than the unmodified flour (table 1). The viscosity of modified flour was about doubled compared to unmodified flour. Previous study showed that along with the increase in acetic acid (but at a lower concentration range) then the viscosity of cassava starch also increased [8]. The ANOVA test showed significant differences between them ($p<0.05$). There was no significant difference in viscosity between the flour produced at drying temperature of 50 and 60°C, but there was significant difference between the flour produced at drying temperature of 40°C with 50°C and 60°C ($p < 0.05$).

The cleavage reaction at α-1,4-glycosidic and α-1,6-D-glycosidic bond increased starch tendency to form gel so that viscosity increased. The hydrolysis using acid produced starches with weaker intermolecular hydrogen bonds so that the starch granules are easier to inflate during heating. Thus, the viscosity increased.

Amylose content in starch may also affect the viscosity. As the higher amylose content, then the viscosity of the starch was also higher. The amylose content of the modified starch was higher than unmodified flour. The results of this research work also demonstrated that at higher drying temperature, the amylose content was also higher (data not shown). Amylose is a fraction starch that role in the formation of gel so the more amylose content in the flour, then the gel was also firmer.

3.4. Swelling power

Swelling power of modified starch flour was higher than unmodified (figure 4.19). This is similar with previous research that found that longer time of hydrolysis result in the increase of swelling power. The hydrolysis by acid weaken the hydrogen bonds that ultimately led to the less dense structure of starch granules. This causes the water molecules is easily to penetrate into the starch granules, so then the starch swell better [9].

Swelling power of modified flour also increased as the increase of drying temperature. The hydrolysis by the remaining acid in the flour allegedly continued during drying, so that at higher drying temperature, hydrolysis reaction rate was also higher. Thus, the structure of the starch granules probably became less dense.

IOP Conf. Series: Materials Science and Engineering **193** (2017) 012024 doi:10.1088/1757-899X/193/1/012024

Swelling power of modified sorghum flour is higher than unmodified flour and even sorghum flour produced by previous study. The ANOVA test showed that there was significant difference between the modified sorghum flour and unmodified sorghum flour. The significant difference in swelling power was also noted between the flour produce at drying temperature of 60°C with 40°C and 50°C. But there was no significant difference at ($p>0.05$) the flour produce at drying temperature of 40°C with 50°C.

3.5. Solubility

The hydrolysis by acid cleave glycosidic bonds in starch to produce simpler compounds include dextrin, maltose and glucose. The increase in solubility is probably due to the formation of reducing sugars which is easily dissolved in water [10, so that sorghum flour modified with lactic acid performed higher solubility than unmodified flour (table 1).

However, the inconsistent results (when the result of starch hydrolysis is considered as the cause of higher solubility), are shown in different drying temperature. The higher drying temperature resulted in a lower solubility. Further observation to the formation complex of lipid-amylose at the higher processing temperature is needed to study this phenomenon.

ANOVA test showed that there was significant difference between the modified sorghum flour with unmodified flour. There was significant difference ($p<0.05$) between the flour produced at the drying temperature of 40°C with 50°C and 60°C, but there was no significant difference ($p>0.05$) between the flour produced at the drying temperature of 50°C with 60°C.

3.6. FTIR

The absorption peaks (figure 1) at 1,651 cm^{-1} were the characteristic of stretching C = O carbonyl vibration, while 2,924 and 3,425 cm^{-1} were the stretching C – H bond and O-H carboxyl vibration, respectively. Starch modification using lactic acid for hydrolysis cleave the glycosidic bond. Termination of α-1,6 glycosidic bond is supposedly easier than α-1,4 glycosidic bond as these branch bonds construct the amorphous regions which are easily attacked by acid. Thus, the modification reduce the amount of glycosidic bond.

In sorghum flour with drying temperature of 60°C, the absorption peak of glycosidic bonds was shifted from 1,334 cm^{-1} (unmodified flour) to 1,373 cm^{-1} (modified flour). Consistent result was shown by modified flour with drying temperature of 40°C, which resulted in the absorption peak associated with glycosidic bond at 1,373 cm^{-1}. The shift of the wave number is also reported in the previous study [11]. The modification by lactic acid reduced the glycosidic bond which the cleavage of glycosidic bond (α-1,6) in amylopectin into short-chain amylose took place during the drying process [12].

4. Conclusion

Higher drying temperature produced sorghum flour with higher viscosity and swelling power, but lower whiteness and solubility. Modified sorghum flour showed higher value than the unmodified flour at all characteristics. Infrared spectroscopy showed that modification shifted absorption peak from 1334 cm^{-1} to 1373 cm^{-1}, but there was no difference between samples dried at different temperatures.

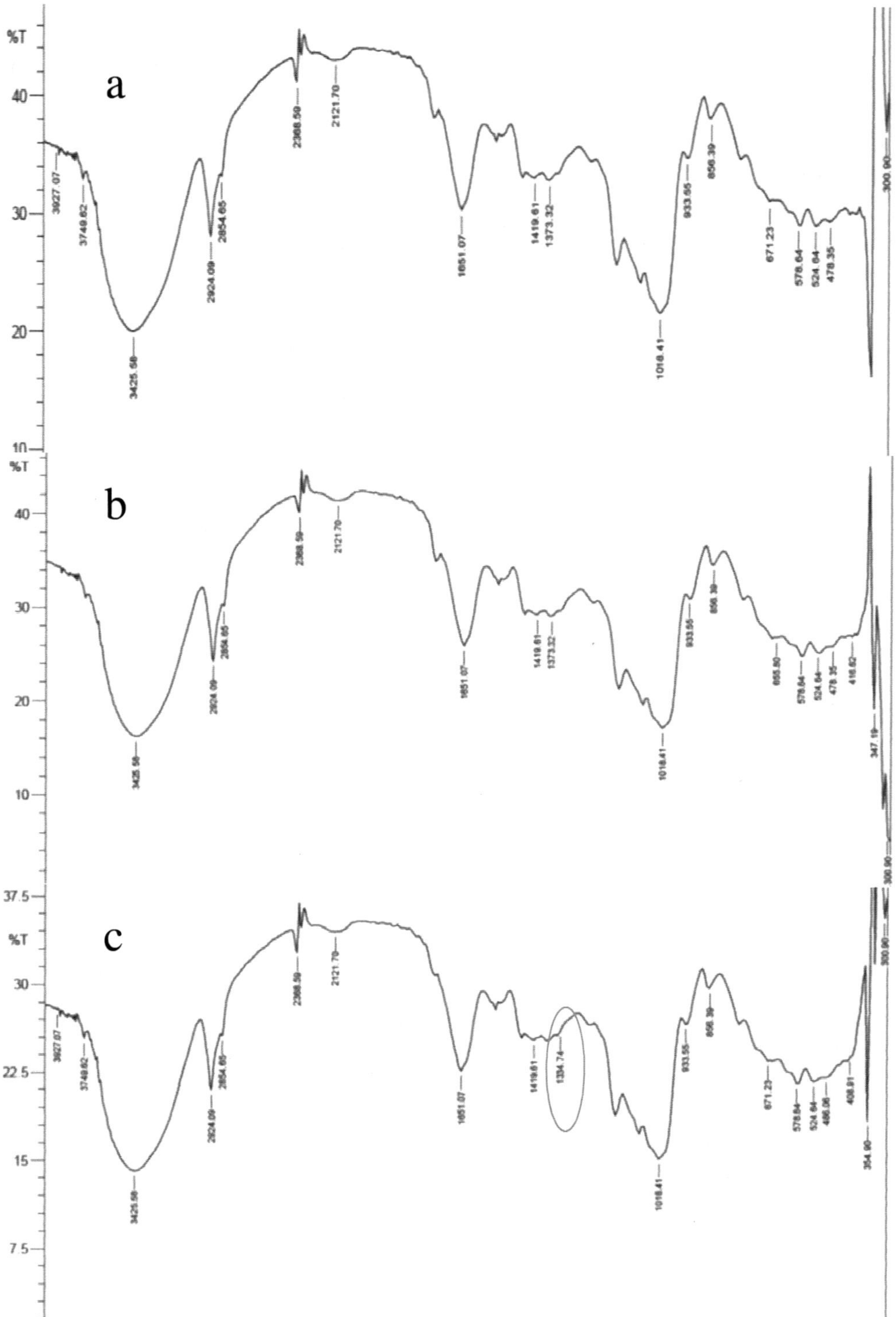

Figure 1. FTIR spectra of modified sorghum flour dried at 40°C (a), 60°C (b), and unmodified sorghum flour dried at 60°C (c)

IOP Conf. Series: Materials Science and Engineering **193** (2017) 012024 doi:10.1088/1757-899X/193/1/012024

References

[1] Munck L 1994 New milling technologies and products: *Whole plant utilization by milling and separation of the botanical and chemical components Sorghum and Millets: Chemistry and Technology ed Dendy D A V* (St. Paul: American Association Cereal Chemistry)

[2] Pessa E Suortti T Autio K Poutanen K Espoo 1992 *Starch/starke* **44** 64-69

[3] Okunlola A and Akingbala O 2013 *Braz. J. Pharm. Sci.* **49** 699-708

[4] Yanuwardana Basito Muhammad D R A 2013 *Jurnal Teknosains Pangan* **2** 1

[5] Codex Standart 1989 *Codex Standart for Sorghum Flour*

[6] Davidek J Velisek J Pokorny J 1990 *Chemical Changes During Food Processing* (Amsterdam: Elsevier Science Publishers)

[7] Salimi Y K 2012 *The role of sorghum (Sorghum bicolor L) extracts and flour in the inhibition of cancer in vitro and in vivo of mice* (Bogor: Institut Pertanian Bogor)

[8] Jyothi A N Sasikiran K Sajeev M S Revamma R Moorthy S N 2005 *Starch/Stärke* **57** 547–555

[9] Leach H W Mc Cowen L D Schoch T J 1959 *Cereal Chem.* **36** 534-544

[10] Subekti D 2008 *Design of vertical screw mixer type machine for maltodextrin production by dry hyidrolysis method using clorida acid* (Bogor: Institut Pertanian Bogor)

[11] Yuanita L 2009 *Jurnal Ilmu Dasar* **10** 49-55

[12] Faridah D N Fardiaz D Andarwulan N Sunarti T C 2010 *Jurnal Teknologi dan Industri Pangan* **21** 135-142

IOP Conf. Series: Materials Science and Engineering **193** (2017) 012025 doi:10.1088/1757-899X/193/1/012025

Shelf Life Prediction for Canned Gudeg using Accelerated Shelf Life Testing (ASLT) Based on Arrhenius Method

R Nurhayati[1,a], E Rahayu NH[1], A Susanto[1], Y Khasanah[1]
[1]Research Unit for Natural Product Technology, Indonesian Institute of Sciences
Jl. Yogya-Wonosari km 31.5 Gading, Playen, Gunungkidul, Yogyakarta, Indonesia.

E-mail: rifa.lipi@gmail.com

Abstract. Gudeg is traditional food from Yogyakarta. It is consist of jackfruit, chicken, egg and coconut milk. Gudeg generally have a short shelf life. Canning or commercial sterilization is one way to extend the shelf life of gudeg. This aims of this research is to predict the shelf life of Andrawinaloka canned gudeg with Accelerated Shelf Life Test methods, Arrhenius model. Canned gudeg stored at three different temperature, there are 37, 50 and 60°C for two months. Measuring the number of Thio Barbituric Acid (TBA), as a critical aspect, were tested every 7 days. Arrhenius model approach is done with the equation order 0 and order 1. The analysis showed that the equation of order 0 can be used as an approach to estimating the shelf life of canned gudeg. The storage of Andrawinaloka canned gudeg at 30°C is predicted untill 21 months and 24 months for 25°C.

1. Introduction

Gudeg is a traditional food from Yogyakarta which is also a spesific food of this region. Gudeg is processed from young jackfruit with the addition of spices and coconut milk in a long of heating time. Based on processing method, there are two types of gudeg, wet and dry gudeg [1]. Wet gudeg obtained by cooking the young jackfruit untill the coconut milk was fully permeated (one time cooking). While the dry gudeg obtained by two times cooking, by sauteing (fried sauteed) of wet gudeg. The water content wet gudeg higher than the dry gudeg. The shelf life wet gudeg was 1-2 days and the dry gudeg 3-4 days [2].

Gudeg has fairly short shelf life. Packaging and preservation technology is needed in order to extend the shelf life and improve the economic value. Canning is a method of preserving food products in hermetic packaging, heated with commercial sterilization process and without additional preservatives.

During storage, canned gudeg changing the quality [3]. To determine the expiry date of canned gudeg, shelf life prediction needs to be done in an easy way and approaching the actual shelf life. Shelf life can be achieved by Accelerated Shelf-Life Testing (ASLT) by storing food products in the specific environment that cause rapid deterioration either in temperature or humidity. Acceleration method can be done in a shorter time with good accuracy [4]. Temperature plays a key parameter to determine the product deterioration. The higher the storage temperature, the faster the food deterioration. The relationship between the temperature at the speed of degradation products can be viewed using the Arrhenius equation.

ASLT method, Arrhenius model, widely used for predicting the shelf life of perishable food by a chemical such as fat oxidation, Maillard reaction, denaturation of proteins, etc. In general, the chemical reaction rate speed faster in higher temperatures, it means the decrease of product quality occur rapidly. The shelf life of food product may be determined by the Arrhenius models, ie: canned food, fruit juice, instant mie, frozen meat and high fat product (fat oxidation), product that contain reducing sugar and protein (browning reaction) [5].

Gudeg has high content of fat (12.74%) [3], thereby potentially damage occurs during storage. Therefore we need a predictive testing period expired canned gudeg using fat damage as a critical parameter. This study aimed to predictions shelf life of canned gudeg using thiobarbituric acid (TBA) value or rancidity test as critical parameters in Arrhenius models.

2. Material and Methods

2.1. Material canned gudeg

Canned gudeg used in this reseach from "Bu Tjitro Andrawina Loka", Laksda adisucipto street 9 Yogyakarta.

2.2. Shelf Life Prediction Test Using Arrhenius Model

Canned gudeg kept at the temperature 37, 50, and 60°C. Observations were made with three replications every 7 days. The parameters analyzed was the number of thio barbituric acid (TBA). TBA value were plotted against time of storage (days) and the linear regression equation obtained. Three equations were retrieved to the three conditions of product storage temperatures. Values ln k plotted versus 1/T (K^{-1}) and obtained the value of the intercept and the slope of the linear regression equation:

$$\ln k = \ln k0 - (Ea / R)(1/T) \quad (1)$$

From the equation obtained exponential factor k0 and the value of activation energy (Ea). The shelf life of canned gudeg calculated with an equation based order reaction kinetics.

2.3. TBA Analysis

TBA analysis following the method Tarladgis [6]. Ten grams sample added with 50 ml distilled water and mash for 2 minutes with a waring blender. Transfer quantitatively to the distillation flask while washed with 47.5 ml of distilled water. Added 2.5 ml of 4 M HCl until the pH to 1.5. Add anti-foaming agent then distilled to obtain 50 ml distillate. Pipette 5 ml of the distillate in a test tube, add 5 ml of reagent TBA, cover, mix evenly and heat for 35 minutes in boiling water. Cool the test tube with water for 10 minutes then measuring the optical density (OD) at 528 nm. TBA value is expressed as mg malonaldehyde per kg sample.

3. Result and Discussion

Shelf life is defined as the period of time under normal storage conditions after which the product will not be fit to eat. The shelf life of canned foods is determined by a variety of factors, related to the deteriorative reaction during storage [7].

Canned gudeg "Bu Tjitro Andrawina Loka" packaged in pull up can size 300 x 205, with 200 g in netto. Nutritional composition of canned gudeg in table 1. Fat content in canned gudeg was high, (15.2 %) if compared with protein (9.53 %) and carbohydrate (3.80 %). Therefore, fat damage is likely to occur during storage of canned gudeg.

The TBA test for malondialdehyde determination is the most frequently used method for the assessment of lipid oxidation in foods because of its sensitivity and relatively simple procedure [8]. TBA test involves the reaction between TBA and malondialdehyde produced from lipid hydroperoxide decomposition to form a pink complex with maximum absorbance at 528 nm [9].

Table 1. Nutrition composition of canned gudeg "Bu Tjitro Andrawina Loka"

Nutrition	Value
Water content	69.9 %
Ash content	1.57 %
Protein	9.53 %
Fat	15.2 %
Carbohydrate *by different*	3.80 %
Calori	190 kal/100 g

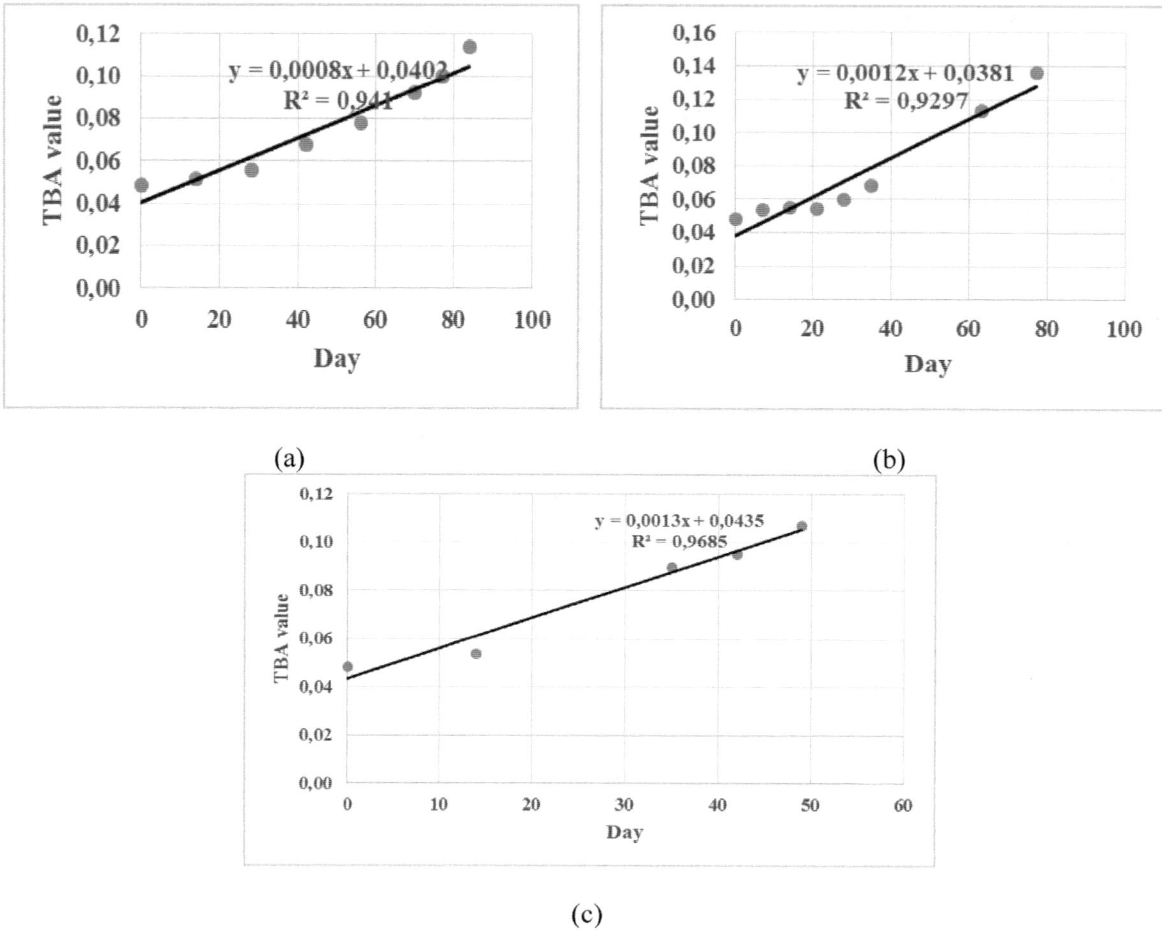

Figure 1. Plot TBA value versus time of storage canned gudeg at 37°C (a); 50°C (b); 60°C (c)

Shelf life test was done by storing the canned gudeg at incubator of 37, 50 and 60°C. TBA test was done every interval 7 days. Plot the data of TBA value versus storage time can be seen in Figure 1. The increase in TBA value during canned gudeg storage follow zero order reactions. This means that the speed of rancidity reaction of canned gudeg was constant. The longer storage, TBA number of canned gudeg was increasing [10,11,12]. The value of TBA for dried shrimp increased during storage at 37°C from 0,81 mg/kg at day 0 to 2,19 mg/kg at day 28 [11].

The TBA value of canned gudeg stored at 60°C changed more quickly than which stored at 37°C and 50°C. TBA value is widely used to measuring malonaldehyde content as a level of lipid oxidation. Malonaldehyde forms from hydroperoxides, which are initial reaction product of polyunsaturated fatty acid with oxygen [13].

The value of ln k versus 1/T (K^{-1}) from each equation Figure 1 was plotted on a graphic and obtained a linear equation ln k = -2239,2 (1/T) + 0,1281 (Figure 2). Energy of activation (Ea) to changed the value of TBA is 4.447,051 cal/mol.

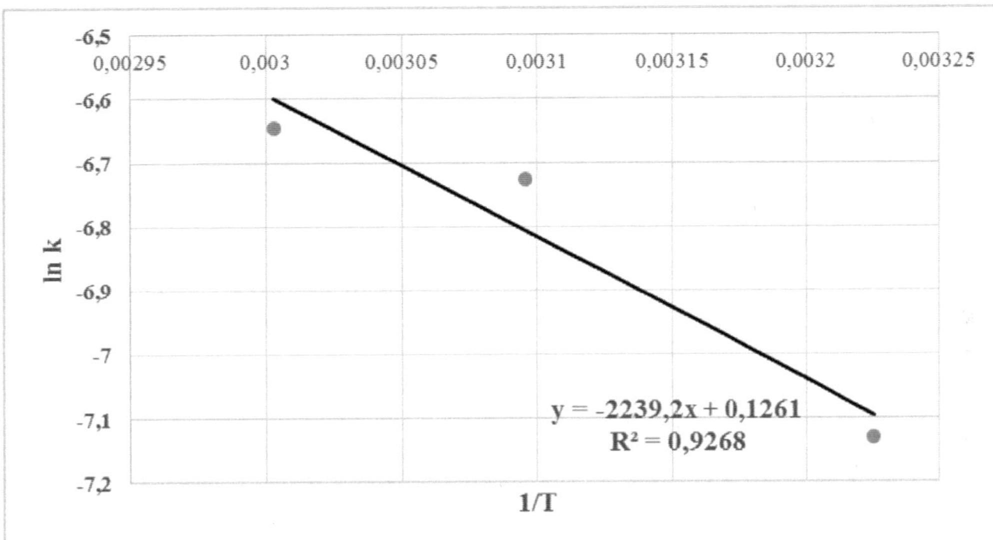

Figure 2. Plot 1/T versus ln k (Arrhenius approach)

Equation ln k = -2239,2 (1/T) + 0,1281 used to calculate the shelf life of canned gudeg at some storage temperature (table 2). The increase in storage temperature may cause a greater reaction speed. This was indicated by the increasing of deterioration constanta value (k). The greater reaction rate of deterioration, the faster the product was damaged, so the shelf life of products was getting shorter.

Table 2. The shelf life prediction of canned gudeg at the storage temperature

Storage Temperature	1/T	ln k	K	Shelf life	
				t (days)	t (months)
30°C	0.0033	-7.2620	0.0007	641.2961	21.3765
25°C	0.0034	-7.3860	0.0006	725.9537	24.1985
4°C	0.0036	-7.9557	0.0004	1283.2444	42.7748

Table 2 showed that the shelf life of canned gudeg at 30°C was 21 months, at 25°C was 24 months, and at 4°C was 42 months. Commercial sterilization of canned gudeg effectively extend the shelf life of gudeg, from 3 days to 641 days. This shelf life prediction accordance with Labuza [14] which states that the shelf life of canned fruits or canned vegetables product range from 1 to probably 3 years, depending on the composition of the product and on the quality aspect being measured. The shelf life of canned "nasi uduk" at 30°C for 9,6 months [15]. The shelf life of sterilized fermented black glutinous rice, traditional food from Mojokerto, Indonesia, at 25°C was 21 months [16].

4. Conclusion

Shelf life predictions of canned gudeg using Accelerated Shelf Life with Arrhenius approach with a number of critical parameters TBA or rancidity. The shelf life of canned gudeg at 30°C was 21.3765 months, 25°C was 24.1985 months and at 4°C was 42.7748 months.

Acknowledgement

We gratefully acknowledge for Mr. Bambang, the owner of Bu Tjitro Andrawina Loka canned gudeg, and Mrs. Ema Damayanti (BPTBA LIPI) for all support.

References

[1] Triwitono P 1993 *Tesis at the faculty agricultural technology* (Yogyakarta: Gadjah Mada University)
[2] Nurhikmat A, Susanto A and Rahayu ENH 2009 *Proceeding of National Conference on Chemistry and Chemical Education* 603-11
[3] Nurhikmat A, Suratmo B, Bintoro N, and Sentana S 2015 *Agritech* **35** 353-358
[4] Labuza T.P 2007 *Department of Food Science and Nutrition* (Minneapolis: University of Minnesota)
[5] Kusnandar F, 2010 (Jakarta: Penerbit Dian Rakyat)
[6] Tarladgis BG, Watts BM, Younathan MT, and Dugan L Jr 1960 *J. Am. Oil Chem. Soc* **37** 44-48
[7] Buculei A, Ionescu V, and Ionescu M 2009 *J. of Agroalimentary Process and Technologies* **15**(1) 140-145
[8] Rahardjo S and Sofos JN 1993 *Meat Sci.* **35** 145-169
[9] Tarladgis BG, Pearson AM, and Dugan L Jr 1964 *J. Sci. Food Agr.* **15** 602-607
[10] Lu F, Zhang JY, Liu SL, Wang Y, and Ding YT 2011 *Food Chem.* **127** 159-168
[11] Kanatt SR, Chawla SP, Chander R and Sharma A 2006 *LWT-Food Sci. and Tech.* **39**(6) 621-626.
[12] Thanonkaew A, Benjakul S, Visessanguan W, and Decker E 2005 *J. Food Sci.* **70**(8) 478-482
[13] Fernandez J, Perez-Alvarez J, and Fernandez-Lopez J 1997 *Food Chem.* **59**(3) 345-353
[14] Labuza 1979 (Washington DC: Government Printing Office Washington)
[15] Kurniadi M, Salam N, Kusumaningrum A, Nursiwi A, Angwar M, Susanto A, Nurhikmat A, Triwiyono, Frediansyah A 2016 *International Conference on Engineering, Science and Nanotechnology*
[16] Haryati, Estiasih T, Heppy F dan Ahmadi Kgs 2015 *Jurnal Pangan dan Agroindustri* **3**(1) 156-165

Preservation Effect of Two-Stage Cinnamon Bark (*Cinnamomum Burmanii*) Oleoresin Microcapsules On Vacuum-Packed Ground Beef During Refrigerated Storage

D Irfiana[1], R Utami[1], L U Khasanah[1], and G J Manuhara[1]
[1] Food Science and Technology, Universitas Sebelas Maret, Surakarta

*E-mail: danitairfiana@yahoo.co.id

Abstract. The purpose of this study was to determine the effect of two stage cinnamon bark oleoresin microcapsules (0%, 0.5% and 1%) on the TPC (Total Plate Count), TBA (thiobarbituric acid), pH, and RGB color (Red, Green, and Blue) of vacuum-packed ground beef during refrigerated storage (at 0, 4, 8, 12, and 16 days). This study showed that the addition of two stage cinnamon bark oleoresin microcapsules affected the quality of vacuum-packed ground beef during 16 days of refrigerated storage. The results showed that the TPC value of the vacuum-packed ground beef sample with the addition 0.5% and 1% microcapsules was lower than the value of control sample. The TPC value of the control sample, sample with additional 0.5% and 1% microcapsules were 5.94; 5.46; and 5.16 log CFU/g respectively. The TBA value of vacuum-packed ground beef were 0.055; 0.041; and 0.044 mg malonaldehyde/kg, resepectively on the 16th day of storage. The addition of two-stage cinnamon bark oleoresin microcapsules could inhibit the growth of microbia and decrease the oxidation process of vacuum-packed ground beef. Moreover, the change of vacuum-packed ground beef pH and RGB color with the addition 0.5% and 1% microcapsules were less than those of the control sample. The addition of 1% microcapsules showed the best effect in preserving the vacuum-packed ground beef.

1. Introduction

Beef is nutritious food. It contains 66 g of water, 18.8 g protein, 14 g fat, 11 mg calcium, 170 mg phosphorus, 2.8 mg iron, and 207 kcal of energy every 100 g [1]. Microbes can cause meat spoilage and decrease its shelf life. *E. coli* is one of microbe which often contaminates the beef [2]. Components of fat in beef can be oxidized on contact with oxygen [3]. This will influence the odor because of the oto-oxidation process on unsaturated fatty acids. Sliced beef has a lower level of damage when compared with ground beef due to its larger surface area [4].

One compound that has the ability to act as an antimicrobial and antioxidant is cinnamaldehyde. Cinnamaldehyde is contained in cinnamon bark [5]. Natural antioxidant activity found in cinnamon extract [6]. Cinnamaldehyde at levels of 50 µl/ml can kill all microbial activity, while at the concentration of 0.78 to 12.5 µl/ml cynamaldehyde can inhibit the growth of microbes such as *Aeromonas hydrophila* and *Enterococcus faecalis* [7]. Cinnamon oleoresin can provide an antimicrobial effect against *Pseudomonas fluorescens* and *Pseudomonas putida* [8]. Cinnamaldehyde of cinnamon bark oleoresin (*Cinnamomum burmanii*) obtained by maceration, contains cinnamaldehyde extract amounting to 65.88% [9]. Oleoresin is sticky and highly concentrated, making difficult to apply to food [10]. Oleoresin also easily loses its constituent components, such as cinnamaldehyde compound that has properties as an antioxidant and antimicrobial properties, so a microencapsulation process is needed. The changes of flavor in the cinnamon oleoresin (*Cinnamomum burmanii*) can be prevented by using the process of microencapsulation [11]. This process can improve the stability of the components of flavor, and make oleoresin into a powder to make it easier to apply to foodstuffs. Cinnamaldehyde in oleoresin microcapsules is equal to 47.95% [11]. It will be used as a preservative.

One method of packaging beef that can be used is vacuum packaging [12]. This packaging can maintain the quality of the beef for 14 days of storage [13]. Vacuum-packed also carried out ground beef with the addition of essential oil and hyssop over 15 days of storage [14]. So in this research the storage of ground beef will be continued for 16 days to discover how the quality of vacuum-packed ground beef is affected by the addition of two-stage cinnamon bark oleoresin microcapsules.

2. Methodology

2.1. Two-Stage Cinnamon Bark Oleoresin Preparation

Two-stage cinnamon bark oleoresin was made by preparation of cinnamon bark size reduction passes 30 mesh and detained 50 mesh. The cinnamon bark was obtained from Bubakan Village, Girimarto, Wonogiri. Cinnamon bark essential oil is produced by the steam distillation method for four hours, and the ratio of cinnamon powder and distilled water is 1: 4 [15]. Having obtained the oil, cinnamon powder residue is dried until it contains a maximum moisture content of 12% (SNI 01-3714-1995). And then, the residue was extracted with ethanol 70% with the ratio of cinnamon powder and ethanol 70% is 1 : 6 at 70-75^{0}C for four hours extraction. This process was followed by evaporation using a rotary vacuum evaporator "IKA Coateng L11" at 80^{0}C and 60 rpm [9]. Having obtained the essential oils and oleoresin cinnamon, both were then mixed together.

2.2. Microencapsulation Process

The microencapsulation process was implemented using a homogenizer "Ultra Turax® Basic® Woke" and followed by spray drying "SD-Basic Lab Plant". Materials that were prepared were distilled water, the coating material, and oleoresin. Then an emulsion was made with a ratio of coating material such as gum arabic and maltodextrin DE 10-12 (1: 3) to distilled water at 20:80. Then two-stage cinnamon bark oleoresin was mixed into the total suspension at 10%. The microencapsulation process was implemented with a spray dryer inlet temperature of 109^{0}C and with a feed rate of 15 or 20 ml per minute [11].

2.3. Application of Two-Stage Cinnamon Bark Oleoresin Microcapsules.

Fresh ground beef was prepared by using a sterile blender (blender sprayed with 96% alcohol and dried). 50 g of ground beef was added with two-stage cinnamon bark oleoresin microcapsules at 0%; 0.5%; and 1%, then mixed together [16]. After that, the ground beef was vacuum-packed in a nylon pack and stored in a refrigerator. Over the next 16 days, TPC testing was conducted [17], as well as TBA [18], pH [19], and color (Lutron Electronic RGB-1002) on days-0,4,8,12, and 16.

3. Result and Discussion

3.1. TPC

Table 1. Analysis of TPC of Ground Beef with the addition of Two-Stage Cinnamon Bark Oleoresin Microcapsules During Refrigerated Storage (4 ± 1^{0}C)

Micro-capsules	TPC (log CFU/g)				
	Day-0	Day-4	Day-8	Day-12	Day-16
0%	$5.27^{Aa} \pm 0.16$	$5.50^{ABa} \pm 0.35$	$5.85^{Bb} \pm 0.02$	$5.89^{Bc} \pm 0.01$	$5.94^{Bb} \pm 0.10$
0.5%	$5.26^{Aa} \pm 0.30$	$5.26^{Aa} \pm 0.05$	$5.28^{Aa} \pm 0.14$	$5.30^{Ab} \pm 0.00$	$5.46^{Aa} \pm 0.00$
1%	$5.00^{Aa} \pm 0.05$	$5.08^{Aa} \pm 0.00$	$5.04^{Aa} \pm 0.06$	$5.16^{Aa} \pm 0.06$	$5.16^{Aa} \pm 0.20$

The figures with the same small superscript letters in the same column and large superscript letters in the same row show no significant difference at the 5% significance level.

TPC with the addition of two-stage cinnamon bark oleoresin microcapsules did not show a significant increase when compared to the control. Essential oil of cinnamon bark contains cinnamaldehyde as an antibacterial agent up to 92.84% [20]. It also contained cinnamon residue oleoresin with 12.22% cynamaldehyde [21]. Cinnamon bark oleoresin microcapsules contains 60.5% cinnamaldehyde [11]. The blending process of the essential oils and oleoresin will provide two-stage cinnamon bark oleoresin with more cinnamaldehyde contents, which will have an inhibitory effect on the number of microbes. The TPC value on day 0 at concentrations of 0%; 0.5%; and 1% were 5.27; 5.26; and 5.00 log CFU/ g respectively. TPC value from this research was almost equal to the value of TPC on day 0 of the study by the other research in which a sample of ground beef TPC control value was 5.525 log CFU/g [22].

Until the 16th day of storage, all samples still had a TPC value below 6 log CFU/g. This is below the maximum standard TPC value. The TPC control sample was higher than the sample with addition two-stage cinnamon bark oleoresin microcapsules. It affects the value of TPC. Cinnamon bark oleoresin microcapsules contains essential oils and contains active compounds such as polyphenols, cinnamaldehyde and antibacterial compounds [23]. Cinnamaldehyde is the main active compounds in cinnamon and plays a role in the inhibition of microbial activity. Cinnamon bark essential oils contains 37.12% cinnamaldehyde and cinnamon bark oleoresin contains 12.22% cinnamaldehyde [15].

The active compound can inhibit the growth of pathogenic microbes or spoiled microbials. It can degrade the cell wall, damaging the cytoplasmic membrane, damaging cellular components, and causes coagulation and malfunction of microbial cells [24]. At low concentrations, cynamaldehyde will inhibit the enzymes that work in cell cytokine interactions and functions of cells. At higher concentrations, cinnamaldehyde will act as an ATPase inhibitor, and at lethal concentrations, cinnamaldehyde will act to damage the cell membrane [25]. Active compounds in cinnamaldehyde can decrease the amount of ATP in intracellular microbes [26]. Cynamaldehyde will increase the permeability of cell membranes and cause cytoplasm leaks [27]. The compound of cinnamaldehyde on the microcapsules will work to reduce the amount of ATP and increases the permeability of cell membranes [26,27].

Essential oils are known to have an inhibitory effect against gram positive and negative bacteria [28]. Cinnamon oil contains (*Cinnamomum zeylanicum*) which is 91.042% cynamaldehyde [29]. Gram-positive bacteria are more sensitive to the essential oil of cinnamon [29]. Some of the oil molecules bind to membrane proteins and inhibit the synthesis of peptidoglycan, therefore the level of inhibition is increased.

3.2. TBA

Table 2. Analysis TBA Ground Beef with the addition Two-Stage Cinnamon Bark Oleoresin Microcapsules during Refrigerated Storage (4 ± 1^{0}C)

Micro-capsules	TBA (mg malonaldehida/kg)				
	Day-0	Day-4	Day-8	Day-12	Day-16
0%	$0.045^{Aa} \pm 0.006$	$0.050^{Ab} \pm 0.005$	$0.050^{Aa} \pm 0.008$	$0.051^{Aa} \pm 0.001$	$0.055^{Ab} \pm 0.003$
0.5%	$0.035^{Aa} \pm 0.003$	$0.037^{Aa} \pm 0.004$	$0.039^{Aa} \pm 0.000$	$0.040^{Aa} \pm 0.002$	$0.041^{Aa} \pm 0.005$
1%	$0.033^{Aa} \pm 0.000$	$0.036^{Aa} \pm 0.001$	$0.037^{Aa} \pm 0.004$	$0.039^{Aa} \pm 0.005$	$0.044^{Aa} \pm 0.000$

The figures with the same small superscript letters in the same column and large superscript letters in the same row show no significant difference at the 5% significance level.

The higher concentration of two-stage cinnamon bark oleoresin microcapsules, will provide an inhibitory effect against oxidation. The addition of 1% microcapsules produce lower TBA values. Cinnamaldehyde as an antioxidant agent can inhibit oxidation reactions, thus rancidity in ground beef can be inhibited and the shelf life is longer. Antioxidants work by reducing the level of rancidity in material. Antioxidants will scavenge free radicals in the material that cause rancidity and act as chelating metals [30]. Two-stage cinnamon bark oleoresin microcapsules will give lower TBA values than the control sample. Cinnamaldehyde include phenolic components, which are phenolic primary antioxidants that prevent the formation of free radicals and turn them into more stable products [30].

The TBA acceptance limit value is equal to 2 mg/kg malonaldehyde. If the TBA values exceed these limits then the meat is not suitable for consumption because it is rancid [31]. TBA value of ground beef with the addition of two-stage cinnamon bark oleoresin microcapsules in this study showed the TBA value to be below 2 mg/kg for 16 days of storage. The beef was still considered safe for human consumption because the value was still below the limit TBA for consumption.

3.3. pH Value During Storage

Table 3. Ground Beef pH analysis with the addition of Two-Stage Cinnamon Bark Oleoresin Microcapsules during Refrigerated Storage (4 ± 1⁰C)

Micro-capsules	Ph				
	Day-0	Day-4	Day-8	Day-12	Day-16
0%	$5.5^{Bc} \pm 0.04$	$5.5^{Bc} \pm 0.00$	$5.5^{Bb} \pm 0.00$	$5.5^{Ba} \pm 0.04$	$5.2^{Aa} \pm 0.04$
0.5%	$5.4^{Bb} \pm 0.04$	$5.4^{Bb} \pm 0.00$	$5.5^{Cb} \pm 0.00$	$5.4^{Ba} \pm 0.07$	$5.1^{Aa} \pm 0.00$
1%	$5.3^{Ba} \pm 0.04$	$5.3^{Ba} \pm 0.00$	$5.4^{Ba} \pm 0.04$	$5.4^{Ba} \pm 0.04$	$5.2^{Aa} \pm 0.04$

The figures with the same small superscript letters in the same column and large superscript letters in the same row show no significant difference at the 5% significance level

Samples with addition of microcapsules have a lower pH value than the control. Maltodextrin made from cassava starch is supplemented with HCl to inactivate the α-amylase enzyme, which does not reverse the neutralization process, so maltodextrin has a low pH [32]. Maltodextrin has a pH of 4.3 for 6.5 to 7.9 DE [32]. From these results, it may be possible that greater addition of two-stage cinnamon bark oleoresin microcapsules will produce a lower pH. The manufacture of maltodextrin with a DE of 17% and 17.4% gives 5 for pH values [33]. If the pH value is low, it would affect the levels of acidity of ground beef. In this study, the pH value of maltodextrin is 4.08, wherein the pH value is quite low.

The pH value of the control samples decreased from 5.5 to 5.2 on the first day and 16th day of storage. The pH value of the samples with 0.5% microcapsules addition decreased from 5.4 to 5.1 from the first day to the 16th day. The pH value of the sample with addition of 1% microcapsules decreased from 5.3 to 5.2 from the first day to the 16th day. The decrease in the pH value of a control sample of ground beef, adding 0.5%, and 1% microcapsules were 0.3; 0.3; and 0.1 respectively. The decrease in the pH value of the sample with additional 0.5% and 1% was lower than the control sample. The addition of 1% microcapsules was more effective because of the higher content of cinnamaldehyde. This is an antibacterial agent that will inhibit the growth of gram-positive bacteria such as *Staphylococcus aureus, Bacillus cereus, E. faecalis,* and *Micrococcus luteus*. The bacteria include lactic acid bacteria in which there is a zone of inhibition of the bacterial activity [7]. It is possible for this compound to inhibit the activity of lactic acid bacteria on meat. So in this study, the addition of 1% microcapsules was found to inhibit the rate of decrease in the pH value due to the activity of lactic acid bacteria.

In the earlier days of storage, there was an increase in the pH value of the sample to which two-stage cinnamon bark oleoresin microcapsules had been added. Samples with addition of 0.5% microcapsules increased from 5.4 to 5.5 on day 8 and sample with addition of 1% microcapsules increased from 5.3 to 5.4 on the eighth day. This is possible due to the growth of bacteria that produce more alkaline metabolites in the samples [34].

3.4. Color Change During Storage

The value of R increasingly rise during storage. This compares to a pH value of meat that still has a low pH value, where low pH values give a bright red meat color [35]. This can be seen from the pH value that is proportional to the TPC. A Higher value of TPC and a lower pH value increase the number of microbes (lactic acid bacteria) growing on ground beef during storage. Fresh beef contains myoglobin pigment which can change color to bright red when a process of oxygenation occurs, and this makes the pigment oksimyoglobin, and vice versa. If the meat is oxygenated for longer, then the color will change to brown or make the metmyoglobin pigment. Brown beef can turn back to red-purple or bright red with oxygen reduction [35].

Table 4. Color analysis (RGB) of Ground Beef with the addition of Two-Stage Cinnamon Bark Oleoresin Microcapsules during Refrigerated Storage (4 ± 1^0C)

	Microcapsules	**Day-0**	**Day-4**	**Day-8**	**Day-12**	**Day-16**
	0%	$240.8^{Aa} \pm 0.4$	$228.5^{Aa} \pm 12.7$	$242.5^{Aa} \pm 1.4$	$291.0^{Ca} \pm 6.4$	$273.0^{Bb} \pm 6.4$
R	0.5%	$225.3^{Aa} \pm 3.9$	$247.0^{ABa} \pm 17.0$	$261.8^{BCb} \pm 5.3$	$287.5^{Ca} \pm 15.6$	$255.8^{Bb} \pm 7.4$
	1%	$229.0^{Aa} \pm 7.1$	$244.3^{Aa} \pm 1.1$	$265.5^{Bb} \pm 7.8$	$279.5^{Ba} \pm 13.4$	$230.8^{Aa} \pm 5.3$
	0%	$124.0^{ABb} \pm 0.0$	$113.0^{Aa} \pm 5.7$	$131.5^{Ba} \pm 9.2$	$163.5^{Ca} \pm 10.6$	$153.5^{Cb} \pm 13.4$
G	0.5%	$116.5^{Aa} \pm 0.7$	$128.5^{ABa} \pm 10.6$	$137.5^{Ba} \pm 2.1$	$166.5^{Da} \pm 0.7$	$151.0^{Cb} \pm 0.0$
	1%	$115.0^{Aa} \pm 1.4$	$128.0^{ABa} \pm 9.9$	$145.0^{Ca} \pm 4.2$	$146.5^{Ca} \pm 7.8$	$136.0^{BCa} \pm 4.2$
	0%	$104.0^{Ab} \pm 1.4$	$99.8^{Aa} \pm 6.7$	$112.0^{ABa} \pm 7.8$	$135.5^{Ca} \pm 10.6$	$128.0^{BCa} \pm 10.3$
B	0.5%	$95.5^{Aa} \pm 0.7$	$106.0^{ABa} \pm 7.8$	$112.8^{Ba} \pm 3.9$	$135.3^{Ca} \pm 0.4$	$126.0^{Ca} \pm 2.8$
	1%	$94.3^{Aa} \pm 1.1$	$108.0^{Ba} \pm 9.2$	$120.8^{Ca} \pm 2.5$	$121.0^{Ca} \pm 3.5$	$112.8^{BCa} \pm 3.1$

The figures with the same small superscript letters in the same column and large superscript letters in the same row show no significant difference at the 5% significance level

All the samples had decreased R value. It is possible, because the response of the environmental conditions are different on ground beef against humidity, room temperature, and light intensity. For example the oxidation of myoglobin decrease red color of ground beef. The decrease of red color is influenced by the texture of the meat, water content, pigments, and activators enzyme that play a role in the formation and degradation of the color of the meat [36]. The control samples tended to have a higher R value when compared with the sample with the added microcapsules. A Higher concentration (0.5% and 1%) give smaller color changes of the meat when viewed from R day 0 to day 16 (difference value R), which means that the color of the meat can be maintained. This is comparable to the TBA value of ground beef, where the value of TBA is lower with the addition of 1% two-stage cinnamon bark oleoresin microcapsules. Contact with oxygen can be inhibited due to the presence of cinnamaldehyde compounds and here is a reduced color change with the addition of two-stage cinnamon bark oleoresin microcapsules. When the packaging is opened, the beef will be in direct contact with oxygen and will have a bright red color. Flesh color will change to a bright red when the meat is exposed to air and a myoglobin oxygenation reaction occurs, forming bright right oksimyoglobin [36].

The value of G (greenness) indicated ground beef metmyoglobin color at a wavelength of 555 nm (the green part of meat color) [37]. On the inside of myoglobin in the meat, green or violet dark colors will increase during storage. The appearance of green or yellow green in the flesh indicates damage to fresh meat. A greenish color can also be caused due to the activity of bacteria in ground beef [38]. A greenish color is comparable to the value of the TPC of ground beef during storage which increases.

Cinnamon bark oleoresin microcapsules affects the quality of the G color in ground beef. The change of green color value of ground beef with the addition of microcapsules is lower than the control sample. This is possible because the active compounds in two-stage cinnamon bark oleoresin microcapsules can inhibit the reduction of myoglobin in meat during storage. So the green color as an indicator of spoiled meat can be inhibited by the addition of microecapsule [38].

Longer storage will give higher G value. There are a correlation with blue color formation and the activity of bacterial. It is possible for yellow color pigments to form during the storage. The yellow color is formed because of the marbling fat content in beef [36]. The yellowness increases with the length of storage time.

Two-stage cinnamon bark oleoresin microcapsule affects the quality of the beef color. The higher of the concentration, the lower of the discoloration of the meat compared to a control sample. The content of the active compounds in two-stage cinnamon bark oleoresin microcapsules will inhibit the reduction of myoglobin in the meat [37], depending on the length of storage.

4. Conclusion

The addition of two-stage cinnamon bark oleoresin microcapsules at concentration of 0.5% and 1% affects the value of TPC, TBA, pH, and the color of ground beef during low temperature storage. It can also inhibit the growth of microbes (TPC) and inhibit the process of fat oxidation (TBA) when compared with control samples. The addition of two-stage cinnamon bark oleoresin microcapsules at concentration of 0.5% and 1% can minimize the changes in pH value and reduce the increase of the color value (R, G, and B) when compared with control samples.

Acknowledgements

This work was financially supported by a research project of PNBP UNS No. 632/UN27.21/LT/2016 from Universitas Sebelas Maret, Indonesia.

References

[1] Sudarisman T dan A R Elvina 1996 *Petunjuk memilih produk ikan dan daging* (Jakarta: Panebar Swadaya)
[2] Sartika R A D, Indrawani Y M, Sudiarti T 2005 *Jurnal Makara Kesehatan* **9** 23-38
[3] Ketaren S 1986 *Pengantar teknologi minyak dan lemak pangan* (Jakarta: UI-Press)
[4] Rahardjo S 2010 *Aplikasi madu sebagai pengawet daging sapi giling segar selama proses penyimpanan*
[5] Kamaliroosta L, Gharachorloo, Kamaliroosta Z, and Zadeh 2012 *J. Med. Plants Res.* **6** 4
[6] Andriyanto, Andriani, dan Widdowati 2013 *Jurnal Teknologi Pangan* **2** 2
[7] Ead-Sanla, Jangchud, Chonhenchob, dan Suppakul 2012 *J. Packaging Tech. and Sci.* 2012 **25** 7-17
[8] Prasetyaningrum, Utami R, dan Anandito R B K 2012 *Jurnal Teknologi Pangan* **1** 1
[9] Widiyanto I, Anindhito B K, dan Khasanah L U 2014 *Jurnal Teknologi Hasil Pertanian*
[10] Rahmawaty T 2014 *Karakteristik fisika dan kimia mikrokapsul oleoresin daun kayu manis (Cinnamomum burmanii) dua tahap dengan variasi rasio bahan penyalut maltodekstrin, gum arab, dan susu skim* (Surakarta: Sebelas Maret University)
[11] Murti L W 2012 *Karakteristik fisik dan kimia mikroenkapsulan oleoresin kayu manis (Cinnamomum burmanii) dengan variasi ratio gum arab dan maltodekstrin sebagai bahan penyalut* (Surakarta: Sebelas Maret University)
[12] Putu I G 2001 Karakteristik daging sapi dikemas dalam kantong plastik hampa udara *Paper of Balai Penelitian Ternak Wartazoa* **11** 2
[13] Smiecinska K, Hnatyk N, Daszkiewiez T, Kubiak D, and Matusevicus P 2015 *J. Vet. Ir Zootechnica* T.0 **92** 1392-2130
[14] Michalczyk M, Macura R, Tesarowick I, and Banas J 2012 *J. Meat Science* **90** 842-850
[15] Yuliarto, Khasanah, dan Anandito 2012 *Jurnal Teknologi Pangan* **1** 1
[16] Juneja V K, and Friedman M 2008 *J. Food Prot.* **71** 1604-11
[17] Fardiaz S 1993 *Analisis Mikrobiologi Pangan* (Jakarta: PT. Raja Grafindo Persada)
[18] Apriyantono A, Fardiaz D, Puspitasari N L, Yasni S, and Budijanto S 1989 *Petunjuk laboratorium analisis pangan* (Bogor: Institut Pertanian Bogor)
[19] Bintoro, Dwiloka, dan Sofyan 2006 *Jurnal Indonesia Trop. Anim. Agric.* **31** 4
[20] Khasanah L U, Baskara K A, Rohula U, Dimas R A M, dan Mukhamad L 2013 *Pemisahan komponen minyak atsiri (Cinnamomum burmanii) dengan destilasi fraksinasi pada berbagai tingkat suhu destilasi Semnas Pros. Agri.* 353-359
[21] Adi D N, Khasanah L U, dan Anandito B K 2012 *Jurnal THP*
[22] Prishandono D, Radiati L E, dan Rosyidi D 2009 *Pengaruh penambahan ekstrak picung (Pangium edule) dengan air dan etanol terhadap recovery Escherichia coli dan Staphylococcus sp serta total mikroba pada daging giling sapi* (Malang: Universitas Brawijaya Malang)
[23] Shan B, Chai Y Z, and Brooks J D 2007 *J. Agric. and Food Chem.* Volume **55** 5484-90
[24] Negi P S 2012 *Intern. J. Food Micro.* **156** 7-17
[25] Nazzaro F, Frattiani F, Martino L D, Coppola R, and Feo V D 2013 *J. Phar.* **6** 1451-74
[26] Oussalah M, Caillet S, and Lacroix M 2006 *J. Food Prot.* **69** 1046–55
[27] Kim J, Marshall M R, and Wei C 1995 *J. Agri. and Food Chem.* **43** 2839–45
[28] Remmal A, Tantaoui E A, Bouchikhi T, and Ettayebi M 1993 *J. Pharm. Belg.* **48** 352-356
[29] Zouheyr H, Rachida A, and Perry M G 2014 *Afr. J. Micro. Research* **8**

[30] Dave D and Ghaly A E 2011 *American J. Agric. and Bio.Sci.***6** 486-510
[31] Campo M M, G R Nute, S I Huges, M Enser, J D Wood, and R I Richardson 2005 *J. Meat Sci.* **72** 303-311
[32] Anwar E 2002 Pemanfaatan maltodekstrin dari pati singkong sebagai bahan penyalut lapis tipis tablet *Makara Sains* **6** 1
[33] Raudah, Kurniasih E, Helmi, dan Dasriani N 2013 *Jurnal Teknik Kimia Politeknik Negeri Lhoksumawe*
[34] Mudawaroch R E dan Zulfanita 2012 *Surya Agritama* **1** 1
[35] Soeparno 1994 *Ilmu dan Teknologi Daging* (Yogyakarta: Gadjah Mada University Press)
[36] Nurdjannah R and Sumarlin R 2010 *Nat. Conf.on Tek. Ptk and Vet.*
[37] Yuniarti F 2011 *Pengaruh penggunaan ekstrak rosella merah (Hibiscus sabdariffa linn) dalam berbagai konsentrasi terhadap residu nitrit dan pewarnaan daging selama proses curing* (Surakarta: Sebelas Maret University)
[38] Buckle K A, Edwards R A, Fleet G H, dan Wotton M 1987 *Ilmu Pangan* (Jakarta: UI Press)

IOP Conf. Series: Materials Science and Engineering **193** (2017) 012027 doi:10.1088/1757-899X/193/1/012027

Exterior and interior physical quality of egg of laying hens fed diets containing different dietary purslane levels

L R Kartikasari[1*], B S Hertanto[1], D Pranoto[1], W N Salim[1] and A M P Nuhriawangsa[1]

[1]Department of Animal Science, Faculty of Agriculture, Universitas Sebelas Maret, Jl. Ir. Sutami No. 36A Surakarta 57126, Indonesia

[*]E-mail: lilikretna@staff.uns.ac.id

Abstract. Purslane is considered a rich vegetable source of alpha-linolenic acid, beta-carotene and various antioxidants. The objective of the study was to investigate the effect of different dietary levels of purslane meal (*Portulaca oleracea*) in the diets of laying hens on physical quality of eggs. A total of 125 Hy-Line Brown hens (54 weeks old) were placed at individual cages and assigned to five dietary treatments. The diets were supplemented with 0, 2, 4, 6 and 8% purslane meal. Laying hens were fed for 5 weeks after a typical period of adaptation (7 days). Water and feed were provided *ad libitum*. A total of 25 egg samples of day 28 and day 35 (n = 5 egg yolks for each treatment) were collected to analyse exterior and interior physical quality of eggs. The data were analysed using ANOVA. Differences between treatment means were further analysed using Duncan's New Multiple Range Test. Results showed that feeding different purslane meal levels in the diets improved egg weight, yolk weight, albumen weight and yolk colour. The highest intensity of yolk colour was obtained with the diet containing 8% purslane meal. However, dietary treatments did not affect egg index, albumen index, yolk index, shell weight, shell thickness and Haugh Unit. It is concluded that including purslane meal to laying hen diets increases the physical qualities of the eggs.

1. Introduction

A number of studies have investigated the effect of dietary omega-3 long chain polyunsaturated fatty acids (n-3 LCPUFA) from marine sources, such as fish meal or fish oil on the fatty acid composition of eggs [1,2]. However, the presence of an off-flavour (fishy off-flavour) in these eggs has been noted in a number of sensory assessments [3,4] and this causes a decrease in consumer acceptability. Therefore, another dietary strategy is needed to provide eggs rich in n-3 PUFA without any off-flavour. A study conducted by Kartikasari et al. [5] showed that n-3 fatty acid content of eggs produced by laying hens fed diets supplemented with n-3 PUFA, alpha-linolenic acid (ALA, 18:3n-3) enhanced without negatively affecting the sensory qualities of the eggs.

Purslane is a green plant that is a source of many nutrients and contains many biologically active compounds, such as β-catotene, alpha-tocopherol, and ascorbic acid. Purslane is one of an herbaceous weed containing high level of n-3 PUFA (ALA), which is essential for normal human growth, health promotion, and disease prevention [6]. Feeding laying hens with plant seeds or oils, rich in the n-3 PUFA, can be one of dietary approach to increase n-3 fat content in eggs. ALA is the n-3 precursor for n-3 LCPUFA. Therefore, the dietary strategy by including ALA-rich plant sources rests on the ability of laying hens to convert ALA to docosahexaenoic acid (DHA). Some researchers reported that the level of n-3 fatty acid of eggs improved by adding plant sources of ALA, such as flaxseed [7] or chiaseed [8]. For example, the inclusion of 10% flaxseed in the diets of laying hens resulted in a higher accumulation of n-3 fatty acids than a control diet [7,9]. In addition, Kartikasari et al. [5] reported that including of flaxseed oil increased the level of n-3 LCPUFA, such as eicosapentaenoic acid (EPA), docosapentaenoic acid (DPA) and DHA without decreasing production performance.

While many numbers of studies reported the effect of plant seeds, oil seeds or vegetable oils rich in ALA on n-3 fat accumulation in eggs, there is little information in relation to the use of plant sources in performance parameters and egg quality. Some researchers found that the including of high levels of dietary flaxseed (10%) have been associated with a decrease in performance parameters such as reducing feed consumption [7] and yolk colour intensity [10] but egg weight, yolk weight, albumen weight, shell weight, and Haugh Unit did not change [7,10]. However, other investigators reported that the dietary inclusion of purslane meal (2%) increased egg weight and egg production [11]. Therefore, the objectives of the current study was to investigate exterior and interior physical quality of eggs of laying hens fed diets enriched with purslane meal, as a source of ALA.

2. Materials and Methods

2.1. Research Materials

A total of 150 *Hy-Line Brown hens* (54 weeks old) were used in this study. The birds were placed at individual cages and assigned to five dietary treatments. The dietary treatments were the corn-soybean based diets containing crude protein 17% and energy 2,925 kkal/kg.

2.2. Research Methods

This experiment was based on a one-way classification design. The variable factor was varying levels of *Portulaca oleracea* meal, as a source of ALA in the diet. There were a total of five diets comprised of a control diet and four experimental diets. The diets were a basal diets supplemented with 0 (R0), 2 (R1), 4 (R2), 6 (R3) and 8% (R4) *Portulaca oleracea* meal (w/w). The level of fat in the diets was held constant at approximately 3.5%. Ingredient composition and nutrient content of experimental diets were presented in table 1. Upon arrival, the hens were immediately weighed and housed one per cage (500 mm wide × 550 mm deep x 500 mm high). Each of the five dietary treatments was replicated five times (n = 5 laying hens for each replication). Laying hens were fed for 4 weeks following a 7-day period of adaptation. Water and feed were provided *ad libitum* for the experimental duration. During the first few days, the laying hens were observed at frequent intervals to ensure that they were comfortable with the environmental conditions and that all had access to adequate feed and water. The data of production performance were taken for a 4-weeks period. Feed intake (FI) was measured weekly and FCR was calculated at the end of the trial (kg feed/kg egg). All hens were weighed individually at the beginning and at the end of the experimental period. A total of 25 egg samples of day 28 and 35 (n = 5 egg for each treatment) were collected to analyse physical quality of eggs.

2.3. Data Analysis

The observed data in this research were egg weight, yolk colour, yolk weight, albumen weight, egg index, albumen index, yolk index, shell thickness, shell weight and Haugh Unit. The obtained data were analysed using analysis of variance (ANOVA). Differences between treatment means were further analysed using Duncan's New Multiple Range Test (DMRT) with significance level of $p<0.05$.

Table 1. Ingredient composition and nutrient content of experimental diets.

Treatment	R0	R1	R2	R3	R4
Ingredient			%		
Yellow corn	53.00	53.00	53.00	53.00	53.00
Rice polish	7.53	6.57	6.15	4.85	2.20
Soybean meal	24.90	24.50	23.71	23.30	23.00
$CaCO_3$	2.15	1.90	1.70	1.46	1.21
Di-calcium Phosphate	1.30	1.34	1.34	1.34	1.34
L-Lysin	0.10	0.10	0.10	0.10	0.10
DL-Metionin	0.15	0.15	0.15	0.15	0.15
Premix	0.30	0.30	0.30	0.30	0.30
Salt	0.15	0.15	0.15	0.15	0.15
Limestone	4.90	4.90	4.88	4.88	4.88
Purslane meal	0.00	2.00	4.00	6.00	8.00
Palm oil	3.00	3.07	3.00	3.20	4.65
Copra meal	1.00	1.00	1.00	1.00	1.00
Filler	1.50	1.00	0.50	0.25	0.00
Vitamin E	0.02	0.02	0.02	0.02	0.02
Total	100.00	100.00	100.00	100.00	100.00
Nutrient value					
DM	79.60	80.05	80.64	80.96	79.73
Crude protein	17.04	17.11	17.06	17.13	17.02
Ether extract	3.37	3.36	3.42	3.37	3.13
Crude fiber	3.34	3.66	4.00	4.33	4.48
ME	2925.40	2925.09	2919.68	2922.87	2852.17
Calcium	3.10	3.10	3.10	3.10	3.10
Phosphorus	0.65	0.65	0.65	0.63	0.60
Available phosphorus	0.37	0.37	0.37	0.37	0.35
Lysine	0.96	0.94	0.92	0.90	0.88
Methionine	0.43	0.42	0.42	0.41	0.40

3. Results and discussion

Adding *Portulaca oleracea* (purslane) meal up to 8% into the diets did not adversely influence egg physical qualities of brown hens including yolk index, albumin index, shell thickness, shell weight and HU (table 2 and 3). Feeding diets enriched with purslane meal for 28 days significantly increased ($p<0.05$) egg weight, albumen weight and yolk weight. Yolk weight of eggs produced by hens at day 35 were consistent with those produced at day 28, which was diets containing purslane meal resulted in heavier yolk compare to those produced from hens fed control diet. However, we observed that egg and yolk weights produced by laying hens at day 35 were relatively similar among the dietary treatments.

Table 2. Interior and exterior physical quality of egg of laying hens fed diets supplemented with *Portulaca oleracea* meal at day 28.

Parameters	R0	R1	R2	R3	R4	*p* Value
Interior						
Yolk colour	8.07^{b}	8.93^{b}	8.73^{b}	9.27^{ab}	10.47^{a}	**
Albumen index	0.073	0.067	0.061	0.067	0.074	NS
Yolk index	0.381	0.388	0.382	0.385	0.390	NS
Albumen weight	32.93^{c}	33.55^{bc}	35.03^{bc}	36.76^{ab}	38.65^{a}	**
Yolk weight	13.88^{b}	14.58^{ab}	14.86^{ab}	15.68^{a}	15.55^{a}	*
Haugh Unit	76.79	73.52	67.80	70.23	77.07	NS
Albumen pH	8.750	8.740	8.757	8.760	8.660	NS
Yolk pH	6.437^{ab}	6.490^{a}	6.373^{b}	6.407^{ab}	6.363^{b}	**
Exterior						
Egg weight	55.89^{c}	57.98^{bc}	59.19^{abc}	62.36^{ab}	64.27^{a}	**
Egg index	0.678^{ab}	0.693^{a}	0.661^{ab}	0.657^{ab}	0.651^{b}	*
Shell weight	5.71	5.86	5.66	5.03	5.90	NS
Shell thickness	0.360	0.359	0.342	0.346	0.349	NS
Specific gravity	1.328	1.284	1.314	1.314	1.327	NS

R0: Basal diet + 0% purslane meal, R1: basal diet + 2% purslane meal, R2: basal diet + 4% purslane meal, R3: basal diet + 6% purslane meal and R4: basal diet + 8% purslane meal.
* significant ($p<0.05$); ** ($p<0.01$); NS= not significant.

Table 3. Interior and exterior physical quality of egg of laying hens fed diets supplemented with *Portulaca oleracea* meal at day 35.

Parameters	R0	R1	R2	R3	R4	*p* Value
Interior						
Yolk colour	8.467^{b}	8.933^{b}	8.867^{b}	9.267^{ab}	10.133^{a}	**
Albumen index	0.066	0.070	0.073	0.068	0.066	NS
Yolk index	0.384	0.399	0.408	0.402	0.393	NS
Albumen weight	34.74	34.55	36.38	37.08	37.98	NS
Yolk weight	13.90^{b}	15.53^{ab}	15.85^{a}	15.69^{ab}	15.26^{ab}	*
Haugh Unit	73.15	73.27	75.21	71.92	72.56	NS
Albumen pH	8.693	8.603	8.657	8.697	8.600	NS
Yolk pH	5.837	5.857	6.150	6.117	6.147	**
Exterior						
Egg weight	54.87	61.23	63.05	62.52	64.44	NS
Egg index	0.662	0.664	0.682	0.659	0.664	NS
Shell weight	5.967	5.763	5.995	5.815	5.729	NS
Shell thickness	0.357	0.344	0.346	0.337	0.338	NS
Specific gravity	1.342	1.277	1.315	1.342	1.318	NS

R0: Basal diet + 0% purslane meal, R1: basal diet + 2% purslane meal, R2: basal diet + 4% purslane meal, R3: basal diet + 6% purslane meal and R4: basal diet + 8% purslane meal.
* significant ($p<0.05$); ** ($p<0.01$); NS= not significant

This could be due to the lipid content of the purslane diets and the control diet was similar (table 1). This is supported by a study conducted by Ayersa and Coates [8], which reported that the increase in weight egg and yolk egg from hens fed chia diets might be due to the higher lipid content of the chia dies compared with the control diet. Some researcher reported that inclusion level of dietary flaxseed (10%) in laying hen diets did not change egg weight, yolk weight and albumen weight [7,10]. In addition, the inclusion of high dietary levels of flaxseed (15%) caused a decrease in egg qualities, such as egg weight, yolk index, yolk colour [9]. In this current study, egg weight, yolk weight and albumen weight, especially from day 28 improved by the supplementation of dietary ALA, suggesting that the use of 8% purslane meal as a source of ALA can be applied without negatively affecting egg quality. In addition, Evaris *et al.* [12] reported that diets supplemented with purslane meal up to a level of 200 g/kg did not affect egg quality, such as egg weight and yolk weight. The increase in egg weight by feeding high ALA diets is in accordance with the previous studies conducted by Aydin and Dogan [11], which reported that adding purslane dried at a level of 10 or 20 g/kg into the diet significantly ($p<0.05$) increased egg weight compared to the control. In addition, a study conducted by Ayersa and Coates [8] reported that brown hens fed diets enriched with chia, a plant source of omega-3, up to a level of 280 g/kg (28%), produced significantly ($p<0.05$) heavier eggs throughout the duration of experimental period (up to day 90) than those fed the control diet. In this case, brown hens were fed chia at a level of 70 g/kg (7%) producing egg yolks significantly heavier than those fed the control diet. However, for white hens, this study did not find any significant differences in egg weight (except for day 58 producing significantly lighter eggs) and yolk weight between white hens fed diets supplemented with chia than those given the control diet up to day 90. This suggested that there was strain effect on egg quality by inclusion plants rich in omega-3 fat. These results also suggested that there might be an interaction between the age and the diets and the differences in maturity of laying hen reproductive system may have led to different response to the dietary alpha-linolenic acid on egg weight and yolk weight.

Diet containing purslane increased yolk colour intensity of eggs. The highest intensity of yolk colour (10.53) was achieved when laying hens were fed diets containing the highest level of purslane meal. Yolk colour depends primarily on xanthophylls and carotene content of diet [13]. Purslane is a plant containing many nutrients, β-catotene and xanthophylls [14,6]. The yolk color intensity of eggs from laying hens fed purslane enriched diet probably resulted from the fact that purslane consumed by these hens were a source of xanthophylls and β-carotene [14,13] for egg yolks. Therefore, increasing the levels of purslane meal in the laying hen diets increased the amount of xanthophylls and vitamin A in the yolks, and this resulted in the increase in yolk color intensity.

4. Conclusions

It was concluded that diets enriched with *Portulaca oleracea (purslane meal)* increased egg weight, yolk weight, albumen weight and yolk colour. The dietary inclusion of *Portulaca oleracea* meal up to 8% can be applied without affecting physical quality of the eggs.

References

[1] García-Rebollar P, Cachaldora P, Alvarez C, De Blas C and Méndez J 2008 *Anim. Feed Sci. Tech.* **140** 337-48
[2] Cachaldora P, García-Rebollar P, Alvarez C, De Blas JC and Méndez J 2008 *Anim. Feed Sci. Tech.* **141** 104 -14
[3] Lawlor JB, Gaudette N, Dickson T and House JD 2010 *Anim. Feed Sci. Tech.* **156** 97-103
[4] Kjos NP, Herstad O, Skrede A and Øverland M 2001 *Can. J. Anim. Sci.* **81** 245-51

[5] Kartikasari LR, Hughes RJ, Geier MS, Bastian SEP, Makrides M and Gibson RA 2014 *Proc. of the 16th Asian-Australasian Assoc. of Anim. Prod. Societies Congress* (Yogyakarta: Gadjah Mada University Press)
[6] Uddin MK, Juraimi AS, Hossain MS, Nahar MAU, Ali ME and Rahman MM 2014 *Scientic World J.* 1-6
[7] Hayat Z, Cherian G, Pasha TN, Khattak FM and Jabbar MA 2009 *J. Appl. Poult. Res.* **18** 541-51
[8] Ayersa R and Coates W 2002 *Br. Poult. Sci.* **43** 283-90
[9] Yassein SA, El-Mallah GM, Ahmed SM, El-Ghamry AA, Abdel-Fattah MM and El-Hariry DM 2015 *Int. J. Res. Stud. Biosci.* **3** 27-34
[10] Cherian G and Quezada N 2016 *J. Anim. Sci. Biotechnol.* **7** 1-8
[11] Aydin R and Dogan I 2010 *J. Sci. Food Agric.* **90** 1759-63
[12] Evaris E, Sarmiento-Franco LA, Segura-Correal JC and Capetillo CM 2015 *Trop. Subtrop. Agroec.* **18** 33-38
[13] Krawczyk J, Sokołowicz Z and Szymczyk B 2011 *Arch. Geflügelk* **75** S151-57
[14] Raju M, Varakumar S. Lakshminarayana R, Krishnakantha TP and Baskaran V 2007 *Food Chem.* **101** 1598-05

Preliminary analysis on hybrid vigor in Indonesian indigenous and crossbred cattle population using data from published studies

S Prastowo[1], TSM Widi[2], N Widyas[1*)],
[1] Animal Science Department, Sebelas Maret University, Surakarta Indonesia
[2] Department of Animal Production, Animal Science Faculty, Gadjah Mada University, Yogyakarta Indonesia

[*)] E-mail : nwidyas@gmail.com

Abstract. Hybrid vigor or heterosis is the phenomenon where a crossbreed progeny has better performance compared to its parents. Heterosis can be quantified relative to the mid-parents value or relative to one of its parent's population average by crossing two breeds. Crossbreeding is aimed to increase the production performance of local breeds. According to the Indonesian government policy, crossbreeding program is one of main strategies to achieve meat self-sufficiency. We explore the possibilities observing of heterosis exhibited by crossing Bali and Peranakan Onggole (PO) cattle as local breed with the exotic breed based on the published data. In this paper, growth and reproductive traits from Bali and PO from year 2000-2010 were used for analysis. Moreover, Limousine and Simmental exotic breed data were collected from official information of artificial insemination (AI) centre. Data in growth trait (chest girth, mature weight, weaning weight and yearling weight) in all breeds and their crosses were then analysed using standard heterosis estimation method. Result, shows that crossbred offspring perform better in the growth trait in relative to Bali and PO as local breed. Specifically in Bali crossed with PO, the offspring shown better estimated heterosis effect in yearling weight compared to both parents. Despite heterosis were observed in some traits, careful planning of crossbreeding program is a must in order to avoid the loss of genetic variance.

1. Introduction

Indonesian government has aimed for meat self-sufficiency in 2026 and one of the strategies is by implementing crossbreeding program [1]. Local or indigenous breeds have lower production performance compared to the exotic ones; however, they excel in the fitness and robustness against local environment and resources [2,3]. Although exotic breeds might originally have higher productive capacity, without the ability to adapt, the desired performance will be far from expectation. Studies had been made to learn about crossbreeding of local cattle (Bali and PO) with exotic *Taurus* and *Indicus* breeds. They reported that crossbred cattle indeed perform better in production traits such as growth traits [4–7] and vital statistics [4,7,8] but they have poor reproductive performance [4,9]. Crossbreeding is expected to be a solution to bring out the advantages of different breeds as well as to overcome within breed weakness.

The phenomenon where a crossbred progeny has better characteristics compared to its inbred parents is called the hybrid vigor or heterosis [10]. This concept has been extensively used to increase production performance in livestock and crops. Although studies revealing the advantages of crossbred in Indonesia are abundant and the concept of heterosis is widely known, very less are concerned about formally exploring them in term of traits heterosis. Heterosis can be quantified relative to the mid-parents value or relative to one of its parent's population average. In a country like Indonesia, where the importance of crossbreeding is to increase the production performance of local breeds, the former type of heterosis is more relevant. Presenting the difference of characteristics among breeds in heterosis term is more feasible for further and advanced analyses from genetics and economic perspectives. This paper explores existing studies on the characteristic of both indigenous and crossbred cattle in Indonesia and presents them in term of heterosis relative to the indigenous parents.

2. Study methods

Mean and standard deviation data are collected from growth and reproduction related traits from studies on Bali and PO cows and their crosses during the year of 2000 – 2010. Presented in table 1 are girth of chest (CG) and mature weight (MW) on PO cows. All crosses were from female PO crossed with *Bos taurus* males using artificial insemination (AI). Semen for AI was produced in Government AI facilities. Information of *Taurus* semen in this paper was obtained from Singosari AI centre in East Java, CG and MW are obtained from official catalogue. The mature weight (MW) variable was estimated based on the Chest girth, thus was named estimated mature weight (EMW). Least square means corrected for the origin of the data effect are presented in table 1 and 2.

Table 1. Phenotype data on PO-cross

Breed	Pedigree	CG (cm)	EMW (Kg)
Limousine	Sire	223.25	912.00
PO	Dam	161.41	329.50
LimPO[1]	Offsp	169.25	402.00
Simmental	Sire	217.65	969.00
PO	Dam	161.41	329.50
SimPO[2]	Offsp	172.94	413.25

[1]Limousine sire crossed with PO dam (via AI)
[2]Simmental sire crossed with PO dam (via AI)

Data from Bali cows on weaning weight (WW) and yearling weight (YW) are presented in Table 2. Data were tabulated from published papers. Bali cows' data are the averages of many observations in different areas around the island of Bali; whereas pure *Taurus* and *Indicus* data were obtained from Singosari AI centre catalogue.

Table 2. Phenotype data on Bali cows and the crosses

Breed	WW (Kg)	YW (Kg)
Bali	86.00	130.00
Brahman	223.50	372.00
Simmental	235.50	430.00
Limousine	225.00	394.50
PO	102.13	134.30
Simbal[1]	131.61	179.21
Limbal[2]	128.75	176.80
Brahbal[3]	115.90	157.60
PObal[4]	111.66	148.25

[1] Cross between Simmental sire and Bali dam
[2] Cross between Limousine sire and Bali dam
[3] Cross between Brahman sire and Bali dam
[4] Cross between PO sire and Bali dam

Considering the purpose of crossbreeding in Indonesia is to increase the production performance of indigenous cattle, we used standard heterosis (= economic/useful heterosis) estimation method as follows:

$$UH(\%) = \frac{\overline{F_1} - std}{std} \times 100 \quad (1)$$

where *UH* is the percentage of standard heterosis, F_1 is the performance of immediate offspring and *std* is the standard breed phenotype. In this study standard breed referred to PO and Bali.

Results and discussions

This paper focused on the crossing of indigenous PO and Bali cows with exotic sires, however, the available information between the crosses differ. Records on chest girth and estimated mature weight were available for PO crosses while for Bali crosses, only data from weaning weight and yearling weights were presented. First step taken in analyzing the data was performance comparison among breeds to confirm the presence of heterosis that is when the crossbred offspring perform better than the standard. In this case, the standard was the performance of local dam populations.

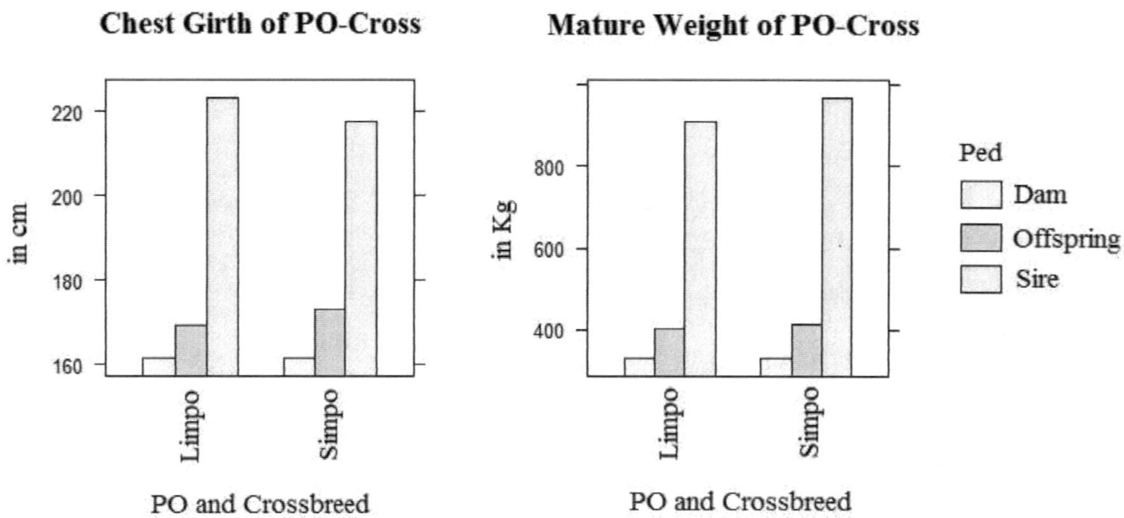

Figure 1. Performance of PO cows and the crosses

Figure 1 presented the performance difference among PO cows and the crosses. Left graph showed the mean chest girth for LimPO dan SimPO group in centimeter unit while the right graph showed the mature weight in kilograms. Each group consisted of three bars. The first bar is the standard or average performance of the dam breed (PO dam), middle bar is the F1 crossbred and the last bar is the performance of exotic sire.

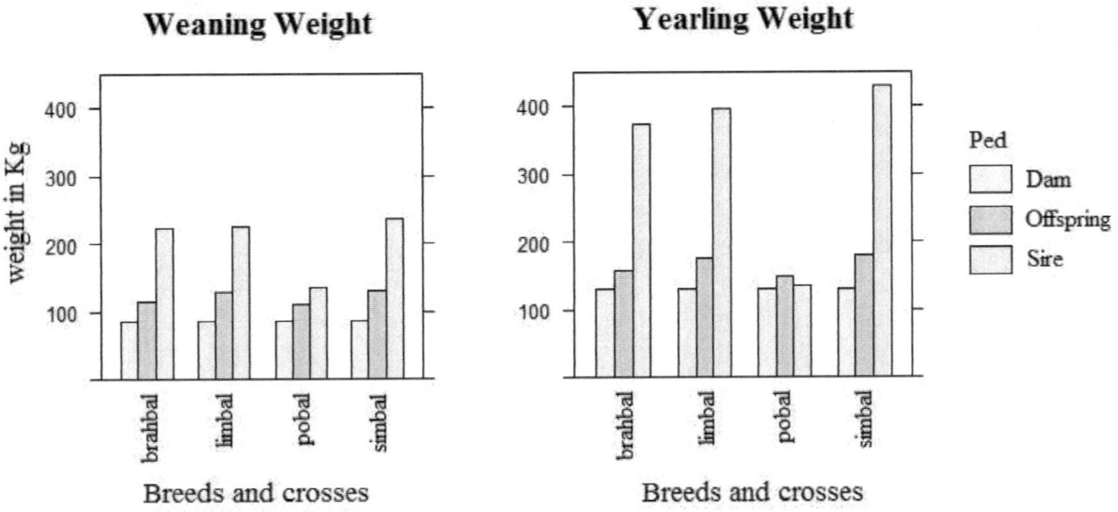

Figure 2. Performance of Bali cows and the crosses

The performance of Bali cows and the crosses are presented in Figure 2 (same explanations as Figure 1 apply). The presence of heterosis is confirmed by the fact that offspring's performance excelled its average dam population in all traits.

Table 3. Usefull (standard) heterosis for both crossbred

Standard	Breed	UH (%)	
PO		CG	MW
	LimPO	4.80	22.00
	SimPO	7.14	25.41
Bali		WW	YW
	SimBal	53.03	37.85
	LimBal	49.70	36.00
	BrahBal	34.76	21.23
	POBal	29.83	14.03

The amount of heterosis varied among breeds and traits. In PO cross, CG trait has lower heterosis compared to MW trait while in Bali cross WW has higher heterosis than YW trait. The degree of heterosis depends on the genetic constitution of the traits. Traits with lower heritability tend to have more benefit from heterosis. Weaning weight on cattle are reported to have lower heritability while yearling weight has moderate heritability [11,12]. On the other hand, the relatedness distance among breed also has an effect in heterosis. The further away parental breeds in the phylogenic tree, the heterosis will be higher [10].

Conclusions

Based on the estimated heterosis values analysis, it's concluded that *Taurus* and *Indicus* cross, perform better growth trait in relative to *Indicus*. However, the use of heterosis doesn't come without cost. Crossbreeding without careful design and program might destroy the genetic constitution in a population or worse, lead to germ plasm extinction. Management cost between indigenous and crossbred animals are also differed. Further analysis on the origin as well as the prediction of the economic value of the heterosis shall be made to understand its genetic background and economic efficiency as to weigh between the lost and benefit of undergoing a crossbreeding program.

References

[1] Menteri Pertanian. 2015. *Rencana Strategis 2015 - 2019.* (Jakarta: Kementrian Pertanian Indonesia)
[2] FAO. 2010. *Breeding strategies for sustainable management of animal genetic resources FAO Animal Production and Health Guidelines.* No. 3. (Rome)
[3] Diwyanto K, Iskandar S. 1999. *Livestock Industries of Indonesia Prior To The Asian Financial Crisis. In: Riethmuller P, editor. Livestock Industries of Indonesia Prior To The Asian Financial Crisis* (Bangkok: FAO Regional Office for Asia and the Pacific) pp. 1–10.
[4] Widi TSM. 2015. *Mapping the impact of crossbreeding in smallholder cattle systems in Indonesia.* (Wageningen: Wageningen University and Research Centre).
[5] Gunawan A, Jakaria. 2011. *Media Peternakan.* **34**: 93–98. doi:10.5398/medpet.2011.34.2.93
[6] Nasution D. 2010. *Agripet.* **10**: 37–41.
[7] Trifena, Budisatria IGS, Hartatik T. 2011. *Buletin Peternakan.* **33**: 11–16.
[8] Dominanto GP, Nasich M, Wahyuningsih S. 2016. *Res Zool.* **6**: 1–7. doi:10.5923/j.zoology.20160601.01
[9] Pribadi LW, Maylinda S, Nasich M, Suyadi S. 2015. *Livest Res Rural Dev.* **27**: 1–10.
[10] Falconer DS, Mackay TFC. 1996. *Introduction to Quantitative Genetics.* 4th ed. (Edinburgh: Prentice Hall)
[11] Cyrill JNSG, Porte J V, Mercadant MEZ, Bonilh SFM, Dia LT, Branc RH. 2014. Estimation Of Genetic Parameters For Body Weight From Birth To 10 Years Old In Nellore Females. *Proc 10th World Congr Genet Appl to Livest Prod.* **5**–7.
[12] Bolton K,. 2002. *Heritability and the beef herd.* (UW-Extension D& LA) p. 44.

IOP Conf. Series: Materials Science and Engineering **193** (2017) 012029 doi:10.1088/1757-899X/193/1/012029

Characteristic of Arenga Starch-Taro (*Colocasia esculanta* L.) Flour Noodle with Addition of Beetroot Extract

E R N Herawati[1]*, D Ariani.[1], Miftakhussolikhah[1], M P Elfanti[2], Y Pranoto[2]
[1]Research Unit for Natural Product Technology, Indonesian Institute of Sciences, Gading, Playen, Gunungkidul, Yogyakarta, Indonesia
[2]Faculty of Agricultural Technology, Gadjah Mada University, Jln. Sosio-Yustisia, Bulaksumur, Yogyakarta, Indonesia

E-mail : ervika.rnh@gmail.com

Abstract. Taro *(Colocasia esculanta L.)* flour can be mixed with arenga starch in noodle making with 25% taro flour and 75% arenga starch as the best proportion. Addition of beetroot extract as red natural color may improve consumer preference and change the noodle characteristic. This research purpose is to know the effect of addition of beetroot extract on the noodle characteristic. Coloring extraction was done using 5 variances of beetroot (0.4; 0.6; 0.8; 1; and 1.2 g (fresh weight/ml water). Then, noodle was made and characteristic including physical and sensory properties were evaluated. The result showed that coloring extracts addition decreased the break compression, elongation, and tensile strength of arenga starch-taro flour sohun. Beetroot extract addition increased the red color and didn't change sensory properties significantly.

1. Introduction

Indonesia has many biodiversity of tuber which contain high amount of carboydrate. One of local tuber that potential and commonly used is taro tuber. Taro tuber contain carbohydrate for 82.15 % [1]. Taro flour can be used for noodle making with addition of arenga starch as amylose source. Arenga starch has been commonly known as a potential raw material to be mixed with tuber in noodle product [2,3]. Arenga starch-taro noodle is an alternative carbohydrate source that can reduce dependency of wheat-based noodle comsumption in Indonesia. Thus, arenga starch-taro noodle can increase the economy value of Indonesian local tuber. From the previous research [1], the best composition of arenga starch-taro flour noodle was made with composition: 75% arenga starch and 25% taro flour. However, arenga starch-taro noodle has a different color with commercial product. The noodle color looks unfamiliar thus making consumers are less interested in trying. The addition of natural color which is safe may change the consumer preference. Addition natural color from plant-based also usually incorporated into noddles to serve as nutrition enhancers or to provide specific physiological functions [4]. One of natural color that usually used in food product is beetroot [5, 6]. The major coloring pigment in beetroot is betalains, whose main pigment is betanin-5-O-β-glucosidase, or betanin, which is the most common betacyanin [7]. Beetroot is commonly used as natural color in food product because it contain polyphenol and antioxidant activity [6,8,9]. The aim of this research is to study the characteristic of arenga starch-taro (Colocasia esculanta L.) flour noodle with addition of beetroot extract.

2. Materials and Methods

2.1. Materials

The materials used in the research were *Colocasia esculanta* L. (local tubers) from Gunungkidul, arenga starch from Klaten, beetroot from Yogyakarta, and commercial noodle products (corn noodle and rice noodle of superior brand that was produced by PT. Tiga Pilar Sejahtera, Solo) for product comparison. The tools used for noodle making were noodle making machine (Food Extruder PD-45N, La Pramigiana), cabinet dryer, tray, gas stove along with boiler, plastic sealer, electric stove, and scales.

2.2. Methods

Research methodologies consisted of (i) extraction of natural color from beetroot [10] with some modifications; (ii) noodle production; and (iii) characterization of physical and sensory properties of noodle. Beetroot was extracted into 5 variances with water as shown in table 1.

Table 1. Variance of beetroot extract

Code	Ratio of fresh beetroot weight : water
B1	0.40 g beetroot/ml water
B2	0.60 g beetroot /ml water
B3	0.80 g beetroot /ml water
B4	1.00 g beetroot/ml water
B5	1.20 g beetroot /ml water

Making of noodle in this research was based on the common practice in a large scale [11] with some modification to be performed in a laboratory scale. The first step was weighing a mixture of arenga starch and taro flour with ratio 75:25. Then, beetroot extract is added to flour mixture with ratio between flour and extract 1:0.5. The mixture was then molded into pellets with a length of 3-5 cm using the extruder with a diameter of 15 mm. Making pellets aims to expand the surface of the dough so as to facilitate the process of gelatinization when steamed. Pellets then steamed for ± 3 min until the surface became shiny. Steaming should not be too long because it is desired to gelatinization partially on the surface of the pellet. While pellets were still hot, they were entered immediately into the extruder to be molded into noodle threads. The mold used in this process had a diameter of 0.7 mm. The obtained raw noodle was then hung on a cart and steamed for ± 20 min until the color of noodle became transparent. The cooked transparent noodle was allowed to stand until it reached room temperature, then separated and dried in a dryer cabinet at temperature of 55 °C for 6 h until the moisture content was about 10 %.

Characterization of the noodle product include of physical and sensory analysis. Physical properties analysis of noodle consisted of color measurement [12], tensile strength [13], elongation [13], and compression test [14]. Sensory properties using hedonic test were determined with attributes: color, taste, odor, and overall liking. The design of the experiment used was a completely randomized design (CRD). Statistical data were analyzed using the software of SPSS version 16 with One Way Anova method with a significant level of 5% in comparison of means using Duncan method.

3. Results and Discussion

3.1 Physical Properties

Physical analysis of noodle i.e color, tensile strength, elongation, and compression test were done. Color is one of important parameters in noodle which influence consumer preferences. Color measurement of noodle product can be seen in table 2.

L-value indicates the lightness or brightness of the noodle. As shown in table 2, the addition of beetroot extract will reduce the brightness of the noodle. Beetroot extract contain red colour, so it decreased L-value of noodle product [5, 15]. The higher score of a value indicates the high intensity of red color in the noodle. The highest a value was B3. Addition beetroot extract increased a value because it contain betanins as red pigment and have strong antioxidant activity [6,8,9]. The score of A value were decreased in B4 and B5 because addition of high amount of beetroot will make blackish-red to the noodle product.

Table 2. Noodle color with addition of beetroot extract

Sample	L	a	B
B0 (control)	51.79^{c}	2.97^{b}	5.69^{bc}
B1	45.80^{b}	7.41^{a}	4.38^{a}
B2	46.50^{b}	8.65^{c}	6.13^{c}
B3	41.56^{a}	10.50^{e}	6.27^{c}
B4	39.92^{a}	9.47^{d}	5.34^{b}
B5	40.31^{a}	7.77^{b}	5.19^{b}

Note: The same superscript symbol in the same column indicates that samples are not significantly different at a significance level of 95%

The other important noodle physical parameters were compression test, elongation, and tensile strength. Compression test was analyse in dried noodle as described in figure 1, while tensile strength and elongation were tested in boiled noodle as shown in table 3. The same superscript symbol in the parameter value indicate that sample are not significantly different at a significance level of 95%

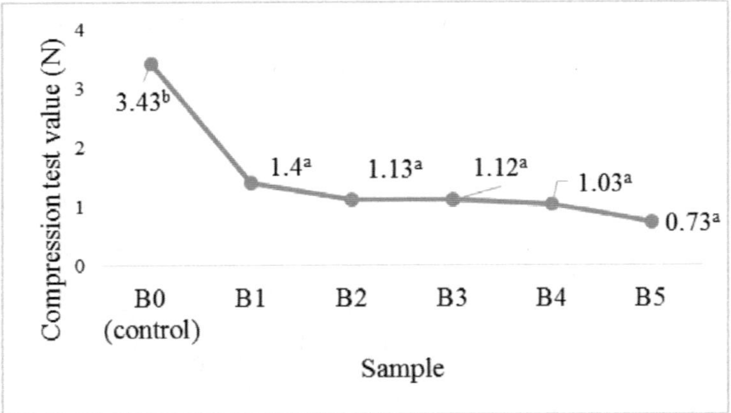

Figure 1. Compression test of dried noodle

Compression test illustrate the noodle power against mechanical (shock) during distribution / storage. The noodle is expected to have a high compression test value so that not easily broken. As shown in figure 1, compression test was decrease with addition of beetroot extract. Compression test was decreased significantly in B1, B2, B3, B4, B5 compared with B0 (control). The more addition of beetroot extract will decreased compression test because addition the particles from beetroot can interfere amylose bond in the noodle. Addition of some plant-based or animal-based tissues and extracts into noodles may negatively affect the sensory, eating, and cooking properties of the final products [16].

IOP Conf. Series: Materials Science and Engineering **193** (2017) 012029 doi:10.1088/1757-899X/193/1/012029

Table 3. Elongation and tensile strength of the noodle with addition of beetroot extract

Sample	Elongation (%)	Tensile strength (N)
B0 (control)	16.08	0.13
B1	n/a	n/a
B2	n/a	n/a
B3	n/a	n/a
B4	n/a	n/a
B5	n/a	n/a

As described in table 3, addition of beetroot extract decreased elongation and tensile strength of boiled noodle therefore it cannot be measure in B1, B2, B3, B4, and B5. Addition of beetroot extract will disrupt the bond between amylose in cellophane so that decrease the elongation and tensile strength of the boiled noodle. Starch amylose content was important parameter which affect textural properties of noodle [17]. Addition of ingredient / extract will make soft product compare to the control and textural properties were decreased [18, 19, 20, 21].

3.2. Sensory Properties

Sensory analysis of noodle with attributes i.e. color, taste, odor, and overall were done as shown in figure 2-5. Sensory analysis using hedonic scale 1-7, which 1 = very dislike; 4 = neutral; 7 = very like. The same superscript symbol in the noodle score indicate that sample are not significantly different at a significance level of 95%

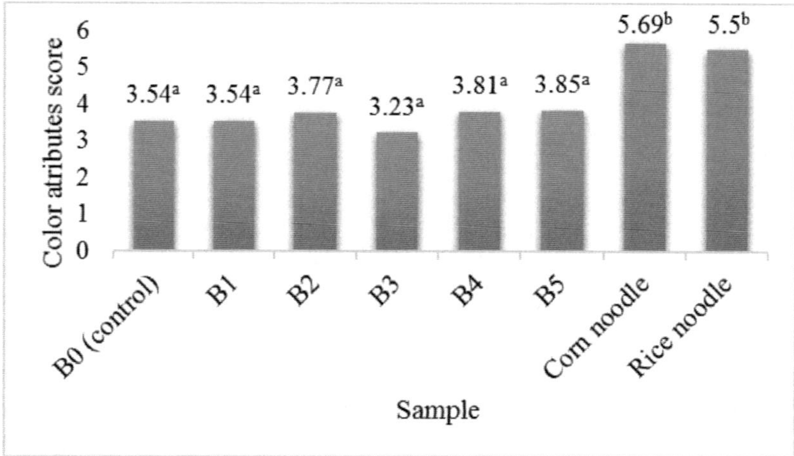

Figure 2. Color atributes value of noodle

For color atributes, the most preferred noodle with addition of beetroot extract was B5, but there not significantly difference between the color of noodle with beetroot extract addition. However, the noodle color with beetroot addition were significantly lower than comparison noodle (corn noodle / rice noodle).

IOP Conf. Series: Materials Science and Engineering **193** (2017) 012029 doi:10.1088/1757-899X/193/1/012029

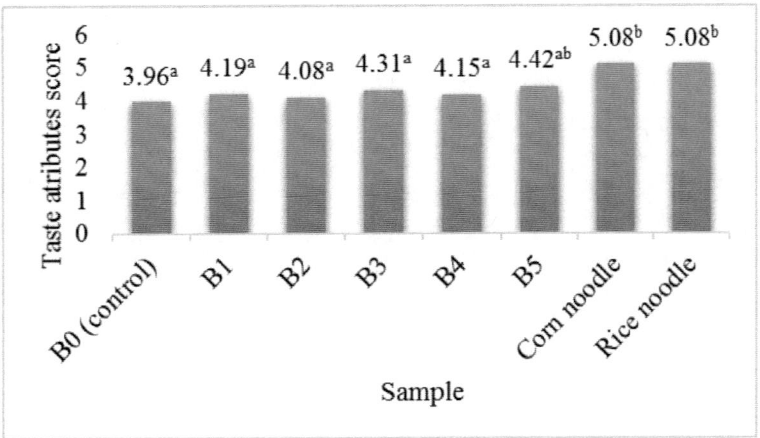

Figure 3. Taste atributes value of noodle

Addition of beetroot extract will increase taste atributes score compared with noodle control. Beetroot extract have sweet taste so can increase the taste of noodle. The most preferred noodle for taste atributes was B5 which highest addition of beetroot extract, and not significantly different with comparison noodle.

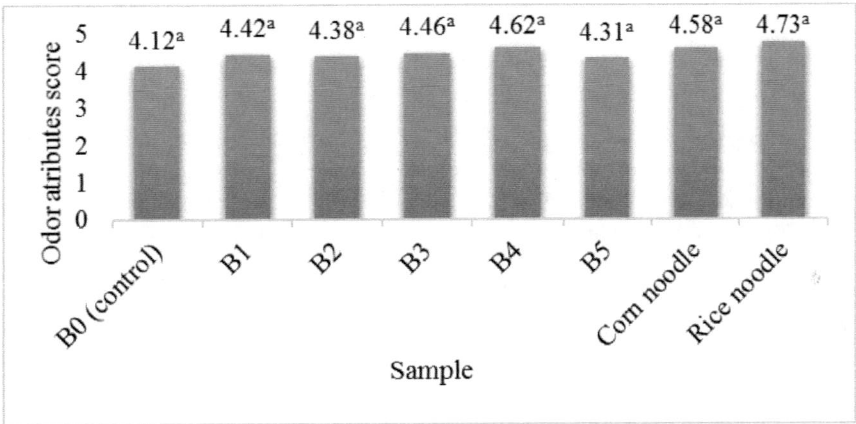

Figure 4. Odor atributes value of noodle

For odor atributes, the most preferred noodle was B4, but there was no significantly difference with the others. This shows that the addition of beetroot extract did not influence the odor of the noodle significantly although beetroot has earthy odor [22].

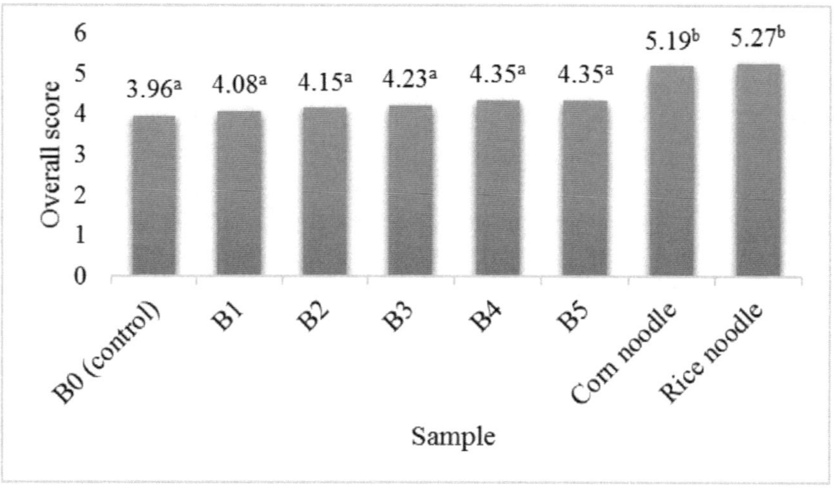

Figure 5. Overall value of noodle

As shown in figure 5, for overall sensory value, the most preferred sample was B4 and B5 but it not significantly different with noodle control and noodle with beetroot extract. This indicates that the addition of beetroot extract did not influence the overall sensory value of the noodle significantly. In wet sago noodles making, addition of beetroot extract also did not influence the overall sensory value [15].

4. Conclusion

Addition of beetroot extract gave an impact on physical properties of arenga starch-taro flour noodle, which increased the color and decreased compression test, elongation, and tensile strength. For sensory analysis, addition of beetroot extract increase the taste attributes scores of the noodle, but it not significantly different in overall attributes of the noodle product.

Acknowledgement

This project was supported by the Competitive Research Grant from Indonesian Institute of Sciences.

References

[1] Susanti Y 2013 *Characteristic of Noodle Made from Taro Flour and Arenga Starch* (Yogyakarta: Gadjah Mada University)
[2] Rahim A 2008 *J. Agroland* **15** 101-105
[3] Herawati E R N, Ariani D, Miftakhussolikhah, Angwar M and Pranoto Y 2015 *Proc. of Int. Conf. Food for a Quality Life* (Bogor : SEAFAST Center)
[4] Li M, Zhu K, Guo X, Brijs K and Zhou H 2014 *Comprehensive Reviews in Food Science and Food Safety* **13** 347-357
[5] Kim M J, Park J E, Park S H, Han J S, Choi J H and Lee H 2015 *J Korean Soc Food Sci Nutr.* **44** (2) 302-306
[6] John F, Li-Shu W, Claudio M, Bethany L, Cassandra H, Colleen M, Kenneth M, Steven J and Gary D 2010 *J. Med. Food* **13** (3) 733-739
[7] Azeredo H, Santos A, Souza A, Mendes K and Andrade M 2007 *Am. J. Food Tech.* **2** (4) 307-312
[8] Mridula D, Gupta R K, Bhadwal, Khaira H and Tyagi S K 2016 *J. Food Sci. and Tech.* **53** (4) 1834–1844
[9] Masum A, Laila K, Khwaja G and Furuta 2010 *Res J Agric Biol Sci.* **6** (2) 186-190
[10] Wisesa T and Widjanarko S 2012 *Jurnal Pangan dan Agroindustri* **2** 88-97
[11] Haryadi 2014 *Noodle Technology* (Yogyakarta: UGM Press)

[12] Aprianita, Todor V, Anna B and Stefan K 2013 *J. Food Sci. Tech.* **51** 3669-3679
[13] Chen, Z L, Sagis, L, Legger A, Linssen J P H, Schols H A and Voragen A G J 2002 *J. of Food Sci.* **67** 3342-3347
[14] Suryani C L 1999 *Whitening and Cross Linking Starch Sagu for Rice Subtitution in Bihun (Thesis)* (Yogyakarta : Gadjah Mada University)
[15] Oktaviani F 2013 *The Effect of Beetroot Addition as Antioxidant and Potassium Source in Wet Sago Noodle : Physico-chemical and Sensory Characteristic (Thesis)* (Semarang : UNIKA Soegijapranata)
[16] Dube M, Schafer C, Neidhart S, Carle R 2007 *Eur. Food Res. Technol.* **225** 287–99
[17] Zhang S B, Lu Q Y, Yang H and Meng D D 2011 *Cereal Chem. J.* **88** 296-301
[18] Song X, Zhu W, Pei Y, Ai Z and Chen J 2013 *J. of Cereal Sci.* **58** (3) 400-407
[19] Nouri L, Nafchi A M and Karim A A 2015 *Int. J. Food Eng.* **11** 221–227
[20] Heo S, Jeon S and Lee S 2014 *LWT-Food Sci. Tech.* **55** (2) 627-631
[21] Chin C K, Huda N and Yang T A 2012 *Int. Food Res. J.* **19** (2) 701-707
[22] Tomezak D W and Zielinska 2006 *PJFNS* **15** (4) 437-444

IOP Conf. Series: Materials Science and Engineering **193** (2017) 012030 doi:10.1088/1757-899X/193/1/012030

The Effect of Biduri (*Calotropis gigantean*) Latex on Meat Quality of Post Laying Hen

A M P Nuhriawangsa[1*], W Swastike[1], B S Hertanto[1], A W[2] and E D Pradisha[2]
[1]Laboratory of Industry and Animal Product Processing, Department of Animal Science, Faculty of Agriculture, University of Sebelas Maret, Jl. Ir. Sutami 36A Kentingan, Jebres, Surakarta, Indonesia +6257126. Phone/Fax. +62271-637457

E-mail: adimagna@staff.uns.ac.id

Abstract. The aim of this research was to investigate the effect of biduri (*Calotropis gigantean*) latex treatment on physical quality and hedonic test of meat of post laying hen. Samples of this research used meat of post laying hen strain Lohman aged 90 weeks. Thigh muscle was used for physical quality and breast muscle for hedonic test. Extract of biduri latex were obtained by cutting and tapping of young stem tissue, and it was centrifuged to obtain crude fluid of biduri latex (supernatant). Variables of this research were pH, cooking loss, tenderness, and water holding capacity (WHC), and hedonic test. Levels of biduri latex were 0, 3, 6 and 9% were measured from weight of the samples and the treatment of biduri latex by smearing on the surface of meat samples. The experimental design of the research used one way randomized design. The data was analyzed by using ANOVA, and differences between treatment means were further analysed using Duncan's New Multiple Range Test. The effect of biduri latex did not affect on pH, but it significantly affected on cooking loss, WHC, tenderness, and hedonic test ($p<0.01$). Cooking loss started to decrease in concentration of 6%. Value of WHC and tenderness went up at concentration of 3%. Increasing of meat quality of tenderness of hedonic test, juicy and overall hedonic test started to increase in concentration of 6%, and also flavor on concentration of 3% and 9%. In conclusion, the treatment of biduri latex with concentration of 3% could increase meat quality of meat of post laying hen.

1. Introduction

Post laying hen has a potential to meet the needs of meat on society, however when its getting older, the meat become tough. So that the social needs of this hen is less in demand. The meat have good protein content and the fatty is low.

Some study prove that tough meat can be tenderize by adding fruit protease enzyme. Tenderness meat of post laying duck [1] and post laying hen [2] can be increase by using papain from papaya fruit. Bromelin from pineapple can be increase to tenderize beef [3,4] and post laying duck meat [5]. The problem is protease source to tenderize meat taken from fuits, while fruits are human food, so there is need to find other source of protease.

One source of protease can be taken from biduri (*Calotropis gigantean*). Moreover there are abundant population of biduri which are optimally untapped [6]. Biduri protease classified as exopeptidase (cut the chain of peptida from outerside) and can hydrolysis fish myofibril and meat myofibril by using hydrolysis with high specific value [7]. Biduri latex also can hydrolysis casein, fibrinogen, fibrin [8] and systein [9]. Biduri's protease enzym activity works at 0,945mg/50mg/hour, better than rennet 0,860mg/50mg/hour [10].

Raw extract biduri latex can be used to raise the quality of meat. The purpose of this research was to understand the effect of raw extract biduri latex and to increase either physical quality and organoleptic quality of meat.

2. Materials and methods

2.1. Materials and samples preparation

Subject on this enzym activity test was latex taken from biduri (*Calotropis gigantean*). Chicken meat taken from post laying hen strain Lohman aged 90 weeks old. The thighs meat used to physical and the breast meat used to hedonic test.Biduri latex obtained from tapping young stem tissue [10] was centrifuged to obtain crude fluid of biduri latex (supernatant) [11]. The thigh and breast meat has been cut with size 5x2x1cm and divided into groups of treatments. Sample were smeary using biduri latex and punctured using needle then incubate for about 30 minutes inside sterile laminar air flow (LAF) in room temperature [12].

2.2. Methods

pH measurement used pH meter [13], cooking loss indicated by how much percentation of meat water loss, tenderness indicated by how strong the meat used warner blatzer method and WHC used Hamm method [12].

Hedonic test used nine point hedonic scale which use numeric data. It consist of value according to [14] in modification description definition score. Score 9 for very much like, 8 for very like, 7 like, 6 for rather like, 5 for neutral, 4 for rather dislike, 3 for dislike, 2 for very dislike, dan 1 for very much dislike, then tabulated and count the average.

2.3. Experimental designs

The design of experiment used One Way Randomized Design [15] with addition of biduri latex in some concentration (0, 3, 6, 9%) from sample weight of meat.

Physical and hedonic test data analyzed using ANOVA. If the difference is significant then use Duncan's New Multiple Range Test [15].

3. Results and Discussion

Physical quality test result of post laying hen with treatment of crude extract latex biduri can be seen in Table 1.

Result shown that there increased of the value of cooking loss and WHC, however there were decreased in tenderness value and pH value did'nt had affected in addition crude extract biduri latex in post laying hen meat. Cooking loss value started to were increased in concentration of 6%, WHC in concentration of 3%. However there was decreased in tenderness value at 3% in addition of crude extract biduri latex.

Table 1. Average and Statistic Test Result of The Physical Quality of Post Laying Hen with Crude Extract Biduri Latex

Variable	Biduri latex addition percentase (%)				P
	0	3	6	9	
Cooking Loss (%)	36.53 ± 1.46^{a}	$35.15 \pm 0.94^{a,b}$	37.68 ± 0.68^{b}	40.59 ± 0.45^{c}	0.001
pH	5.78 ± 0.06	5.78 ± 0.05	5.71 ± 0.09	5.72 ± 0.07	0.576
WHC (%)	28.43 ± 0.68^{a}	38.00 ± 2.38^{b}	46.26 ± 3.95^{c}	42.38 ± 1.82^{c}	0.000
Tenderness (kg/cm^2)	4.24 ± 0.18^{a}	3.72 ± 0.15^{b}	3.64 ± 0.15^{b}	3.58 ± 0.13^{b}	0.010

Description: ^{a,b}Different superscript in the same row indicates significantly

There are similarity and difference beetwen predecessor study and this study. The addition of raw extract protease taken from *Calotropis procera* up to 0,5% will affect to decrease the value of tenderness, WHC and cooking yield, however increase the value of poultry meat pH [16]. Addition of 6×10^3 unit activity/100g from the total of protease meat from *Calotropis procera*, papaya and bromelin can decrease pH and tenderize chicken, beef, catfish meat [17]. The use of bromelin pineapple crude extract increase tenderness, cooking loss and WHC, however decrease in pH for beef meat [3].

Actin and myosin having proteolysis because of protease from *Calotropis procera*, it becomes protein with molecul weight fewer on chicken [16], beef, catfish and squib meat [18]. This also happen when pineapple protease is used [3]. Then protein myofibril hydrolized by protease, it cause meat fragmentation and the establishment of microstructure space, so therefore water inside meat can move and loss while being cooked. These reason induced in increasing cooking loss value, WHC and decreased tenderness value, based on this study.

Regardless of amino acid by bromelin proteolitic activity in protein meat, causing the decrease of pH [3]. This action did'nt happen in this study, it might caused by meat pH still in normal condition, based on [19] in about 5,1-7,2. Meat with normal pH which close to meat isoelectric pH, therefore ionic protein molecul in the meat is well-balanced. This statement in accordance with [20], pH in isoelectrical point consist of positive ion and negative ion in protein molecul when they are balance. This made meat pH no differently in this study.

Hedonic quality test result of post laying hen with crude extract biduri latex can be seen in Table 2.

Table 2. Average and Hedonic Statistic Test Result of Post Laying Hen with Crude Extract Biduri Latex

Variable	Biduri latex addition percentase (%)				P
	0	3	6	9	
Tenderness	6.12 ± 0.91^a	6.44 ± 1.33^a	5.28 ± 1.56^b	4.60 ± 1.72^b	0.000
Flavor	4.32 ± 1.32^a	5.36 ± 1.35^b	6.20 ± 1.32^c	4.48 ± 1.55^a	0.000
Juicy	6.68 ± 1.05^a	5.00 ± 1.77^b	5.04 ± 1.78^b	4.56 ± 1.65^b	0.000
Overall Taste	6.20 ± 1.79^a	4.56 ± 1.42^b	4.48 ± 1.36^b	4.32 ± 1.76^b	0.000

Description: ^{a,b}Different superscript in the same row indicates significantly

Hedonic analytic result shown was decreased in tenderness, juicy and overall taste value, therefore increased in flavor when crude extract latex biduri added. Tenderness taste decreased in 6%, juicy and favorite decreased in 3%, while flavor increased in 3% then decreased in 9% of levels of biduri latex.

Predecessor study concerning favorite rate by hedonic in protease meat shown different result. Addition of biduri hay (*Calotropis procera* SW.) in feed decrease juicy, flavor and overall taste from *semimembranosus* (SM) sheep meat [21]. Addition of protease biduri latex concentration have no effect on aroma, taste and flavor of windu shrimp [22]. The addition of papain doses injection up to 3mg/kgBB can increase tenderness, decrease juicy and increase overall taste of post laying hen meat [2].

Degradation occur in endometrium layer by gap formation in meat bundle which circling muscle, degradation happen by the provision of protease from papaya latex and *C. procera* latex. It means there are protease activity in protein myofibril connective cell in meat [17]. Higher concentration of protease result in higher hidrolysis rate, therefore more chain loss of molecule in colagen protein [23].

Tenderness taste decreased at 6% levels of crude biduri latex. This case may happen caused by the provision of 6% biduri latex therefore when the meat is cooked, structural protein denaturated and the water went out eksudatively, it caused muscle shortening in protein structure so the meat seems toughing. Myofibrilar protein such as actin and myosin denaturated by heating, resulted in the meat tenderness quality [24]. Heating on meat can make the water dipped from the meat and myofibril degenerate in diameter and shrink, therefore affect the tenderness, WHC and juicy value [25].

Flavor in addition biduri latex increased at 3% and 6% otherwise did'nt in 9% levels of crude biduri latex. This happen because fatty acid aroma when cooked and accumulate in bigger microstructure gap and will induce flavor in response of degradating meat microstructure in crude biduri latex 3% and 6%. Addition of bromelin caused increasing pH and increasing cooking loss [5], therefore meat water increase and so the flavor. Based on [26], intramucular fat will melting when cooked and going to fiber muscle and stimulate salivary gland when eating, so that resulting in flavor, taste, juicy and meat aroma. At 9% there are decreased of flavor taste because fat comes out with meat water eksudatively, because microstructure meat gap cant be maintained, therefore flavor taste decrease in this study.

Juicy flavor and overall taste decreased in addition biduri latex at 3% otherwise did'nt at 6% and 9%. The panel preferred juicy taste than overall taste. Sensation taste decreased in addition biduri latex. Protease biduri latex hydrolysis protein from long chain protein to short chain protein or hydrolisat protein in fish [27]. Cooking process can make the water form water gap in meat microstructure and when it repeat the water exsudative from microstructure [12]. Cooking process can make the water content decrease, coagulate and denaturate of protein [28]. Structure change in meat caused by biduri protease and cooking process can make water inside meat eksudated and become dry, therefore juicy rate decreased. At 3%, 6%, and 9% addition of biduri latex did'nt make difference in juicy taste and favorite taste, this may caused by water eksudative rate inside meat resulted the same on this study. Based on [12] water eksudative meat happen in freely water and immobilization water inside meat. Moreover water eksudative meat formed when protein which binding water either enzymate or heating are degradated. Eksudation will maximally work when all water bound freely and immobilization is up.

4. Conclusions

The treatment of biduri latex with concentration of 3% could increase meat quality (cooking loss, WHC, tenderness of hedonic, juicy, flavor, overall taste) of meat of post laying hen.

References

[1] Nuhriawangsa A M P dan Pudjomartatmo 2002 *Kegunaan Enzim Papain dan Pemanggangan untuk Meningkatkan Kualitas Daging Itik Afkir* (Surakarta: Fakultas Pertanian Universitas Sebelas Maret)

[2] Widiastuti A, Pudjomartatmo dan Nuhriawangsa A M Patriadi 2012 *Sains Pet.* **10** 100

[3] Chaurasiya R S, Sakhare P Z, Bhaskar N and H Umesh H 2015 *J. Food Sci. Techno.* **52** 3870

[4] Ketnawa S and Rawdkuen S 2011 *Food Nutr. Sci.* **2** 393

[5] Utami D Putri, Pudjomartatmo, Nuhriawangsa A M, dan Patriadi 2011 *Sains Peternakan* **9** 82

[6] Witono Y 2009 Spesifitas dan stabilitas protease dari tanaman biduri (Calotropis gigantean) *Proseding Seminar Nasional Peran Ilmu dan Teknologi Pangan dalam Mewujudkan Ketahanan Pangan* (Padang: Fakultas Teknologi Pertanian Universitas Andalas)

[7] Saputri D S 2011 *Spesifitas Katalik Protease Biduri (Calotropis gigantea)* (Indonesia: Universitas Jember)

[8] Joshi H, Gururaja M P and Suares D 2011 *Intl. J. Pharm. Res.* **3** 975

[9] Rajesh R, Gowda C D R, Nataraju A, Dhananjaya B L, Kemparaju K and Vishwanath B S 2005 *Toxicon* **46** 84

[10] Nuhriawangsa A M P, Swastike W, Cahyadi M dan Gunawan D 2013 Aplikasi ekstrak kasar tanaman biduri (Calotropis gigantea) sebagai pengganti rennet terhadap kualitas keju susu

sapi *Seminar Nasional Peternakan Berkelanjutan ke-5* (Bandung: Fakultas Peternakan Universitas Padjajaran)
[11] Wang D I C, Cooney C L, Demain A L, Dunnill P, Humphrey A E and Lilly MD 1979 *Fermentation and enzyme technology* (New York-Chichester-Brisbane-Toronto-Singapore: A Wiley-Interscience Pub. John Wiley & Sons)
[12] Soeparno 2005 *Ilmu dan teknologi daging* (Indonesia: Gadjah Mada University Press Yogyakarta)
[13] Bouton P E, Harris P V and Shorthose W R 1971 *J. Food Sci.* **36** 435
[14] Lawless H T, Popper R and Kroll B J 2010 *Food Qual. Prefer.* **21** 4
[15] Steel R G D dan James H Torrie 1995 *Prinsip dan Prosedur Statistik* (Indonesia: PT Gramedia Pustaka Utama Jakarta)
[16] Rawdkuen S, Jaimakreu M and Benjakul S 2013 *Food Chem.* **136** 909
[17] Rawdkuen S and Benjakul S 2012 *African J. Biotech.* **11** 14088
[18] Rawdkuen S, Pintathong P, Chaiwut P and Benjakul S 2012 *Food Bioprod. Process.* **89** 73
[19] Lawrie R A 2003. *Meat Science* (Oxford-NewYork-Seoul-Tokyo:Pergamon Press)
[20] Tein M Lin and Jae W Park 1998 *J. Food Sci.* **63** 215
[21] Costa R G, da Silva N V, de Azevedo P S, de Medeiros A N, de Carvalho F F R, Queiroga R C R E and de Medeiros G R 2011 *R. Bras. Zootec.* **40** 1266
[22] Hardi J dan Diharnaini 2014 *J. Nat. Sci.* **3** 39
[23] Doneva M, Miteva D, Dyankova S, Nacheva I, Metodieva P and Dimov K 2015 *Biotech. Anim. Husbandry* **31** 407
[24] Martens H, Stabursvik E and Martens M 1982 *J. Texture Stud.* **13** 291
[25] Lawrie R A and Ledward D A 2006. *Meat and Human Nutrition. In: Lawrie's Meat Science* (Cambridge: Woodhead Publishing, Limited)
[26] Schlesinger C and Willoughby J 2010 *How to Cook Meat* (New York: HarperCollins Publishers Inc)
[27] Witono Y, Taruna I, Widrati W Siti dan Ratna A 2014 *JITP.* **25** 140
[28] Toldrá F 2010 Thermal Processing *Handbook of Meat Processing* (Iowa: Blackwell Publishing)

IOP Conf. Series: Materials Science and Engineering **193** (2017) 012031 doi:10.1088/1757-899X/193/1/012031

Sensory Attributes and Preliminary Characterization of Milk Chocolate Bar Enriched with Cinnamon Essential Oil

A Ilmi, D Praseptiangga, D R A Muhammad

Department of Food Science and Technology, Sebelas Maret University (UNS). Jl. Ir. Sutami 36A Surakarta 57126, Central Java, Indonesia

E-mail:dpraseptiangga@yahoo.com

Abstract. Cocoa (*Theobroma cacao*) is one of Indonesia's main commodities with annually increasing production. Chocolates are semi-solid suspensions of fine solid particles in a continuous fat phase. Primary chocolate categories are dark, milk, and white that differs in content of cocoa solid, milk fat, and cocoa butter. Milk chocolate bar is one of the most popular processed cocoa products in Indonesia. Widely cultivated in Indonesia, cinnamon is potential to be developed and is expected to add flavor and taste as well as enhance functional properties of milk chocolate, since it is well-known of its high antioxidant properties. The aim of this study was to determine the effect of cinnamon essential oil addition on the sensory attributes and physicochemical properties of milk chocolate bar. Three formulas of milk chocolate bar with an addition of cinnamon essential oil (0.1%, 0.3%, and 0.5%) were evaluated in this study. Panelists acceptance level decreased with increasing concentrations of cinnamon essential oil added, while moisture content and color analysis results did not show any significantly different for each formula, suggesting that milk chocolate bar with the addition of 0.1% of cinnamon essential oil had the highest level of acceptance and preferences for some of properties evaluated.

1. Introduction

As the third largest producer of cocoa (*Theobroma cacao* L.) in the world, the production of Indonesian cocoa continuously increases which were 740.513 tonnes, 777.539 tonnes, 817.322 tonnes in 2012, 2013, and 2014, respectively, based on the data of Directorate General of Estate Crops on 2014. Cocoa beans production increased annually, in 2012 recorded 53.300 ton cocoa beans was produced, in 2013 55.500 ton and in 2014 57.800 ton [1], suggesting that Indonesia has a great opportunity to be the centre of cocoa agroindustry in the world.

Milk chocolate is one of confectionary products that consist of cocoa mass, cocoa butter, sugar, and milk powder with or without the addition of other food ingredients and food additives such as lecithin [2]. Although the market of milk chocolate has been well established, some studies are still required to improve the quality and functionality of the chocolate. In the last decade, studies about processing and manufacturing chocolate products are linked to the biological active compound contained in cocoa, such as source of antioxidants, vitamins, minerals, and to the prevention of diseases [4]. In order to improve potential health benefit of chocolate, some studies have been conducted to enrich the chocolate by using functional material derived from other plant substrate. For an instance, the addition of dried fruits, including prunes, papaya, apricots, raisins, and cranberries have been reported to be able to improve sensory characteristics and bioactive compounds in dark and milk chocolates [5].

Widely cultivated in Indonesia, *Cinnamomum burmannii*, one of cinnamon species, has an excessive potency as an antioxidant and flavoring agent. They can be further processed by changing the crust into a powder, or by performing an extraction process to obtain cinnamon extract or cinnamon essential oil. Indonesia is one of the largest cinnamon producers in the world, in which approximately 89.500 tonnes per year [1]. The term of essential oils is understood to mean blends of secondary metabolites obtained from plant volatile fraction by distillation. Essential oil is widely used as a flavor agent because of its volatile compound [6]. Cinnamon essential oil is potential to improve flavor characteristic and to enhance functional properties of milk chocolate, due of its high antioxidant activity. The addition of cinnamon essential oil (5%) has been reported to increase antioxidant activity on salted egg [7]. Thus, this study aimed at determining the effect of cinnamon essential oil addition on the sensory attributes and physicochemical properties of milk chocolate bar.

2. Experimental

Cocoa liquor and cocoa butter derived from Forastero beans were used in this study. The beans were harvested from Kaliwining Jember (fermented for 2x48 hours). Full cream milk powder (Indomilk), sugar, lecithin were provided by the Indonesia Coffee and Cocoa Research Institute (ICCRI), Jember, Indonesia. Cinnamon essential oil (*Cinnamomum burmannii*) was obtained from CV. Anugrah Alam Abadi, West Sumatra.

The chocolate was produced by mixing of 496.5 g/500 g (consisting of 125 g/500 g of cocoa liquor, 137.5 g/500 g of cocoa butter, 112.5 g/500 g of milk powder and 121.5 g/500 g of sugar), and the additional ingredients of 3.5 g/500 g (consisting of 1.5 g/ 500 g lecithin, 1.5 g/500 g of soda, and 0.5 g/ 500 g of vanilla). Cinnamon essential oil at the concentration of 0.1%, 0.3% and 0.5% were chosen according to some previous experiment trials.

Cocoa liquor, cocoa fat, milk powder, sugar, soda, vanilla, were mixed first in a mixer, then 0.3% lecithin was added. The mixture was then refined in ball miller refiner to gain particle size up to 40-50 μm (50°C, 80 rpm, 20 hours). Conching was done to improve and enhance the flavor and viscosity of the end of the milk chocolate paste. Conching process was done using conche rotations at 60°C for 16 hours. The second refining step was conducted to mix 500 grams of milk chocolate with cinnamon essential oil in which F1 with 0.1% addition of cinnamon essential oil, F2 with 0.3% addition of cinnamon essential oil, and F3 with 0.5% addition of cinnamon essential oil. This step was conducted in 10 minutes by using ball miller with a rotary speed of 80 rpm at 50°C. Tempering process was done by melt down milk chocolate bar with a temperature less than 45°C. 2/3 part of the melted milk chocolate then stirred and sliding on marble table for 10-15 minutes until the temperature stabilized at 25°C. When the stable temperature was reached, 2/3 part of tempering results poured back into the bowl containing 1/3 part of untempered milk chocolate, then stirred for a few minutes until the temperature stable at ± 30°C, and then followed by moulding into chocolate plastic moulds. After 20 minutes, the moulded chocolates were de-moulding and were wrapped in the aluminium foil and stored at refrigerator temperature until used for further analysis.

Cinnamon milk chocolates were then evaluated using Hedonic methods [8]. The experimental chocolates were subjected to sensory evaluation by a group of panelists, comprising 30 random panelist members varying in age between 18-22 years. Five attributes (color, aroma, taste, appearance, overall acceptabillity) were evaluated. Three samples of chocolate with different codes were presented in a single serving plate with crackers and mineral water for rinsing between samples. Panelists were asked for not comparing the samples with other samples. The sensory properties were presented on a five point scoring scale 1) Dislike, 2) Dislike moderately, 3) Neither like nor dislike, 4) Like moderately, 5) Like very much).

Color measurement of milk chocolate bar was conducted using Konica Minolta Sensing Chromameter inc. Results obtained expressed in L^* (lightness, $L^* = 0$, black; $L^* = 100$, white), a^* ($a^* =$ 0-100, red; $a^* = 0 - (-80)$, green), b^* ($b^* =$ 0-70, yellow; $b^* = 0 - (-70)$, blue) [9]. While the value of °Hue represented the dominant wavelength that will determine whether the colors of red, green, or yellow

[10]. °Hue value calculated based on the value of a* and b* obtained by the formula: °Hue = $\tan^{-1}(b^*/a^*)$ [4].

Moisture content of chocolate was determined using Moisture Meter CA-200 Mitsubishi Chemical Analytech by Karl Fischer titration method. In the automatic system a pre-weighed amount of chocolate is dispersed in a mixture of formamide, chloroform, and methanol in the reaction vessel. This vessel is then sealed to prevent further moisture coming in from the surrounding air. The mixture was then continuously monitored by two platinum electrodes that are placed in it. If any free iodine is present, this will depolarise the cathode and so the current will stop flowing. The potential difference across the electrodes is therefore used to control the titration. Initially the Karl Fischer reagent (which includes pyridine) is added slowly via a peristaltic pump, which is able to record accurately the amount that passes through it. When the reaction is complete and no moisture remains, the current will stop flowing. The instrument can then calculate the percentage of water present, based on the sample weight and the amount of reagent used [11].

Completely Randomized Design (CRD) with one factor was used in this study. Data from the analysis were statistically analyzed using One-Way ANOVA. If there was a difference, Duncan's Multiple Range Test (DMRT) with a significance level of $\alpha = 5\%$ was conducted. From the sensory test results, the best formula will be obtained.

3. Results and Discussion

3.1. Sensory attributes

Sensory analysis is a process of identification, scientific measurement, analysis and interpretation of product attributes through the five human senses. Sensory analysis also involves a measurement that can be quantitative or qualitative. The sensorial quality of food products play an important role in the choice of food. Preference test or hedonic test aims to identify the level of preference and acceptance of a product [12]. The preference test result with a scoring method on milk chocolate bar with the addition of cinnamon essential oil shown in Table 1

Table 1. Level of panelists acceptance of milk chocolate with cinnamon essential oil addition

The addition of cinnamon essential oil	Attributes				
	Color	Aroma	Taste	Appearance	Overall
0.1% (F1)	$4.63^a \pm 0.56$	$4.13^b \pm 0.90$	$4.43^c \pm 0.77$	$4.60^b \pm 0.56$	$4.57^c \pm 0.57$
0.3% (F2)	$4.43^a \pm 0.82$	$3.77^{ab} \pm 1.00$	$3.87^b \pm 1.00$	$4.23^b \pm 0.93$	$4.00^b \pm 1.02$
0.5% (F3)	$4.27^a \pm 0.78$	$3.50^a \pm 0.97$	$3.03^a \pm 1.33$	$3.77^a \pm 1.07$	$3.20^a \pm 0.92$

[a]Within a row, means with the same superscript letters are not significantly different from one another ($\alpha > 0.05$)
ANOVA, analysis of variance (One-way)

3.1.1. Color attributes

Determining quality of food in general depends on several factors, including taste, color, texture and nutritional value [10]. Change in color on chocolate is due to reactions based on a class of chemicals found in cocoa called tannins (polyhydroxyphenols). These are made up of epicatechin molecules, which during the different fermenting, drying, and roasting stages may join together, oxidise or react with other chemicals within the cocoa. This increases the number of color-giving molecules and makes the cocoa much darker [11]. In general, milk chocolate bar has a lighter color than dark chocolate bar due to the presence of milk. Table 1 showed that that F1, F2, and F3 have the same letter notation. This means that the variation of the addition of cinnamon essential oil into milk chocolate was not significantly different to the colour of the three samples. Cinnamon essential oil has a light yellow color [13]. Milk chocolate bar with the addition of cinnamon essential oils up to 0.5% did not significantly affect the panelists' acceptance in the parameter of color.

3.1.2. Aroma attributes

The results showed that milk chocolate with 0.1% cinnamon essential oil is the most preferred product. Aroma of chocolate can be affected by several factors. Free amino acid is one of the essential components of the chocolate aroma precursors [11]. The characteristic smell of chocolate can also be produced by the reaction of amino acids such as leucine, threonine and glutamine with glucose, when heated until 100°C. Higher temperatures will produce a much more penetrating/pungent smell. Best chocolate aroma is usually produced from the seeds of high levels of free amino acids [14]. In addition, roasting cocoa beans also affects the aroma of chocolate. High temperature caused unpleasant aroma. The main components of cinnamon essential oil are cinnamaldehyde, eugenol, acetogenin and other aldehydes in small quantities. Cinnamon essential oil also contains methyl-n-amyl ketone [15]. The cinnamaldehyde has a spicy perception. Therefore, it can be expected that enriching 0.3% and 0.5% of cinnamon essential oils resulted in too much spicy taste, and then decreased consumer acceptance to the product.

3.1.3. Taste attributes

In chocolate, fermenting and roasting the cocoa beans has a significant impact on the formation of the chocolate flavor. Over fermented cocoa beans will cause a sour taste in chocolate. Similarly, if the roasting process is carried out at temperatures that are too high and too long will caused the chocolate has a bitter taste. Adding flavor agent is an alternative way to improve the taste attribute of chocolate. Due to its potency as flavor agent, cinnamon essential oils may be able to improve the taste of chocolate. Based on Table 1, three samples of milk chocolate with the addition of 0.1%, 0.3% and 0.5% of cinnamon essential oil have a different notation. Essential oil has a bitter taste, sometimes sharp, biting, gives the impression of warmth depending on the type of their components [16]. Therefore, panelists' acceptance will decrease with increasing concentration of cinnamon essential oil. The most preferred formula by panelists was milk chocolate with 0.1% addition of cinnamon essential oil.

3.1.4. Appearance attributes

Tempering is important for glossiness, a key quality attribute in chocolate. In under-tempered chocolates light scattering is caused by reductions in surface regularity. Over-tempering causes significant increase in product hardness, stickiness with reduced gloss and darkening of product surfaces. Under-tempering induced fat bloom in products with consequential quality defects on texture, color and surface gloss [17]. The results of sensory analysis showed a 0.1% addition of cinnamon essential oil was not significantly different with the addition of 0.3% cinnamon essential oil, while the addition of 0.5% cinnamon essential oil significantly different than the other formulas.

3.1.5. Overall attributes.

Consumers' assessment of a food product is not only influenced by one factor, but also a combination of several factors considered in their entirety. In this case, panelists rate the overall of milk chocolate bar in terms of color, aroma, taste, and appearance. Table 1 showed that the overall level of acceptance in the milk chocolate bar between samples with the addition of 0.1% of cinnamon essential oil, 0.3% of cinnamon essential oil and 0.5% of cinnamon essential oil were significantly different. Milk chocolate bar with 0.1% addition of cinnamon essential oil is the most preferred by panelists. Therefore, it can be concluded that in overall attributes, milk chocolate bar with the addition 0.1% cinnamon essential oil has the highest level of acceptance from panelists. The aroma and taste parameter are the main factor of the cinnamon enriched-chocolate acceptability.

3.2. Color measurement

According to Table 2, milk chocolate bar with addition 0.1% and 0.3% of cinnamon essential oil were in range of red color. While the milk chocolate bar with the addition of 0.5% cinnamon essential oil was in range of red-yellow color.The color of chocolate bar is in range of red and red-yellow color [4]. It was shown that the color of milk chocolate bar was getting brighter with increasing concentration of cinnamon essential oil. The color of cinnamon essential oil is light yellow [13]. It is explaining that with the addition of essential oil of cinnamon will cause the color of milk chocolate bar into red-yellow range.

Table 2. Color attribute of cinnamon essential oil-enriched milk chocolate bar

The addition of cinnamon essential oil	L	a^*	b^*	°Hue
0.1 % (F1)	35.10^a±0.14	9.68^b±0.11	11.45^a±0.14	49.78^a±0.02
0.3% (F2)	37.17^c±0.19	10.25^b±0.41	13.53^c±0.23	52.86^b±0.64
0.5% (F2)	36.15^b±0.12	7.36^a±0.03	12.22^b±0.00	58.92^c±0.12

To determine the color of milk chocolate bar range based °Hue value can be determined by Table 3

Table 3. Value of °Hue and color range[a]

°Hue	Color range
18°-54°	Red
54° -90°	Red-Yellow
90° -126°	Yellow
126° -162°	Yellow-Green
162° -198°	Green
198° -234°	Green-Blue
234° -270°	Blue
270° -306°	Blue-Magenta
306° -342°	Magenta
342° -18°	Magenta-Red

[a]Source: Etikawati [18]

3.3.Moisture Content

Table 4. Moisture content analysis on milk chocolate bar with addition cinnamon essential oils

The addition of cinnamon essential oil	Moisture content (%wb)
0.1 % (F1)	2.32^a±0.07
0.3% (F2)	2.33^a±0.35
0.5% (F3)	2.37^a±0.00

Water is an important component in food ingredients, since water can affect the appearance, texture and flavor of food. Water content can determine acceptability, freshness, and shelf life of a food stuff [10]. Based on Table 4, moisture content of milk chocolate was not significantly different. Chocolate generally has low moisture content. When moisture content of chocolate more than 2% caused the chocolate has lower shelf life, and the quality of the chocolate texture will also decrease. In addition, the high moisture content in chocolate will induce fat bloom development [11].

4. Conclusion

This study evaluated the consumer acceptance of milk chocolate with cinnamon essential oil addition. In terms of sensory attributes, milk chocolate bar with the addition of 0.1% of cinnamon essential oil has the highest level of panelists' acceptance. Panelists acceptance level decreased with increasing concentrations of cinnamon essential oil added, while moisture content and color analysis results did not show any significantly different for each formula. The results provided additional information in developing Indonesian chocolate signature products, particularly in development of milk chocolate with an addition of cinnamon essential oil.

5. Acknowledgements

This study was financially supported by the research grant of PNBP UNS No. 501/UN27.11/PN/2014 and the authors wish to express their gratitude for this research support. The authors also wish to thank Mrs. Dwi Puji Astutik and Dr. Sukrisno Widyotomo (Indonesian Coffee and Cocoa Research Institute (ICCRI)), Jember, Indonesia for their helps in the preparation and production of milk chocolate bar.

6. References

[1] Central Bureau of Statistic 2015 Monthly Production of the Indonesian Plantation 2012-2014
[2] Badan Standar Nasional 1996 *Cokelat Susu* SNI 01-4293-1996
[3] Taylor M 2011 RPT-FEATURE-Indonesia's Chocoholics Lead an Asian Quality Wave *Article*
[4] Misnawi 2011 *Pelita Perkebunan* **27**(3) 216-229
[5] Komes, Draženka, Ana B, Svjetlana Š, Aleksandra V, and Arijana B 2013 *Food Sci. Tech.* **53** 360-369
[6] Fitriana N, Rumayati, Nelvira S, Afghani J, Syaiful and Harliya 2014 *Jurnal Aplikasi Teknologi Pangan* **3(1)** 12-15
[7] Andriyanto A, Andriani M, and Widowati E 2013 *Jurnal Teknosains Pangan* **2(2)**13-20
[8] Kartika B, Hastuti B and Supartono W 1988 *Pedoman Uji Inderawi Bahan Pangan* (Yogyakarta: PAU Pangan dan Gizi UGM)
[9] Hadnađev, Tamara D, Ljubica P , Miroslav S, Milica M, Slađana M and Aleksandra M 2013 *Food and Feed Res.* **40(2)** 101-108
[10] Winarno F 1984 *Kimia Pangan dan Gizi* (Jakarta: Gramedia Pustaka Utama) 41-43
[11] Beckett S T 2008 *The Science of Chocolate* 2nd *Edition* (UK: RSC Publishing) 169-170
[12] Setyaningsih D, Anton A and Maya P 2010 *Analisis Sensori* (Bogor:IPB Press) 59
[13] Badan Standar Nasional 2006 *Minyak Kulit Kayu Manis* SNI 06-3734-2006
[14] Haryadi and Supriyanto 2012 *Teknologi Cokelat* (Yogyakarta: Gadjah Mada University Press) 144
[15] Fitriyeni I 2011 Study of Cinnamon Processing Industry Development in West Sumatra Sekolah Pascasarjana Institut Pertanian Bogor
[16] Inna M, Atmania N and Prismasari S 2010 *J. Dentistry Indonesia* **17(3)** 80-86
[17] Afoakwa E, Peterson A, Fowler M and Vieira J 2008 *J. Food Eng.* **89** 128–136
[18] Etikawati E C 2007 *Effect of Passing Treatment,* Na_2CO_3 *Concentration and Water Content Of Physical Quality on Corn Noodle Using Screw Extruder Cookers* (Thesis) (Bogor: Department of Food Science and Technology)

Evaluation of Cooking Quality of Chromium Fortified – Parboiled Rice

W A Yulianto[1], C L Suryani[1] and S Ulfah[1]
[1] Faculty of Agroindustry, University of Mercu Buana Yogyakarta, Jl. Wates Km 10, Argomulyo, Bantul, Yogyakarta 55753

E-mail: wisnuadi@mercubuana-yogya.ac.id

Abstract. Parboiled rice was developed to produce rice that has low glycemic index, especially for diabetics. Yet, parboiled rice is not enough because diabetics also lack of chromium. The purpose of this experiment is to investigate the influence of paddy soaking time and chromium fortification on cooking quality of chromium fortified - parboiled rice (Cr -PR). The paddy was soaked for 2.5, 3.0, and 3.5 hours with chromium fortification at 0, 1.08, 3.42, 5.40 and 7.47 mg/L. The paddy soaked in different time and chromium fortification level did not affect the alkali spreading value and cooking time of CR-PR. The elongation of Cr-PR was from 1.39 to 1.48. Water uptake ratio increased due to addition of chromium, but the more addition of chromium (> 1.08 mg/L) the less water uptake ratio increase. The rice solid loss declined and it was influenced by the length of soaking time. The longer the soaking time and the higher the chromium addition, the harder the rice texture generated.

1. Introduction

The prevalence and incidence of diabetes mellitus has increased drastically in the new industrial countries and developing countries, including Indonesia. The prevalence of diabetes for all age-group worldwide was estimated to be 2.8% in 2000 and 4.4% in 2030. The number of people with diabetes is projected to rise from 171 million in 2000 to 366 million in 2030. In that year, the number of diabetics in Indonesia is predicted to reach 21.3 million, an increase of 150 percent from 8.43 million people in 2000 [1]. Globally, at least one in 10 deaths among adults aged 35-64 years is attributable to diabetes, rising to a quarter of all deaths in some parts of the world [2]. International Diabetes Federation (IDF) states that two individuals develop diabetes every 10 seconds and two individuals die of diabetes related conditions every 10 second worldwide [3]. One strategy that can be applied to manage diabetes is to consume foods that raise blood sugar slowly, but can provide the satisfaction of satiety. The trick is to consume food products that have a low glycemic index (GI). The research showed that traditional parboiled rice has IG 46 and can significantly lower the blood sugar profile of type 2 diabetics compared with non-parboiled rice is consumed (GI 55) [4]. The low GI values can be caused by high levels of resistant starch.

Yet, parboiled rice is not enough because diabetics also lack of chromium [5]. The mineral deficiency cause high blood sugar levels. Signs of Cr deficiency had similar metabolic syndrome with high blood sugar, high triglycerides, low density lipoprotein, and hypertension. Supplemental Cr has been shown to improve all these signs in human subjects [6].

In producing parboiled rice, the first step is soaking process. This stage plays an important role in hydration and pre-gelatinization process, whereas gelatinization process itself will take place during steaming. With the fortification of chromium in the parboiled rice processing, it is expected to be a functional food for diabetics. Iron fortification during parboiling of paddy has reported by Prom-u-thai et al. [7-9]. The soaking stage and the addition of chromium can affect the quality of physical, chemical and cooking quality of parboiled rice. Therefore, the research needs to be done to find out the effect of soaking time and chromium concentration to cooking quality of Cr-PR.

2. Experimental

2.1. Production of Chromium Fortified- Parboiled Rice (Cr-PR)

Five kilograms of Ciherang variety rice was washed to remove its chaff and other impurities. After that, soaking was done in hot water with a temperature of 65 °C for 2.5; 3; and 3.5 hours with the addition of chromium 0; 1.08; 3.42; 5.40 and 7.47 mg / L, then it was drained and heated/steamed for 25 minutes, and cooled at 0°C for about 14 hours and later it was dried. Grain drying was repeated 2 times, first it was drained until the water content reached 15% - 16%, while the second drying was conducted until the moisture content was 13% - 14%. The drying was done by using a cabinet dryer at a temperature of 50°C and then it was continued to the hulling. The result of parboiled rice then was tested its cooking quality and preference level of Cr-PR.

2.2. Alkali Spreading Value

Alkali spreading value (ASV) was used for predicting gelatinization temperature (GT). The spreading degree of individual milled rice kernel in a weak alkali solution (1.7% KOH) at room temperature (32±2°C) was evaluated on a 7- point numerical scale [10]. Each test was conducted three times. Each time, 10 intact milled grains were placed on a Petri dish to which 15 ml of 1.7% KOH was added. The grains were carefully separated from each other and incubated at ambient temperature for 23 hours to allow spreading of the grains. Grains swollen to the extent of a cottony center. A cloudy collar was given an alkali spread value score 4. Grains that were unaffected were given ASV of 2, and grains that were dispersed and disappeared completely were given a score of 7. A low ASV corresponds to a high gelatinization temperature. Assessment results are evaluated with a scale of 2-7, which is 2 to high GT, 3-5 to intermediate GT and 6-7 to low GT [10].

2.3. Cooking Time

Rice samples (10 g) were individually taken and cooked around 97 - 99°C with 70 ml distilled water in a 100 ml beaker glass which placed in a boiling water bath. Sampling was done 20 minutes after cooking. Cooking time was determined by parallel plates method in which cooked rice was pressed between 2 glass plates every 1 minute until at least 90% no longer has opaque centre or till no white core was left [11].

2.4. Elongation Ratio

Rice was cooked in excess water. Two grams rice was cooked around 97 - 99°C with 20 ml distilled water in a 100 ml beaker placed in a boiling water bath [12]. Samples were removed at the cooking time to measure the length and width (before and after cooking) and calculated by equation:

$$\text{Elongation ratio} = \frac{\text{length of cooked rice (mm)}}{\text{length of raw rice (mm)}} \qquad (1)$$

2.5. Water Uptake Ratio

Rice was cooked in excess water. Two grams rice was cooked around 97 - 99°C with 20 ml distilled water in a 100 ml beaker placed in a boiling water bath [12]. Samples were removed at the cooking time to weigh and calculated by equation:

$$\text{Water uptake ratio} = \frac{\text{weight of cooked rice (g)}}{\text{weight of raw rice (g)}} \qquad (2)$$

2.6. Solid Loss and Texture
Rice was cooked in excess water. Two grams rice was cooked around 97 - 99°C with 20 ml distilled water in a 100 ml beaker placed in a boiling water bath [12]. Five-ml sample of the cooked rice water (previously well agitated) was taken at the cooking time and placed in a pre-weighed flask and air- dried in an oven at 105°C for 1 h. After that, the sample was fully dried at the same temperature until a constant weight. The amount of solids leached was calculated as kilogram of solids per kilogram of dry grain [12]. Measurement of texture of the cooked chromium fortified – parboiled rice was used Zwick (Type DO-FBO. 5TS).

2.7. Statistical Analysis
The research used completely randomized design (CRD) with 2 factorial patterns; hydration time (2.5, 3 and 3.5 hours) as the first factor and the addition of $CrCl_3$ (0; 1.08; 3.42; 5.40 and 7.47 mg/L) as the second factor. Analysis of variance (ANOVA) and test of significance were performed using SPSS (Statistical Analysis System Software) version-19 with confidence level of 95 % (significance level $P < 0.05$). The samples were randomized for all the analyses describe above.

3. Results and Discussion
3.1. Alkali Spreading Value
Different soaking time and chromium concentration do not affect the alkali spreading value (ASV) of the Cr-PR as exhibited in table 1. Cr-PR has alkali spreading value of 3 - 5. According to Bergmen et al. [13], the data indicated that the rice is at the medium / intermediate level of GT (Gelatinization Temperature) which means the Cr-PR can undergo gelatinization of starch at the temperature of 70 – 74°C. It is the temperature at which 90% of rice starch granules swell irreversibly in hot water with loss crystalline structure and birefringence [13]. Dipti et al. [14] reported ASV of six fine grain rice varieties (Superfast, Basmati 4488, Khaszar, Basmati PNR, Badshabhog, and BRRI dhan 28) range from 3.0 – 3.9. GT was indexed by alkali spreading test. ASV corresponds to the following GT range: low = < 70°C (ASV = 6-7), intermediate = 70 - 74°C (ASV = 3 – 5), and high = > 74°C (ASV = 2) [10]. A low ASV corresponds to high GT, conversely, a high ASV indicates a low GT.

Table 1. Alkali spreading value of Cr-PR during soaking.

Soaking Time (hours)	$CrCl_3$ (mg/L) Concentration				
	0	1.08	3.42	5.40	7.47
2.5	3	4	4	5	5
3.0	4	5	5	5	5
3.5	3	4	4	4	4

Note: Not significant

3.2. Cooking Time
Cooking time for the rice is until 90% of the starch in the grain no longer shows opaque center when it is pressed between two glass plates. Rice differs in optimum cooking time in excess water between 15 – 25 minutes without pre-soaking [15]. The treatment with various lengths of soaking time and concentrations of chromium do not affect cooking time of Cr-PR as shown in table 2. This result is in line with the value of ASV as shown in table 1. The present study indicated that ASV of Cr-PR cooked in excess water range from 23 to 25 minutes and on the average 23.80 ± 1.16 minutes. This length is faster than the cooking time of rice without parboiling process (28 minutes) because of pre-gelatinization process (hot soaking) during the manufacturing of parboiled rice.

Table 2. Cooking time (minutes) of Cr-PR during soaking.

Soaking time (hours)	$CrCl_3$ (mg/L) concentrations				
	0	1.08	3.42	5.40	7.47
2.5	24	25	24	23	24
3.0	24	24	24	24	24
3.5	24	24	23	23	23

Note: Not significant

3.3. Elongation Ratio

Different soaking time and chromium concentrations made a significant effect on Cr-PR elongation after cooking process. Results are shown in table 3. Elongation is one of the important factors for the cooking rice. If the elongation of the rice reaches the appropriate length, it means the better the rice appearance. Based on the data in table 3, it is noted that the elongation ratio of parboiled rice with fortification chromium was ranged from 1.39 to 1.48. It is likely that the longer the soaking time, the longer the elongation ratio of Cr-PR. On the other hand, Cr addition (1.08-7.47 mg/L) in the same cooking time did not affect the Cr-PR elongation ratio. Elongation of rice can be influenced by both the length (l) / breadth (b) ratio and the amylose contents. Grain elongation ratio of *Ofada* rice (*Oryza sativa* L.) ranged from 1.24 – 1.75 [17]. A positive correlation was also recorded by both amylose content and l/b ratio in relation to elongation of rice [16].

Table 3. Elongation ratio (mm/mm) of Cr-PR during soaking.

Soaking time (hour)	$CrCl_3$ (mg/L) Concentration				
	0	1.08	3.42	5.40	7.47
2.5	1.39^{a}	1.43^{bc}	1.44^{bcd}	1.41^{ab}	1.43^{bc}
3.0	1.42^{abc}	1.44^{bcd}	1.41^{ab}	1.41^{ab}	1.44^{bcd}
3.5	1.41^{ab}	1.48^{e}	1.45^{cde}	1.45^{cde}	1.48^{e}

Note: The number followed by different letter indicates significant different (P<0,05)

Table 4. Water uptake ratio (g/g) of Cr-PR during soaking

Soaking Time (hour)	$CrCl_3$ (mg/L) concentration				
	0	1.08	3.42	5.40	7.47
2.5	2.90^{g}	3.06^{i}	2.93^{h}	2.88^{g}	2.73^{cd}
3.0	2.81^{f}	2.87^{g}	2.77^{e}	2.77^{e}	2.75^{de}
3.5	$2,60^{a}$	$2,71^{c}$	$2,68^{b}$	$2,66^{b}$	$2,60^{a}$

Note: The number followed by different letter indicates significant different (P<0,05)

3.4. Water Uptake Ratio

Water uptake ratio is important parameter in cooking rice. If the bulk density is higher, then correspondingly water uptake ratio will also be high. This has been attributed to the compact structure of a rice variety. Water uptake ratio shows a positive significant influence on grain elongation. Water uptake ratio of *Ofada* rice varied between of 1.74 – 2.11 [17]. The results of Cr-PR water uptake ratio measurement are presented in table 4. Based on the results of statistical analysis, it is known that the soaking time and the chromium concentration significantly affected water uptake ratio. The longer the soaking time and the greater the concentration of chromium (> 1.08 mg/L), the less the water absorption in the parboiled rice would be. In this study, it was observed water uptake ratio to be highest (3.06 g/g) for soaking time of 2.5 minutes and chromium concentration of 1.08 mg/L. The reduction of water absorption capacity in Cr-PR had caused changes in the structure of starch granules as the reaction of heat (hot soaking and steaming) during the proses of parboiling and retrogradation during cooling. Therefore, it can produce a sturdy structure. Disorganized cellular structure can enhance the probabilities for high water absorption during cooking and can contribute to longer cooking time [16].

3.5. Solid Loss
Solid in cooking water may be correlated with amylose content and may be related to stickiness of cooked rice. The data in table 5 shows that different soaking time and chromium concentration influenced the solid loss of Cr PR. Loss of solids in cooking water or the amount of solid released into cooking water was related to the amount of water that is absorbed. Low solid contents in the cooking gruel can be attributed to fact that the surface area in contact with water is smaller, resulting in a lower l/b ratio [16]. Rice varieties with higher amylose content are more prone to leaching out into the cooking water as starch grains expand during cooking [18]. This statement is contrary to results of this study that shown in table 5. It can be meant that the less amylose, the more solids separated at the cooking time, and the higher amylose content the less solids were separated.

Table 5. Solid loss (g/100g) parboiled rice during soaking

Soaking Time (hour)	$CrCl_3$ (mg/L) Concentration				
	0	1.08	3.42	5.40	7.47
2.5	5.82^{k}	5.44^{ij}	5.46^{j}	5.36^{i}	4.88^{ef}
3.0	5.10^{h}	4.92^{fg}	4.76^{d}	4.62^{c}	4.36^{b}
3.5	5.12^{h}	4.98^{g}	4.94^{fg}	4.82^{de}	4.22^{a}

Note: The number followed by different letter indicates significant different (P<0,05)

3.6. Rice texture
Based on the results of statistical tests, it was obtained that the interaction between soaking time and the chromium concentration significantly affected the texture of CR-PR (table 6). The longer the soaking time, the harder the rice grain become. The hardness of the rice texture was caused by chromium fortified-parboiled rice amylose content which is in line with the analysis results of amylose content on a variety of soaking time and concentration of chromium (data not showed). The higher the amylose content of rice, the harder its texture was. The greater force (N) produced the harder texture of the rice. Such condition is in line with the higher level of amylose in the rice. There was a negative correlation between cooked rice firmness with moisture content of rice during cooking, that provides more evidence of the significant role moisture uptake plays in determining cooked rice firmness [19].

Table 6. Texture (Newton) of Cr-PR during soaking

Soaking time (hour)	$CrCl_3$ (mg/L) concentration				
	0	1.08	3.42	5.40	7.47
2.5	3.52^{b}	3.26^{a}	3.65^{c}	3.86^{d}	3.91^{e}
3.0	4.12^{f}	4.14^{f}	4.38^{h}	4.44^{i}	4.45^{i}
3.5	4.20^{g}	4.18^{g}	4.50^{j}	4.84^{k}	4.83^{k}

Note: The number followed by different letter indicates significant different (P<0.05)

4. Conclusion
From the research results of research and discussion, it can be concluded that treatment of soaking time and the chromium fortification did not affect the alkali spreading value and cooking time. The elongation of Cr-PR was from 1.39 to 1.48. The addition of chromium had increased water uptake ratio, but the greater concentration of chromium, the less water uptake ratio became. The longer soaking times, the

lower its solid loss would be. The longer the soaking time and the addition of chromium, the harder the rice texture generated.

Acknowledgements
The author would like to thank the Directorate General of Higher Education, Ministry of Education and Culture, Indonesia, which has provided research funding through the National Strategic Research Program.

References

[1] Wild S, Roglic G, Green A, Sicree R and King H 2004 *Diabetes Care* **27** 1047–53
[2] Roglic G and Unwin N 2005 *Diabetes Voice* **50** 33–4
[3] International Diabetes Federation 2007 *World diabetes media kit: every 10 seconds 1 person dies of diabetes* (Belgium: International Diabetes Federation Brussels)
[4] Larsen H N, Rasmusse O W, Rasmussen P H, Alstrup K K, Biswas S K, Tetens I, Thilsred S H and Hermansen K 2000 *EJCN* **54** 380–5
[5] Anderson R A 2008 *Proc. of Nutrition Society* **67** 48–53
[6] Cefalu W T and Hu F B 2004 *Diabetes Care* **27** 2742–51
[7] Prom-u-thai C, Fukai S, Godwin D I, Rerkasem B and Huang L 2008 *Food Chem.* **110** 390–8.
[8] Prom-u-thai C, Glahn R P, Cheng Z, Fukai S, Rerkasem B, and Huang L 2009 *Food Chem.* **112** 982–6
[9] Prom-u-thai C, Rerkasem B, Fukai S and Huang L 2010 *Food Chem.* **123** 628–34
[10] Khush G S, Paule C M and Delacruz M N 1979 *Proc. of the workshop on chemical aspect of rice quality* 21–31
[11] Juliano B O 1982 *An International survey of methods used for evaluation of cooking and eating qualities of milled rice* (Los Banos Laguna: Philippines IRRI)
[12] Sareepuan K, Siriamornpun S, Wiset L and Meeso N 2008 *WJAS* **4** 409–15
[13] Bergman C J, Bahattacharya K R and Ohtsubo K 2004 Rice end-use quality analysis *Rice: Chemistry and Technology* ed Champagne E T (Minnesota: America Association of Central Chemists Inc)
[14] Dipti S S, Hossain S T, Bari M N and Kabir K A 2002 *Pakistan J Nutr* **1** 188–90
[15] Raghavendra R S N and Juliano B O 1970 *J. Agric. Food Chem.* **18** 289–94
[16] Thomas R, Wan-Nadiah W A and Bhat R 2013 *IFRJ* **20** 1345–51
[17] Danbaba N, Anounye J C, Gana A S, Abo M E and Ukwungwu M N 2011 *IFRJ* **18** 629–34
[18] Juliano B O 1971 *Cereal Science Today* **16** 334–8
[19] Saleh M and Meullenet J F 2013 *IFRJ* **20** 1337–44

IOP Conf. Series: Materials Science and Engineering **193** (2017) 012033 doi:10.1088/1757-899X/193/1/012033

Utilization of Bagasse Fly Ash to Remove the Unpleasant Odor of Stevia Extract and Soy Milk

Ashadi[1], M Masykuri[1], Haryono[1]
[1]Department of Chemistry Education, Teacher Training and Education Faculty, Universitas Sebelas Maret, Surakarta - Indonesia.

E-mail: ashadi.uns.ac.id@staff.uns.ac.id

Abstract. Stevia is a safe natural sweetener that, but has a slightly unpleasant odor. Soy milk is undoubtedly high nutritional value, soy milk slightly unpleasant odor. Bagasse Fly Ash (BFA) is a sugar factory waste which is abundant, not widely used yet, and allowed to accumulate around the sugar factory. BFA can be activated with a solution of NaOH become adsorbent. Utilization of activated BFA to remove the odor of stevia extract and soy milk means the utilization of a waste to reduce other waste. Deodorizing done by batch system. Before being used as adsorbent, BFA characterized using SEM, XRD, FTIR, and AAS. Odor and color analysis conducted by organoleptic. The results shown activation increases the cavity, BFA containing SiO_2 and Al_2O_3, does not contain Pb, Cr, Cd. The results shown that the BFA can reduce odor of stevia from a scale of 4 to 2, the color becomes more clear, unpleasant odor of soy milk is also reduced.

1. Introduction

Stevia leaves contain stevioside and rebaudioside A, the level of sweetness up to 300 times that of sucrose, beneficial to health, because the calorie content is low, has an antioxidant, anti-fungal and non-carcinogenic. This plant comes from Paraguay that has spread to several countries in Asia, Eropa and Canada, cultivation of commercially in Karanganyar- Tawangmangu Central Java [1]. Stevia also developed in Cibodas, from the area of 1 hectare (population 40,000 plants), generated from 3.6 to 7.2 tons of dry stevia leaves. This amount is far higher than the productivity of other varieties of stevia which is only 3-6 tons per year at a price of dry stevia reached Rp 25,000 per kg [2]. Stevia extract from stevia leaf powder can be done by methanol, ethanol or water, 20 grams of powdered leaves stevia with 200 ml of methanol was obtained 0.2985 grams of stevia crystals [3]. The problem to consume stevia is the stevia extract has a slightly unpleasant smell.

Soy milk is a high nutritional beverage, rich of protein and vitamin. Similar with stevia, this milk has a rottenunpleasant smell. To overcome the problems there are two manner to reduce the odor, first use an adsorbent, second mask the odor by essence. For the first manner, we must choose an adsorbent that is cheap and abundant. Fly ash known as bagasse Bagasse Fly Ash (BFA) is a sugar mill waste that has not been widely used, is available in abundance, which can be activated into adsorbent can be selected. Abundane of bagasse in Indonesia is quite large. Increasing production of cane in 1999 totaled 2,270,623 tons, thus bagasse produced ranged from 340 593 tonnes to 711 614 tonnes [4]. One of Sugar Industry Kremboong can grind 2,700 tons of cane per day, every hour produces 2.8 tons of bagasse, in one year can reach 5103 tons. Indonesia has many sugar industries spread in Java, Sumatra and Sulawesi, so availability of bagasse is very much.

Regarding the effectiveness of the BFA as absorbents have been studied, including disappearances cadmium and nickel from waste water using bagasse fly ash, equilibrium adsorption is affected by the initial pH (pH_0) of the solution. The $pH_0 \approx 6.0$ is found to be the optimum for the individual removal of Cd(II) and Ni(II) ions by BFA. The pH of the system, however, increases during the initial sorption process for about 20 min and, thereafter, it remains constant. The equilibrium adsorption data were obtained at different initial concentrations (C_0 = 10–100 mg/l), 5 h contact time, 30 °C temperature, BFA dosage of 10 mg/l at pH_0 6. Desorption with various solvents showed that the hydrochloric acid is the best solvent; the maximum elution being about 65% for Cd(II) and about 42% for Ni(II). **[5]** Removal by adsorption of phenol from synthetic aqueous solutions by bagasse fly ash (BFA) was investigated at 303^0 K under dynamic conditions in a packed bed. The effects of sorbent bed length (Z = 40–90 cm), flow rate (Q = 0.01–0.04 dm^3/min), bed diameter (D = 2–4 cm), and initial concentration (C_0 = 50–500 mg/dm^3) on the sorption characteristics of phenol were investigated at an influent pH of 6.5. More than 99.5% of phenol was removed in the column operated at C_0 =100 mg/dm^3 of phenol. The column performance improved with increasing Z and decreasing Q. The sorptive capacity of BFA for phenol was found to be 9.93 mg/g [6] brown color of stevia extract contain tannin can be removed by activated carbon under optimum condition using 200°C of temperature and using 6 of pH operation. Concentraion of tannin measure by spectrophotometer using 725 nm of wavelength.

Study examines the adsorption of pyridine (Py) from aqueous solutions, using BFA as an adsorbent have been performed to evaluate the influence of various parameters, such as initial pH (pH_0), adsorbent dose (m), contact time (t), initial concentration (C_0), and temperature (T), on the removal of Py from the aqueous solutions, batch adsorption studies. The maximum removal of Py is determined to be 99% at lower concentrations (<50 mg dm^{-3}) and 95% at higher concentrations (600 mg dm^{-3}), using a BFA dosage of 25 kg m^{-3} at normal temperature. The desorption of Py from Py-loaded BFA with several solvents shows that only 68.70% and 51% of Py could be recovered, using ethyl alcohol and 0.1 N H_2SO_4, respectively [8].

The adsorption of 2-picoline from aqueous solutions onto bagasse fly ash (BFA) have been studied The influence of various parameters like initial pH (pH), adsorbent dose (m), contact time (t), initial concentration (C_0) and temperature (T) on the adsorption of 2-picoline from the aqueous solutions were studied using batch adsorption experiments. The maximum removal of 2-picoline is found to be 98% at lower concentrations (<50 mg dm^{-3}) and 49% at higher concentrations (600 mg dm^{-3}) using 5 kg m^{-3} of BFA dosage at normal temperature and natural pH_0 (~6.5) [9]. Base on the capacity of BFA to adsorb some chemical compound, BFA bill try to remove unpleasant odor in Stevia extract and soy milk.

2. Methodology

2.1. Activation and characterization:

Equipments required: Oven, Reflux apparatus, SEM, XRD, FTIR. Materials: Bagasse Fly Ash, NaOH. Procedure: (a) washing and drying the BFA, (b), do reflux BFA with a solution of NaOH 2.5 M at a temperature of 100 ° C for 3 hours (c) washing the BFA to neutral and drying, (d) characterization BFA with SEM, XRD and FTIR,

2.2. Removing unpleasant odor from Stevia extract:

Equipments required: Hotplate, filter paper whatman 42, shaker. Material: activated BFA, stevia leaf, aquadest. Procedure:(a) Extract 1g of stevia leaf by with 100 mL hot water of distilled water. (b) when cool add 1g of BFA, shake for 0.5 hour, (c) filter with whatman 42. Do organoleptic test against the filtat of stevia before and after added BFA

2.3. Removing unpleasant odor of Soy milk:

Equipment required: Hope plate, beaker glass, blender and shaker. Materials:soy bean, aquadest, activated BFA. Procedure: (a) soak 1kg of white Soybean Overnight with 500 mL of aquadest (b)

IOP Conf. Series: Materials Science and Engineering **193** (2017) 012033 doi:10.1088/1757-899X/193/1/012033

boiled for 1 hour, (c)lend until Smooth, (c) Filter with a clean cloth, to produce Soy Milk (d) take 100 ml of Soy Milk, (e) Add 1 g of BFA, filter with Whatman 42. (f) DO organoleptic test against the filtrate before and after added BFA.

3. Result and Discussion

Activation make expanding pore, shown by SEM data figure 1. Activation also reduce crystalinity of BFA.

Before activation After activation

Figure 1. SEM image of BFA before and after activation

Main ingredients BFA is SiO_2 and Al_2O_3, as shown in figure 2. There are peaks characteristic of SiO_2 on the value of 2θ = 26.77°, 20.95° and 50.46° corresponding to the standard JCPDS No.88-2302. Al_2O_3 peaks characteristic at 2θ = 34,74°, 42.58°, and 64.22° according to the standard (JCPDS No. 88-0107). The activation process caused crystallinity of BFA is reduced and more amorph, that mean increasing the capacity of absorption.

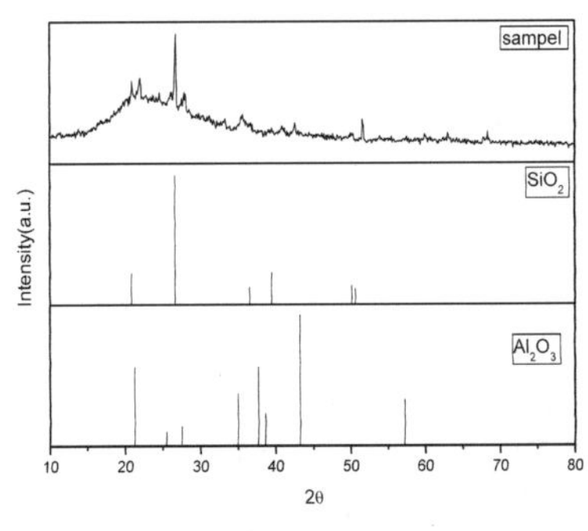

Figure 2. XRD data of BFA

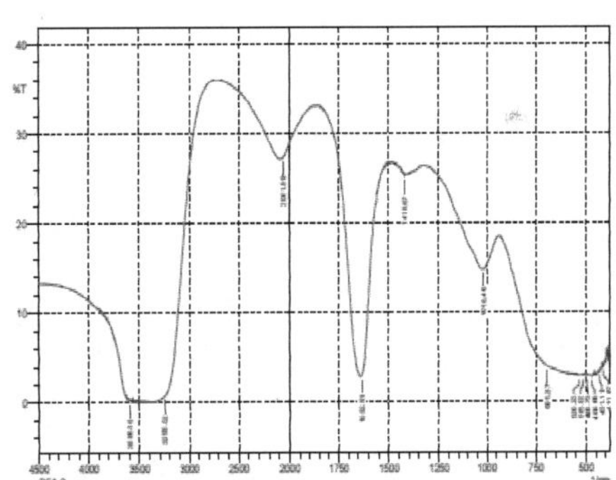

Figure 3. FTIR data

FTIR Data provides their information some functional groups, that may form a chemical bonds, beside to van der Waals bond as shown in figure 3. BFA reduce unpleasant odor of stevia from a scale of 4 to 2, the color becomes more clear, unpleasant odor of soy milk is also reduced

4. Conclusion

Bagasse Fly Ash (BFA) effective to eliminate odor and color of stevia extract. BFA reduce odor of Soybean Milk.

Recommendation
Although BFA effective to reduce the color and odor of stevia extract and soy milk, but not immedietly can be sugestion to use it, more need further study to ascertain whether the using of BFA as adsorbent meet the requirements of health drinks criteria.

References
[1] Djajadi 2014 *Perspectif* **13** 25-33
[2] Fajar Riyadi (2015, Juni 2) *Stevia, Si Manis Rendah Kalori* Berita Pertanian Indonesia
[3] Ratnani RD dan R Anggraeni 2005 *Momentum* **1 (2)** pp 27-32
[4] Subiyono (2015, November 14) *Ampas Tebu Diubah jadi Listrik: Energi Terbarukan* konfrontasi.com
[5] Srivastava VC, Prasad B, Mishra IM, Mall ID, and Swamy M. 2003 *Elsevier Water Res.* **37** (16) 4038-44
[6] Srivastava VC, Prasad B, Mishra IM, Mall ID, and Swamy M. 2008 *Eng. Chem. Res.* **47** (5) 1603–13
[7] Ibrahim G, Bayu W dan Abdullah 2013 *Jurnal Teknologi Kimia dan Industri* **2** (4) 198-204
[8] Lataye D, Indra H, Mishra M, and Mall ID 2006 *Ind. Eng. Chem. Res.* **45** (11) 3934–43
[9] Daware, GB, Babu PV, and Bhausaheb L 2014 *Desalination Water Treat* **52** 31-33

Dietary Protected Feed Supplement to Increase Milk Production and Quality of Dairy Cows

A Pramono[1*], E Handayanta[1], D T Widayati[2], P P Putro[3] and Kustono[2]
[1]Department of Animal Science, Faculty of Agriculture, Universitas Sebelas Maret, Surakarta
[2]Faculty of Animal Science, Universitas Gadjah Mada, Yogyakarta
[3]Faculty of Veterinary Medicine, Universitas Gadjah Mada, Yogyakarta

*E-mail: ahmad_pram@staff.uns.ac.id

Abstract. The efforts to improve and optimize productivity of dairy cows require sufficient availability of nutrients, especially high energy in the early period of lactation. Increasing energy intake in dairy cows can be conducted by increasing the density of energy. The research aimed to evaluate dietary protected feed supplement on milk production and quality, including: fat, protein, and lactose content of Friesian Holstein dairy cow milk. Protected feed supplement was produced from sardine fish oil, through saponification and microencapsulation protection methods. The experiment consists of two treatments i.e. P0: basal diet (control) and P1: basal diet + 3 % protected feed supplement. Each treatment was repeated 15 times. Data were analyzed by independent samples t-test analysis. Results showed that supplementation of protected sardine fish oil had no effect on lactose content, but increased milk yield production ($p<0.01$), milk fat content ($p<0.05$), and protein content ($p<0.05$).

1. Introduction

Milk production and composition of dairy cattle is influenced by the nutrients consumed. Nutrients are chemical substances in feedstuffs that can be mobilized and utilized for basal maintenance, production, and reproduction [1]. Nutritional factors are an important aspect in the lactation management of dairy cows. The efforts to improve and optimize the production performance of dairy cows require sufficient availability of nutrients, especially high energy in the early period of lactation. Increasing the energy intake for dairy cows can be conducted by increasing the energy density of feed by using fat (oil). One of potential energy source that is widely available and cheap is sardine fish oil [2]. Sardine fish oil contains high unsaturated fatty acids. Unsaturated fatty acid supplementation is proven to be able to increase energy efficiency through increasing energy density and sustain the efficiency improvement of milk protein synthesis through increasing the flow of nitrogen (N) non-ammonia into the duodenum [3].

Protected fat supplementation in dairy cows is expected to be able to fulfill the energy needs, especially in early lactation in order to gain optimum milk production, healthy, and especially dairy cattle conditions related to ideal Body Condition Score [4]. Protection treatment is needed to gain the real benefit from the presence of energy source supplementation or protein in feed. In the unsaturated fatty acid supplementation, protection is required to prevent unsaturated fatty acids from ruminal microbes biohydrogenation. Protection is also useful to eliminate the negative effects of high concentration of unsaturated fatty acid supplementation, and reduction in fibre degradability. Therefore the aim of current study was to evaluate the protected feed supplementation on milk production and quality, including: fat, protein, and lactose content of Friesian Holstein dairy cows.

2. Materials and Methods

2.1. Sample preparation

Thirty of Friesian Holstein cows in the Dairy Cattle Breeding Centre, Ministry of Agriculture Baturraden, Indonesia were employed in this study with an average lactation period of 75 days and an average body weight of 575 kg. They were weighed initially to determine their daily feed requirement. Body condition score (BCS) of dairy cows used in this study was 3 ± 0.2 (scale 1 –5). This study used 30 individual (2.5x1.5) m^2 cages, the feed facility consisted of a concentrate feed, forage and drinkers. The feed treatment was conducted for 1 month. Protected feed supplement was produced from sardine fish oil, through saponification and microencapsulation protection methods. The experiment consists of two treatments i.e.: basal diet (P0; control diet) and basal diet + 3 % protected feed supplement oil (P1). Each treatment was repeated 15 times. All animals were given feed according to their individual requirement, based on dry matter intake [5].

$$\text{Dry Matter Intake (DMI)} = (0.372 \times FCM + 0.0968 \times BW^{0.75}) \times (1 - e^{(-0.192 \times (WOL + 367)}) \quad (1)$$

Where,

FCM : fat corrected milk
BW : body weight
WOL : week of lactation

Table 1. Feedstuffs composition and nutrient content of treatment diets

Item	P0	P1
Feedstuff composition, %		
Elephant grass	50	50
Concentrate 1	40	37
Concentrate 2	10	10
Protected sardine fish oil	-	3
Total amount	100	100
Analyzed nutrients content, %		
Crude Protein	14.29	14.24
Extract Ether	4.57	4.69
Crude Fibre	23.45	22.99
Ash	13.06	13.30
Total Digestible Nutrient	61.60	62.44

2.2. Measurement of fat, protein and lactose

Fat, protein, and lactose content were analyzed in the laboratory using milk analyzer Lacto Scan MCC50.

2.3 Data analysis

Data obtained in current study were analyzed by independent Student's t-test using SPSS 22.

3. Result and Discussion

Milk production and quality of dairy cows milk are shown in table 2.

Table 2. Milk production and composition of dairy cows used in the study

Variables	P0	P1
Milk productions (liters)	17.29^{d}	20.18^{c}
Milk fat (%)	3.41^{b}	3.66^{a}
Milk protein (%)	2.76^{b}	2.81^{a}
Milk lactose (%)^{ns}	3.96	4.04

P0 = basal diet without supplementation, P1 = basal diet with protected feed supplement; ^{ab}Different superscript in the same row indicates significant difference (p<0.05); ^{ns}Non-significant different.
^{cd}Different superscript in the same row indicates highly significant difference (p<0.01)

Result showed that dietary supplementation of protected feed supplement increased milk yield (p<0.01). The enhancement in milk production in this study might be attributed to the presence of protected sardine fish oil supplementation, which improves the energy density of the diets, which in turn improve milk production. [6] stated that fat (oil) can be used to fulfill energy requirements in early lactation period in dairy cows. In addition, [7] observed that milk production of dairy cows can be pursued with increased energy density. Results of this study were in line with previous findings of [7] (2005) and [8] that stated supplementation of protected fat in the diet increased milk production of dairy cows. However, result of this study differs from previous studies [9]; [10] that reported non-significant milk production improvement following dietary supplementation of protected fat.

Supplementation of protected feed supplement increased milk fat content (p<0.05). This might showed that protected fatty acids in the diet can be transfered directly into milk fat formation. [11] stated that total milk fat production was depend on the balance transfers of long-chain fatty acid ration to milk fat and milk fat synthesis in the mammary gland. This result was consistent with finding of [7] which reported that protected fat supplementation increased fat content of milk. Different result was reported by [12] which indicated reduction in milk fat content following fish oil and canola oil supplementation.

Protein content of the milk in supplemented cattle (P1) was higher than P0 (p< 0.05). This might be caused by the increased of energy availability due to the supplementation of protected fat that stimulate milk protein synthesis. Protected fat is an alternative for energy source that has a role in the synthesis of milk protein. The high milk protein in P1 probably is due to the coating material used in the microencapsulation process used to protect fat, calculating 50% soybean meal of the total amount of raw materials used or 1 : 1 sardine fish oil used. The result was in line with the report of [13] that noted combination of fat supplementation and high quality protein stimulated milk production and protein.

There was no different found in lactose content of milk from both groups. It might be because protected fat added in the diets as energy source is used more for milk production, milk fat synthesis, and milk protein, rather than for lactose improvement. [14] stated that, glycerol is a substrate for the glucose synthesis in gluconeogenesis in lactating cattle that was generally used for energy rather than be used as a precursor of milk lactose. There fore, dietary fat addition did not give significant effect on lactose content of milk.

4. Conclusion

Results showed that supplementation of protected sardine fish oil had no effect on lactose content, but increased milk yield production, milk fat content, and protein content. Supplementation of protected sardine fish oil in dairy cows was able to increase milk production and quality of dairy cows.

References

[1] Pramono A, Kustono and Hartadi H 2010 Reproductive performance of dairy cows in Yogyakarta province based on balanced ration given. *Proc. The 5th International Seminar on Tropical Animal Production Community Empowerment and Tropical Animal Industry* (Yogyakarta)

[2] Pramono A, Kustono, Putro P P, Widayati D T and Hartadi H 2014 Dietary Supplementation of Protected Sardine Fish Oil on Milk Production and Quality of Dairy Cows. *Proc. of The 16th Asian-Australasian Association of Animal Production Congress* (Yogyakarta)

[3] Johnson K A, Kincald R L, Westberg H H, Gaskins C T, Lamb B K and Conrath J D 2002 *J. Dairy Sci.* **85** 1509–15

[4] Pramono A 2014 *Pengaruh Sinkronisasi Estrus Intrauterin dan Intramuskular Dengan Pgf2α Dan Gnrh serta Suplementasi Lemak Terproteksi Terhadap Respon Estrus, Profil Hormon Estrogen dan Progesteron serta Conception Rate* (Disertasi) (Yogyakarta: Universitas Gadjah Mada)

[5] NRC 2001 *Nutrient Requirements of Dairy Cattle* 7th Revised Ed (Washington, D. C: National Academy Press)

[6] Santos J E P, Bilby T R, Thatcher W W, Staples C R and Silvester F T 2008 *Reprod. Domest. Anim.* **43** 23–30

[7] Leonardi C S, Bertics and Armentano L E 2005 *J. Dairy Sci.* **88** 2820–2827

[8] Lohrenz A K, Duske K, Schneider F, Nürnberg K, Losand B, Seyfert H M, Metges C C and Hammon H M 2010 *J. Dairy Sci.* **12** 5867–5876

[9] Cortes C, da Silva-Kazama D C, Kazama R, Gagnon N, Benchaar C, Santos G T D, Zeoula L M and Petit H V 2010 *J. Dairy Sci.* **93** 3146–3157

[10] Hristov A N, Domitrovich C, Wachter A, Cassidy T, Lee C, Shingfield K J, Kairenius P, Davis J and Brown J 2011 *J. Dairy Sci.* **94** (8) 4057–4074

[11] Palmquist D L 1984 *Use of fat in Diets for Lactating Dairy Cows. In: FATS in Animal Nutrition* ed J. wiseman (London: Butterworths)

[12] Vafa T S, Naserian A A, Moussavi A R H, Valizadeh R and Mesagaran M D 2012 *Asian-Aust. J. Anim. Sci.* **25**(3) 311–319

[13] Chan S C, Huber J T, Theurer C B, Wu Z, Chen K H and Simas J M 1997 *J. Dairy Sci.* **80** 152–159

[14] Hames, D and Hooper N 2005 *Bios Instan Notes, Biochemistry* 3rd ed (New York: Taylor and Francis Group)

IOP Conf. Series: Materials Science and Engineering **193** (2017) 012035 doi:10.1088/1757-899X/193/1/012035

Starch Product of Wild Plants Species Jalawure (*Tacca leontopetaloides* L.) Kuntze as The Source of Food Security in The South Coastal West Java

Wardah[1], E N Sambas[1], Ridwan[1], and D Ariani[2]

[1] Research Center for Biology-LIPI, Cibinong Science Center, Jl. Raya Jakarta- Bogor, KM 46 Cibinong.

[2] Research Unit for Natural Product Technology-LIPI, Jl. Raya Yogjakarta-Wonosari Km 31,5 Ds. Gading Kec. Playen, Wonosari.

E-mail : wardah_etnobio@yahoo.com

Abstract. Majority of people of South coast of West Java, from Sukabumi, Cianjur, Garut are fishermen. Natural conditions are very dry and the area of land for agriculture, particularly rice cultivation is minimal. So that the condition of the society is more directed to high enough levels of food insecurity. Because coastal areas tend to have a longer dry season from rainfall. Results of research conducted in the years 2013 - 2016 in the area of Pelabuhan Ratu, Cidaun (Cianjur), Coastal area of Jayanti, Ranca Buaya, Mekar Mukti, and along the coast until Pameungpeuk, Leuweung Sancang, is known that jalawure plant which grows wild at South-coast region of West Java is precisely the alternative solution to address food insecurity. The results of the starch flour is a source of carbohydrate that is high enough to be used as a substitute for rice and wheat. Another potential source of jalawure nutrition is also recommended for diabetics consumption.

1. Introduction

Indonesia as an agricultural country and maritime has a wealth of natural resources potential, it is natural to be able to meet food needs for the population, because food affects the community, the nation and the state, both in the economic, political, social, cultural and defense. Although the rice self-sufficiency had been achieved in 1984, but with the climate disruptions and change the orientation of economic development, so Indonesia again become a net rice importer. Associated with food needs in the broad sense, Indonesia also has not been able to achieve self-sufficiency, especially when associated with the fulfillment of carbohydrates, protein, fat, vitamins and minerals. The number of people are getting serious to press the availability of natural resources that can be used to provide food, which is already very limited. Indonesia is estimated to have population growth in 2035 will be double the current amount, to less than 400 million inhabitants. As a result, a period of 35 years Indonesia needs additional supply of food more than twice the current supply [1], so that efforts to achieve food security is a challenge that must be given priority to the welfare of the nation. National food production needs to be significantly improved so that domestic demand can be met, while food development in Indonesia still faces many problems. The problem is that food consumption of Indonesian is fixated on rice consumption, and government policies by freedom duties on rice imports in order to meet the availability of rice, but the implications for rice producers is deteriorated.

Though Indonesia has high biodiversity, the irony will be food shortages. Foodstuffs derived from tubers carbohydrates is still widely spread in all corners of Indonesia, which can be a source of carbohydrates as food alternative. In this case the Research Center for Biology LIPI as the agency in charge of increasing the basic knowledge of science, its role in enhancing the country's food security is expected. One of the activities conducted is looking for wild plants as an alternative source of food. Data and information obtained from this study are expected to be the foundation in the development of the use of plants by other competent agencies in the field. In this case the research is directed at the species of minor tubers relatively unnoticed by the other institutions to be developed. Therefore *Tacca leontopetaloides* is one kind of tubers that became the main target of the study.

Tacca leontopetaloides plant is known locally different in each region, kecondang, condang (Java), labin (Madura), kecondang (Seribu Islands). Also in Tahiti is known as arrowroot or Polynesian arrowroot [2-3]. Jalawure (*Tacca leontopetaloides*) is known by the coastal communities in Sukabumi, South Cianjur and South Garut as a plant that has been used by local communities since their ancestors is hereditary in overcoming conditions of famine and food shortages that often occur in their area. This is due to the westerly winds that blow away the heat, which causes the fishermen not to fish and harvest failures. Various studies on this species have been done such as on physiological aspect [4] and ethnobotanical aspect [5].

Tubers of jalawure (*Tacca leontopetaloides*) is processed by the local community in the research sites to serve as a substitute for rice starch or flour to overcome food shortages and famine [6]. The results of the analysis carried out in the Department of Health and Nutrition Research Center on Post Harvest Lab Department of Agriculture on Tacca tuber starch turns that carbohydrate content of the starch flour is quite high i.e., 83.07% -89.4% and storability of flour could reach 1 to 1,5 years [7]. This species is found growing wild in the coastal areas of West Java. Regarding this species in Indonesia is still rarely carried out, both from the aspect of ethnobotany, cultivation and other aspects, including the excavation efforts in exploiting their local wisdom. The existence of this plant begins rarely found in nature, this is due to iron sand mining and shrimp ponds in coastal areas of research location. Therefore, this study aimed to determine the indigenous knowledge of local communities, the importance of this plant as a food source of carbohydrates that can be used as an alternative solution in facing famine and food insecurity in order to feed themselves for food self-sufficiency.

2. Materials and Methods

2.1. Time and Research Location

The study had been conducted in 2011 to 2015 in the coastal areas of South West Java. The research locations were in Kertajaya village, Simpenan sub-district, Sukabumi regency; Cidamar village (Jayanti beach), Cidaun sub-district, Cianjur regency and to the regency of Garut (Rancabuaya, Cicalobak; Cigadog, Cikelet, Santolo beach, and Sayangheulang). Climatic conditions at the study sites according to the classification of Schmidt and Ferguson, including climate type B, i.e. the wet type with value Q (Quotient) amounted to 24.19% where Q is the percentage ratio between the average number of dry months with the average number of wet months . Duration needed to get to the location of the study in normal conditions is about 12 hours, but when the bad weather, the rain and fog, can reach 15 hours.

2.2. Working Procedures

Data was collected through interviews with members of the local community, village head, village chief, sub-district, Gapoktan, KWT, and the people who use the plants, especially on jalawure (*Tacca leontopetaloides*). Interviews were conducted openly and freely. Direct field observations in a way to traverse along the coast based on information from the local community. Starting from the coast of South Cianjur to the coast in South Garut in West Java by purposive sampling (where plants were available). Besides interviews there were also collecting plant samples for specimen sample (voucher), and plant parts such as tubers and flour used for proximate analysis and mineral.

Figure 1. Location of the study in the Coastal of West Java

3. Results and Discussion

3.1. Results

3.1.1. General Situation of Research Sites

Pamipiran, Sangrawayan, and Citamiang sub-villages are administratively included in Kertajaya village, Simpenan sub-district, Sukabumi regency and located between 7 ° 04'50.8 "S, 106 ° 32'46.5" E, research is located in the coastal hamlet, a population of approximately 175 people consisting of 56 families. Conditions of hamlet is directly facing the Indonesia Ocean with a height of 20 m-50 m above sea level. The research location is hilly topography and steep cliffs up ramps. Population is Sundanese, majority of Islam religion, and the main livelihood of the fishermen is catching fish in the sea. They do rice paddy farmland on low sloping section, and the results are not enough to feed them one month. Because the land with an area slightly too infertile, they are planting other tubers but oftenly consumed by wild boar. Therefore, people who live in this area often had a bad season and food insecurity.

Cidamar Village, Cidaun sub-district, South Cianjur, Cianjur regency is located between coordinates 7 ° 29'13.1 "S 107 ° 21'18.4" E. The total area of about 1.89 million ha is located on the south coast ± 10 meters above sea level. With rainfall ± 2000 mm / year and a rainy months for 6 months, Cidamar village has a daily average temperature ranges between 29^{0}C - 32^{0}C. Main livelihoods are fishing and farming.

Rancabuaya beach is located in the Purbayani village, Caringin sub-district, Garut regency, West Java. North side of this site is adjacent to the Caringin village, south adjacent to the Indralayang village, the east by the Sinarjaya village, and west by the Indonesian Ocean at coordinates 7 ° 31 ' 42 "S, 107 ° 28'46" E. Other sites are Cicalobak beach of Mekarmukti sub-district; Village of Cigadog, Santolo beach, sub-district of Cikelet is located at 7 ° 39 '40' 'S and 107 ° 41' 11 " E; and Sayangheulang beach, Village of Mancagahar, Pameungpeuk sub-district is at 7 ° 40'6 "S, 107 ° 41'34 "E, and Nature Reserves of Leuweung Sancang 7 ° 41'48 "S, 107 ° 52'18" E are all in Garut regency. The total area of 2248.83 km2, or approximately 73.37% of the current Garut regency with a population density of 1,171,846 inhabitants (census of 2010), or about 43% of the total population of the Garut regency today. The main livelihoods are farming and fishing communities to catch fish in the sea.

3.1.2. Botany Information

According to [8], the genera *Tacca* consists of 8 species including *Tacca leontopetaloides* (L.) O. K., synonym of *Tacca pinnatifolia* Gaertn. This plant is herb, growing wild, up to 2 meters. Tuber rounded, flattened or elliptic, thin-skinned brown when young and dark brown when a bit old, inside the tuber white milk, grows under the ground to a depth of about 50 cm. Midrib rounded leaves and stems widen the breech eggs grooved and dark green fruit which generally consists of 20-40 fruits (see figure 2).

Figure 2. Plants and development of *Tacca leontopetaloides*

Inflorescence has flowers 1:25; 20-40, the flowers are protected by the green outer involucrum sometimes edged violet, or yellow, grow upright, can reach 40 cm long. Fruit dark green, green, rounded, rounded eggs or elliptic. Seeds many, varied forms of its size.

Tacca leontopetaloides on the coast of West Java, is known by the local name "jalawure" (South Cianjur and South Garut), in Sukabumi the name is "jalamure". In Central Java, on the island of Karimun Jawa is known by the name "kecondang", on the island of Madura is known as the "lorkong". In Central Java and Madura this plant is used as food only at certain times only. In Indonesia, it is found in such as the island of Madura, Krakatau, Karimunjawa islands and several other islands. In Garut, jalawure is found on the beach in the coastal of Cikelet sub-district. Jalawure grows at the seaside on a sandy soil under the shelter of a number of plant species such as pandanus , thistle and daffodil. [9] and [10], state that jalawure is a tropical savanna plant that can live in the dry season because it has a tuber that can store water.

3.2. Discussion

So far we only know rice as a staple food source of carbohydrates for satisfy nutrition for the community. But the price of rice is now perceived to be quite high for the people who live in rural areas, particularly those living on the seashore of the least fertile land and rainfed. Therefore, since 1962 jalawure has been used by people who live in the sub-village Cicadas of Cigadog village, South Garut as food alternatives to cope with the famine and food shortages that hit the village. Jalawure (*Tacca leontopetaloides*) is processed to be used as food ingredients with variations depending on the capabilities of each member of the family. The role of jalawure for people on the south coast of West

Java greatly assists government programs in terms of food self-sufficiency. This is apparent in difficult conditions, the people in this region can overcome its own problems by making jalawure as being substitutes of rice [7]. At first, Cigadog village has a population of poor and food-insecure around 60%, through the guidance and direction, the village eventually could reduce the population living in poverty to 20% (data of Cigadog village office 2014). This is because people try to cope with the crisis-prone fed and how to process local food sources that exist around them into local food instead of rice. Jalawure as one species of local minor tubers grow wild in the coastal is processed into foodstuffs. The work done by the village of Cigadog is awarded with Adhikarya Food Nusantara in 2011 from the government. Because of Cigadog villagers creativeness to process a variety of local food into food of carbohydrates substitute.

Results of research conducted in communities from South Cianjur area to Pameungpeuk of Garut regency show that the community in this area has been utilizing Tacca since their ancestors as a food substitute in famine conditions. However, as the transportation is smooth to the southern coastal areas, local wisdom in utilizing existing vegetation around them began to erode with the inclusion of a variety of food from outside the area which instantly can be eaten. Besides a disturbance at the plant habitats with the exploitation of iron sand mining and aquaculture ponds in coastal is already started to spread along south coast of Garut make communities that had been easy to get this plant is now getting hard to find it. This research is done in order to encourage the community to preserve the resources of Jalawure to remain a source of local food by the way it is grown, without depleting the habitat.

3.2.1. Potential tuber starch of Jalawure (Tacca leontopetaloides)

Results of research conducted in the Caringin , sub-district of Cidaun, South Cianjur, in the coastal communities Tacca tuber is used in a simple way, processed as starch flour for daily meal together with burned small fish. This shows the poor life condition of fishermen in this region. This is due to the people in this area who depend on the goodness of nature for fishing. Starch of Tacca tubers has a very important role in the lives of people in this area as a food ingredient to overcome food shortages.

Communities in South Garut from Purbayani village, coastal of Rancabuaya, Cikelet village, the village of Cigadog and until Pameungpeuk utilize starch from tubers of jalawure is more advanced than the people in South Cianjur. In addition to consume as a food substitute for rice and wheat flour, they process them into various foods for the big days such as the Feast and at weddings. The role of this plant in the community is very large, but the process of making flour starch is still indispensable hygiene processing technology and quality. Because of the way of processing technology that can improve the quality of the flour is also the quality of food products produced.

Research conducted by LIPI in South Garut from 2011 to 2016 has made a considerable improvements to the coastal communities of this area. *Tacca leontopetaloides* plants that had only been utilized by the communities living on the seashore, now almost all the people in the study site knowing this plant without worrying in to use them as food. *Tacca leontopetaloides* is a plant that is still wild scattered along the south coast of West Java, and is starting to do its development. Making a pilot demonstration plot has been done so that local communities can cultivate without having to take the wild continuously. The development of this species is done in cooperation with the local government area of the Garut regency to see the potential in the future to replace the consumption of wheat flour into *Tacca leontopetaloides* flour.

Propagation is done by intercropping with maize, peanut or also with green beans, for around 8-9 months of age Tacca to be harvested. So tuber harvesting is done after 3 times harvesting of corn. Because the harvesting season of the tubers falls during the dry season and other crops can not be harvested except Tacca tubers. At that time the tubers are old and starches is produced quite a lot [6].

3.2.2. The nutritional value of tuber starch flour of Tacca leontopetaloides

Result analysis of the nutrient content of tuber starch flour of jalawure (*Tacca leontopetaloides*) per 100 gram portion which can be consumed, it contains the potential to be developed as an alternative food ingredient see table 1. Apparently the results of the analysis shows that starch flour has the potential to develop into an alternative foodstuffs and to make starch from wild tubers Tacca is similar with the flour-starch from other minor tubers that have been cultivated and developed. The chemical composition of starch tubers of wild Tacca for the content of protein, fat, carbohydrates, can compare the nutrient content for the types of tubers such as tapioca starch, white sago flour, tubers of *Dioscorea esculenta*, black potato tubers, canna tubers etc. (Table 2).

Table 1. Nutrient content of *Tacca leontopetaloides* starch per 100g

Analysis	Nutrient content /100g
Protein	5.23
Fat	0.785
crude fiber	1.85 %
Carbohydrate	80.11 – 88.07 %
Energy	334-369.165kcal/100g
Mg	221mg/100g
Fe	6.185mg/100g
Ca	283.2mg/100g
K	616.635mg/100g
P	233.105mg/100g
Water content	15.65g/100g
ash content	1.3g/100g
amylopectin	53.28 %
amylase	23%

Source: [7, 11]

The value of protein and fat of Tacca flour is lower when compared to rice, but with the addition and processing can improve the nutritional value of Tacca and this can be proven that over the years the people in the South shores of Garut can survive in conditions of famine and food insecurity. Table 1 shows that the Tacca flour content of carbohydrates is high compared with other tubers (table 1 and 2)and also contains vitamin C, fat and protein (tabel 1.). Content of this flour included in classification of Low Glycemix Index, this product is suitable for development as the possibility of consumption for diabetics. Unlike the case with natural Glycemix carbohydrates with high indexes, such as rice and corn [12]. Another advantage of the wild tuber starch flour of Tacca is based on research results is the amylose content of the starch is high (31.08%) and amylose content in a starch said to be high if it is> 25%. Amylose content of the Tacca starch is high to cause swelling of the starch granules and starch limited solubility. So it is suitable for food processing of some kind of noodles, because it will produce a strong high breakage noodle while still raw, cooking loss of solids due to the low, low adhesiveness and high tensile strength. High amylose content in Tacca starch, also demonstrated the potential Tacca starch as resistant starch. Starch with high amylose content, crystalline structure will tend to be more difficult to digest. Besides acute toxicity tests have been conducted on the Tacca flour that shows practically non-toxic [13].

Table 2. Content of nutrients from minor tubers per 100g

Types of Samples	Fat (g)	Protein (g)	Carbohy drate (g)	Ca Mg/100g	P Mg/100g	Fe Mg/100g	Vit.C mg/100g	Ash content (g)	Water content (g)	Energy (Kkal)
Wheat flour	1.3	8.9	77.3	16	**106**	1	1			365
White sago(*Maranta arundinacea)*	0.2	0.7	85.4	8.3	23	1.45	-	-	-	346.2
Cilawu tuber	0.7	1.8	27.9	30	49	0.7	-	-	-	125.1
Manggu cassava	0.6	2.8	78.5	33	56	0.55	-	-	-	330.6
Legi tuber	0.1	1.1	2.1	-	40	0.2	-	-	-	89
Gadung	20	0.2	23.1	79	0.6	9.	0.1	0.9	73.5	101
Kimpul	26	0.4	34.2	54	1.4	-	-		63.1	145
Dioscorea esculenta	14	0.1	20.4	49	0.8	4	0.5		75	98
Canna	21	0.11	22.6	70	1.9	10	0.1		76	95
Black potato	34	0.4	33.7	75	0.2	38	0.02		64	142
Taro	61	28	1.9	61	1	17	0.13		73	23.7

Source: [14]

Selection of the types of food and consumption patterns that are less appropriate can cause various diseases, including diabetes mellitus, hypertension, heart disease, and cancer. Carbohydrates consumed from a food will be digested and absorbed by the body. The higher the faster or a starch digestibility, the more glucose is generated to cause an increase in blood glucose levels. Carbohydrates are slow digesting power to be consumed by diabetics because of the increase in blood glucose becomes slow. The starch digestibility of each food is different, and can be detected through approach of glycemic index (GI) value. GI value of a food is influenced by many factors. One of the main factors that affect the value of GI is the absorption rate of carbohydrates by the body of a food [15].

3.2.3. *The process of making flour tuber starch of Tacca*

Processing technology of starch from tubers of Tacca wild plants which still done by the people is very simple, but to get a quality starch people need guidance. Guidance and awareness has been done to farmers Gapoktan, KWT, the PKK and the surrounding coastal communities to cultivate the fresh Tacca tuber hygiene and quality. The current result of Tacca starch quality when compared to starch that has been set by SII (Standard Industrial Indonesia) indicate that Tacca starch can achieve these standards. Judging the results of the starch flour which has almost the same level of whiteness with tapioca flour.

The processing of tubers into flour starch is done by a simple technology. To get a good quality flour is firstly in terms of raw materials (the tubers). Tubers are taken on the right harvest time which is during dry season, brown tuber skin without injuring oldness and harvesting the tubers. Because the Tacca tubers are very easily injured if exposed to scratches, streaks will cause brown color and that part will be rotten, so the flour quality is not good [6-7]. After the tubers are harvested then : a) soaking the tubers while cleaning the dirt from sticking; b) followed by removing the outer skin layer tuber with a brush or knife by means of scraping, the outer skin is easy to clean because it is soft after soaking ; c) tubers already scraped then soaked with clean water; d) scarring the tubers; e) then filtered to obtain a starch; f) starches that have been obtained then re-soaking up water from starch flour, shall be considered the work is completed; g) then the starch is to dry (figure 3).

Figure 3. The processing of starch Tacca

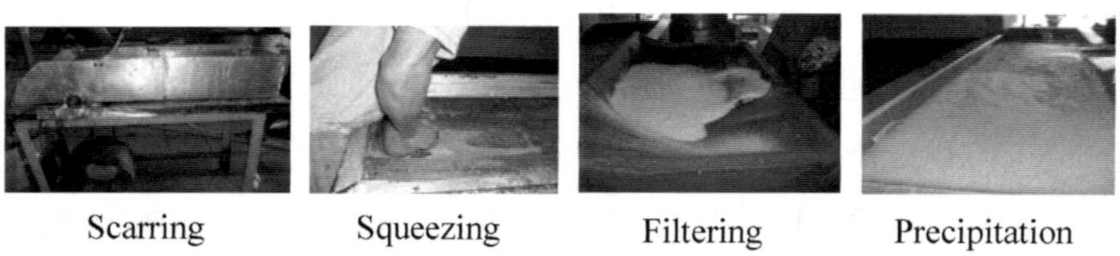

Figure 4. The process of manufacture of starch with mechanical technology

Processing to produce starch either traditional or managed using the same means of mechanical technology - can produce the same quality of starch. The difference is if the management of starch in the scale of home industry can be done in a simple, but if in a large scale rather difficult to do because it uses energy scarring. The resulting starch yield levels did not differ with producing tapioca starch. The yield of tuber starch flour of Tacca is quite high if the raw materials in a timely harvest is about (30% -35%) [16] and tapioca flour around (25%) [17].

3.2.4. *Prospects of tuber starch of Tacca*

Tacca tuber starch is proved to have good prospects, and could be a wise choice to meet the needs of local food-based raw materials with consideration of its main raw material of jalawure (*Tacca leontopetaloides*). Because the development on the local habitat, with most of the agro-climatic coastal land stretching, the low cultivation cost, productivity is high, produce large enough tubers. The execution management is easy enough that a housewife could manage their own needs, do not require a mechanical technology that will spend money on the process. Cassava growers are also planting in this area, the constraints is on the time of harvesting, cassava prices is low due to unstable price fluctuations, thus allowing it to be pursued. Finally, the low sale price of cassava, cassava farmers are forced to sell at low prices. To be processed into flour can not be done in the home industry, as cassava is difficult to do scarring.

Development of the current Tacca starch can improve the lives of farmers, because of Tacca starch can be used as raw material to be processed into various food products, see figure 5. One of the advantages of the Tacca tuber starch flour can survive long enough around 1.5 years [16]. Therefore, the flour has a fairly large prospects as food raw materials to tackling food shortages and famine.

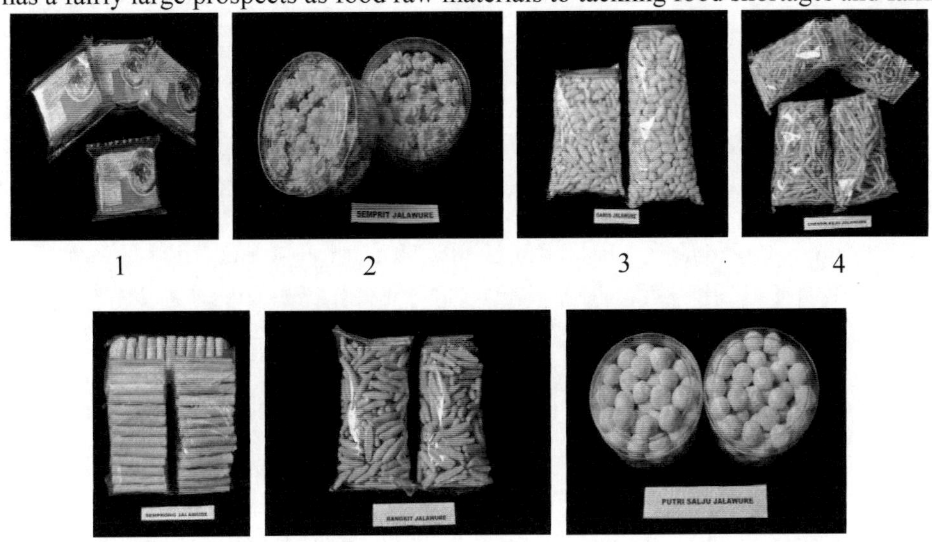

1 2 3 4 5 6 7

Figure 5. A wide variety of products that can be produced from starch tubers of Jalawure. Description: 1. A noodle product produced from starch of jalawure; 2. Jalawure syringe Cake; 3. Jalawure cork cake; 4. Cheestick of jalawure; 5. Jalawure semprong cake; 6. Jalawure rose cake; 7) Snow princess cake of jalawure.

In the industrial manufacture of noodles, biscuit or flour-based bakery, materials and equipment required is relatively simple. If during the many small industries that can operate with flour raw material, why we do not try to replace it with tuber starch flour of Tacca. This would mean at the present time, especially since it would reduce imports of wheat flour as a raw material. In addition to the number of companies that lay off employees right now, people will have difficulty finding job. With the industrial manufacture of jalawure tuber flour and dairy products will absorb a certain amount of labor. This will at least help a little on economic difficulties that so far have not passed in our country. Moreover, because the price of raw materials is cheaper, the price of flour produced from jalawure tubers will also be cheaper. The benefits arising from the utilization of sweet potato flour efforts is the emergence of the processing industry allows absorption of surplus labor generally found in rural areas, the processed food industry to reduce the cost of production and its dependence on wheat, the country could save foreign exchange through a reduction in imports of wheat. It is expected that Tacca tuber flour and dairy products can be accepted by consumers to further the success of diversification as one of the government programs. The success of this food diversification efforts will increase the economic value. Besides efforts to disseminate the use of jalawure tuber flour and processed results is also expected to achieve food security in Indonesia. Currently the manufacture of noodle products from raw materials of Tacca starch tubers are in the process which is more geared to functional food.

4. Conclusion

Tacca leontopetaloides is one kind of minor tubers that are wild and not yet widely cultivated. Apparently the result of research conducted in the coastal areas of West Java, namely in the village of Kertajaya, sub-district of Simpenan; Cidamar village of Cidaun sub-district; areas along from

Rancabuaya to Pameungpeuk sub-district, tubers of *Tacca leontopetaloides* have been used since time immemorial for generations by the community to replace rice as the main source of food. Starch flour is used as a substitute for rice or starch flour to overcome the famine and food shortages that often occur. Knowledge of Tacca plant can lift local communities to improve their living standards by creating a wide variety of food products to their market. Its starch flour is also recommended for functional food ingredients.

Acknowledgement
Research is done on the south coast of West Java since 2011 until 2016, through the National Priority Program DIPA program (PN 9) of the Food, The Research Center for Biology, Competitive program LIPI and we thank the Chairman of the Research Center for Biology- LIPI and Local Government of Garut Regency, Regional Secretary Asisstant of Garut , Head of Food Security of Garut regency, and Head of Food Diversification of Garut, Chairman of Gapoktan Cigadog village, sub-district and villagers of Cikelet, and Office of P3K Cikelet who had helped in the implementation of research in this area.

References

[1] Husodo 2001 *Kemandirian di Bidang Pangan Kebutuhan Negara Kita* (Makalah Seminar) (Semarang: PATPI)
[2] Heyne K 1987 *Tumbuhan Berguna Indonesia* Jilid I Cetakan 1(Jakarta: Badan Litbang Kehutanan)
[3] Jukema and Paisooksantivatana. 1996 *Tacca leontopetaloides In*: M. Flanch and F. Rumawas (Eds.). Plants yielding non-seed carbohydrates (Bogor: PROSEA)
[4] Syarif F, Lestari P dan Wawo A H 2014 *Berita Biologi* **13** (2) 161-171
[5] Susiarti S 2015 *Berita Biologi* **14** (1) 97-103
[6] Wardah 2012 *Laporan Penelitian karakterisasi tumbuhan berpotensi sebagai sumber bahan pangan lokal di Kabupaten Garut* (Cibinong : Puslit Biologi LIPI)
[7] Wardah 2011 *Analisis kandungan gizi jalawure (Tacca leontopetaloides*) (Cibinong : Puslit Biologi LIPI)
[8] Drenth E 1976 *Flora Malesiana* **7** (2) 375-387
[9] Fitter A H and Hay R K M 1981 *Enviromental Physiology of Plants* an Indonesian Translation (Yogyakarta: Gadjah Mada University Press)
[10] Ukpabi U J, Ukenye E and Olojede A O 2009 *J. Food Technol.* **7** 135-138
[11] Wardah 2013 *Jalawure sumber karbohidrat berpotensi sebagai ketahanan dan kemandirian pangan pada masyarakat di Garut Selatan* (Laporan Triwulan) (Cibinong : Puslit Biologi LIPI)
[12] Budijanto and Slamet 2009 *Majalah Pangan Media Komunikasi dan iInformasi* **54** 55-66
[13] Wardah and Ariani D 2013 *Wild tuber of jalawure (Tacca leontopetaloides) is an alternatif to overcome food insufficiency of sosiety in Cikelet Subdistrict, South Garut.* (Nusa Dua ASIA HORCs General meeting)
[14] Nutrition Directorate of the Ministry of Health Indonesia 1981 *Daftar Komposisi Bahan Makanan* (Jakarta: Bhratara Karya Aksara)
[15] Zabidi and Noor (2009) *Partial Substitution of Wheat Flour with Chempedak (Artocarpus integer) Seed Flour in Bread* http://books.google.co.id.
[16] Wardah 2014 *Laporan perjalanan sosialisasi jalawure dan pengembangannya pada masyarakat Garut Selatan dan koordinasi dengan PEMDA Garut* (Cibinong : Puslit Biologi LIPI)
[17] Suprapti L M 2005 *Tepung tapioka pembuatan dan pemanfaatannya* (Yogyakarta: Penerbit Kanisius)

IOP Conf. Series: Materials Science and Engineering **193** (2017) 012036 doi:10.1088/1757-899X/193/1/012036

Physical Characteristics of White Sweet Potato (*Ipomoea batatas* L.), Rice (*Oryza sativa* L.), and Tapioca (*Manihot esculenta)* Flours - Based Seasoning Composite Flour

NNA Alfani, D Ishartani, C Anam, D Praseptiangga, G J Manuhara
Department of Food Science and Technology, Universitas Sebelas Maret (UNS), Jl. Ir. Sutami 36A Kentingan Surakarta 57126, Central Java, Indonesia

E-mail: dishtani@yahoo.com

Abstract. The objective of this study was to determine the physical characteristics of seasoning composite flour that made from white sweet potato, rice, and tapioca flours, and determined the best formula of seasoning composite flour. A completely randomized design (CRD) with formula as the single factor was used. The data were analyzed by one-way ANOVA method and followed by Duncan Multiple Range Test (DMRT) at significance 5% if there was a significant difference. The best formula of seasoning composite flour was 30% tapioca flour, 30% rice flour, and 40% white sweet potato flour. The physical characteristics of the best formula were 5.689 ml/g of swelling power, 2.681 g/g of water absorption capacity, 0.887 ml/g of oil absorption capacity, and 22.03% cooking loss. Physical characteristics of the best seasoning composite flour were significantly different from the commercial seasoning flour and showed a better cooking loss, oil absorption capacity, and swelling power than commercial seasoning flour.

1. Introduction

Seasoning flour is a food material mixed from flour and seasoning with or without an addition of other food materials and permitted additives. It has a good market trend. This is because the seasoning flour simplifies and accelerates the preparation of fried products such as fried chicken, fried tempeh, peanut brittle, etc. Today's seasoning flour is made from 50% wheat flour that contains gluten [1]. Gluten-free seasoning flour is a promising food material since gluten was associated with Autism Syndrome Disorder and Celiac Disease [2]. However, research on product development of gluten-free or wheat-free seasoning flour is still limited.

Tuber and cereal flour such as tapioca, rice flour, and white sweet potato flour are very potential to be used as raw material gluten-free seasoning flour, since the abundant availability. The world rice and cassava production always increases for last two decades. Rice production was more than 700 million tons and cassava production was more than 250 million tons in 2014. While, sweet potato production remains stable around 100 million tons for last eight years [3].

Tuber and cereal flour could be applied as composite flour. Composite flour is a mixture of several flours. Food product development based on composite flour has been applied on some products, such as composite flour of semolina durum with canna starch for spaghettini [4]; composite flour of canna starch, cassava starch, and modified cassava flour to substitute material on corn noodles instant product [5]; composite flour of arrowroot starch, cassava flour and gotu kola for flakes [6], composite flour of from red bean flour and canna starch for infants' instant porridge [7], and composite flour of cassava starch waste, maize starch and mungbean flour for fettucine [8]. However, composite flour based on tapioca, rice, and white sweet potato flours, and its application in the seasoning flour has not been studied yet. Thus, physical characteristics of formulation and physical characterization of seasoning composite flour from tapioca, rice, and white sweet potato flours was conducted in this study.

2. Experimental

Tapioca (Rose Brand), rice flour (Rose Brand) and white sweet potato flour (CV. KUSUKA UBIKU) were obtained from local market. Those three raw materials were sifted using a 100 mesh sieve and then mixed into composite flour. There were nine formulations of seasoning composite flour that are shown in table 1.

Table 1. Seasoning Composite Flour Formulation

Tapioca Flour (%)	Rice Flour (%)	White Sweet Potato Flour (%)
30	70	0
30	65	5
30	60	10
30	55	15
30	50	20
30	45	25
30	40	30
30	35	35
30	30	40

Composite flours were further mixed with blended spices consisting of refined salt, pepper powder, onion, garlic, and candlenut. Mixing is done using a mixer at medium speed for 20 minutes.

2.1. Characterization of Seasoning Composite Flour

Determination of swelling power, 0.1 gram sample was mixed with 10 ml of distilled water in a centrifuge tube and heated at 60°C for 30 minutes. The mixture was being shaken during the heating process. When the heating was done, the suspension was centrifuged at 3000 rpm for 30 minutes.

$$\text{Swelling power}=\frac{\text{sendiment weight}}{\text{dry sample weight}} \quad (1)$$

Determination of water absorption capacity (WAC), 3 grams sample was placed on a filter paper, and then it was added with warm water (in 40°C) by 13 grams and allowed for 3 minutes. The dripped water was collected and pondered.

$$\text{WAC}=\frac{\text{Water weight at the onset-dipped water weight}}{\text{sample weight}} \quad (2)$$

Determination of oil absorption capacity (OAC), 1 gram sample was added by 10 ml of vegetable oil into a 25 ml centrifuge tube. The mixture was centrifuged for 30 minutes at 2000 rpm.

$$\text{OAC (\%)}=\frac{\text{supernatant weight-sample weight}}{\text{sample weight}} \quad (3)$$

Determination of cooking loss, Fillet-chicken breast was coated with sample batter followed by sample flour, and then it was fried until its color changed into golden brown using deep frying method at the temperatures of 175 - 180^{0}C. Sample batter was prepared by mixing 1 portion water to 1 portion sample.

$$\text{Cooking loss (\%)}=\frac{\text{raw coated meat weight-fried meat weight}}{\text{raw uncoated meat weight}} \quad (4)$$

The best formulation was determined by de Garmo analysis with a modification. The best formulation is determined from the highest value of the results value (RV). RV calculation was started by determined Variable Score (VS), followed by Normal Score (NS) calculation, Effectiveness Value (EV) calculation and ended by Result Value (RV) calculation. Variable Scores (VS) are sorted by priority of physical characteristic that give higher contribution to the sample quality. VS of sample were determined as follows: swelling power 0.5; WAC 0.5; OAC 0.5; cooking loss 1.0. Then Normal Score (NS), Effectiveness Value (EV), and Results Value (RV) were calculated using the formula:

$$NS=\frac{VS}{\Sigma\, VS};EV=\frac{\text{parameter value-worst value}}{\text{best value-worst value}};RV=NS\times EV \quad (5)$$

3. Results and Discussion

3.1. Swelling power

The ratio of rice flour and white sweet potato flour had a significant effect on the swelling power of the seasoning composite flour. White sweet potato flour portion had positive correlation with the swelling power of the seasoning composite. The seasoning composite flour with the lowest swelling power was the seasoning composite flour of 30% tapioca flour; 70% rice flour; 0% white sweet potato flour. The seasoning composite flour with the highest swelling power was the seasoning composite flour of 30% tapioca flour; 30% rice flour; 40% white sweet potato flour. Seasoning composite flour with 0-35% white sweet potato flour had a lower swelling power than the commercial seasoning flour (controls = K1 and K2) while the seasoning composite flour with 40% white sweet potato flour had not significant different with controls.

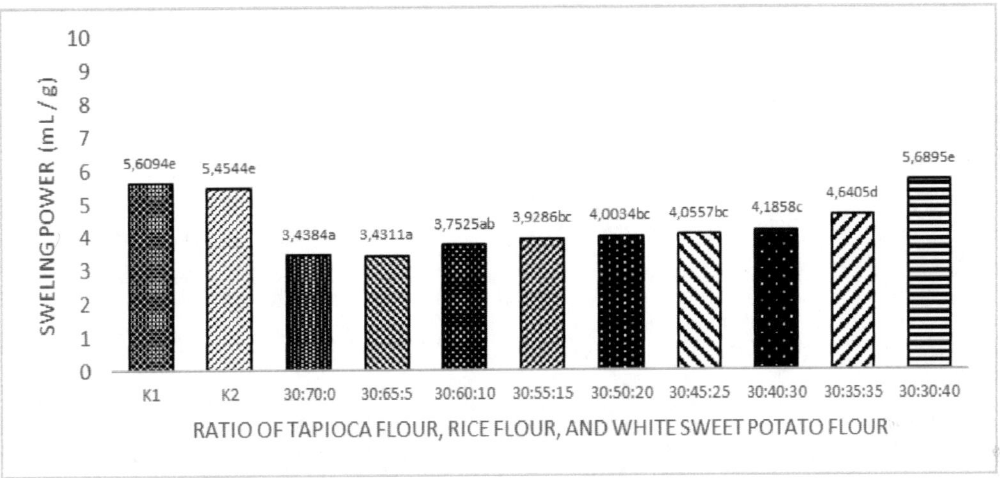

Figure 1. Sweeling Power of Seasoning Composite Flour

3.2. Water Absorption Capacity

The ratio of rice flour and white sweet potato flour affected the water absorption capacity of the seasoning composite flour significantly. The greater the ratio of white sweet potato flour was, the higher the water absorption capacity of the seasoning composite flour would be. The water absorption capacity of seasoning composite flour with the portion of 0-10% white sweet potato flour was not significantly different from the controls' while the water absorption capacity of the seasoning composite flour with the portion of 15-40% white sweet potato flour was greater than that of the controls.

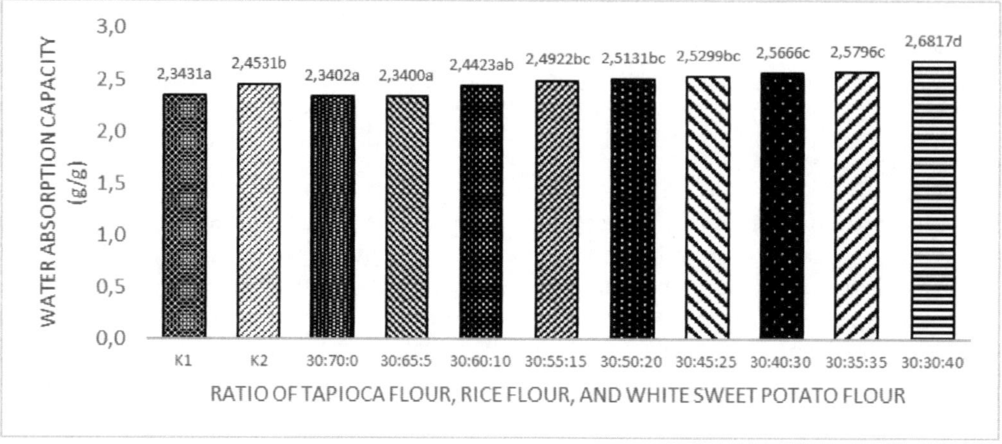

Figure 2. Water Absorption Capacity of Seasoning Composite Flour

3.3. Oil Absorption Capacity

The oil absorption capacity of the seasoning composite flour was affected significantly by the ratio of rice flour and white sweet potato flour. Sweet potato flour portion gave inverse proportion to the oil absorption power of the seasoning composite flour. The portion of 0-35% white sweet potato flour had a higher oil absorption capacity than that of the controls while the portion of 40% white sweet potato flour had not significant different with controls.

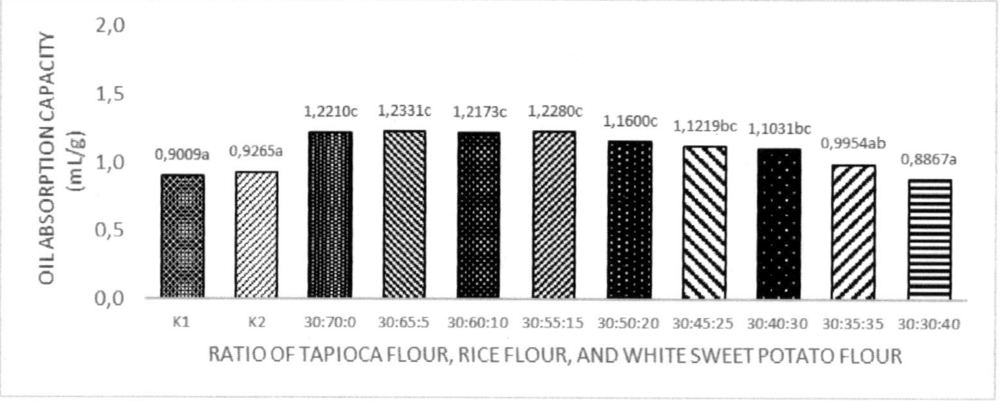

Figure 3. Oil Absorption Capacity of Seasoning Composite Flour

3.4. Cooking loss

The result showed that the cooking loss of the seasoning composite flour decreased as the increase of white sweet potato portion. The controls'cooking loss was not significant different with the seasoning composite flour's using 0-20% white sweet potato flour portion. The factor which affected the cooking loss was viscosity of the used flour. The higher the viscosity value of the used flour was, the lower the cooking loss value would be [12].

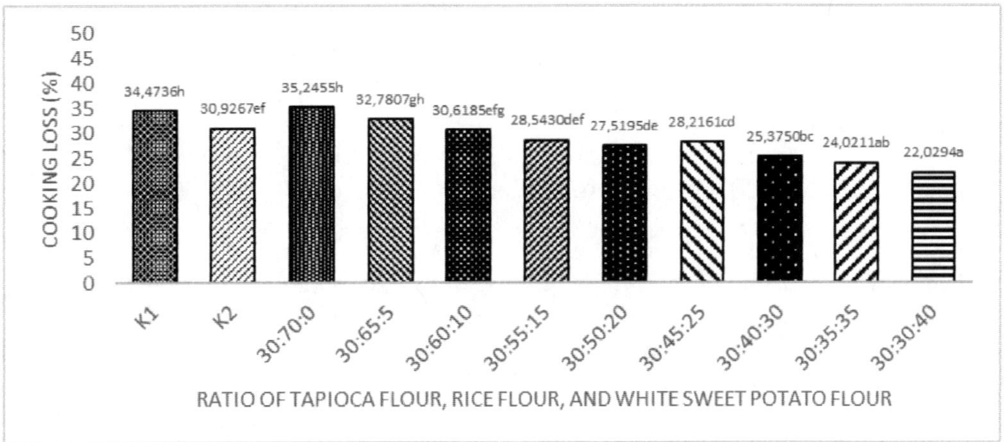

Figure 4. Cooking Loss of Seasoning Composite Flour

3.5. Determination the best formula

Based on the de Garmo analysis result (table 2), formula that is composed by 30% tapioca flour, 30% rice flour and 40% white sweet potato flour was determined as best formula. Compared to controls, seasoning composite flour had Result Value (RV) highers than control.

Table 2. Determination the best formula of seasoning composite flour

Sample	Result Value (RV)				
	CL (%)	WAC (g/g)	OAC (ml/g)	SP (ml/g)	Total
K1	0.0235	0.1748	0.1917	0.1929	0.1233
K2	0.1306	0.1163	0.1773	0.1791	0.6034
30 : 70 : 0	0.0000	0.1157	0.0070	0.0006	0.1233
30 : 65 : 5	0.0746	0.1156	0.0000	0.0000	0.1902
30 : 60 : 10	0.1400	0.1409	0.0091	0.0285	0.3185
30 : 55 : 15	0.2029	0.1532	0.0029	0.0441	0.4030
30 : 50 : 20	0.2338	0.1584	0.0422	0.0507	0.4851
30 : 45 : 25	0.2128	0.1716	0.0642	0.0553	0.5039
30 : 40 : 30	0.2987	0.1748	0.0751	0.0668	0.6154
30 : 35 : 35	0.3397	0.2000	0.1372	0.1071	0.7841
30 : 30 : 40	0.4000	0.0000	0.2000	0.2000	0.8000

4. Conclusion

The physical characteristics of seasoning composite flour include the following: swelling power ranged from 3.4385 ml/g to 5.6895 ml/g; the water absorption capacity ranged from 2.3402 ml/g to 2.4480 ml/g; the oil absorption capacity ranged from 0.8867 ml/g to 1.2210 ml/g; and the cooking loss ranged from 22.03% to 35.25%. The best formula of seasoning composite flour were made from 30% tapioca flour, 30% rice flour, and 40% white sweet potato flour. The physical characteristics of the best formula were swelling power 5.689 ml/g, water absorption capacity 2.681 g/g, oil absorption capacity 0.887 ml /g, and cooking loss 22.03%. The physical characteristics of the seasoning composite flour significantly different from the commercial seasoning flour. Seasoning composite flour was better than commercial seasoning flour for parameters cooking loss, oil absorption capacity, and swelling power.

References

[1] Bagja JS, Yuwono SS and Widyaningtyas D 2015 *J. Food and Agroindustry* **3** 1627–36
[2] Lau N M, Peter H R, Green, Annette K, Taylor, Hellberg D, Mary A, Caroline Z.T, Barry E. Kosofsky, Joseph J H, Anjali M, Rajadhyaksha, Armin A 2013 *J. Ploss One* **8** 1–6
[3] FAO (2017, January 24) http://www.fao.org/faostat/en/#data/QC/visualize
[4] Harmani SA, Haryadi, Cahyanto MN and Yudi P 2016 *AGRITECH* **36**
[5] Indrianti, N, Rima K, Riyanti E and Doddy A D 2013 *AGRITECH* **33**
[6] Sianturi DP and Marliyati SA 2014 *JGP* **9** (1) 15-21
[7] Yustiyani and Setiawan B 2013 *JGP* **8** (2) 95-102
[8] Sunyoto, M, Tati N, Rima R K, Deddy M 2016 *J. Food* **25** 43–50
[9] Ocloo, FCK, Bansa D, Boatin R, Adom T and Agbemavor WS 2010 *Agric. Biol. J.N. Am.* **1** (5) 903–8
[10] Lidiasari, E, Merynda I S and Friska 2006 *J. AGRIPLUS* **16** 1–7
[11] Beuchat LR 1977 *J. Agric.Food Chem.* **25** 111–120
[12] Mukprasirt A, Herald TJ, Boyle DL, and Boyle EAE 2001 *Poultry Science Journal* **96** 1–9

IOP Conf. Series: Materials Science and Engineering **193** (2017) 012037 doi:10.1088/1757-899X/193/1/012037

Rooms for genetic improvement in Indonesian Bali cattle population

N Widyas, T Nugroho, S Prastowo
Animal Science Department, Sebelas Maret University, Surakarta Indonesia

E-mail : nwidyas@gmail.com

Abstract. Bali cattle is a species of *Bos javanicus d'Alton*, a local cattle in Indonesia. They are loaded with potential as meat producer and well adapted to tropical climate and limited feed resources. Studies have been made to characterize the species. This paper presents a rough estimate of the opportunity to improve the Bali cattle population genetically. Our aim is to endorse that the Bali cattle could be both superior and efficient as tropical meat producer cattle. Results shows that Bali cattle population size is decreasing for the last years with a possibility to be accompanied by genetic quality decline. However, there is hope in improving Bali cattle genetically by a proper breeding strategy. This could also be an answer to the challenge of climate change which leads to global warming; where species adaptable to such environment is more beneficial in the future.

1. Introduction

Bos javanicus d'Alton is an indigenous species in South-East Asia. It is known for its robustness against tropical environment. Despite the harsh habitat with low quality feed resources, they manage to have distinguishable performances. Studies have been made to document their performances on high fertility [1–4], good growth and excellent carcase quality [2,5–11]. In Indonesia, *Bos javanicus d'Alton* spread all over the archipelago, they, however, have an exclusive habitat in the island of Bali. These animals reside and evolved in Bali Island over a long period of time. They had undergone domestication in this area since ten to five thousand years ago [3] and thus are specifically called the Bali cattle.

Conservation program has been established for Indonesian Bali cattle in the island of Bali. The rules prohibited the intrusion of new genetic resources from outside Bali while selling out cattle to other part of the country is allowed. Because better cattle make better money, the conservation policy leads to selection against genetically superior individuals in Bali cattle population [4]. Not only did the Bali cattle population decrease in number [12], we suspected that the genetic quality of the cattle population itself is also deteriorating. On the other hand, Indonesian government fixed a plan to obtain meat self-sufficiency and endorses mass production of meat from local breeds [13,14]. In order to achieve long term success of the program, superior local genetic resources are needed. Hence, attempts are needed to be made in order to improve the genetic status of Bali cattle; an option is by implementing a proper and well assessed breeding program.

Bali cattle is evolutionary distant from both *Bos taurus* and *Bos indicus* [15], underlying the species unique characteristics which have not been much studied. Their crosses with *Bos taurus* and *Bos indicus* are potential of exhibiting heterosis on production traits [5,16,17], but other studies revealed poor reproductive performance on the F1 crosses [1,3,4] to the extreme that male crosses are infertile. This fact provides reason that straight breeding program is favoured. In this paper, we presented opportunities in genetically improve production traits of Bali cattle within the juridical area of Bali Province. Further we predicted the coming year's performance increase due to the genetic improvement through straight breeding program.

2. Study methods

This paper used published data on Bali cattle population inside and outside the Province of Bali [2,6,8,10,12,18–21]. The information originated within Bali were grouped into two populations: 1) Balai Pembibitan Ternak Unggul (BPTU) is a population inside the Pulukan breeding centre; and 2) Population outside BPTU or from over the island. We focused our analysis on body weight traits with the following variables: birth weight (BW), weaning weight adjusted for 205 days (WW), yearling weight adjusted for 365 days (YW) and mature weight (MW) which is the weight of mature cattle above 2 years of age. Least square means and standard deviation of the pooled data along with the heritability of the traits is presented in table 1.

Table 1. Means, standard deviation and heritability of Bali cattle production traits

Location	Weights (Kg)			
	BW	WW	YW	MW
Mean ± sd[1]				
BPTU Bali	17.63 ± 0.64	92.67 ± 2.30	138.00 ± 2.82	338.00 ± 97.89
Bali Province	16.77 ± 0.94	92.10 ± 5.75	140.63 ± 9.69	330.76 ± 88.57
Outside Bali	12.79 ± 0.88	84.69 ±15.19	134.17 ± 26.39	271.17 ± 21.86
All population	13.46 ± 1.81	85.61 ± 14.30	135.09 ± 24.22	278.62 ± 29.21
Heritability[2]	0.245	0.43	0.565	0.65[3]

[1]Means and standard deviations summarized from various studies
[2]heritability of BW, WW and YW are averaged across studies in Bali cattle
[3]Heritability for MW is taken from reference [22] for beef cattle, as no specific information available

Genetic gain or response to selection for each trait was estimated as the parameter describing the genetic improvement of each trait after selection. Assumptions and variables for simulation were based on the previous published studies and scenario following progeny testing program held in BPTU Bali cattle. Annually, BPTU kept around 90 male candidates for progeny test where only four will be chosen as elite bulls. In addition, around 900 females are kept and 120 were chosen as elite dams. Following this scenario, we obtained selection intensity of 2.15 for males and 1.76 for females. Estimated breeding value accuracy of 70% was obtained from progeny test analysis [22] with 30 offspring each candidate, while female accuracy was obtained based on mass selection as much as the square root of the traits' heritability ($\sqrt{h^2}$). Phenotypic standard deviation for each trait is presented in table 1 and average generation interval is 5 years for beef cattle. Genetic gain is estimated with the following formula:

$$\Delta G = \frac{i_m \cdot r_m \cdot \sigma_g + i_f \cdot r_f \cdot \sigma_g}{L_m + L_f} \quad (1)$$

where *m* and *f* subscript indicate male and female selection part, i is selection intensity, r is the accuracy, σ_g is the additive genetic standard deviation for each traits obtained from $\sqrt{\sigma_P^2 \times h^2}$ and L is the generation interval. We assume linear genetic gain over years.

3. Results and discussions

Conservation policy of Bali cattle in the island of Bali had backfired the local population with selection towards less superior individuals; where the local performance decrease as people sold the best cattle. The Bali cattle population had undergone decreasing trend over recent years of time (Table 2) while the genetic impact of this statement needs further justification.

Table 2. Bali cattle population from 2003 - 2015

Year	District								
	Klungkung	Gianyar	Bangli	Badung	Buleleng	Jembrana	Kr.asem	Tabanan	Denpasar
2003	43315	0	0	42526	0	0	0	0	0
2004	41822	55324	0	43413	0	0	0	0	0
2005	43688	56168	0	45844	0	0	0	0	0
2006	43943	56279	0	49450	0	0	0	0	8599
2007	44059	56433	0	55092	0	0	147388	0	9311
2008	44272	57236	93725	65671	144821	0	148385	67037	9370
2009	44664	57815	95818	65118	142736	0	152437	68157	9393
2010	44724	57022	97830	67390	150857	0	149268	67027	9812
2011	46615	47282	94058	48049	136189	54114	135506	66333	8710
2012	46905	53809	96733	49051	145600	54306	136957	59616	7990
2013	33914	40302	74327	36662	120127	52014	132231	53800	7586
2014	37250	42456	75164	37862	121613	52306	122369	52916	7241
2015	38732	44615	72880	30771	119473	51825	121637	51567	6689

Source: BPS Bali

We suspected that not only the population is decreasing but the genetic quality of the cattle also deteriorated. The decrease in population's genetic quality could be caused by inbreeding and/or as an impact of selection against superior genetic resources within the species. Precautions need to be made by designing a proper genetic improvement program. We estimated genetic improvement of production traits using response to selection.

Scenario implied for the genetic gain simulations are based on the condition picturing the progeny test program in BPTU. Following the scenario, we could obtain an increase of 0.21 Kg of birth weight every year after the first generation; followed by the increase of 2.48; 5.08 and 6.90 Kg of weaning weight, yearling weight and mature weight, respectively. These estimates only hold if we were to keep the superior animals to disseminate the genetic improvement achieved by selection into the population inside the BPTU.

Table 3. List of variables used in the simulation

Simulation variables	BW	WW	YW	MW	Total
Selection intensity (*i*)					
Male					2.15
Female					1.76
Phenotype sd (σ_g^2)	1.81	14.30	24.22	29.21	
Heritability (h^2)	0.25	0.43	0.56	0.65	
Generation interval (*L*)					
Male					5
Female					5
ΔG/Year (Kg)	0.21	2.48	5.08	6.9	

The estimated improvements are quite remarkable; however, selection was done in a very strict manner that the size of elite (breeding) herd is very small. This is where the effect of inbreeding kicked in. According to the number for breeding population, the rate of inbreeding will reach 15.5% per generation [22,23], which is very high. This simulation was based on the population in BPTU while the number of actual population is a lot higher of more than 500,000 animals. Nevertheless, this paper was written in order to give an overview that there are rooms for genetically improving Bali cattle population into more efficient beef cattle breed through a properly measured breeding program. Further observations and analysis need to be done while accommodating all aspects in a breeding program to get more detailed and accurate insights for genetic improvement program in Bali cattle.

4. Conclusions

Bali cattle population size in Bali had decreased for the past seven years (2008 – 2015), which might be accompanied with decline in genetic quality due to selection against superior genetic resources. However, based on our simulated estimation, there is still room to improve the genetic quality of Bali cattle by properly assessed breeding program.

References

[1] Lenstra J and Bradley D 1999 *Genet. Cattle CABI* 1–14
[2] Talib C 2001 *Wartazoa* **12** 100–107
[3] Copland JW 1996 *Bali Cattle: Origins in Indonesia* In: Wilcox GE, Soseharsono S, N. DDM, Copland JW, editors. Jembrana Disease and the Bovine Lentiviruses (Canberra: ICAR)
[4] Sutarno dan Setyawan ADWI 2015 *Biodiversitas* **16** 327–354
[5] Dominanto GP, Nasich M dan Wahyuningsih S 2016 *Res. Zool.* **6** 1–7
[6] Entwistle K C T 2001 *Preliminary Report* 31–39
[7] Hartati, Muladno, Priyanto R, Gunawan A dan Talib C 2015 *JITV* **20** 168–174
[8] Prahani L 2004 *Genetic Evaluation for Growth Traits, Reproductive Performance and Meat Tenderness in Beef Cattle* (Florida: University of Florida)
[9] Gunawan A Jakaria 2011 *Media Peternakan* **34**: 93–98
[10] Warmadewi DA, Oka L, Sudyadnya P dan Sudana LB 2014 *J Anim. Sci. Udayana Univ.* **3** 1–9
[11] Lindell IC 2013 *Phenotyping of Bali cattle and interviewing farmers in Indonesia - a minor field study* (Swedish University of Agricultural Sciences Faculty)
[12] Bali BPS 2016 *Populasi ternak Sapi Potong Menurut Kabupaten/Kota di Bali* (Denpasar: Badan Pusat Statistik)
[13] Menteri Pertanian 2015 *Rencana Strategis 2015 - 2019* (Jakarta: Kementrian Pertanian Indonesia)

IOP Conf. Series: Materials Science and Engineering **193** (2017) 012037 doi:10.1088/1757-899X/193/1/012037

[14] Menteri Pertanian 2010 *Pedoman Umum Program Swasembada Daging Sapi Tahun 2014* (Jakarta: Kementrian Pertanian)
[15] Mohamad K, Olsson M, van Tol HTA, Mikko S, Vlamings BH and Andersson G, *et al* 2009 *PLoS One* **4** 1–6
[16] Pribadi LW, Maylinda S, Nasich M, and Suyadi S 2015 *Livest. Res. Rural. Dev.* **27** 1–10
[17] Nasution D 2010 *Agripet.* **10** 37–41
[18] Yusuf M, Syamsu JA, Rahim L, and Ali HM 2010 *Studi Uji Performa Ternak Sapi Bali Di Kabupaten Barru, Sulawesi Selatan (Preliminary Study)* (Jambi: Universitas Jambi)
[19] Sampurna IP 2013 *Menentukan standar sapi bali berdasarkan pola pertumbuhan dan kedekatan hubungan dimensi tubuhnya* (Denpasar)
[20] Suranjaya IG, Ardika IN and Indrawati RR 2010 *Majalah Ilmiah Peternakan* **13** 83–87
[21] Panjaitan T, Fordyce G and Poppy D 2003 *JITV* **8** 183–188
[22] Falconer DS and Mackay TFC 1996 *Introduction to Quantitative Genetics* 4th ed (Edinburgh: Prentice Hall)
[23] Crow JF 1986 *Basic concepts in population, quantitative, and evolutionary genetics* (New York: W.H. Freeman and Company)

IOP Conf. Series: Materials Science and Engineering **193** (2017) 012038 doi:10.1088/1757-899X/193/1/012038

Identification of Morphological Character and Esterase Isozyme Pattern in Second-Generation Black Rice Plant Irradiated to Gamma Rays

R S Hartanti[1], T A N Putri[1], F Zulfa[1], Sutarno[1]*, and Suranto[1]

[1]Department of Biology, Faculty of Mathematic and Natural Science, Sebelas Maret University, Jl. Ir. Sutami 36A Surakarta, Central Java 57126 Indonesia

*E-mail: nnsutarno@staff.uns.ac.id

Abstract. Black rice is one of the functional foods due to its high anthocyanin content. Black rice grain was irradiated by gamma rays with a dose of 200 Gy and 300 Gy. The main purpose of this irradiation is to induce mutation to the black rice plant in order to achieve the improved organism. This study was undertaken to elucidate the morphological character and esterase isozyme pattern of black rice plant after irradiated by gamma rays. There were morphological differences on leaves, stems and grains between irradiated and non irradiated black rice plant. Gamma radiation dose of 200 Gy showed the significant influence of the length of the stem, number of internodes, and length of leaves. The radiation dose of 300 Gy showed the significant influence of the decrease value of diameter of 3rd internodes, number of branches and width of leaves. Flowering time is getting faster as increasing radiation dose. At the age of 74 days after planting there are 9.15% plants of 200 Gy radiation dose that have flowered faster than normal plants. This value increased into 11.45% at the dose of radiation 300 Gy. There were differences in the esterase banding pattern between radiation dose of 200 Gy and 300 Gy than the control plants, indicated that randomly mutation has occurred.

1. Introduction

Black rice is potential as functional food because it contains of vitamins, minerals and antioxidant [1]. This rice has been recorded to have high flavonoids contents, and one of the flavonoid type contained in the black rice is anthocyanin. Anthocyanins are color pigments that have antioxidant activity and it has proved to be an effective antioxidant against the negative effects of free radicals [2]. Anthocyanins identified in the black rice variety were cyanidin-7-O-galactoside, cyanidin-3-Oglucoside, cyanidin-3'-O-glucoside, cyanidin-5-O-glucoside, cyanidin-3, 7-O-diglucoside, cyanidin-3, 5-O-diglucoside and peonidin-4'-O-glucoside [3]. The antioxidant activity of the extract of black rice has been reported to prevent or reduce the risk of many degenerative diseases such as cancer, hypertension, and diabetes. They also can reduce the risk of hepatic steatosis, lowering LDL cholesterol (Low Density Lipoprotein) or the bad cholesterol in the blood and may help fight heart disease [4, 5]. Black rice also has fibers which are higher compared to most other types of rice. Dietary fiber is very good for the body, plays a role in lowering cholesterol. They could decrease blood sugar levels, and has a chemo-protective effect on colon [6].

Besides of the advantages, black rice plant has shortage of challenges in the business of its cultivation, such as planting time which is longer than white rice and the habitus is very tall, therefore the plant so easily collapse [7]. Therefore, it needs plant breeding efforts to improve the quality of black rice plants in order to obtain the superior black rice. One way of plant breeding could do is by inducing mutation. Plant synthetizes the new protein and DNA under radiation stress as an adaptation to environment [8]. This mutation induction can be done using a physical mutagenic agents such as radiation of gamma rays.

Rice grains irradiated with gamma rays yielded by core of Cobalt 60. Gamma rays radiation dose used in this study are 200 Gy and 300 Gy, respectively. Those doses have been used because they are the most effective dose in rice breeding [9, 10]. Those doses caused little physical damage but produce higher mutation probability in the plant [9]. Indications biochemical changes in seed after irradiation with gamma rays such as changes in enzyme activity and respiration rate, and chromosome damage. Radiation is expected to induce mutations that lead to the improvement of plant characters such as shorter in stem length and flowering time. Furthermore, identification of the morphological and molecular alteration is done to ensure that the changes occur in plants as the result of mutation induction by gamma rays radiation. Isozyme analysis can be used as molecular markers to determine differences between mutant and control plants [11].

2. Material and Methods

2.1. *Plant Samples Material*

This study was conducted used 2000 irradiated seeds of rice which were cultivated in field. Control plants were also plotted side by side in order to look at wheter any difference between the treated seed and the control one. These plants were then used for electrophoretic purposes as well as for morphological observations. The morphological character observed when plants aged 4 months, counted from planting time. For molecular identification, the fresh leaves of *Oryza sativa* L. were used for electrophoretic procedures and esterase staining were conducted according to Suranto's [12].

2.2. *Observation of morphological character*

The length of the stem was measured from the base of the stem to the tip. The number of internodes was calculated from the base of the stem to the tip. The length of internode and stem diameter was measured at the third internode from the base. The number of branches on the main stem also calculated. The length and width of the first, second, and third leaves were measured with a ruler. Flowering time was determined by the interval time from the planting time to the first flower appeared. The length and width of the seeds were measured following the procedure from Chang [13]. The number of seeds per panicle is whole calculated.

2.3. *Identification of esterase isozyme pattern*

All methods used in the analysis of isozyme banding pattern was following the procedure of Suranto's [12, 14]. In this study, isozyme banding pattern was detected using polyacrylamide gel electrophoresis method. Staining process of electrophoresis results for esterase isozyme banding pattern, and gel visualization used dye solution with the composition as follows: 0.0125 g of α-naphthyl acetate dissolved in 2.5 ml acetone, then added with 50 ml of 0.2 M phosphate buffer pH 6.5 and 0.0125 g of Fast Blue BB Salt.

3. Result and Discussion

3.1. *Morphological Analysis*

3.1.1. Stems

Table 1. The average length of stem and internodes of black rice after radiated using different gamma rays

Parameters	Dose of Radiation (Gray)($\bar{x}\pm$ SD)		
	0	**200**	**300**
Length of stem (cm)	$94{,}04^{a} \pm 0{,}48$	$106{,}61^{b} \pm 1{,}66$	$93{,}66^{a} \pm 1{,}01$
Number of internodes	$5{,}16^{a} \pm 0{,}17$	$6{,}05^{b} \pm 0{,}21$	$5{,}43^{a} \pm 0{,}38$
Length of 3rd internodes (cm)	$18{,}07^{a} \pm 0{,}74$	$17{,}04^{a} \pm 0{,}70$	$17{,}92^{a} \pm 1{,}08$
Diameter of 3rd internodes (cm)	$0{,}57^{b} \pm 0{,}06$	$0{,}54^{ab} \pm 0{,}03$	$0{,}50^{a} \pm 0{,}02$
Number of branches	$1{,}07^{a} \pm 0{,}07$	$1{,}17^{a} \pm 0{,}13$	$1{,}70^{b} \pm 0{,}50$

Note: Value with same small letter annotation means has no significant difference in 95% DMRT (Duncan Multiple Range Test)

The average length of stems employed the 3 doses of radiation were 0 Gy is 94.04 cm, 200 Gy is 106.61 cm, and 300 Gy is 93.66 cm. The radiation dose of 200 Gy showed a significant difference compared to the control plant (0 Gy), whereas the radiation dose of 300 Gy showed decreasing the stem, although the shorter growing than the control plant was not significantly different. The average number of internodes resulting from the 3 doses of radiation were 0 Gy was 5.16, 200 Gy (6.05), and 300 Gy (5.43) respectively. It was interested to note that using radiation dose of 200 Gy the result again showing a significant different number of internodes compared to the control, meanwhile no significant difference were recorded when radiation doses of 300 Gy was applied. The average diameter of the third segment on the radiation dose 0 Gy was 0.57 cm, 200 Gy was 0.54 cm, and 300 Gy was 0,50 cm. These results confirmed that treated seed have no effect on the development of stem diameters. The number of branches has increased along with increasing radiation dose. The average number of branches on the control plant was 1.07, and 1.17 after a treatment with dose of radiation 200 Gy has increased significantly to 1.70 after dose of radiation 300 Gy was conducted.

3.1.2. Leaves

Table 2. Morphological character of leaves of black rice plant

Parameters	Dose of Radiation (Gray)($\bar{x}\pm$ SD)		
	0	**200**	**300**
Length of 1st leaf (cm)	$29{,}88^{a} \pm 2{,}44$	$32{,}77^{b} \pm 0{,}77$	$29{,}48^{a} \pm 1{,}67$
Length of 2nd leaf (cm)	$41{,}38^{b} \pm 3{,}77$	$46{,}47^{c} \pm 0{,}52$	$37{,}19^{a} \pm 1{,}18$
Length of 3rd leaf (cm)	$45{,}28^{a} \pm 2{,}46$	$49{,}88^{b} \pm 0{,}83$	$43{,}45^{a} \pm 1{,}40$
Width of 1st leaf (cm)	$1{,}42^{b} \pm 0{,}04$	$1{,}50^{b} \pm 0{,}04$	$1{,}26^{a} \pm 0{,}09$
Width of 2nd leaf (cm)	$1{,}29^{b} \pm 0{,}05$	$1{,}33^{b} \pm 0{,}03$	$1{,}15^{a} \pm 0{,}06$
Width of 3rd leaf (cm)	$1{,}17^{b} \pm 0{,}04$	$1{,}26^{c} \pm 0{,}06$	$1{,}07^{a} \pm 0{,}06$

Note : Value with same small letter annotation means has no significant difference in 95% DMRT (Duncan Multiple Range Test)

Observations on the morphological character of leaves were the length and width. These parameters were opserved at the third leaf from the top.The radiation dose of 200 gy create a longer leaf with length of 1st leaf is 32.77 cm, length of 2nd leaf is 46.47 cm and length of 3rd leaf is 49.88 cm, while the control plant have length of leaf in a row are 29.88 cm, 41.38 cm and 45.28 cm respectively, while the radiation dose of 300 Gy have almost the same length leaves of control that are 29.48 cm, 37.19 cm and 43.45 cm. Beside the significant difference in leaf length, the width of leaf result the opposite instead. There is no significant difference found in the width of leaf where the value appears to be in a close range each other.

3.1.3. Grains

The parameters used in measuring the grains were length and width of grains, shape of grain (ratio between length and width of grain) and number of grains per panicle. The data of morphological difference of grains that was analyzed with SPSS (Statistical Package for the Social Science) software were presented at table 3.

Table 3. Morphological character of grains of black rice plant

Parameters	Dose of Radiation (Gray)($\bar{x}$± SD)		
	0	**200**	**300**
Grain Length (cm)	$0{,}98^a \pm 0{,}00$	$0{,}98^a \pm 0{,}00$	$0{,}98^a \pm 0{,}00$
Grain Width (cm)	$0{,}30^a \pm 0{,}00$	$0{,}3\ ^a \pm 0{,}00$	$0{,}30^a \pm 0{,}00$
Grain Shape (L:W)	$3{,}27^a \pm 0{,}06$	$3{,}28^a \pm 0{,}06$	$3{,}26^a \pm 0{,}05$
Number of grains per panicle	$126{,}21^a \pm 12{,}19$	$130{,}23^a \pm 8{,}10$	$124{,}73^a \pm 10{,}07$

Note : Value with same small letter annotation means has no significant difference in 95% DMRT (Duncan Multiple Range Test)

Gamma radiation seems to have no effect on grains morphology. Where the results appeared no one show the significant differences among all the parameters of measurement. However, radiation dose of 200 Gy produce the highest number of grains per panicle. There were no significant difference on the treated seeds and the control one. And this may be due the existance of seed coats.

3.1.4. Flowering time

Flowering time was measured from the day of first flower appears. The data shown below is flowering time of black rice plant irradiated by gamma rays. The control plant doesn't include in the table because they flowers later. So this table just recorded the flowering time that appear faster than normal plant (c.a. 80 days).

Table 4. Flowering time of black rice plant

Age of plant	Number of flowers appear per 2000 plants	
	200 Gy	**300 Gy**
54-60	16	41
61-67	95	114
68-74	72	74
Total flowers appear	183	229

There are remarkable results of observation that radiation could initiate the flowering of black rice plant. From total amount of 6000 black rice plant, 412 plants could flower faster than normal plants (80 days after planting). The radiation dose of 300 Gy have more influence on the early time plant flowering than the 200 Gy. This could be proved by many number of flower appear faster than the normal black rice plant. In the age of 54-60 days after planting, while the radiation dose of 200 Gy produce only 16 flowers while using radiation dose of 300 Gy 41 flowers have been produced. Interestingly, more flowers (114) have been produced when 300 Gy was applied compared to the 200 Gy (95). The radiation dose of 300 Gy obtained the number of plants are flowering faster than the control are 229 plants from a total amount of 2000 plants. The radiation dose of 200 Gy obtained the number of plants that flowering faster than the control plants are 183 plants from a total amount of 2000 plants. The percentage (%) of early time of flowering plants were quite high (9.15 %) at the dose of radiation 200 Gy. This value increased into 11.45 % at the dose of radiation 300 Gy. This indicated that both two different doses of gamma radiation (200 and 300 Gy) had resulted a number of plants flowering early. We do believe that this early flowering time was significantly difference to control plants. In this prelimenary report showed very promising molecular evidence that those radiation to have affects on the alteration of enzyme, protein or DNA of the plants as also shown by De Britto [8].

3.2. *Molecular Analysis*

Figure 1. Zymogram of esterase isozyme patterns of the black rice after radiated using different gamma rays. Gel is loaded by samples in following orders; lane 1 and lane 10 were samples from control plants, lane 2 to 5 were samples from plants with radiation dose of 200 Gy, and lane 6 to 9 were samples from plants with radiation dose of 300 Gy.

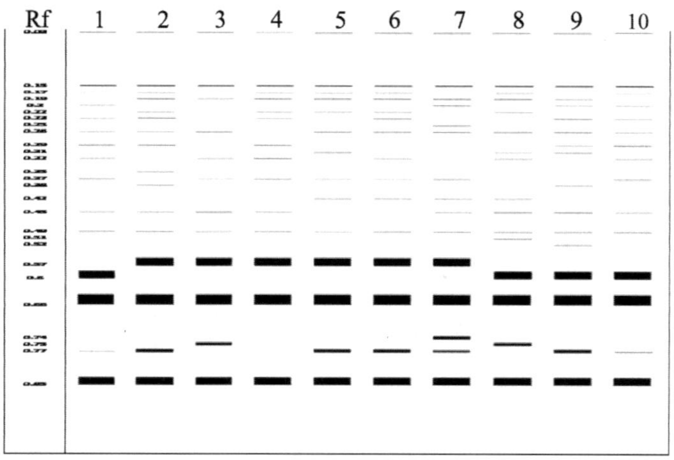

Figure 1. Zymogram of esterase isozyme patterns of the black rice after radiated using different gamma rays.

Result of isozyme banding pattern analysis show 27 bands. Those 27 bands located in the Rf between 0.03 and 0.85 i.e. 0.03; 0.15; 0.17; 0.19; 0.20; 0.22; 0.23; 0.25; 0.26; 0.29; 0.31; 0.32; 0.35; 0.37; 0.38; 0.42; 0.45; 0.49; 0.51; 0.52; 0.57; 0.60; 0.66; 0.74; 0.75; 0.77 and 0.85. There were 12 bands appear from sample of mutated plant that do not exist in the sample of normal plant. From those 12 bands, there was one (1) band (i.e. Rf = 0.35) appears only in sample of mutated plant with 200 Gy radiation dose, 4 bands (i.e. Rf= 0.25; 0.51; 0.52; 0.74) appear only in sample of mutated plant with 300 Gy radiation dose, and 7 bands (i.e. Rf = 0.19; 0.22; 0.31; 0.39; 0.42; 0.57; 0.75) appear in both samples. There were 2 bands (i.e. Rf = 0.20; 0.60) appeared in sample of normal plant and mutated plant with 300 Gy dose of radiation. Besides, there was one (1) band (i.e. Rf = 0.77) showed different thickness among other band in the same line. This band showed thicker band in both sample of mutated plant. This very preliminary evident could prove that radiation leads to mutations that trigger the alteration in isozyme banding pattern. And this could be appeared as new isozyme bands or different thickness of the bands. With increasing radiation dose, it could potentially produce more additional bands or possibly causing a few bands disappeared.

The result of this study was consistent with the previous study which showed addition in band number of protein pattern and DNA pattern of *Zea mays* L. [2]. Exposure of radiation with different dose and time leads to alteration of isozyme banding pattern, and it could be increased or decreased the number of band. This was happen because radiation may resulted mutation which could activated some gene expressions or distract the expression of the gene [15, 11].

4. Conclusion

This preliminary study showed the differences in the morphology of rice plants after treated with gamma rays. That could be noted on stem length, number or internodes, number of branches, leaf length, and flowering time. The dose of radiation 200 Gy have better value than 300 Gy. This first evident of isozyme banding pattern of irradiated plants which differ from the control plant would possibly to do further test molecularly (DNA and protein). And hopefully the induction of gamma rays would give another clue that mutation could occur in mutated rice plants.

Acknowledgements

We would like to thanks to National Nuclear Energy Agency of Indonesia for the support. And our thanks went to Mrs. Atik and Mrs. Nina (Molecular Biology Lab) for the technical assistance.

References

[1] Kristamtini, Taryono, Basunanda P, Murti R H dan Supriyanta 2012 *J. Agric. Biol. Sci.* **7** 982
[2] Miksusanti, Elfita dan Hotdelina 2012 *Jurnal Penelitian Sains* **15** 60
[3] Ali H A, Cho I K, Kim S J, Kim S N, Kim S Y, Cho Y S, Baek H J and Kim B J 2011 *Korean J. Environ. Agric.* **30** 111
[4] Ling W H, Cheng Q X, Ma J and Wang T 2001 *J. Nutr.* **131** 1421
[5] Ling W H, Wang L L and Ma J 2002 *J. Nutr.* **132** 20
[6] Santoso A 2011 *Magistra* **75** 35
[7] Kristamtini 2009 *Tabloid Sinar Tani* (Jakarta : Sinar Tani)
[8] De Britto A J, Jeevitha M and Raj T L S 2011 *J. Stress Physiol. Biochem.* **7** 232
[9] Sobrizal 2008 *Jurnal Ilmiah Aplikasi Isotop dan Radiasi* **4** 1
[10] Utami S 2013 *Agrium* **18** 158
[11] Talukdar D 2010 *Int. J. Genet. Mol. Biol.* **2** 112
[12] Suranto 2001 *Biodiversitas* **2** 85
[13] Chang T T, Bardenas, Eliseo A, Rosario D and Arnulfo C 1965 *Technical Bulletin* (Manila: The International Rice Research Institute)
[14] Suranto 2002 *Biodiversitas* **3** 257
[15] Hamouda M, Saad-Allah K M and Kasim W A 2014 *Int. J. Agric. Res.* **4** 46

IOP Conf. Series: Materials Science and Engineering **193** (2017) 012039 doi:10.1088/1757-899X/193/1/012039

Physical Characteristics of *Coleus tuberosus* Flour and Noodle in Various Arenga Starch Substitution

Miftakhussolikhah[1], D Ariani[1], M Angwar[1], J Kevin[2]
[1]Research Unit for Natural Product Technology, Indonesian Institute of Sciences, Jl. Jogja-Wonosari km 31,5
[2]Faculty of Agricultural Technology, Gadjah Mada University

E-mail: miftalipi@gmail.com

Abstract. *Coleus tuberosus* is one of local commodities in Indonesia which contains high carbohydrate. However, its utilization does note maximally. Therefore, *C. tuberosus* made into flour by grater method, and then used for noodle making with arenga strach as substituting material. The aim of this study was to determine the effect of *C. tuberosus* flour substitution in noodle preparation from arenga starch on its textural and sensory properties. In this study, noodle was made in some variations which were 100% arenga starch; 75% arenga starch:25% *C. tuberosus* flour; 50% arenga starch:50% *C. tuberosus* flour; 25% arenga starch:75% *C. tuberosus* flour and 100% *C. tuberosus* flour. Characterization of noodle were investigated including water content, strength, tensile strength, strain at break and stickiness. Sensory evaluation was conducted to analyse consumers acceptance. Noodle was compared with two commercial products. The result showed that arenga starch substitution in *C. tuberosus* noodle affect textural properties of noodle. The higher concentration of *C. tuberosus* flour caused tensile strength and strain at break getting low. The water content and stickiness were increased as the *C. tuberosus* flour substitution ratio increase. None of the noodle resulted from *C. tuberosus* flour and arenga starch mixture was exactly as same as maize and rice commercial noodles. However 25% of *C.tuberosus* noodle has better characteristics than other *C.tuberosus* noodles.

1. Introduction

Diversification of staple food consumption is not intended to replace rice in total but changing people's food consumption patterns so that people will consume a wider variety of food and get better nutrition. Various tubers have a prospect for development as a rice substitute and processed into prestigious food.

Coleus tuberosus is one of carbohydrate sources which potential as wheat or rice substitute. In taxonomy, this tuber is classified in family *Lamiaceae*. High content of flavonoid and phenolic in this tuber, makes it potential as a food with functional properties. Nugraheni et al [10] mentioned that *C. tuberosus* had component which potential as antioxidant and anticancer. High potential of carbohydrate content and functional properties in *C. tuberosus* was not used yet as a processed food.

High carbohydrate content in *C. tuberosus* was not used as a processed food. High content of flavonoids and phenolic in this tuber, makes it potential as a food with functional properties. While the raw material for noodle production were still imported. Utilization *Culous tuberosus* as a noodles was expected could increase economic and benefit value of this tuber.

Noodle is one of very popular food in Indonesia, because its taste and practical in preparation. According to BPS [2], noodle consumption in Indonesia reached 80 grams/capita/week both in rural and urban areas. Indonesia is the 2nd largest noodle consuming country in the world after China. While raw

IOP Conf. Series: Materials Science and Engineering **193** (2017) 012039 doi:10.1088/1757-899X/193/1/012039

material for noodle production were still imported. Utilization *C. tuberosus* as a noodles was expected could increase economic and benefit value of this tuber. The aim of this study were to determine characteristic of *C. tuberosus* flour and to determine effect of arenga starch substitution on textural characteristic of *C. tuberosus* noodle.

2. Experimental

2.1. Material

C. tuberosus tubers obtained from Mertelu Village, Gedangsari, Gunungkidul, Yogyakarta. *C. tuberosus* was made into flour by peeling the skin, slicing the tuber, drying and then milling. Arenga starch was bought from home industry at Klaten, Central Java.

2.2. Swelling Power and Solubility of C. tuberosus flour

Swelling power and solubility of *C. tuberosus* flour were conducted according to Adebowale, *et al* [1]. *C. tuberosus* flour (1 gram) was added into 50 ml of distilled water and stirred until homogeneous. Then the slurry was inserted into a centrifuge tube, and heated to temperature of 60, 70, 80 or 90°C for 30 minutes in the water bath. The samples were cooled to room temperature and centrifuged at ± 3000 rpm for 15 minutes. Supernatant was dried until obtain constant weight (W2). The pellet was moved into a bottle and was weighed (N1), then the pellet was dried to obtain a constant weight (N2).

$$\text{Solubility (\%)} = \frac{\text{W2 (g)}}{\text{W1 (g)}} \text{x}100\% \qquad (1)$$

$$\text{Swelling power (g/g)} = \frac{\text{N1 (g)}}{\text{N2 (g)}} \qquad (2)$$

2.3. Noodle making

Noodle-making was done by modifying the noodle-making method on a large scale according Haryadi [5]. Five variations flour mixture *i.e.* 100% arenga starch; 25% *C. tuberosus* flour:75% arenga starch; 50% *C. tuberosus* flour:50% arenga starch; 75% *C. tuberosus* flour:25% arenga starch; and 100% *C. tuberosus* flour made into dough by adding water to flour with ratio mixture flour and water was 1:0.5. Dough was molded into pellets using Food Extruder PD-45N, La. Pramigianaa 15 mm diameter mold. Pellets steamed for 4 minutes to gelatinize starch. Gelatinization occurs only on the surface of pellet so that it can serve as an adhesive when it was made as noodle. Pellets then molded into noodle with diameter 0.7 mm and obtained raw noodle strands. Raw noodle steamed in a steamer for ±20 minutes. Noodle then dried in the cabinet dryer at 60°C for 6 hours until water level was ± 10%.

2.4. Tensile strength and strain at break of noodle

Tensile strength and strain at break was done according to Chen *et al* [4]. Noodle samples that have been cooked in the optimum cooking time is clamped in the universal testing machine tools, then pulled with 5 kg force speed 1.00 mm / sec to drop. Maximum force at the time of breaking noodle expressed as tensile strength, while the strain at break at break compared beginning noodle expressed as strain at break percent.

2.5. Stickiness of noodle

Noodle stickiness determined according to Chen *et al.* [4] by taping two strands of noodle that has been cooked and removed from water, then pulled to separate them each other using universal testing machine. Measurements were made with speed 1.00 mm / sec and 5 kg load.

2.6. F break of dry noodle

The strength of dry noodle (F break) was conducted according to Suryani [13]. Dry noodle was cut into 1 cm length, then placed and clamped on measurement table Lloyd Instrument in perpendicular position to the discounter. The sample was then pressed with a cylindrical probe (diameter 3.5 cm) with a speed of 1 mm/second using 10kg heavy load until the sample broke. Maximum force when dry noodle fracture was calculated as strong fracture of dry noodle.

3. Results and Discussion

3.1. Swelling power and Solubility

The results of swelling power and solubility analysis of *C. tuberosus* flour can be seen in table 1. From these results it can be seen that the swelling power and solubility increase as the temperature increase. The starch granules will continue expand at temperature above the gelatinization temperature of flour [12]. The solubility of starch are also influenced by the presence of amylose-lipid complexes. Amylose-lipid complexes are insoluble in water and require a higher temperature to dissociate the compound [12]. Swollen starch granules will become increasingly vulnerable to shear forces resulting in disintegration and will release the material dissolved when the starch granules were destroyed [12].

Table 1. Physical characteristics of *C. tuberosus* flour

Temperature (^{O}C)	Swelling power (g/g)	Solubility (%)
60	4.83^{a}	6.67^{a}
70	5.35^{a}	7.57^{ab}
80	8.60^{b}	8.54^{b}
90	13.62^{c}	10.44^{c}

The values with different superscripts within a column are significantly different ($p < 0.05$) by Duncan's multiple range test.

Swelling power and solubility is influenced by the amylose content in starch. The higher amylose content, the lower the swelling power. This is caused by molecules of linear amylose thus reinforcing its internal network [7]. Swelling power of starch is highly dependent on the strength of inter and intra-molecular starch granules. According to Leach [7], that binding is influenced by various factors including: comparison of amylose and amylopectin, amylose and amylopectin molecular weight, molecular weight distribution, degree of branching, and the length of the outer branches of amylopectin molecules that may play a role in bindings. When the granules swell, the amylose molecules will separate from the network and leaching into the liquid [7]. Swelling power and solubility is proportional untilthe equilibrium is reached. Equilibrium is reached when the concentration of soluble starch and expanded starch granules had a same amount [7].

From statistical analysis of data with $p < 0.05\%$, solubility *C.tuberosus* flour at 60°C and 70°C were not significantly different. At a temperature 60-70°C starch granules still not absorb water quickly. When starch is heated at temperature 80-90°C, the starch getting closer to gelatinization temperature so the solubility increase rapidly and demonstrated significantly different results. Similar results were shown for the swelling power, at 60°C and 70°C did not differ significantly and at temperature 80°C and 90°C has significantly different swelling power.

3.2. Textural characteristics of noodle

Tensile strength is one of the physical properties of noodle which shows maximum force required to break the noodle. Strain at break shows the change in length of noodle acquiring the maximum tensile force to break up compared to the initial length. The higher tensile strength and strain at break will have better noodle quality, and it is expected by the consumer.

Noodle strength is influenced by the amylose content of flour used [3]. Granular structure and noodle will be more compact because of the increased bonding between amylose molecules that form crystalline clusters [9]. Tensile strength is directly proportional to strain at break, it could be due to more compact noodle, then noodle will be more easy to pull and more difficult to break at withdrawal. Statistical analysis was done with $p < 0.05\%$ showed that tensile strength were significantly different for all samples noodle. Arenga starch and *C. tuberosus* mixing will give real effect to noodle tensile strength. Samples 100% *C. tuberosus* flour has a tensile strength that is not significantly different from the commercial product B, and samples with *C. tuberosus* 25% are not significantly different from the commercial product A. Noodle made with *C. tuberosus* ratio of 0% in the flour mixture has strain at break which is not significantly different from the commercial product A and product B, tensile strength value of *C. tuberosus* noodle was lower than that of from rice noodle [14]. Higher concentration of *C. tuberosus flour* in flour mixture,will made lower amylose content of flour mixture. The bond between amylose molecules causing declining compactness of noodle granular. The force required to break the noodle will be smaller due to increased concentration of *C. tuberosus* flour. Compare to noodle from rice noodle [15], strain at break of *C. tuberosus* flour was much higher than rice noodle.

F break of noodle illustrates the strength of mechanical treatment (shock) during distribution/storage which caused the destruction of dried noodle/before cooking. The greater force needed to break the noodle, the more powerful of resulting noodle. Noodle is expected to have a strong value for fracture (N) which is great because it is not easily broken. Table 2 displays test results noodle textural properties.

Table 2. Textural characteristics of noodle

Noodle (*C.tuberosus* flour : arenga starch)	Tensile strength (N)	Strain at break (%)	F break (N)	Stickiness (N)
0% : 100%	0.19^{e}	54.39^{c}	3.67^{g}	0.01^{ab}
25% : 75%	0.12^{d}	46.36^{bc}	3.19^{f}	0.01^{b}
50% : 50%	0.10^{c}	35.43^{abc}	2.50^{e}	0.02^{c}
75% : 25%	0.08^{b}	20.07^{ab}	2.16^{d}	0.04^{d}
100% : 0%	0.06^{a}	15.38^{a}	1.79^{c}	0.05^{e}
Maize commercial	0.03^{d}	20.50^{c}	1.08^{b}	0.01^{a}
Rice commercial	0.05^{a}	59.52^{c}	0.79^{a}	0.01^{b}

The values with different superscripts within a column are significantly different ($p < 0.05$) by Duncan's multiple range test.

Strength of dry noodle influenced by bonds between molecules in the starch granules. Suryani [13] states that the higher the number of levels of high amylose starch in the flour mixture, the more crosslinking on the starch granules so that the bonds between the starch molecules stronger so it is not easily broken. High amylose content which causes retrogradation perfect so that the structure of the resulting instant starch noodle tougher, stronger, and compact [11]. Noodle made from materials that yield high amylose, has a tensile strength, and strong texture / compact [8]. Good quality of noodle should have high value of tensile strength, strain at break and F break. In the other hand, stickiness value of noodle should be lower.

Excessive starch leaching (high cooking loss) will make the noodle surface becomes sticky. Noodle sticky surface will be positively correlated to its solubility. When noodle has high adhesiveness, it will make noodle strands become stuck together and difficult to be separated and it is inevitable happened because it is not preferred by consumers. Noodle made from starches with low amylose content such as sweet potato starch, potato starch and cassava starch very easy to stick and be difficult to separate between noodle strands with one another during the drying process [4].

From table 2, it can be seen that stickiness of noodle increased along with the increase of *C. tuberosus* flour. From statistical analysis with $p < 0.05\%$, noodle with 100% arenga starch has a stickiness that is not significantly different from commercial products A and B, while made with *C. tuberosus* concentration of 25% was not significantly different from the product B. For other commercial samples have values significantly different adhesiveness. Black potato starch mixed with starch sugar will affect the stickiness of noodle. Black potato starch has sticky properties for mengdanung banyk amylopectin. Starch sugar which has a high amylose content will make noodle produced has a low adhesiveness. Amylose can increase the bond strength between molecules capable of forming crystalline clusters so as to prevent the occurrence of amylose leaching and decrease the stickiness [6].

4. Conclusion

The higher concentration of *Coleus tuberosus* flour caused F break, tensile strength and strain at break getting low. The water content and stickiness were increased as the *Coleus tuberosus* flour substitution ratio increase. Noodle from 25% *C.tuberosus* flour and 75% arenga starch has better characteristic than other *C.tuberosus* noodles.

References

[1] Adebowale KO, Afolabi TA, and Lawal, O.S 2002 *Food Chem.* **78** 305–311
[2] Anonym 2013 *Pengeluaran untuk Konsumsi Penduduk Indonesia* (Jakarta: Badan Pusat Statistik)
[3] Chansri R, Puttanlek C, Rungsadthogy V and Uttapap D 2006 *J. Food Sci.* **70** (5) 337–342
[4] Chen Z L, Sagis L, Legger A, Linssen JPH, Schols HA and Voragen AGJ 2002 *J. Food Sci* **67** 3342-47
[5] Haryadi 2006 *Teknologi Pengolahan Beras* (Yogyakarta: UGM Press)
[6] Kim YS, Wiesenborn DD, Lorenzem JH, and Berguland P, 1996 *Cereal Chem.* **73** (3) 302-308
[7] Leach HW 2009 Gelatinization of Starch. In: Whistler, R.L. and Paschall, E.F. (eds). *Starch: Chemistry and Technology*, Vol I. Industrial Aspects. 3rd Edition (New York and London: Academic Press)
[8] Li JH and Vasanthan T 2003 *J. Food Res. Int.* **36** 381-386
[9] Lii CY and Chang Y H 1981 *J. Food Sci* **46** 78-81
[10] Nugraheni M, Santoso U, Suparmo, and Wuryastuti 2011 *J. Food Res. Int.* **18** 1471-80
[11] Rahim A 2008 *J. Agroland* **15** (2) 101 – 105
[12] Singh N, Singh J, Kaur L, Sodhi NS, and Gill SB 2003 *Food Chem.* **81** 219-231
[13] Suryani CL 1999 *Pemutihan dan Pengikatan Silang Pati Sagu untuk Substitusi Beras pada Pembuatan Bihun* (Yogyakarta: Universitas Gadjah Mada)
[14] Loubes MA, Flores SK and Tolaba MP 2016 *Food Sci. Tech.* **69** 280-286
[15] Ye X and Sui Z 2016 *Int. J. Biol. Macromolecule* **84** 428-433

IOP Conf. Series: Materials Science and Engineering **193** (2017) 012040 doi:10.1088/1757-899X/193/1/012040

Chemical changes associated with lotus and water lily *natto* production

S D Lestari[1*], N Fatimah[1] and R Nopianti[1]
[1] Fisheries Product Technology Study Program, Sriwijaya University, Ogan Ilir, Indonesia 30662

*E-mail: shantidwita_thi@unsri.ac.id

Abstract. *Natto* is a traditional Japanese food made by fermenting whole soybean seeds with pure culture of *Bacillus subtilis* subsp. *natto*. The purpose of this study was to investigate the suitability of lotus (*Nelumbo nucifera*) and water lily (*Nymphaea stellata*) seeds as the raw materials for *natto* production. Chemical (proximate, amino acids and minerals) changes were observed on raw, steamed and fermented seeds. Proximate compositions of all samples were calculated in both wet basis and dry basis. In wet basis calculation, steaming and fermentation tended to lower the carbohydrates, ashes, fats and protein content which were attributed to the increase of moisture. The total amino acid, iron and magnesium contents of raw lotus seeds were 24.29%, 5.08 mg 100g-1 and 174.23 mg 100g-1 dry matter, respectively. After a 24h-fermentation at 40°C, the total amino acids decreased while iron and magnesium contents increased significantly reaching, in respective order, 9.9 mg 100g-1 and 411.36 mg 100g-1 dry matter. Changes in chemical composition after fermentation were more pronounced in lotus seeds than water lily seeds indicating that their nutrient composition were more suitable to support *Bacillus subtilis* growth.

1. Introduction

Natto is a traditional Japanese fermented soybean prepared commercially using a pure starter culture of a spore-forming Gram-positive bacterium, *B. subtilis*. There are some other unsalted fermented soybean foods resembling *natto*, for instance Himalayan *kinema* [1], Korean *chongkukjang* [2] and Thai *thua nao* [3]. *Kinema* is traditionally made by wrapping lightly crushed boiled soybeans in local fern leaves inside a bamboo basket to provide a source of *Bacillus subtilis* bacteria and let to ferment at warm temperature for 1-3 days. The same steps applied to *natto* and *chongkukjang* beside the use of rice straws instead of fern leaves. As for *thua nao*, the boiled soybean is wrapped in bamboo baskets lined and covered with banana leaves. All of these products do have similar properties for instance distinct flavor and stickiness. The nutritional properties of the raw materials used are improved while the texture and wide variety of flavors develops during fermentation. *Natto* fermentation also adds the functional properties of soybean by the presence of nattokinase, polyglutamic acid and dipicolinic acid [4]. Nattokinase is a fibrinolytic enzyme shows ability to break the cross-linking of fibrin *in vitro* [5]. Research conducted by [6] indicated that the enzyme has optimal pH value and temperature of 8.0 and 40 °C, respectively and shows high thermostability at 30 to 50 °C.

Other raw materials have been used to replace soybean, for instance chickpeas (*Cicer arietinum*) [7] [8] and pigeon pea [9] based on their nutritional value and acceptability as a food. Lotus (*Nelumbo nucifera*) and white lily (*Nymphaea stellata*) are underutilized water plants which seeds have been found to be highly nutritious [10] [11]. The seeds of *N. nucifera* are rich in protein, unsaturated fatty acids, minerals and starch [12]. Several bioactive compounds have also been derived from lotus, including alkaloids, flavonoids, glycosides, triterpenoids, and vitamins which all have therapeutic impact [13]. Isoflavone, tannins, saponins and lignan are also detected in both plants [14], adding their benefits to human's health. Based on their richness in nutritive and functional components, lotus and water lily seeds stand as promising raw materials for the preparation of fermented food products, including *natto*. However, their suitability as raw materials of solid-state *natto* fermentation has not been reported. Therefore, the current research focused on bringing to light their potentials as *natto* raw materials by evaluating the amino acids profile, proximate and mineral changes of lotus and water lily seeds during *natto* production.

2. Experimental detail

2.1. Starter preparation

The overnight culture of *natto* starter culture (*Bacillus subtilis*) on NB medium was streaked sterilely into NA medium and incubated at 37 °C for 24h. Actively growing colonies on NA were diluted to 0.85% saline water in order to create turbidity equaled to Mc Farland standards 1.

2.2. Seeds preparation

Mature lotus (*Nelumbo nucifera*) and water lily (*Nymphaea stellata*) fruits were collected from several locations of swamp waters in Banyuasin and Ogan Ilir districts, South Sumatera. Ripen lotus seeds with black color were separated from the fruit's pods. Water lily fruits sinking below the water were collected, cut, slightly crushed and washed thoroughly to separate the small dark green ripen seeds. Both types of seeds were then sun-dried and subjected to hard shell removal leaving brown-colored seeds which were used as *natto* raw materials.

2.3. Natto preparation

The method used to prepare *natto* was similar to the current common practice in Japanese households in respect to the relative proportions of seeds and water as well as the fermentation time and conditions. Lotus and water lily seeds were soaked in tap water for 24h at 29±1 °C with a ratio of 1:3 w/v and during the process the soaked water was changed once. At the end of soaking period, the water was discarded. After dehulling, the lotus seeds were rinsed and cut in two pieces in order to remove the bitter green embryos inside. Both type of seeds were washed thoroughly, drained and weighed. Steaming was done for 10 minutes at 121 °C in an autoclave by placing the seeds inside the stainless steel baskets. After steaming, both seeds were placed in a sterile polypropylene container, cooled down until the temperature reached 50 °C and directly inoculated with 4mL of *Bacillus subtilis* in saline solution (0.85% NaCl) per 100 g steamed seeds. The cell density of the inoculums was approximately 3 x 10^8 CFU/mL, attained by adjusting the turbidity of the cells with McFarland standards 1 (Himedia). After mixed thoroughly, the containers were covered with sterile cheese clothes and their lids were put just on top of cheese clothes for air circulation. Samples were incubated in the water bath (Memmert) for 24h at 40 °C and 99% relative humidity. After fermentation, samples were stored in a refrigerator at a temperature of 5±1 °C for 12h.

2.4. Analytical methods

The nutritional compositions (ash, fat, moisture and protein) of the fermented lotus and water lily seeds were evaluated according to the Indonesian National Standard (SNI) 01-2891-1992 method. The carbohydrate content was estimated using by-difference calculation. The nutritionally essential elements including Fe and Mg were determined using Atomic Absorption Spectrophotometer (AAS) and Inductively Coupled Plasma Optical Emission Spectrometry (ICP-OES). The energy contents of the samples were obtained by multiplying crude protein, crude fat and carbohydrate by factors of, in respective order, 4, 9 and 4. Amino acids analyses were performed by using Waters Acquity UPLC H

Class Amino Acid Analysis System following the method outlined in [15]. Amino acids were identified by comparing retention times with a standard mixture of 15 authentic amino acids. The content of each amino acid was calculated on the basis of the standard curve of each authentic amino acid standard.

3. Results and discussion

Chemical changes during the preparation of *natto* were assessed in relation to the raw materials. The chemical changes that took place during the preparation of *natto* are given in table 1.

Table 1. Changes in chemical characteristics of lotus and water lily seeds during *natto* preparation

Parameters	Lotus seeds			Water lily seeds		
	Raw	Steamed	Fermented	Raw	Steamed	Fermented
Moisture (%)	11.18	73.8	74.92	13.58	54.25	71.67
Dry matter (%)	88.82	26.2	25.08	86.42	45.75	28.33
Ashes (% DM)	4.40	4.08	5.02	0.79	0.74	0.78
Total Fat (% DM)	2.09	2.44	3.39	1.13	1.18	1.98
Protein (% DM)	27.18	29.96	34.09	11.18	11.82	11.79
Total Carbohydrates (% DM)	66.33	63.51	57.49	86.90	86.23	85.46
Total energy (Kal)	392.86	395.88	396.85	402.52	402.84	406.78
Energy from fat (Kal)	18.85	21.98	30.50	10.21	10.62	17.79
Soluble dietary fiber (% DM)	5.92	Nd	Nd	5.38	nd	nd
Total sugar	6.98	Nd	Nd	0.86	nd	nd
Fe (mg/100g)	5.08	0.77	9.90	3.54	2.82	7.18
Mg (mg/100g)	174.22	226.03	411.36	61.50	83.61	70.03
Vitamin K_1 (mcg/100g)	<2	<2	<2	<2	<2	<2

nd not determined

The initial proximate composition of lotus seeds used in this study was 11.18% moisture, 3.81% ashes, 1.86% fats, 24.14% proteins and 58.91% total carbohydrates (wet basis) with the total energy of 348.94 Kal/100g. The nutritional value of lotus seeds had also been evaluated by [10]. In 100g, lotus seeds contains 63-68% of carbohydrate which mostly in the form of starch, 17-18% of protein, 1.9-2.5% of fat, and the remainders are water (about 13%) and minerals (mainly sodium, potassium, calcium, and phosphorus) providing about 350 Kal of energy. The author also found out that lotus seeds were low in fiber and not a good source of vitamins.

Water lily seeds have lower protein and total sugar content than lotus seeds, accounted for only 11.18% and 0.86% dry basis compared to 27.18% and 6.98% dry basis respectively. Soybean, which is the common raw material for *natto* production, has 44.19% protein [16]. In Japan, small sized and white colored Nattoshoryu soybean cultivar with 42.6% protein and 30.4% carbohydrates (dry basis) was preferred for *natto* production [17].

During the fermentation of *natto*, the biochemical processes that cause typical changes in chemical and physicochemical characteristics of the raw ingredient take place. Those changes are induced by sugars, proteins and fats decomposing-enzymes which present in the tissue or secreted by microbes. Soluble sugars, especially sucrose, are the major source of energy for fermentation [18]. The amount of sucrose in soybean ranges from 1.5 to 10.2% of total dry matter [19]. As the free sugars composition of soybean influences the effective fermentation, increased total sugar and sucrose content is important to achieve the proper rate of fermentation during the *natto*-making process and to improve the taste and flavor since sucrose is consumed faster than other types of sugars such as raffinose and stachyose on the fermentation process [17]. Enzymatic degradation of proteins is also a key process in *natto* fermentation. In general, the proteins are hydrolyzed into peptides, amino acids, and ammonia. All enzymatic activities in the process are expected to contribute to distinct attributes of *natto* including texture, appearance, flavor, and aroma.

Good quality *natto* products should be covered with a white colored mucous substance, has a slimy appearance, softer texture and are able to generate silky and sticky mass when mixed/stirred with a pair of chopsticks [20]. Based on those criteria, only lotus seeds *natto* gave satisfactory result. No formation of white-colored viscous substance was observed in water lily seeds after a 24h-fermentation at 40°C. The changes in sensory attributes were in line with its chemical properties. From Table 1, it could be noticed that in dry basis analysis, fermentation process lowered the total carbohydrates of lotus seeds, while the ashes, protein, and fats increased. The proximate value of soybean and soybean *natto* had been compared and it was revealed that fermentation increased total fat, slightly lowered the protein and ashes and considerably decreased the total carbohydrates [16]. Increase in fats, moisture and carbohydrates followed by decrease in proteins of cooked soybeans in the fermentation of *chongkukjang* for 2-4 days at 40-43°C were also reported [2].

Some of the carbohydrates were used up as carbon source by the fermenting organism and converted to microbial biomass. It contributed to the decrease in carbohydrate content of lotus seeds after fermentation. The increase in the protein content of lotus seeds after fermentation in this study might be due to the secretion of extracellular enzymes by *Bacillus subtilis* which consequently increased the protein content of the fermented product, as well as the microbial biomass. The protein values of raw, steamed and fermented lotus seeds samples which determined based on the total N were, in respective order, 27.18%, 29.96% and 34.09%. The amino acids profiles of those three products are presented in table 2. The quantities of total amino acids analyzed were found to be 24.294%, as dry basis, in raw lotus seeds. Steaming slightly increase the total amount of fifteen amino acids to 25.799% while fermentation decreased them to 22.237%. The gaps between total protein and total amino acids were accounted for other amino acids, such as methionine, tryptophan, sulfurous amino acids and probably other non protein nitrogen compounds.

Table 2. Changes in amino acids composition of lotus and water lily seeds during *natto* preparation

Seeds type	Lotus			Water Lily			Soy bean [16]	*Thua Nao* [16]	*Natto* [16]
Processing step	Raw	Steamed	Fermented	Raw	Steamed	Fermented			
Amino acids									
L-Asp	2.337	2.626	1.876	0.648	0.665	0.619	4.842	5.006	4.335
L-Glu	4.958	5.674	5.403	1.760	1.868	1.790	8.525	8.740	9.363
L-Ser	1.917	1.876	1.173	0.822	0.885	0.879	2.304	2.353	1.967
Gly	1.312	1.292	1.209	0.346	0.380	0.364	2.025	2.134	1.804
L-Ala	1.052	1.288	1.308	0.565	0.609	0.592	1.920	2.220	1.668
L-Pro	0.786	0.882	0.697	0.259	0.269	0.241	2.047	2.297	1.990
L-Tyr	0.664	0.655	0.792	0.390	0.402	0.331	1.441	1.894	1.903
L-His*	0.705	0.651	0.658	0.366	0.375	0.315	1.012	1.026	1.055
L-Arg*	2.508	2.125	1.539	1.391	1.398	1.203	3.533	3.297	2.601
L-Ile*	1.008	1.092	0.924	0.496	0.505	0.467	2.099	2.186	1.879
L-Leu*	1.746	1.878	1.561	1.000	1.011	0.933	3.564	3.680	3.300
L-Lys*	1.892	2.342	1.722	0.272	0.270	0.285	2.751	2.628	2.699
L-Val*	1.180	1.293	1.119	0.556	0.570	0.533	2.054	2.136	1.971
L-Phe*	1.117	0.978	1.202	0.647	0.626	0.456	2.455	2.384	2.213
L-Thr*	1.112	1.146	1.039	0.494	0.497	0.471	1.742	1.849	1.593
Total	24.294	25.799	22.237	10.013	10.331	9.480			

*essential amnino acids
Data expressed as % sample dry matter

Amino acid analyses are usually feasible for the expression of total amino acids in various foods. Their amino acid content and pattern are important from nutritional point of view. Changes of fifteen amino acids composition were analyzed in every step of *natto* processing. Predominant amino acids in lotus *natto* were glutamic acid, followed by aspartic acid, lysine, leucine, arginine, alanine, glysine and phenylalanine. The predominant amino acids compared favorably with those of soybean thua nao and *natto* [16] despite a slight difference in pattern was noted. In both soybean products, the order was glutamic acid, aspartic acid, leucine, arginine, lysine and serin. Some hydrophobic amino acids, such as alanine and phenylalanine increased after fermentation, while histidine remained at the same level in all processing steps. The same trend was also observed by [16]. The decrease in the rest of amino acids suggests that they were consumed or transformed somehow at a greater rate than they were formed by proteolytic activity.

Amino acids, peptides and proteins have direct contribution to the flavor of the foods. L-glutamic acid is known for its flavor-enhancing properties. According to [21], pure amino acids have their own taste properties. L-tryptophan, L-phenylalanine, L-tyrosine and L-leucine are amino acids which contribute to product's bitterness. Two out of these were presented in lotus *natto* in considerable amounts and constitute *natto*'s prominent amino acids. Lotus *natto* was also rich in amino acids which tastes are barely perceptible, for instance L lysine and L arginine. The fermentation process of lotus *natto* contributed to the increase of some sweet-tasted amino acids such as L-alanine and glysine.

4. Conclusions

Based on chemical assessment on both lotus and water lily *natto*, this research confirmed the suitability of lotus seeds as promising *natto* raw material. Decomposition of carbohydrates took place during the fermentation of lotus seeds, the same trend was also observed on soybeans by other authors. Very low total sugar content in water lily seeds compared to lotus seeds and soybeans was probably the reason of their unfitness as *natto* raw material.

Acknowledgments

The authors would like to thank Sriwijaya University for funding through SP DIPA Number 042.01.2.400953/2016 for SAINTEKS Research Scheme 2016.

References

[1] Sarkar P, Tamang JP, Cooks PE and Owens JD 1994 *Food Microbiol.* **11(1)** 47–55
[2] Shin D and Jeong D 2015 *J. Ethn. Foods* **2** 2-7
[3] Chukeatirote E 2015 *J. Ethn. Foods* **2** 115-18
[4] Fujiwara K, Miyaguchi Y, Feng XH, Toyoda A, Nakamura Y, Yamazaki M, Nakashima K and Abe H 2008 *Asian-Aust. J. Anim. Sci.* **21(12)** 1766-72
[5] Fujita M, Nomura K, Hong K, Ito Y, Asada A and Nishimuro S 1993 *Biochem. Biophys. Res. Commun.* **197** 1340-7
[6] Wang C, Du M, Zheng D, Kong F, Zu G and Feng Y 2009 *J. Agric. Food Chem.* **57(20)** 9722–9
[7] Xiao P, Yao S, Rehman RU, Kang R and Wang Y 2014 *Adv. J. Food Sci. Technol.* **6(12)** 1294-1300
[8] Wei X, Luo M, Xie Y and Liu H 2012 *Appl. Biochem. Biotechnol.* **168(7)**
[9] Jin S, Feng C, Luo M, Wang W, Xiaa XX, Zua YG, Lic LP and Fua YJ *2015 Innov. Food Sci. & Emerg. Technol.* **31** 160-9
[10 Dharmananda S 2002 Lotus seed: food and medicine *Itmonline.org*. Web. 16 Oct. 2016.
[11] Musa A, Birnin-Yauri UA, Muhammad C and Umar A 2012 *Afr. J. Food Sci* **3(7)**
[12] Wu JZ, Zheng YB, Chen TQ, Yi J, Qin LP, Rahman K and Lin WX 2007 *Food Chem.* **105** 540–7
[13] Mukherjee PK, Mukherjee D, Maji AK, Rai S and Heinrich M 2009 *J. Pharm. Pharmacol.* **61** 407–22
[14] Baehaki A, Lestari SD, Apriyanti W 2015 *J. Chem. Pharm. Res.* **7(11)** 221-224
[15] Aristoy MC and Toldra F 2013 Amino acids *Food analysis by HPLC 3rd Edition* ed LM Nollet and F Toldra (Boca Raton Florida: CRC Press)
[16] Nikkuni S, Karki TB, Vilkhu KS, Suzuki T, Shindoh K, Suzuki C and Okada N 1995 *Food Sci. Technol. Int.* **1(2)** 107-11
[17] Taira H 1990. *Jpn. Agric. Res. Q.* **24** 224–30

[18] Hou A, Chen P, Alloatti J, Li D, Mozzoni L, Zhang B and Shi A 2009 *Crop Sci.* **49** 903–12
[19] Hymowitz T and Collins FE 1974 *Agron. J.* **66** 239–40
[20] Wei Q, Wolf-Hall C and Chang KC 2001 *J. Food Sci.* **66(1)**
[21] Solms J 1969 *J. Agric. Food Chem.* **17(4)** 686–8

IOP Conf. Series: Materials Science and Engineering **193** (2017) 012041 doi:10.1088/1757-899X/193/1/012041

Physical and Chemical Characterization of Greater Yam (*Dioscorea alata*) and Jack Bean (*Canavalia ensiformis*) - Based Composite Flour

D R Affandi, D Praseptiangga, F S Nirmala, B Sigit Amanto, W Atmaka

Department of Food Science and Technology, Faculty of Agriculture, Universitas Sebelas Maret, Jl. Ir. Sutami 36 A, Kentingan, Surakarta 57126, Indonesia

E-mail : d_affa@hotmail.com

Abstract. Indonesia is a tropical country that has great potential in agriculture. Tubers and legumes as examples of the potential commodities are needed to be more developed. Flour production is one of the best alternatives to be chosen as the downstream stage of the tubers and legumes utilization. Greater yam *(Dioscorea alata)* and jack bean *(Canavalia ensiformis)* were used in this study. This study was conducted to determine the best formula of composite flour based on physical, chemical, and functional characterization of composite flour produced. Variations of formula used was the ratio of greater yam flour and jack bean flour, which were 85:15 (F1), 70:30 (F2), 55:45 (F3), respectively, and this study was conducted using completely randomized design (CRD). The formula variations did not show any significant effect on the water absorption capability, water holding capacity (WHC), oil holding capacity (OHC), swelling power, and starch content of the composite flour. However, the formula variations had a significant influence on the colour, proximate parameters, amylose and amylopectin content, resistant starch content, dietary fibre, total phenol, and antioxidant activity of the composite flour produced. Considering the results of physical, chemical, and functional characteristics of composite flour, formula (F1) was selected as the best composite flour developed from greater yam and jack bean flours.

1. Introduction

Indonesia is rich in legumes and tubers. Local tubers have been consumed as a source of carbohydrates from earlier times and proved to have a bit trigger for a degenerative disease in the past [1]. Legumes is a cheap source of protein, it also contains carbohydrates, cholesterol, high-fiber, low in fat, and high concentration of unsaturated fatty acids [2].

Greater yam is one kind of potential tubers. The productivity is 40 tons per hectare, it could grow above the sea level to more than 1500 above sea level and from the moist soil (swamp) to marginal land [3]. Heretofore greater yam only considered as a wild plant that is limited use [4]. Greater yam contain nutrients such as carbohydrates are quite large, which is 24.6% [5]. Most carbohydrates are in the form of starch consisting of amylose and amylopectin. Amylose content in greater yam is in the range of 19-20% [6]. Greater yam also contains micronutrients such as flavonoids, saponins, and phenolic compounds that are beneficial to the body so that it can be considered as selected tuber for daily consumption [7]. Greater yam has a low glycemic index and fiber content is high [8]. Despite having the functional potential and good nutritional value, but greater yam contain only small content of protein in the amount of 0.32 to 2% [9].

Jack bean is one of the local legumes that can be used as a source of vegetable protein with a carbohydrate content of 55% and 24% protein [10]. Jack bean also contains anti nutritional compound in the form of phytic acid and glucocyanide toxin. Jack bean can be processed into food products such as flour and the derivatives products such as cakes, cookies and other bakery products, crackers, tempeh and some other products. Moreover, jack bean is a commodity available in large quantities on the market that in the year 2010-2011 was recorded on the land area of 24 hectares, 12 districts in Central Java has produced 216 tons of jack bean in every harvest [11].

Functional potential owned by greater yam and jack bean made the commodity can be used as an alternative to a functional food that can be consumed by the public. Many previous studies about substitution of wheat flour as the main purpose of composite flour's production, but not much is discussed about the functional potential of composite flour. Therefore, this study was conducted to learn more about the functional potential possessed by composite flour from greater yam (*Dioscorea alata*) and jack bean (*Canavalia ensiformis*) based on their physical and chemical characteristics. Then, the best formula based on physical and chemical characteristics (functional charateristics was included) would be determined.

2. Materials and Methods

2.1. *Making Composite Flour*

In the making of greater yam flour, there was a pretreatment where the sliced greater yams were soaked in a solution of Na-metabisulfite 0.3% for 20 minutes to prevent browning. The sliced greater yams that have been soaked then washed, dried in a cabinet dryer (temperature of 50°C for 24 hours), powdered, and sieved (80 mesh). In the making of jack bean flour there was a pretreatment where the beans were soaked in water for three days with water changes every 12 hours, aiming to reduce the levels of HCN. Jack beans that have been soaked and then boiled, peeled, chopped, dried in a cabinet dryer (temperature of 60 ° C for 10 hours), powdered, and sieved (80 mesh).

The formula used in the making of composite flour was the ratio between greater yam flour and jack bean flour, ie 85:15 (F1), 70:30 (F2), and 55:45 (F3). The three formulas of composite flour were made by mixing greater yam flour and jack bean flour in a plastic jar. Then a plastic jar was shaken up until greater yam flour and jack bean flour mixed homogeneously.

2.2. *Method of Analysis*

Several analysis were conducted on composite flour. Physical properties tests were conducted on water absorption capacity / WHC by filtration method [12], water holding capacity by centrifugation method [13], oil holding capacity [14], swelling power by centrifugation method [15], and color test with chromameter [16]. Chemical analysis were conducted on water content by thermogravimetric method [17], ash content in the dry method [17], protein content by Kjeldahl method [17], fat content by Soxhlet extraction method [17], carbohydrate content by difference, starch content [18], amylose and amylopectin content by the spectrophotometric method [18], and resistant starch [19]. Analysis of functional properties were conducted on antioxidant activity by DPPH radical-scavenging activity [20], dietary fiber by multienzyme method [21], and total phenolic by Folin Ciocalteu method [22].

3. Results and Discussion

3.1. *Physical Characteristics*

3.1.1. *Color.*

Color analysis conducted using chromameter to get the value of L *, a *, and b *. Then the °Hue was calculated with the formula °Hue = arc tan (b * / a *).

Table 1 Lightness (L*), Yellowness (b*), Redness (a*) and °Hue of Greater Yam (*Dioscorea alata*) and Jack bean (*Canavalia ensiformis*) - Based Composite Flour

Parameter	Composite Flour**		
	F1	F2	F3
L*	70.78 ± 0.03^a	71.47 ± 0.08^b	72.23 ± 0.05^c
a*	2.75 ± 0.005^a	2.61 ± 0.05^a	2.62 ± 0.06^a
b*	14.79 ± 0.01^a	15.01 ± 0.03^b	15.33 ± 0.08^c
°Hue	79.49 ± 0.01^a	80.16 ± 0.15^b	80.30 ± 0.17^b

Value in the same line followed by different letter was significantly different (p=0,05)
** Proportion of greater yam flour:jack bean flour were F1=85:15, F2=70:30, F3=55:45

Table 1 shows the test result of each color composite flour formula. F3 has the highest lightness level (L*), while F1 had the lowest lightness level. The highest level of b * owned by F3 because the proportion of jack bean flour most compared to other formulas. Jack bean flour had more white color (b * higher) than greater yam flour. Highest °Hue owned by F3, meaning the intensity of the yellowish red getting stronger. Comparison of the three color composite flour with greater yam flour and jack bean flour could be seen in Figure 1.

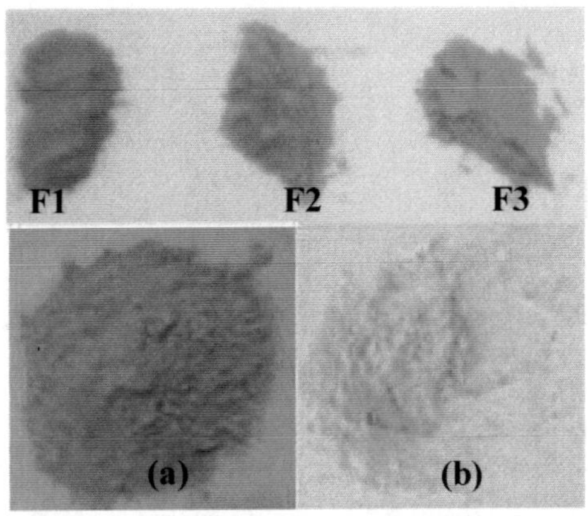

Figure 1 The Color of Composite Flour (F1, F2, F3), Greater Yam Flour (a) and Jack Bean Flour (b)
(Proportion of greater yam flour:jack bean flour are F1=85:15, F2=70:30, F3=55:45)

Figure 1 shows that greater yam flour and jack bean flour had a different color with composite flour. Greater yam flour had yellowish red color that appear as shades of brown. Greater yam flour color was darker than jack bean flour that had a bright yellow color. The more proportion of greater yam flour in composite flour caused the color to become darker. F1 with the most proportion of greater yam flour had the darkest color (low of L* and low of °Hue) compared to F2 and F3. Table 1 showed that the °Hue of F1, F2, and F3were in the range of 50°-90°, indicated that the composite flour had a yellowish red color range. Increasingly approaching 90° then the color of flour produced would be more bright (yellow).

3.1.2. Water Absorption Capability (WAC), Water Holding Capacity (WHC), Swelling Power, and Oil Holding Capacity (OHC)

Table 2 shows that the variation in the proportion of greater yam flour and jack bean flour had no effect on WAC, WHC, OHC, and swelling power. However, when seen in Figure 2,3 and 4, WAC likely to increase from F1 to F3, as well WHC and swelling power. Instead OHC value tend to decline as more proportion of jack bean flour.

Table 2 Water Absorption Capacity (WAC), Water Holding Capacity (WHC), Oil Holding Capacity, and Swelling Power of Greater yam (*Dioscorea alata*) and Jack bean (*Canavalia ensiformis*) - Based Composite Flour

Flour*	WAC (%)	WHC (g/g)	OHC (%)	Swelling Power (g/g)
F1	77.79 ± 0.52[a]	2.40 ± 0.02[a]	1.08± 0.006[a]	4.99 ± 0.06[a]
F2	78.04 ± 6.34[a]	2.54 ± 0.05[a]	1.07± 0.02[a]	5.31 ± 0.16[a]
F3	78.32 ± 0.32[a]	2.56 ± 0.07[a]	1.01± 0.04[a]	6.06 ± 0.52[a]

Value in the same line followed by different letter was significantly different (p=0,05)
* Proportion of greater yam flour:jack bean flour were F1=85:15, F2=70:30, F3=55:45

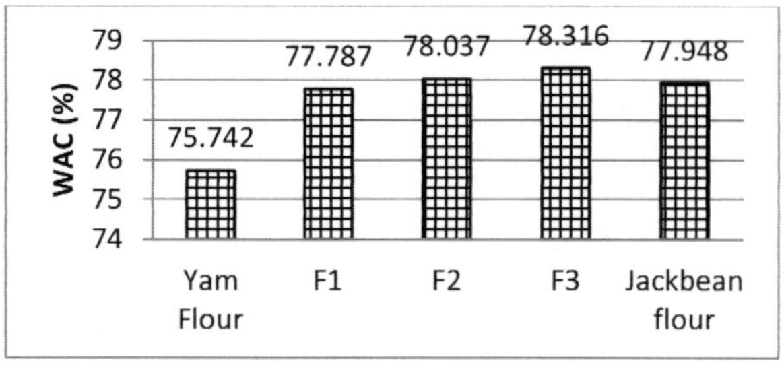

Figure 2 Water Absorption Capacity (WAC) of Greater Yam Flour, Jack Bean Flour, and Composite Flour
(Proportion of greater yam flour:jack bean flour were F1=85:15, F2=70:30, F3=55:45)

Figure 2 and Figure 3 show that the water absorption and water holding capacity of jack bean flour was higher than greater yam flour. The high water absorption of jack bean flour was caused by the high protein content of jack bean flour (19.796%) which was higher than greater yam flour (7.410%), according to preliminary research. There were several factors that could affect the WAC and WHC a food ingredient. Besides the presence of protein in a food, the WAC is affected by the composition and physical properties of the starch granules after adding water [23]. Based on preliminary research, jack bean bean flour had a higher protein content than greater yam flour, whereas the starch content of greater yam flour was higher than jack bean flour. Protein is a factor which could determine the ability of flour to bind water [24]. The higher the protein content of a food will increase the value of water holding capacity (WHC). A low fiber and starch content may cause low absorption ability of water into the flour [25]. WAC and WHC composite flour was higher than basic flour essentially thought to be caused by the high protein of jack bean flour and higher starch content of greater yam flour.

Figure 4 shows that the swelling power of composite flour was likely to increase as more and more jack bean flour. Swelling power is a property that is influenced by amylopectin. The more branch chain of amylopectin are contributing to increase the swelling power [26]. Based on this explanation, it can be said that the swelling power will increase as the higher amylopectin content or lower amylose content. Amylose content of greater yam flour was higher (28.072%) than jack bean flour (22.885%), based on preliminary research, so that the more proportion of greater yam flour would increase amylose content in composite flour which had decreased the swelling power.

Figure 5 shows that oil holding capacity (OHC) of greater yam flour higher than jack bean flour. The high OHC of greater yam flour affected by amylose content of greater yam flour that was higher than jack bean flour (preliminary research). OHC of composite flour was lower than the OHC of basic flour. This presumably because the protein from leguminoceae flour was hydrophilic. Other studies showed that the composite flour oil absorption capacity would decrease with the addition of a legume flour [27]. It is also reinforced with a statement saying that the flour made from the legume naturally have a protein with a hydrophilic side which is more dominant than the hydrophobic, so that the oil absorption capacity flour made from legume becomes lower [28]. Thus, the addition of jack bean flour in the production of composite flour would further lower the value of oil holding capacity.

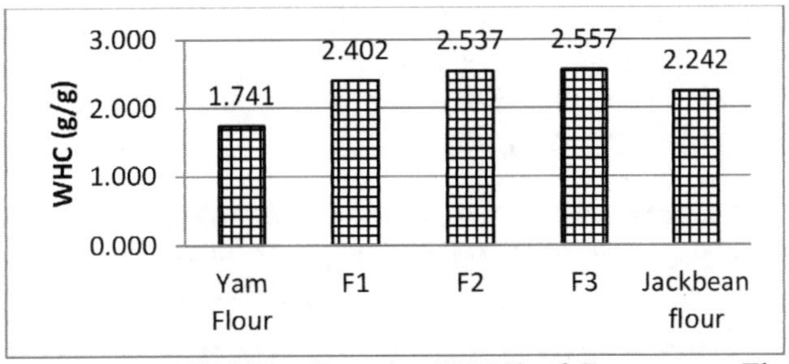

Figure 3 Water Holding Capacity (WHC) of Greater yam Flour, Jack bean Flour, and Composite Flour

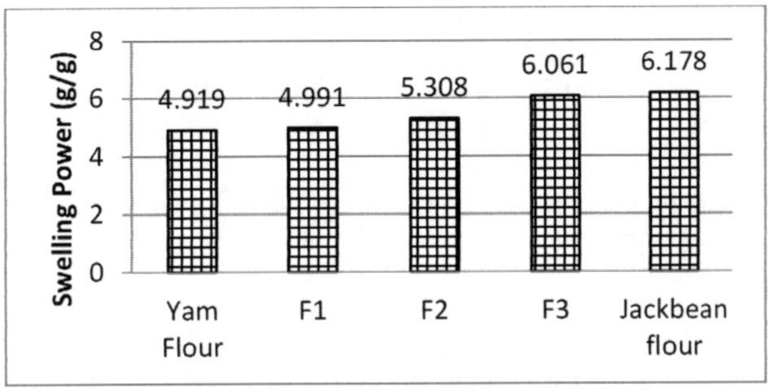

Figure 4 Swelling Power of Greater yam Flour, Jack bean Flour, and Composite Flour

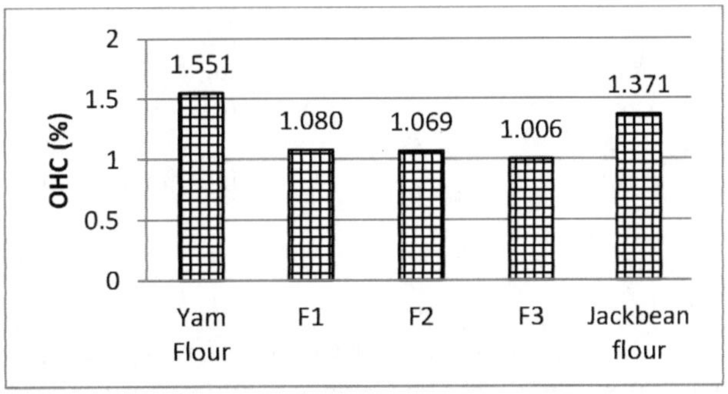

Figure 5 Oil Holding Capacity (OHC) of Greater yam Flour, Jack bean Flour, and Composite Flour
(Proportion of greater yam flour:jack bean flour were F1=85:15, F2=70:30, F3=55:45)

3.2. Chemical Characteristics

The moisture content of the composite flour was affected by the moisture content of basic flour. Soaking was one of steps in the making of greater yam flour or jack bean flour, so that water was not only naturally derived from the material but also from water immersion. Jack bean flour added in the making of composite flour increased the moisture content (Table 3), it was supposed that the protein of jackbrean flour was hydrophilic. Composite flour from legumes have a protein with hydrophilic dominant side [28].

The ash content and carbohydrate content of composite flour decreased from F1 to F3 (Table 3) . These because the ash content of jack bean flour was lower (1.261%) than greater yam flour (4.294%), and carbohydrate content of jack bean flour was lower (66.567%) than greater yam flour (82.143%).

The protein and fat content of composite flour increased from F1 to F3 (Table 3). This because the protein content of jack bean flour was higher (19.795%) than greater yam flour (7.41%), and fat content of jack bean flour was higher (4.463%) than greater yam flour (0.431%). The proximate data of basic flour was based from preliminary research.

Table 3 Chemical Characteristics of Greater yam (*Dioscorea alata*) and Jack bean (*Canavalia ensiformis*) - Based Composite Flour

Parameter	Composite Flour*		
	F1	F2	F3
Moisture (%)	8.78 ± 0.02^b	8.51 ± 0.02^a	10.03 ± 0.02^c
Ash (%)	4.05 ± 0.01^c	3.53 ± 0.01^b	2.93 ± 0.02^a
Fat (%)	1.14 ± 0.02^a	1.70 ± 0.03^b	2.22 ± 0.01^c
Protein (%)	11.34 ± 0.02^a	14.14 ± 0.02^b	16.06 ± 0.006^c
Carbohydrate (%)	74.69 ± 0.03^c	72.11 ± 0.04^b	68.77 ± 0.05^a
Amylose (%)	20.38 ± 0.07^b	21.25 ± 0.09^c	17.23 ± 0.09^a
Amylopectin (%)	48.84 ± 0.41^b	46.09 ± 0.49^a	51.07 ± 0.43^c
Starch (%)	69.22 ± 0.34^a	67.34 ± 0.40^a	68.30 ± 0.51^a

Value in the same line followed by different letter was significantly different (p=0,05)
*Proportion of greater yam flour:jack bean flour were F1=85:15, F2=70:30, F3=55:45

Amylose and amylopectin content of greater yam flour were higher than jack bean flour, based on preliminary research. The addition of greater yam flour would increase the level of amylose and amylopectin content. However, amylose content of F1 with the more proportion of greater yam flour was even lower than the amylose content of F2. The amylose content of F3 was the lowest of which it was appropriate to amylose ratio of the constituent (basic flour). In general, Table 3 showed that formula variation caused decreasing in amylose content of composite flour compared to the basic flour.

The starch contents of three composite flour were higher than the greater yam flour (64.913%) and jack bean flour (54.033%). The production of composite flour was done mechanically and did not have the factors that caused increasing level of starch. Therefore it is necessary to do further research, especially concerning the molecular structure of starch, to find out how increasing level of starch could occur.

3.3. Functional Characteristics

Resistant starch (RS) content of greater yam flour was slightly higher (6.26%) than jack bean flour (5.975%), so the addition of greater yam flour would increase resistant starch content of composite flour. Resistant starch content of composite flour was appropriate to the resistant starch content of basic flour (Table 4).

Resistant starch content of composite flour was higher than the RS content of basic flour. Improved RS can occur due to heating and chemical modification [29], the crystallization of amylose (Eerlingen et.al., 1993), and bonding between amylose and lipid [30]. High RS content in the composite flour thought to cause high dietary fiber content of composite flour. Resistant starch is included in the dietary fiber [31]. Dietary fiber in the composite flour is dominated by insoluble dietary fiber. Insoluble dietary fiber is known to dominate on seven different types of legumes [32], and the lima bean [33]. Total dietary fiber and insoluble dietary fiber content of composite flour were higher than in the basic flour. However, increasing levels of resistant starch and dietary fiber in this study was still unclear.

Soluble dietary fiber slows gastric emptying time, whereas insoluble dietary fiber has an influence on human faecal mass. The content of dietary fiber in food, either soluble or insoluble, may be a prebiotic. Many studies reported data that supported the role of dietary fiber in triggering the growth of lactic acid bacteria (*Lactobacillus*) having metabolic properties such as bifidobacteria to produce short-chain fatty acids and repair the immune system [34]. The content of dietary fiber in the composites flour (Table 4) showed that both had the potential to be a prebiotic flour and to help improve metabolism.

Table 4 Functional Characteristics of Greater Yam (*Dioscorea alata*) and Jack Bean (*Canavalia ensiformis*) - Based Composite Flour

Parameter	Composite Flour*		
	F1	F2	F3
Resistant starch (%)	19.22 ± 0.04^{b}	19,10 ± 0.04^{b}	18.11 ± 0.04^{a}
Soluble dietary fiber (%)	2.31 ± 0.02^{a}	2,50 ± 0.02^{b}	2.57 ± 0.02^{b}
Insoluble dietary fiber (%)	41.59 ± 0.12^{c}	40,64 ± 0.17^{b}	39.63 ± 0.26^{a}
Dietary fiber n(%)	43.90 ± 0.11^{b}	43,14 ± 0.15^{b}	42.20 ± 0.24^{a}
Total Phenolics (%)	1.58 ± 0.005^{c}	0.99 ± 0.003^{b}	0.79 ± 0.001^{a}
Antioxidant activity (% / sample weight)	49.10 ± 0.15^{c}	47.40 ± 0.19^{b}	45.82 ± 0.45^{a}

Value in the same line followed by different letter was significantly different (p=0,05)
*Proportion of greater yam flour:jack bean flour were F1=85:15, F2=70:30, F3=55:45

Total phenolics content of greater yam flour was 1.75% and total phenolic of jack bean flour was 0.13% (preliminary research). Based on the total phenolics content of basic flour, the addition of greater yam flour to the composite flour would increase the total phenolics content (**Table 4**). Phenolic compounds known to contribute significantly to its antioxidant activity, the higher content of the phenolic group compound, the greater the antioxidant activity is [35]. Antioxidative mechanism of phenolic compounds is the provision of an electron donor to oxygen and alkyl radicals to form stable radical phenocyl [34].

The antioxidant activity of greater yam flour amounted to 52.93% and jack bean flour antioxidant activity amounted to 15.63% (preliminary research). The more greater yam flour on composite flour increased antioxidant activity (Table 4). The antioxidant activity of composite flour included in the scale of moderate, it was between 20-50% [36].

4. Conclusion

In consideration of the purpose of study, based on physical, chemical, and functional characteristics, then F1 with the proportion of greater yam flour : jack bean flour = 85 : 15, assessed as the best formula of composite flour.

References

[1] Sari I P, Lukitaningsih E, Rumiyati, Setiawan I M 2013 *Jurnal Trad. Med.* **18** 127-131
[2] Retna G, Affandi D R, dan Ishartans D 2013 *Jurnal Teknosains Pangan* **2**
[3] Frech, 2006 cit Fahmi A, dan Antarlina SS 2007 Ubi Alabio Sumber Pangan Baru dari Lahan Rawa (Sinar Tani)
[4] Harzau, Hazzizah, dan Estiasih T 2013 *Jurnal Pangan dan Agroindustri* **1** 138-147
[5] Njie D N, Rumsey T R, Singh R P 1998 *J. Food. Eng* **37** 63-76
[6] Martin F. and Rubeste R. 1976. cit Ezeocha V C & Ojimelukwe P C. 2012. *J. Stored Prod. Postharvest Res.* **3**
[7] Mishra, T, Goyal A K, Middha S. K A, & Sen, A 2011 *IJNPR* **2** 315-321
[8] Sari I P, Lukitaningsih E, Rumiyati, & Setiawan I M 2013 *Jurnal Trad. Med.* **18** 127-131
[9] Prawiranegara D 1996 *Daftar Komposisi Bahan Makanan* (Jakarta: Direktorat Gizi Departemen Kesehatan RI)
[10] Windrati W S, Nafi A, Augustine P D 2010 *Agrotek.* **4** 18-26
[11] Wahjuningsih S B, Saddewisasi W 2013 *Riptek* **7** 2

[12] Elly, 1990. cit Lidiasari E, Syafutri M I, Syaiful F, 2006 *Jurnal Ilmu-Ilmu Pertanian Indonesia* **8** 141-146
[13] Subagio A, Windrati W S, & Witono Y 2003 *Jurnal Teknologi dan Industri Pangan* **14** 2
[14] Sosulski F W, Humbert E S, Bui K, and Jones J O 1976 *J. Food Sci.* **41**
[15] Leach H W, Mc Cowen L D, and Schoch T J 1959 *Cereal Chem.* **36**
[16] Hutching J B 1999 *Food Colour and Appearance* (London: Chapman and Hall Publisher)
[17] AOAC 2005 *Official Methods of Analysis of Association of Official Analytical Chemist* (Virginia: Inc. Arlington)
[18] Fardiaz D, Apriyantono A, Yasni S, Budiyanto S, dan Puspitasari N L 1986 *Penuntun Praktikum Analisa Pangan* (Bogor: Institut Pertanian Bogor)
[19] Goni I, Garcia-Diz L, Manas E, and Saura-Calixto F 1996 *Food Chem.* **56**
[20] Molynoeux P 2004 *Songklanakarin J. Sci. Tech.*
[21] Asp N G, Johansson C-G, Hallmer H, and Siljestrom M, 1983 *J. Agric. Food Chem.* **31**
[22] Rusydi M & Azrina A 2012 *Int. Food Res. J.* **19**
[23] Elliason A C 2004 *Starch in Food: Structure, Function, and Application* (New York: CRC Press)
[24] Raikos V, Neacsu M, Russel W, and Duthie G 2014 *Food Sci. Nutr.* **2**
[25] Ahmed J, Al-Attar H, and Arfat Y A 2015 *Food Hydrocol.*
[26] Li J Y and Yeh AI 2001 *J. Food Eng.* **50**
[27] Menon L, Majumdar S D, Ravi U 2014 *IJNPR* **5**
[28] Bhatt S M and Gupta R K 2015 *J. Phar. Phytochem.* **4**
[29] Wulan S N, Widyaningsih T D, Ekasari D 2007 *Jurnal Teknologi Pertanian* **8** 69-74
[30] Evans C, Yakubu E G, Tawakalitu B 2013 *Annals Food Sci. and Tech.*
[31] Lattimer, James M, Mark D H 2010 *J. Nutr.* **2** 1266-1289
[32] Lee B W, Andrews K W, Pehrsson P R 2002 *J. Food Comp. Anal.* **15** 715-723
[33] Ahmed S, Hasan M S, Mahmood Z A 2015
[34] Kusharto, Ciara M 2006 *Jurnal Gizi dan Pangan* **1**
[35] Arief 2008 cit Hardiana R, Rudiyansyah, Saharah T A 2012 *JKK* **I**
[36] Wulansari dan Chaerul 2011 cit Saefudin, Marusin S, Chairul 2013 *Jurnal Penelitian Hasil Hutan* **31**

IOP Conf. Series: Materials Science and Engineering **193** (2017) 012042 doi:10.1088/1757-899X/193/1/012042

Physical Properties of Milk Cincau Curd on Different Concentrations of Green Cincau Leaf (*Cyclea barbata L.Miers*)

B S Hertanto[1*], L R Kartikasari[1], Winny Swastike[1], M Cahyadi[1], A Yuliani[2], and A M P Nuhriawangsa[1]

[1]Laboratory of Industry and Animal Product Processing, Department of Animal Science, Faculty of Agriculture, University of Sebelas Maret, Jl. Ir. Sutami 36A Kentingan, Jebres, Surakarta 57126. Phone/Fax. 0271-637457

[2]Graduated Student of Department of Animal Science, Faculty of Agriculture, University of Sebelas Maret, Jl. Ir. Sutami 36A Kentingan, Jebres, Surakarta 57126. Phone/Fax. 0271-637457

*E-mail: bayuhertanto@uns.ac.id

Abstract. The objective of this study was to determine the effect of cincau leaf (*Cyclea barbata L.Miers*) on the physical properties of milk cincau curd. The materials of this research were milk cow of Local Friesian Holstein and leaves of cincau. This research used one way randomized design. The treatment of this research was concentration ratio between cincau leaf and cow milk (w/v): A1 = 10%:90%; A2 = 20%:80%; A3 = 30%:70%. The data was analyzed using ANOVA, and differences between treatment means were further analysed using Duncan's New Multiple Range Test. Our study revealed that different concentrations of cincau leaf significantly affected cohesiveness, chewiness, hardness, gumminess, springiness, pH and syneresis ($p<0.01$). However, it did not affect adhesiveness. In addition, the level of 30% of cincau leaf increased cohesiveness, and the level of 20% increased chewiness, hardness, gumminess, springiness, pH. On the other hand, syneresis decreased at the level of 20%. It can be concluded that the addition cincau leaf up to a level of 20% improved the physical properties of milk cincau curd.

1. Introduction

Green cincau leaf (*Cyclea barbata L.*Miers) is local name in Java Island. It is a vine growing in Asia, Africa and South America that can produce a gel-forming prepared by mixing the leaves with water [1]. Extracts of green cincau leaf contain pectin as a source of dietary fiber polysaccharides and the cause of the gel [2]. Chemical composition of the gel from green cincau leaf has been elucidated and classified as pectin consisting of the partly methylated and acetylated polygalacturonic acid backbone with neutral sugar side chains [3]. Pectin from green cincau leaf has unique characteristics because a gel occurs spontaneously in a short period of time at room temperature in the absence of neither sucrose nor calcium, and at the pectin concentration of as low as 0.5% w/v [4].

Application of pectin is not only in water but also in other solvents such as milk. Milk is a rich source of nutrients (fat, protein, carbohydrates, minerals and vitamins), protective compounds, enzymes, growth factors and components, making it very useful for humans [5]. Pectin has been used to produce sour milk (yogurt) in Iran with addition of salt and flavoring [6]. Pectin is used as a stabilizer in sour milk beverage by absorbing κ-casein from casein micelles [7]. Therefore, future application of green cincau leaf into milk can be furthered by investigating the physical quality of the milk curd.

2. Materials and Methods

The materials of this research were leaves of green cincau and milk cow of Local Friesian Holstein. Leaves of green cincau were collected in sub district of Imogiri, Yogyakarta Province, sub districts of Jepon and Mertoyudan, Central Java Province. Milk cow were obtained from small holder dairy farm in Karanganyar District, Central Java Province.

2.1. Sample preparation

The treatment of this research was concentration ratio between cincau leaf and cow milk (w/v): A1 = 10%:90%; A2 = 20%:80%; A3 = 30%:70%. Milk from local dairy farmer was pasteurized at 62-63^{0}C for 30 minutes, and it was cooled to 20^{0}C. Sample preparation of cincau refers to [8]. Green mature leaves were trimmed and washed in water, cut into small pieces, and mixed well together.

Sample of green cincau leaves was put in pasteurized milk and squeezed for 10 minutes. Thereafter, the mixture was incubated in room temperature for 30 minutes until forming a curd.

2.2. Measurement of pH value

Value of pH was measured in accordance with AOAC method [9]. As many as 5 g of sample were crushed and added 5 ml of distilled water and stirred until evenly. After that, the sample was measured by using instrument of pH meter.

2.3. Syneresis measurement

Syneresis measurement refers to [10]. Gel sample was inserted to plastic cup with the same weight for each treatment. Then, it was stored for a few days in a refrigeration temperature. After that, the level of syneresis gel was observed by taking water separated from the gel, and it was weighed and calculated by the formula:

$$\text{Syneresis level}=\frac{\text{initial weight-final weight}}{\text{initial weight}}\text{x } 100\% \qquad (1)$$

Where, initial weight is gel weight in plastic and final weight is gel weight in plastic after separating water from gel.

2.4. Measurement of adhesiveness, chewiness, hardness, gumminess and springiness

To measure the variables of adhesiveness, chewiness, hardness, gumminess, springiness by using texture analyzer by brand *LFRA texture Analyzer* and refers to [11]. In principle, the value obtained represents the measured variables. Preparation of sample based on height of curd of milk cincau approximately 3 – 5 cm as a requirement, and then the samples were analyzed by texture analyzer.

2.5. Data analysis

This research used one way randomized design. The data was analyzed using ANOVA, and differences between treatment means were further analysed using Duncan's New Multiple Range Test.

3. Result and Discussion

The treatment did not affect on adhesiveness, but significantly affected on chewiness, hardness, gumminess, springiness, pH and syneresis ($p<0.01$). The addition concentration of green cincau leaf of 30% was able to increase the value of cohesiveness. In addition, the level of 20% could increase chewiness, hardness, gumminess, springiness, pH, but the level of 20% decreased syneresis (table 1).

The addition of 0.1% pectin can increase the value of adhesiveness and whey syneresis [12]. Samples of low fat cheese by adding pectin did not affect on hardness, gumminess, chewiness and adhesiveness [13]. This difference was influenced by the stability of the interaction between pectin and casein in milk caused by the difference in pH [14], sensitivity of Ca [15], water content of the milk and content of the buffer in the milk serum [16]. Increasing the value of chewiness, hardness, gumminess and springiness in this study were caused by pectin in leaves of green cincau. Pectin can separate the casein in milk, thus forming a gel [17]. According [4] pectin as a maker of gel contained in green cincau leaf (*Cyclea barbata L. Miers.*). Separation of whey from the milk sour milk is influenced by the concentration of pectin [18] and the concentration of protein in milk [15]. Precipitation of gel formation is due to the hydrophobic molecular bonding with a methyl group on the side poligalaktoronik in pectin [4]. Proteins and polysaccharides covalently bonded or ionic in emulsion form [19]. Casein molecules will interact in the presence of Ca ions, resulting in sedimentation that facilitate the separation of the serum of milk [20]. Concentration of green cincau leaf of 20% was increase the amount of curd of milk cincau because hydrophobic group of casein has a maximum bonded to methyl group in poligalaktoronik of pectin.

Table 1. The concentration of green cincau leaf on physical quality of cincau milk curd

Parameters	A1 (10%:90%)	A2 (20%:80%)	A3 (30%:70%)	*P*
Chewiness (gmm)	80.511 ± 10.808^{x}	143.128 ± 10.850^{y}	157.553 ± 3.227^{y}	0.001
Adhesiveness (gs)	67.684 ± 0.040	86.780 ± 101.936	8.027 ± 2.619	0.277
Hardness (g)	99.167 ± 8.845^{x}	146.667 ± 7.576^{y}	133.111 ± 5.445^{y}	0.007
Gumminess (g)	24.622 ± 3.010^{x}	35.378 ± 2.531^{y}	37.057 ± 1.477^{y}	0.005
Springiness (mm)	3.364 ± 0.178^{x}	3.510 ± 0.086^{y}	4.198 ± 0.149^{y}	0.001
pH	5.578 ± 0.050^{x}	5.722 ± 0.039^{y}	5.422 ± 0.000^{z}	0.001
Syneresis (%)	44.324 ± 1.204^{y}	39.318 ± 1.871^{x}	35.389 ± 2.194^{x}	0.004

^{x,y} Different superscript in the same line showed significantly different.

The value of pH was enhancement by addition of green cincau leaf at the level of 20%, but decreased at 30% (table 1). The addition of pectin concentration of pill *Citrus sinensis* on a yoghurt fermented for 2 and 4 hours can demote the pH [12]. Commercial skim milk added pectin of 0.2% did not change the pH [21]. The addition of green cincau leaf up to 20% increase pH on milk cincau curd [8]. Changes in pH was caused by complex formation of carbohydrate-protein influenced by changing in the isoelectric point of the protein (constituent amino acid groups) and the number of ions in the carboxyl group of carbohydrates [22].

Syneresis was decrease with increasing concentrations of green cincau leaf (table 1). The addition of carrageenan and carbohydrates can decrease syneresis [23]. The higher addition of green cincau leaf (*Cyclea barbata L. Miers*.) resulted in lower grades of syneresis [24]. Pectin added from *Citrus sinensis* pills of 0.6% can reduce syneresis of whey in yogurt [12]. Impairment syneresis of this study was caused by an increase in the amount of pectin. Pectin was able to bind water in the presence of carboxyl acid groups [25].

4. Conclusion

The concentration of 20% increased chewiness, hardness, gumminess, springiness and pH, but decreased syneresis. The concentration of green cincau leaf of 20% can raise the physical quality of cincau milk curd.

Acknowledgment

The authors gratefully acknowledge to Sebelas Maret University for funding the research through maintenance research group (No. 632/UN27.21/LT/2016).

References

[1] Smitinand, T and Larsen K 1991 *Flora of Thailand* (Bangkok: Chutima Press)

[2] Nurdin S U, Suharyono A S dan Rizal S 2008 *JTHP* **13** 4

[3] Voragen A G. J, Pilnik W, Thibault J-F, Axelos M A V and Renard C M G C 1995 *Pectins: Food Polysaccharides and Their Applications*, ed A M Stephen (New York: Marcel Dekker)

[4] Arkarapanthu A, Chavasit V, Sungpuag P and Phuphathanaphong L 2005 *J. Sci. Food Agr.* **85** 1741

[5] Huppertz T and Kelly A L 2009 *Properties and Constituents of Cow's Milk In: Milk Processing and Quality Management*, ed A. Y. Tamine (UK: Wilew-Blackwell Limited Publisher)
[6] Kiani H, Mousavi M E, Razavi H and Morris E R 2010 *Food Hydrocoll.* **24** 744
[7] Tuinier R, Rolin C and de Kruif C G 2002 *Biomol.* **3** 632
[8] Hertanto B S, Nuhriawangsa A M P, Kartikasari L R, Cahyadi M, Swastike W, Gunawan D dan Raharjo N P 2015 *Pros. Seminar Nasional II Pengembangan Ternak Lokal* (Padang: Universitas Andalas)
[9] AOAC 1995 *Official Methods of Analysis* ed 16th (Washington D.C: Association of Official Analytical Chemist)
[10] Imeson A 1992 *Thickening and Gelling Agent for Food* (London: Blackie Academic and Professional)
[11] Bourne M C 2002 *Food Texture and Viscosity In: Concept And Measurement: Practice of Objective Texture Measurement* (London: Academic Press)
[12] Arioui F, Saada D A and Cheriguene A 2016 *Food Sci. Nutrition* **2016** 1
[13] Liu H, Xu X M and Guo S D 2008 *Int. J. Food Sci. Technol.* **43** 1581
[14] Pereyra R, Schmidt K A and Wicker L 1997 *J. Agr. Food Chem.* **45** 3448
[15] Laurent, M A and Boulenguer P 2003 *Food Hydrocol.* **17** 445
[16] Jensen S, Claus Rolin and Ipsena R 2010 *Food Hydrocol.* **24** 291
[17] Matia-Merino L and Singh H 2007 *Food Hydrocol.* **21** 765
[18] Lucey J A, Tamehana M, Singh H and Munro P A 1999 *J. Texture Stud.* **30** 305
[19] Dickinson E 2009 *Food Hydrocol.* **23** 1473
[20] Françoise K A, Kablan T, Kamenan A and Lagaude A 2009 *Eur. J. Sci. Res.* **25** 584
[21] Karimi N and Nateghi L 2015 *Int. J. Biol. Pharm. Appl. Sci.* **4** 6745
[22] Ye A 2008 *Int. J. Food Sci. Technol.* **4** 3406
[23] Verbeken D, Bael K, Thas O and Dewettinck K 2006 *Int. Dairy J.* **16** 482
[24] Manurung S I, Erline dan Hardoko 2008 *Itepa.* **6** 127
[25] Endress, H U and Mattes F 2009 *Pectin In:Fiber Ingredients: Food Applications and Health Benefits, ed S. S. Cho and P. Samuael* (New York: CRC Press)

IOP Conf. Series: Materials Science and Engineering **193** (2017) 012043 doi:10.1088/1757-899X/193/1/012043

Preparation and Characterization of Nano Chitosan from Crab Shell Waste by Beads-milling Method

E Rochima[1,3], S Y Azhary[1], R I Pratama[1] , C Panatarani[2,3], and I M Joni[2,3,*]
[1]Dept. of Fishery Processing Technology, Faculty of Fishery and Marine Science, Universitas Padjadjaran, Jl. Raya Bandung-Sumedang, Jatinangor 45363, West Java, Indonesia
[2]Dept. of Physics, Faculty of Mathematics and Natural Sciences, Universitas Padjadjaran, Jl. Raya Bandung-Sumedang, Jatinangor 45363, West Java, Indonesia
[3]Nanotechnology and Graphene Research Center, Universitas Padjadjaran, Jl. Raya Bandung-Sumedang, Jatinangor 45363, West Java, Indonesia

E-mail: imadejoni@phys.unpad.ac.id

Abstract. This research aimed to produce well-dispersed chitosan nanoparticles suspension from a crab shell waste by beads-milling method. Tween 80 surfactant was applied as a dispersing agent to improve the dispersion stability. A mono-dispersed chitosan nanoparticle suspension was obtained at milling times 90 and 120 minutes by adjusting their pH values. The characterization was performed to obtain the size and size distribution, the particle stability, the characteristics of the functional groups and the degree of deacetylation. The results showed that chitosan nanoparticle was achieved in the milling process 90 minutes and low pH dispersion medium. The average particle size of the as-prepared suspension is 232.5 nm with polydispersity index of 0.452. Beads milling process did not change the glucosamine and N-acetylglucosamine content on chitosan structure indicated by degree of deacetylation higher than 70%. Based on the FTIR investigation the mechanism of surface modification of chitosan due to milling process was proposed. It was concluded that beads milling process can be applied to prepare chitosan nanoparticles by proper adjustment of the milling time, pH and appropriate dispersing agent type and amount.

1. Introduction

Chitosan nanoparticles (CS-NPs) have received a great attention because of their nano-size, large surface area, biodegradable, and good biocompatibility [1]. These characteristics favour the employment of chitosan nanoparticles in a wide range of various applications, including drug delivery systems, tissue engineering, sensors, packaging materials, protein carriers, removal of pollution and catalysts [1,2]. Thus, developments of CS-NPs are potential to be applied in various areas of modern industries. Chitosan $[(C_6H_{11}NO_4)n]$ is a natural, cationic aminopolysaccharide (pKa 6.5) copolymer of glucosamine and N-acetylglucosamine obtained by the alkaline and partial deacetylation of chitin [3]. It is a natural biopolymer derived from crustacean shells such as crabs, shrimps and lobsters [4].
Recently, various efforts applied for the development of chitosan through chemical or physical modification of chitosan, i.e. surface modification or functionalization and reducing particle size or grain [5]. The synthesis methods for chitosan nanoparticles has been a major challenge to meet the

commercial production requirement that the process should simple, low cost and high yield, i.e. bottom and top down approaches [6]. CS-NPs are prepared using numbers of different strategies.

These include cross linking, ionotropic gelation, microemulsion, emulsification solvent diffusion and polyelectrolyte complexation [3,4]. All these techniques adopt a "bottom-up" approach where nanoparticles are generated from individual macromolecules through self-assembly [7]. Extensive work had been carried out by adopting bottom-up strategies to prepare chitosan nanoparticles, however research carried out using top down technologies is almost insignificant [8]. Top down synthesis approach is a particle synthesis process starting from bulk material and pass through mechanical grinding or comminuting to form particles either submicron and/or nano-sized particles [9,10]. Beads-mill is an instrument of top-down process to synthesis nanoparticles and capable to produce well-dispersed suspension since allows using smaller beads around 0.015 to 0.5 mm and engineering the dispersing agent [10-15]. This paper reports the preparation of chitosan nanoparticles by beads milling process aim to obtain well-dispersed suspension and to investigate the surface properties CS-NPs. In addition, the mechanism of surface modification of chitosan due to milling process was proposed.

2. Material and Methods

The raw material of chitosan received from commercially available flakes chitosan with the degree of deacetylation higher than 80%. A nonionic surfactant of Polysorbate 80 (Tween 80) was used as a dispersing agent to prevent particles agglomeration. The beads used for the wet milling process was zirconia with 30-μm in size, while the media of the suspension was distilled water. Chitosan flake was crushed into micron-size chitosan sample and then was completely dried in oven at 60°C for 5 days. Then the dried chitosan powder was pulverized for 5 minutes to obtain fine powder. The obtained powder was selected to obtain more uniform micron size (around 37 μm) by using sieves at mesh -400 and subjected for beads milling process.

The vessel was filled with beads to 70% capacity. The slurry was prepared by mixing the chitosan powder and distilled water. The mixing chitosan was stirred using magnetic stirrer for 30 minutes before beads milling. The Tween 80 as a dispersing agent was added after the mixing time 15 minutes. The slurry of the chitosan was pumped into the vessel (with an optimized recirculation mass flow rate of 8 L/min) which contained zirconia beads and impeller operated at a speed of 4070 rpm. The beads were agitated in the lower portion of the vessel (dispersing section) to break up the aggregate to avoid agglomeration of chitosan nanoparticles in the suspension. After dispersion, the suspension was pumped from the dispersing section to the separation region where centrifugal force was used to separate the zirconia beads from the particle suspension. The chitosan particle suspension was then recycled back to the dispersing section. In order to keep the temperature of the system constant, the vessel was supported by a water jacket system and was completely sealed.

The chitosan concentration in suspension determined in weight % in which a ratio of the weight percentage of chitosan to the total weight of media (distilled water solution), chitosan powder, and dispersing agent. The wt % of dispersing agent was the concentration of Tween 80 compared with the wt % of chitosan concentration. The total solution was 575 g, and received 1 wt % of chitosan (5.75 g), 40% of dispersing agent (2.3 g), and 566.95 g of the distilled water. The milling time operations were 90 minutes and 120 minutes where samples denoted respectively as CS-NPs 90 and CS-NPs 120.

The proximate analysis on chitosan flakes was carried out to determine parameters such as; ash content, moisture content and nitrogen content of the chitosan (AOAC, 1990). A Delsa™ Nano Series instrument (Beckman Coulter, Inc) used to measure the particle size distribution, the polydispersity index (PI), and zeta potential. The functional group of chitosan, before and after bead milling, were also observed by FTIR (Fourier Transform Infra-Red) ranging from 400 to 4,000 cm^{-1} (Nicolet™ iS™ 5 Thermo Fisher Scientific Inc).

IOP Conf. Series: Materials Science and Engineering **193** (2017) 012043 doi:10.1088/1757-899X/193/1/012043

3. Results and Discussion

In this section we discuss how to format the title, authors and affiliations. Please follow these instructions as carefully as possible so all articles within a conference have the same style to the title page. This paragraph follows a section title so it should not be indented.

3.1. Proximate Analysis of Chitosan Flakes

The moisture, ash, and nitrogen content of the chitosan flakes were 12.45%, 1.33%, and 36.57% respectively. The moisture content of chitosan flakes was found to be higher than the standard chitosan by Protan Laboratories. This could be attributed to the effectiveness water content in chitosan flakes where dried to constant weight before the deacetylation process [20]. Chitosan with high moisture content was indicated that chitosan more hydrophilic than chitin [16]. Nitrogen content of the chitosan flakes was slightly higher than that of chitosan showed acceptable effective deproteinization step of as-received chitosan flakes. Nitrogen content of chitosan flakes was attributed to the high degree of deacetylation of the flakes chitosan [20,21].

Table 1. The results of proximate test chitosan flakes

Items	Test Results	Protan Laboratories Standard
Moisture content	12.45%	≤ 10%
Ash content	1.33%	≤ 2%
Nitrogen content	5.85%	≤ 5%

3.2. Preparation of Nano Chitosan by Beads Milling

Figure 1 showed the chitosan suspensions after beads milling at different milling time (CS-NPs 90 and CS-NPs 120) at 1 wt % chitosan content. This result indicated that the chitosan successfully reduced into fine suspended powder; but poorly dispersed respectively. The poorly dispersion and large agglomerated particles, which was confirmed by a poor degree of transparency [11,13]. Selected dosages of dispersing agent (Tween 80) and dispersant did not significantly prevent dispersed particles from reagglomeration because less ionic and steric density at the surface promoted the Van der Waals forces to dominate. Thus, the CS-NPs 90 and CS-NPs 120 suspension was optimized by adjusting pH value. Figure 1.showed the corresponding pictures of chitosan suspension for different pH values. The decrease of pH, increasing the transparency of chitosan suspension.so the ionic strength increased and gained an electrostatic.

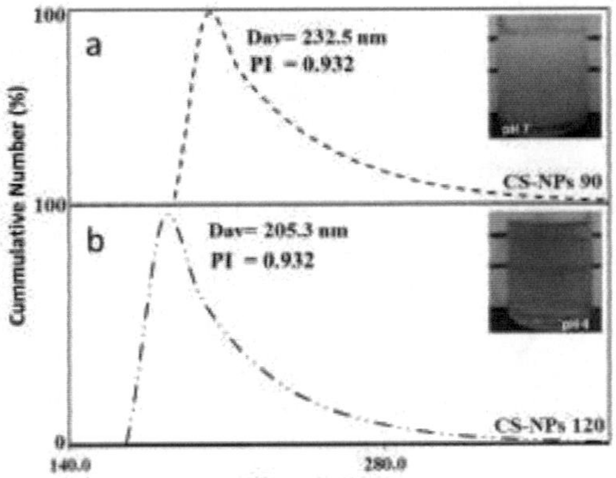

Figure 1. The particle size distribution, the polydispersity index (PI) of chitosan suspension

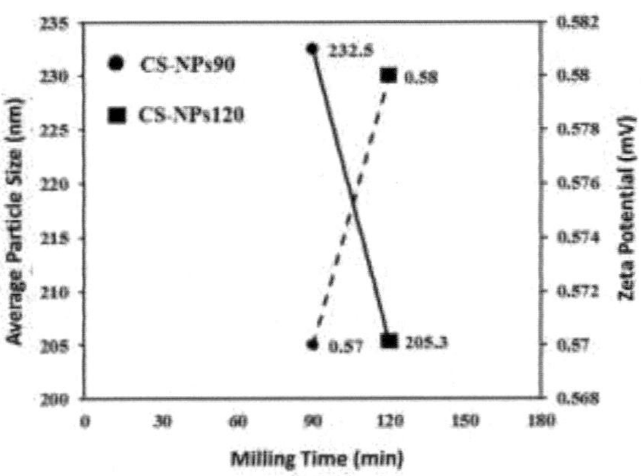

Figure 2. Average particle size and zeta potential of chitosan suspension

The size distribution of CS-NPs 90 and CS-NPs 120 suspension were well-dispersed with an average size of 232.5 nm and 205.3 nm after optimizing pH values. The main peak of the CS-NPs 120 size distribution shifted to small size, which showed successful in size reduction. The effective size reduction was occurred because of the surface modification using optimize pH value effectively prevented formation of aggregates or agglomerates of chitosan particles. However, the PI value of CS-NPs 120 was 0.932 iindicates a very broad distribution of particle sizes (PI > 0.7). In contrast the PI value of CS-NPs 90 was 0.452 indicating a narrow and favourable particle size distribution (PI < 0.5) [21].

3.3. Dispersion Stability of Chitosan Suspension

Figure 2 showed the average size and zeta potential of chitosan suspension in water after optimized by pH versus milling time. The zeta potential of CS-NPs 90 and CS-NPs 120 was 0.57 mV and 0.58 mV, respectively. These results showed that the effect of the decrease of pH, increase of concentration of ion in medium not significantly. Ionic environment in the vicinity of colloid and electrical repulsive forces occur, could be influence the observed zeta potential. In contrast showed that the decrease of pH, increasing the clarity of chitosan suspension. Thus the Twin 80 as a dispersing agent promoted a steric interaction at the surface of the particles and stabilized the dispersion in aqueous media. The decrease pH 4, increasing the positive charges of the chitosan particles surfaces. The positive chitosan sites at the surface can act as counter ions, causing Tween 80 adsorption onto chitosan surface. Thus, this result was an evidence for the existence of the electrosteric stabilization in chitosan suspension.

3.4. Characteristics nano chitosan after beads mill process

Figure 3 showed that FTIR spectra of chitosan particles before beads milling process resulted a peak absorption at 3352.01 cm^{-1} wave that showed the OH stretching vibration. Absorption peaks at wave number 3289.17 cm^{-1} was NH-CO stretching vibration that indicated the presence of an acetyl group. The absorption peaks at wave number 2872.17 cm^{-1} was due to the presence of methylene and methyl groups in the chitosan structure. Absorption at wavenumber 1644.49 cm^{-1} indicate the C=O stretching vibration and wave number 1317.86 cm^{-1} showed the C-N stretching vibration in which also indicated the presence of an acetyl group. Wavenumber 1590.72 cm^{-1} which was the vibration absorption NH (R-NH) bending and 1149.03 cm^{-1} indicated that the vibration absorption CN (NH_2) stretching as an evidence of amine groups was formed. Absorption at wave number 1022.62 cm^{-1} showed a peak vibration of C-O-C stretching symmetric showed that commercial chitosan formed through glycosidic bonds [22].

Figure 3 showed that after milling process (CS-NPs 90 and CS-NPs 120), the resulting spectrum was slightly changed in the position and intensity of the pattern. However, there is no significant different pattern between CS-NPs 90 and CS-NPs 120. A broad absorption wavelength range of 3330 cm^{-1} was the collective absorption of OH and NH in chitosan nanoparticles. Broad absorption indicates that intramolecular hydrogen bonds in the structure were very strong. Absorption at a wavelength of 1636.80 cm^{-1} and 1636.75 cm^{-1} showed the present of acetyl groups at chitosan nanoparticles. In addition, the present of acetyl group was also clearly observed at peak absorption of 1318.29 cm^{-1}. In contrast, the Amine group was observed for the CN absorption at a wavelength of 1154.52 $cm^{-1.}$ The percentages of glucosamine and N-acetylglucosamine on chitosan structure after the milling process were decreased.

Based on observations on FTIR spectrum, the surface functionalization mechanism of chitosan particles before and after milling were illustrated in figure 4. Chitosan particles before beads milling was a copolymer composed N-acetylglucosamine and D glucosamine with relative proportions (figure 4a). This caused CH_3, NH_2 and OH bond associated to surface charge chitosan. After the beads milling chitosan copolymer transformed into smaller size due to the high energy mill. However, because the surface area of the particles of chitosan was high, then the particles tend to agglomerate (figure 4b) since the ionic electrostatics stabilization could not prevent agglomeration. On the decrease of pH of chitosan suspension the steric strength increased due to association of the dispersing agent, in turn received a gain on the steric stabilization (figure 4c). In addition the ionic charges of the NH_3^+ and OH acted as the surface charges of chitosan suspension. Thus the stabilization of chitosan was received both ionic and steric stabilization usually named electro-steric stabilization.

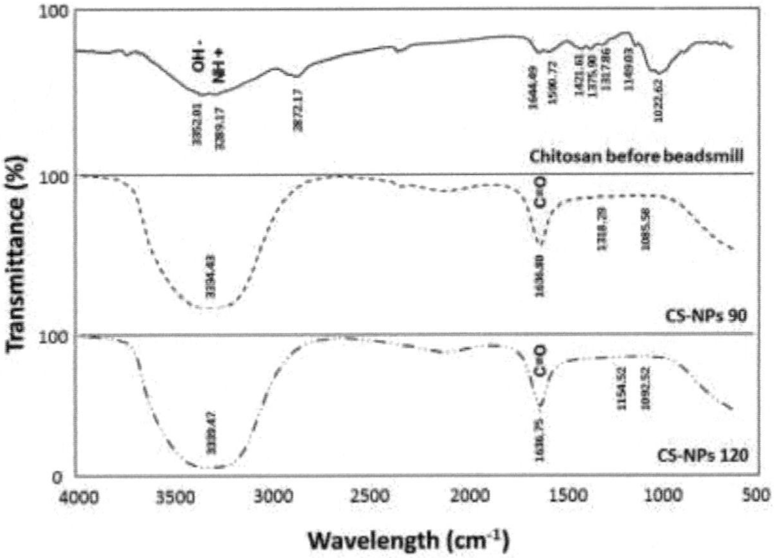

Figure 3. FTIR spectrum of chitosan particles before and after beads milling

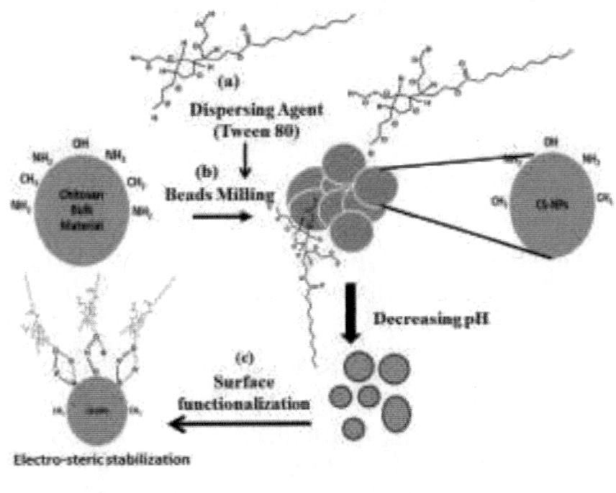

Figure 4. The surface functionalization mechanism of chitosan nanoparticle with dispersing agent and optimized by pH value

Degree of deacetylation (DDA) is one of the chitosan chemical characteristics affect the performance of chitosan in many application [23]. DDA also represents the number of amino groups contained in the structure of chitosan. DDA of chitosan raw materials was 88.20%. After the milling process 90 minutes and 120 minutes were respectively 81.18% and 78.88%. According to the Protan Laboratories Standard of Japan, DDA of the chitosan was above 70%. This result showed that as prepare well-dispersed chitosan nanosuspension were acceptable quality because of an acetyl group contained in chitosan were small in quantities.

4. Conclusion

Well-dispersed chitosan nanoparticles suspension successfully prepared by using a Beads-Mill method at milling time 90 and 120 minutes. As-prepared chitosan suspension resulted average diameter of

IOP Conf. Series: Materials Science and Engineering **193** (2017) 012043 doi:10.1088/1757-899X/193/1/012043

232.5 nm and 205.3 nm after optimized by pH respectively for CS-NPs 90 and CS-NPs 120. It was concluded that the stability of the suspension received from the contribution of electronic and steric stabilization after adjusting the pH of the suspension. In contrast, beads milling process preserved the percentage glucosamine and N-acetylglucosamine on chitosan structure indicated by the degree of deacetylation higher than 70%.

References

[1] Safari J, Zarnegar Z, Jazini Z A and Sadeghi M 2016 *Catal. Commun.* **77** 108
[2] Khan Y H, Islam A, Sarwar A, Gull N, Khan S M, Munawar M A, Zia S, Sabir A, Shafiq M and Jamil T 2016 *Carbohydr. Polym.* **146** 131
[3] Sonia T A and Sharma C P 2011 *Adv. Polym. Sci.* **243** 23
[4] Tiyaboonchai W 2003 *Naresuan Univ. J.* **11** 51
[5] Suptijah P, Jacoeb A M dan Rachmania D 2011 *Jurnal Pengolahan Hasil Perikanan Indonesia* 14 78
[6] Yadav T P, Yadav R M and Singh D P 2012 *Nanosci. Nanotech.* **2** 22
[7] Iqbal P, Preece J A and Mendes P M 2012 *John Wiley & Sons Ltd.* **8** 3589
[8] Wijesena R N, Tissera N, Kannangara Y Y, Lin Y, Amaratunga G A J and Silva K M N A 2014 *Carbohydr. Polym.* **117** 731
[9] Joni I M, Panatarani C, Hidayat D, Serianto, Wibawa B M, Rianto A and Thamrin H 2013 *AIP Conference Proceedings* **1554** 20
[10] Joni I M, Panatarani C and Maulana D W 2016 *AIP Conference Proceedings* **1712** 050019
[11] Joni I M, Purwanto A, Iskandar F and Okuyama K 2009 *Ind. Eng. Chem. Res.* **48**(15) 6916
[12] Joni I M, Nishiwaki T, Okuyama K, Isoi S and Kuribayashi R 2010 *Powder Tech.* **204**, 145
[13] Joni I M, Balgis R, Ogi T, Iwaki T and Okuyama K 2011 *Colloids and Surfaces A: Physicochem. Eng. Aspects* **388** 49
[14] Joni I M, Ogi T, Purwanto A, Okuyama K, Saitoh and Takeuchi K 2012 *Adv. Powder Technol.* **23** 55
[15] Joni I M, Ogi T, Iwaki T and Okuyama K 2013 *Ind. Eng. Chem. Res.* **52** 547
[16] Ahmad L O, Permana D, Wahab, Sabarwati S H, Ramadhan L O A N and Rianse U 2015 *Adv. Environ. Geological Sci. Eng.* **38** 373
[17] Hussain M R, Iman M and Maji T K 2013 *Int. J. Adv. Eng. Appli.* **2** 4
[18] Abdulkarim A, Isa M T, Abdulsalam S, Muhammad A J and Ameh A O 2013 *Civil Environ. Res.* **3** 108
[19] Julian G D and George A F R 1985 *Macromol. Chem. Phy.* **186** (8) 1671
[20] Isa M T, Ameh A O, Tijjani M, Adama K K 2012 *Int. J. Biol. Chem. Sci.* **6** 446
[21] Dounighi N M, Eskandari R, Avadi M R, Zolfagharian H, Sadeghi A M M and Rezayat M 2012 *J. Venom. Anim. Toxins. Incl. Trop. Dis.* **18** 44
[22] Gyliene O, Razmute I, Tarozatte R and Nivinskiene O 2003 *Chemija (Vilnius)* **14** (3) 121
[23] Khan T A 2002 *J Pharm Pharmaceut Sci.* **5**(3) 205

IOP Conf. Series: Materials Science and Engineering **193** (2017) 012044 doi:10.1088/1757-899X/193/1/012044

The influence of composition and final pyrolysis temperature variations on global kinetics of combustion of segregated municipal solid waste

Pranoto[1], D A Himawanto[2] and N A Arifin[3]

[1] Department of Chemistry, Faculty of Mathematic and Natural Science, Sebelas Maret University
[2] Mechanical Engineering Department Faculty of Engineering, Sebelas Maret University
[3] Program of Masters studies of Environmental Science, Sebelas Maret.University

[1]pak_pran@yahoo.com

Abstract. The combustion of segregated municipal solid waste (MSW) and the resulted char from the pyrolysis process were investigated in this research. The segregated MSW that was collected and used can be divided into organic and inorganic waste materials. The organic materials were bamboo and banana leaves and the inorganic materials were Styrofoam and snack wrappings. The composition ratio of the waste was based on the percentage of weight of each sample. The thermal behaviour of the segregated MSW was investigated by thermo gravimetric analysis. For the pyrolysis process the prepared samples of 200gram were heated from ambient temperature until a variance of final pyrolysis temperature of 550°C, 650°C and 750°C at a constant heating rate of 25°C/min. It was found that the highest activation energy of the raw materials is achieved from sample CC1 (Char with 100% inorganic materials). The activation energy of the raw materials is relatively lower than that of the char. The higher the final pyrolysis temperature, the lower the calorific value of char. The calorific value gradually increases with the amount of inorganic materials.

1. Introduction

An increasing world population with new developments in technology and economy increases the demand in resources such as food and other living essentials. These developments result in an increase in the amount of waste that is being produced. USEPA defines solid waste as any garbage, refuse, sludge from a wastewater treatment plant, water supply treatment plant, or air pollution control facility, and other discarded material, including solid, liquid, semisolid, or contained gaseous material, resulting from industrial, commercial, mining, and agricultural operations and from community activities [1,2]. In the city Surakarta, often called Solo the growing economy is also noticeable. This is a city in Central Java, Indonesia with a total population around 588.110 people (in 2011) with an average density of 13.354 people/km^2. According to the Department of Sanitation in Surakarta the waste generation in Surakarta has been estimated at around 250 tons/day.

IOP Conf. Series: Materials Science and Engineering **193** (2017) 012044 doi:10.1088/1757-899X/193/1/012044

In the last decade adopting new renewable energy resources have increased although it will take a lot more years to decrease the dependency of the use of fossil fuels. Turning MSW into a renewable energy source can be a beneficial solution for both decreasing the amount of waste and for creating a new energy source.

Biomass is any type of organic material that is available on a renewable or reoccurring basis, and includes such things as agricultural crops and waste, wood and wood wastes, animal wastes, aquatic plants, and organic fractions of municipal and industrial waste [3,4]. Depending on the operating condition, pyrolysis can be classified into three main categories: conventional, fast and flash pyrolysis.

Combustion of biomass or waste in combination with a base fuel is a simple and economically way to replace fossil fuels by biomass and to utilize waste. The burning characteristics of biomass may vary considerably depending on the composition of the raw material used [5,6].

The kinetics of the In this research the global kinetics of combustion of municipal solid waste (MSW) is being studied to have a better understanding of the thermal degradation of MSW. Segregated MSW that has no economic value yet on the market is used in this research and have been collected from the TPA. The original condition of the selected samples of waste is that it already has been mixed with other MSW. The composition of the segregated MSW can be divided into two main groups: organic and inorganic MSW.

Combustion characteristics are influenced by different factors. In this research the composition and the final pyrolysis temperature will be varied in order to identify the influences on the global kinetics. The problem statement in this research can be formulated into two questions: The variations of the composition of waste influence the global kinetics on combustion of segregated MSW ; The variations of the final pyrolysis temperature influence the global kinetics on combustion of segregated MSW.

2. Research Methods

The equipment used in this study consists of test pyrolysis equipment, combustion equipment, supporting equipment and measuring equipment.

2.1. Characteristics of slow pyrolysis equipment:

The slow pyrolysis equipment is a model fixed bed that is equipped with a thermo controller that goes until 1000 ^{0}C

2.2. Characteristics of combustion equipment:

The combustion equipment is in the form of a cylindrical tube with a diameter of 96 mm that is getting heat from the coil heating element with heat settings up to 1000°C using a thermo-controller.

2.3. Supporting equipment:

2.3.1. Crusher equipment

This tool is used to cut the materials into a size of 20 mesh. This smoothing tool type disk mill is powered with 1 HP that rotates the inside blade. After the material has been cut, it is sieved with an output of a size of 20 mesh.

2.3.2. Measuring equipment are a digital scale with an accuracy of 0,01gram, thermocontroller and thermocouple type-K.

2.3.3. Analyzing equipment

The Thermogravimetric Analyzer (TGA) is an essential laboratory tool used for material characterization. Thermogravimetric Analysis is a technique in which the mass of a substance is monitored as a function of temperature or time as the sample specimen is subjected to a controlled

IOP Conf. Series: Materials Science and Engineering **193** (2017) 012044 doi:10.1088/1757-899X/193/1/012044

temperature program in a controlled atmosphere. In other words, TGA is a technique in which, upon heating a material, its weight increases or decreases.

2.3.4. Software

The software that is being used to record the data is WinCT that records the mass change and Adam.NET utility records the data temperature. In the program WinCT the Rs Key is included. The Rs Key is a program that is used to record data scales during the pyrolysis and combustion process. The obtained data is stored in MS Excel format that subsequently will be process into a graph.

2.4. Sample of the research

The materials that will be used in this research are segregated municipal solid waste collected in Surakarta, Central Java, Indonesia. These waste are mainly waste that are not being collected by the waste collectors or have no value for them. The beginning weights of the samples are each 20 grams. The organic materials that are being used are banana leaves and bamboo. The inorganic materials are styrofoam and snack wrappings.

3. Result and Discussion

3.1. TGA Analysis

The combustion process of the raw MSW materials in this research is accordance to the theory of solid fuels [7,8] where there are 3 physical stages that can be identified that includes drying, de-volatilization and the stage of char combustion where then ash will be left over. The drying stage can be characterized by the slow mass losses in the beginning. The burning profile of the samples can be divided into 4 different temperatures points. The first is the volatile matter initial temperature (ITVM) where the mass starts to decrease. The second is the fixed carbon initial temperature (ITFC) when the rate of the mass loss accelerates due to the onset of the char combustion. The third is the peak temperature (PT) when the rate of the mass loss has reached a maximum value. The last important point is the burn out temperature (BT) when the mass becomes constant at the completion of burning. The ITVM, ITFC, PT . The fluctuations in the temperature in the samples can be caused by a restricted airflow into the inside of the samples; the ash covers the sample.

The combustion profiles of 0%, 10%, 20%, 70% and 100% organic materials are given below in the form of graphics. In the combustion of char, the boundary of the combustion stages where unclear so that the ITVM cannot be identified of sample CC10, CC11, CC12 and CC21 (see graphs in Appendix 1). The BT stage of sample CC13, CC14, CC15, CC21 and CC24 could also not be identified. The BT could not be identified for these samples due to incomplete combustion. Inorganic materials tend to cover the organic materials during combustion. The samples should be mixed thoroughly to get even layers for a complete combustion process.

Comparing these results to a previous research done by Himawanto et. al in 2013, the ITFC, PT and BT are relatively similar while the ITVM differs with more than 100°C. A possibility of this difference can be the moisture content of the organic material of this research is lower than that [9]. The combustibles of a solid fuel are volatile matter and fixed carbon. In the combustion of char, it was shown that the drying zones occurred in a shorter time than the combustion of raw materials and the ITVM occurred relatively above 100°C. The fluctuations in the remaining mass may be caused by the layer of ash covering the sample inhibited the diffusion of oxygen into the sample and the combustion gases out of the sample. In the raw materials the PT was higher than that of the char combustion. ITFC of all the samples are relatively the same within the temperature range 200-400°C.

IOP Conf. Series: Materials Science and Engineering **193** (2017) 012044 doi:10.1088/1757-899X/193/1/012044

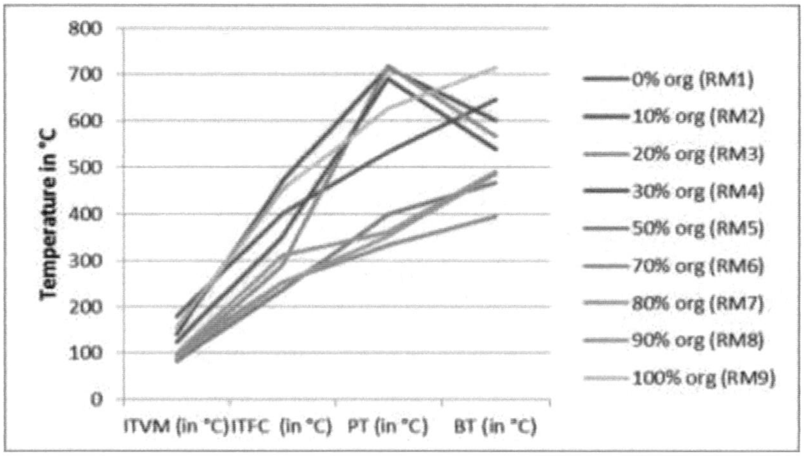

Figure 1. Combustion profile of raw materials.

At the point that intersects the line drawn parallel to the gradient of the dots until you get to the point of zero mass fraction values decline. From that point again draw a straight line down to cross the x axis.

For the combustion of raw materials MSW the decomposition zone was easier to analyze than that of the char MSW. Taking the pyrolysis reactions into account, the degradation reaction was accompanied by the release of water, gaseous products and eventual formation of char. With the presence of oxygen the char residues underwent ignition during the combustion reaction resulting in an appearance of the second decomposition zone.

For 20%, 50% and 90% organic materials the Tonset gradually decreases with the final pyrolysis temperature variation. For 90% organic materials the Tonset,h combustion, the weight loss of cellulose (Tshoulder) started at higher temperature and finished at lower temperature. This behavior is attributed to the fact that the cellulose composes of only one simple repeating unit, cellubiose. Therefore, cellulose decomposes through a rapid depolymerization process by cleavage of glycosidic bonds during the combustion process. Tshoulder is relatively the same for all this composition with different final pyrolysis temperature. The Toffset is the highest for the sample with the highest final pyrolysis temperature (CC24). The fluctuations in the decomposition of char graphs can be explained by the amount of oxygen that is still available at which point the char residues underwent ignition. The decomposition of hemicellulose started at a higher temperature for the combustion of the raw materials than that of the combustion of char MSW. The breakdown of cellulose also started at a higher temperature and finished at a lower temperature for the combustion of char MSW.

It takes more energy to burn the maximum amount of MSW (Tpeak- maxium mass loss) in the samples with a higher amount of inorganic material than in the raw waste material samples. The Toffset temperatures are lower than the Tpeak of all the samples with an exception of a few. A higher peak temperature indicates that the sample reacts slower. Identifying the devolitilization zones in the combustion process were more difficult due to the influences of oxygen and the surrounding air that fluctuate the combustion process.

3.2. Activation Energy

The activation energy indicates the amount of energy where the sample can be burned or ignited at which point a chemical reaction starts. The higher the activation energy, the more difficult the char is ignited. The highest activation energy is achieved from sample CC1 (Char MSW with 100% inorganic materials). The activation energy is the lowest with 100% organic materials. The activation energy of the raw waste materials is all under 130kJ/mol. The highest activation energy of the raw materials is 130,3kJ/mol (20% organic) and the lowest is 45,59kJ/mol (10% organic).

The highest activation energy of char with a final pyrolysis temperature of 550°C is 643,36 kJ/mol (0% organic) followed by 225,35kJ/mol (30% organic). The highest activation energy of char with a final pyrolysis temperature of 650°C is 288.81kJ/mol (90% organic) followed by 238.44kJ/mol (30% organic). The highest activation energy of char with a final pyrolysis temperature of 750°C is 155,56 kJ/mol (10% organic) followed by 121,8 kJ/mol (50% organic). According to these results, the higher the final pyrolysis temperature, the lower the activation energy will be.

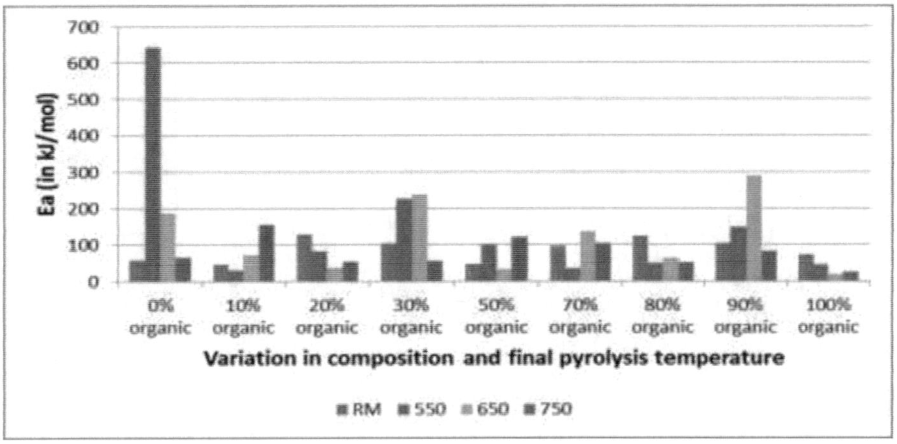

Figure 2. Activation energy.

3.3. Calorific Value Analysis

The calorific value of inorganic raw materials (snack wrappings and styrofoam) is higher than that of organic raw materials (banana leaves and bamboo). The char MSW with the composition of 100% inorganic materials has also the highest calorific value. The composition of 70% organic and 30% inorganic raw materials, which composes the rough calculation of the MSW composition in Surakarta, has a calorific value of 21.49 MJ/kg (final pyrolysis temperature of 550°C), 19.55MJ/kg final pyrolysis temperature of 650°C) and 18.46MJ/kg (final pyrolysis temperature of 750°C) [11]. All mixed compositions reaches the minimum calorific value similar of conventional biomass. The samples that underwent a higher final pyrolysis temperature have lower calorific values. A higher calorific value is obtained at a final pyrolysis temperature of 550°C. The higher the final pyrolysis temperature, the lower the calorific value of char MSW. The calorific value gradually increases with the amount of inorganic materials.

4. Conclusions

The variation in composition of MSW influences the global kinetics on combustion of segregated MSW. The highest activation energy is achieved from sample CC1 (Char MSW with 100% inorganic materials) and the lowest with 100% organic materials. The activation energy of the raw materials is relatively lower than that of the char. The higher the final pyrolysis temperature, the lower the activation energy of the char is.

The combustion process of the raw MSW materials in this research is accordance to the theory of solid fuels where there are 3 physical stages that can be identified that includes drying, de-volatilization and the stage of char combustion where then ash will be left over. The drying stage can be characterized by the slow mass losses in the beginning. The burning profile of the samples can be divided into 4 different temperatures points: ITVM, ITFC, PT and BT. In the combustion of char MSW the drying zones occurred in a short time. The fluctuations in the remaining mass may be caused by the layer of ash covering the sample inhibited the diffusion of oxygen into the sample and the combustion gases out of the sample. The decomposition of hemicellulose started at a higher temperature for the combustion of the raw materials than that of the combustion of char MSW. The breakdown of cellulose also started at a higher temperature and finished at a lower temperature for the combustion of char MSW.

It takes more energy to burn the maximum amount of MSW (T peak- maxium mass loss) in the samples with a higher amount of inorganic material in the raw waste material samples. The higher the final pyrolysis temperature, the lower the calorific value of char MSW. The calorific value gradually increases with the amount of inorganic materials.

5. REFERENCES

[1] Claudia Juliana Gómez Díaz 2006. *Chemical Engineer Understanding Biomass Pyrolysis Kinetics: Improved Modeling Based on Comprehensive Thermokinetic Analysis* (Barcelona : Universitat Politècnica de Catalunya)
[2] Giudicianni,P., G. Cardone, R. Ragucci, Cellulose 2013 *J. Anal. Appl. Pyrol.* **100** 213-222
[3] Mark Shaw 2006 *Annual Conf. Edmonton* **06-105**
[4] Hosoya T., H. Kawamoto, S. Saka 2007 *J. Anal. Appl. Pyrol.* **80** 118-125
[5] Rhen, C., Ohman, M., Gref, R., Wasterlund, I 2007 *Biomass and Bioenerg.* **31** 66-72
[6] Wang, S.R., X.J. Guo, K.G. Wang, Z.Y. Luo 2011 *J. Anal. Appl. Pyrol.* **91** 183-189
[7] Ragland Kenneth W., Bryden Kenneth M. 2011. *Combustion Engineering, second edition.* ISBN-13: 978-1-4665-0001-3
[8] DA Himawanto, I Indarto, H Saptoadi, TA Rohmat 2013 *Modern Applied. Sci.* **7** 33
[9] DA Himawanto *IPTEK* **24** 1
[10] Carlo S. Alburo, Radzwell H. Conje, Maria Gracelda B. Pino,Engr. Patrick U. Tan. 2010. *Calorific Values and Proximate Analysis of Sargassum spp. And Ulva spp.* (Cebu City : University of San Carlos)

IOP Conf. Series: Materials Science and Engineering **193** (2017) 012045 doi:10.1088/1757-899X/193/1/012045

Effect of Air Cooling and Vacuum Cooling Storage on the β-Carotene Content and Proximate Analysis (Water Content, pH, Total Protein and Content of Sugar) in Carrot

T Kusumaningsih*, T Martini, K S Rini, and L Okstafiyanti
Department of Chemistry, Faculty of Mathematics and Natural Science, Sebelas Maret University, Surakarta, Indonesia

*E-mail: triana_kusumaningsih@staff.uns.ac.id

Abstract. The study of air cooling and vacuum cooling storage effect on the β-carotene content and proximate analysis in carrot has been studied. The aim of the research to determine the effective storage in carrot to improve the quality and the shelf life. Parameters measured during the 12 weeks of storage process were β-carotene, pH, water, sugar and protein content. Validation analysis for β-carotene method showed a good linearity (r^2= 0.997) in a range of 0-8 mg/L and (r^2=0.999) in a range of 0-1 mg/L. The precision was exemplified by %RSD of 0.88%-7.48%. Mean recovery was 100.66% during accuracy studied. UV analysis revealed the LOD values were 0.009 mg/L and LOQ values were 0.032 mg/L. The decreased content of β-carotene, water, protein, and pH from carrot during vacuum cooling storage were higher than in the air cooling storage period. The sugar content for air cooling storage increased up to eight weeks and decreased at the end of storage while the vacuum cooling storage decreased from the beginning of the storage period. All the data indicates that the air cooling storage was more effective storage techniques for extending the shelf life of carrot compared to the vacuum cooling storage.

1. Introduction

Deficiency of vitamin A is major cause of premature death in developing country, particularly among children. According to the WHO around 250.000-500.000 the children not growing up because lack of vitamin A. Vitamin A has many systemic functions in humans can be produced within the body from carotenoids, notably β-carotene [1]. Carrot has high β-carotene about 80% of total carotenoid contain [2] which will converted into vitamin A. The wrong post-harvest handling can reduce its quality, nutrition and the shelf life of carrot because of physical, chemical, and microbiological effects [3]. One of chemical effect is a respiration process [4], which makes improvement in decay process of carrot due to activity of microorganism increase.

Therefore, a storage technique is needed to improve the quality and the shelf life of carrot. A storage techniques has been do are air cooling and vacuum cooling storage, because low temperature can extend shelf life and maintain the nutritional quality of carrots, by inhibiting the maturation process [4-6]. Vacuum storage tends to reduce the growth of aerobic bacteria during storage period because its cannot do respiration proses due to lack of O_2 [7, 8]. In the vacuum cooling storage, the selection of the best packaging material is a crucial point for extend the food shelf-life. Therefore, this research used polivinyl chloride (PVC) and nylon plastic for vacuum process because those are have a low permeability to H_2O so it's hard to absorb water from outside to the system [9]. Parameters measured from the both of storage that may occur during 12 weeks the storage period are β-Carotene content and proximate content include water content, pH, protein content and sugar content.

2. Experimental

2.1 Proximate Analysis

Proximate analysis for water content using oven method based on SNI-01-2891-1992, sugar content using Lufft Schrool method based on SNI 01-2892-1992, pH test based on SNI-01-2891-1992 and protein assay using semi micro kjeldahl method based on AOAC .

2.2 β-Carotene Analysis

β-Carotene analysis was carried out using UV-Vis spectrophotometer by modified procedure from Karnjawanipagul research [10]. Briefly, 40 g of carrot was blended with 4 g anhydrous sodium carbonate and mixed with a mechanical blender. 5 g of the mixture was transferred into a centrifuge tube, added with 10 mL tetrahydrofuran and mixed for 24 hours at 5°C. The mixture was centrifuge at 1000 rpm for 30 min and the supernatant was collected. Extraction was performed by adding 7.5 mL dichloromethane and 7.5 mL of 10% w/v NaCl into the supernatant and shaken for 2 min. The extraction was repeated twice, organic layer was collected and evaporated under nitrogen steam. The residue was kept at 5°C, 0.1 mL of residue reconstituted with 5 mL DCM for UV measurements.

2.3 Method of Validation

Validation method was evaluated in terms of linearity, accuracy, precision, limits of detection (LOD) and limits of quantitation (LOQ). Linearity of the method was performed in a range of 0-8mg/L and 0-1mg/L standard β-carotene solution which diluted in dichloromethane solvent. Precision was determined from repeatability, intra-day and inter-day precision and relative standard deviation (RDS) was calculated. Repeatability was from repetitive UV measurement of standard β-carotene solution at 4 mg/L (n=6). Intra- and inter-day precision was determined from UV measurement of standard β-carotene solution at 4 mg/L on the same day (n=10) and on different days (n=6). Accuracy was performed by standard addition method and recovery (R) was calculated. %R was calculated from (amount found/amount added) x 100. LOD and LOQ were calculated from (3 x SD)/s and (10 x SD)/s, where SD and s were standard deviation of blank measurement (n=7) and slope of calibration curve, respectively.

3. Result and Discussion

3.1 Analysis Proximate

3.1.1. Physical Changes during the Storage Period

During the both of storage physical changes that occur were a decrease the intensity of color and texture of carrot (Table 1). In the air cooling storage a black rot was occur at the surface skin of carrot in week 12, while in the vacuum cooling storage has occurred in week 4 of the storage period.

3.1.2. Water Content of Carrot during Storage Process

The water content in the air cooling storage decreased by 2.67% over 12 weeks of storage period (Figure 1) due to microbial unable to do respiration process and transpiration process especially water at fiber tissue, because this type of water is easy to evaporated [11]. This storage inhibited activity of pathogenic bacteria *L.monocytogenes, Salmonella enteritidis,* and *C.sakazakii* [12]. While in the vacuum cooling storage decreased by 0.6% over the four weeks of storage and had increased 0.57% for twelve weeks of storage. That showed vacuum process only inhibited aerobic respiration for 4 weeks of vacuum cooling storage and had been working out as usual caused by the activity of *Pseudomonas sp*, because these bacteria keep running its activities in low temperatures [13].

Table 1. Physical changes of carrot during the storage period.

	0weeks	4 weeks	8 weeks	12 weeks
Air Cooling Storage	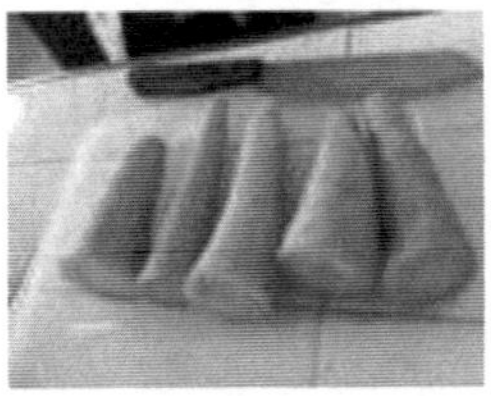	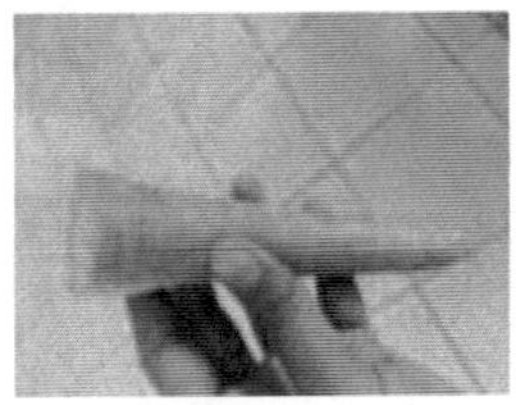	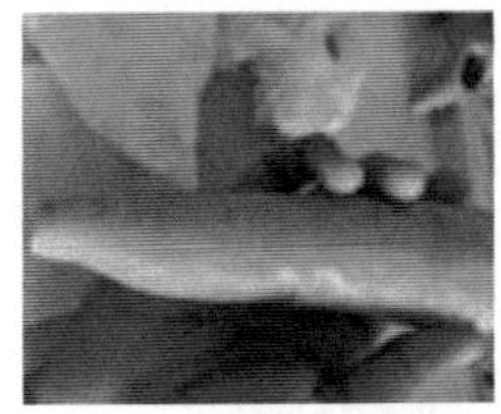	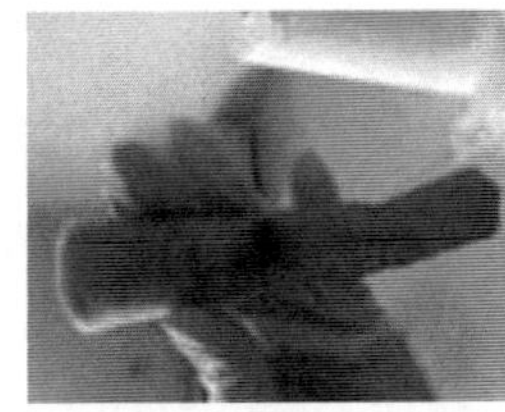
Vacuum cooling Storage	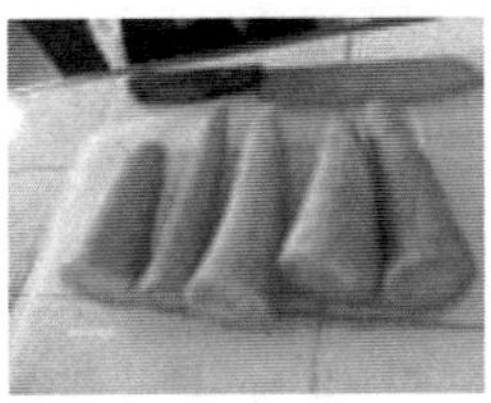	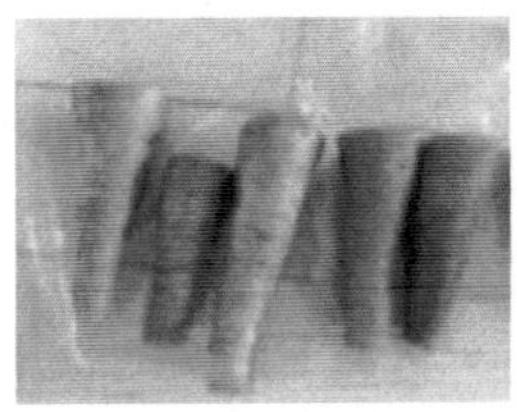	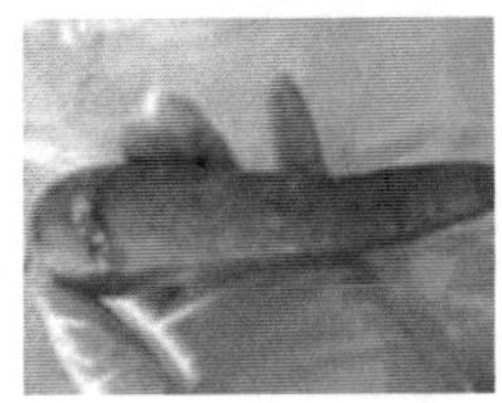	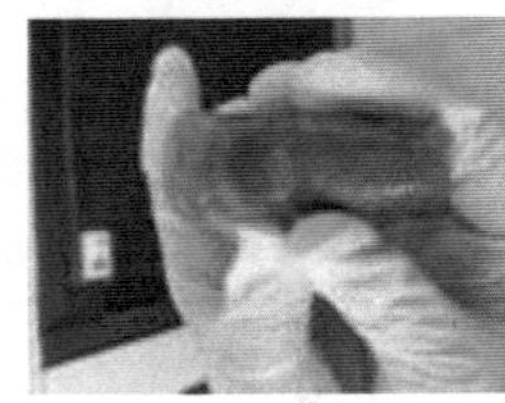

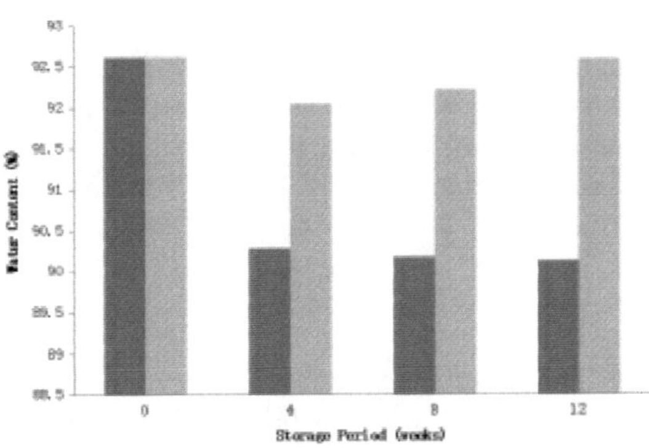

Figure 1. Water content of carrot during air cooling (red) and vacuum cooling (green) storage period.

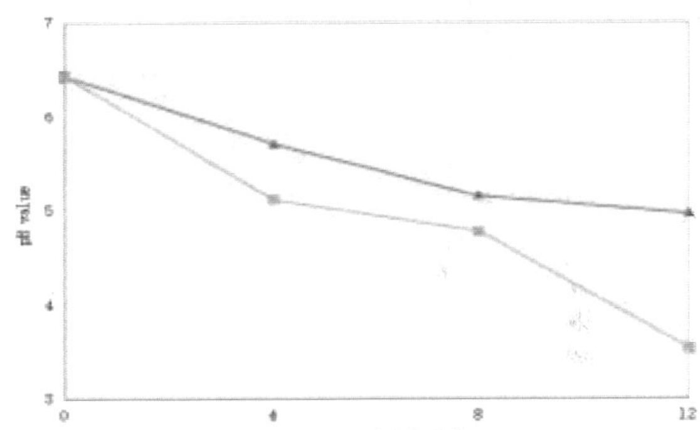

Figure 2. pH value of carrot during air cooling (red) and vacuum cooling (green) storage period

3.1.3. pH value of Carrot during Storage Process

pH value of carrot during storage process showed at figure 2. The decrease of pH during the both of storage is estimated because of anaerobic respiration process by bacteria *L.mesenteroides*, these bacteria can do anaerobic respiration at a temperature of 2 °C [14]. The decreased of pH value from carrot on 12 weeks storage period vacuum cooling were higher than the carrot in the air cooling storage, on the air cooling storage was 6.4 to 4.9 and on the vacuum cooling storage was 6.4 to 3.5. That happened because vacuum condition by lowering the concentration of O2 will increase the growth of fungus [15] in carrots and cause anaerobic respiration which produces excess CO2 and lactic acid that caused decrease in the pH value.

3.1.4. Protein Content of Carrot during Storage Process

Protein content of carrot during air cooling storage decreased by 75.39% while in the vacuum cooling storage decreased by 95.17% (Figure 3). Protein content during the both of storage decreased caused by

IOP Conf. Series: Materials Science and Engineering **193** (2017) 012045 doi:10.1088/1757-899X/193/1/012045

coagulation and denaturation process due to changes in pH and cold temperatures. Denaturation occur low temperature and lower pH will improve the process of protein denaturation [16].

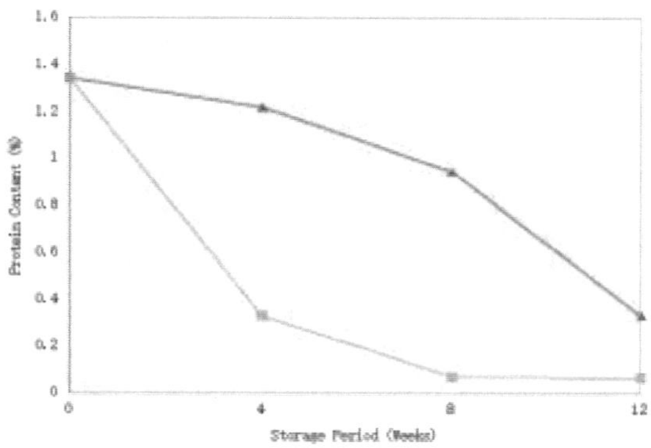

Figure 3. Protein content of carrot during air cooling (red) and vacuum cooling (green) storage period.

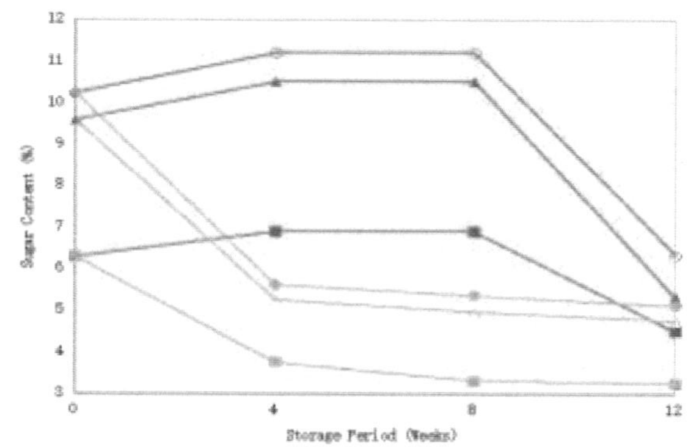

Figure 4. Sugar Content (circle maltose, triangle lactose, rectangular glucose and fructose) of Carrot during air cooling (red) and vacuum cooling (green) storage period.

3.1.5. Sugar Content of Carrot during Storage Process
Sugar content of carrot during storage process showed at figure 4. The sugar content increased up to 8 weeks at the air cooling storage because of starch hydrolysis process into constituent compounds enzymatically [17] and decreased at the end of storage because the starch content is already used and enzyme activity has decreased simultaneously with the sugar content [18]. While at vacuum cooling storage decreased from the beginning of the storage period because maturation process not occur in this storage and the starch contained in carrot be used to anaerobic respiration.

3.2 Identification of Maximum Wavelength
UV spectrum of β-carotene was scanned from 350 to 800 nm and maximum absorption was obtained at 461 nm (figure 5). This was in good agreement with that reported in literatures [10]. UV spectrum of carrot extract and the standard showed the same pattern UV absorption at the same wavelength (Figure 6). This result confirmed that the extraction procedure was valid and the extract contained β-carotene.

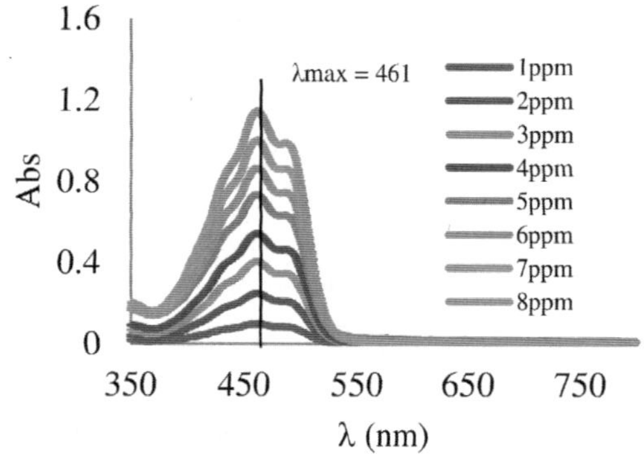

Figure 5. UV-Vis spectra of 1-8 mg/L standard β-carotene in dichloromethane.

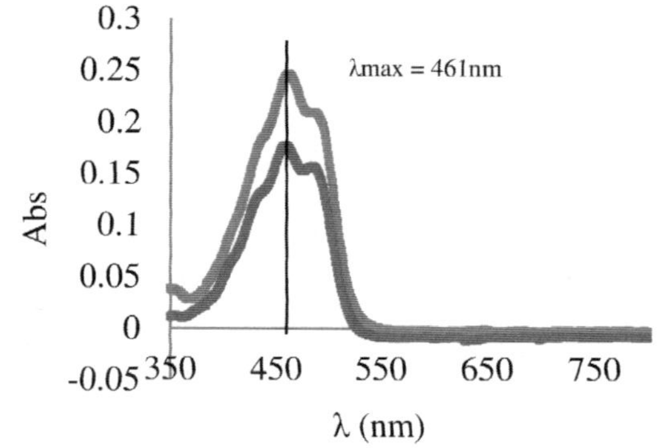

Figure 6. UV-Vis spectra of standard β-carotene (green) and β-carotene from carrot extract (red).

3.3 Method Validation

Linearity of the method was performed in a higher range of 0-8 mg/L showed good linearity with a regression of $y = 0.147x - 0.031 (r^2 = 0.997)$ (Figure 7), where x and y were β-carotene concentration and UV absorption at 461 nm, respectively. At the lower range, the method also provided acceptable linearity with regression of $y = 0.222x - 0.002$ ($r^2 = 0.999$) (Figure 8).

Method repeatability showed RSD of 0.880%. Intra-and inter-day precision revealed RSD of 7.477% and 3.564%, respectively. RSD values of Intra-day were slightly high since β-carotene was unstable and easily degraded at room temperature. However, the other one were in acceptable ranges [19]. Method accuracy showed %Recovery was 100.66%, that showed the procedure is valid because it is not disturbed by the sample matrix and has an acceptable range [19]. LOD and LOQ were 0.009 mg/L and 0.032mg/L, respectively. That indicates that the instrument cannot measure concentrations below 0.032 mg/L. Validation data indicated that the proposed method showed good linearity, precision, accuracy and sensitivity, which could be used for determination of β-carotene in carrot.

3.4 β-Carotene Content of Carrot during Storage Process

A decrease of β-carotene content during the both of storage presented at figure 7.

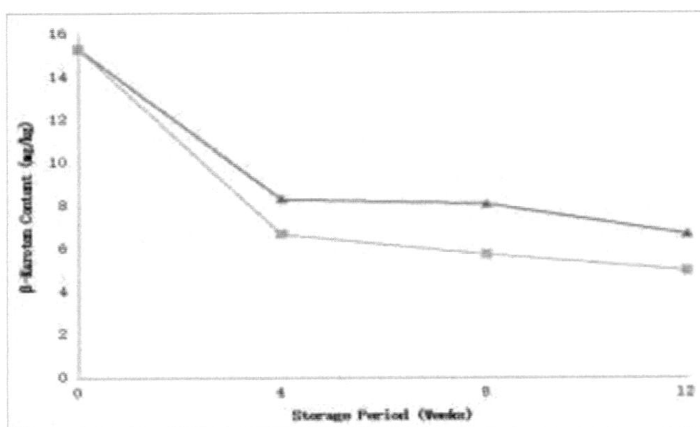

Figure 7. β-carotene content of carrot during air cooling (red) and vacuum cooling (green) storage.

From figure 7, the decrease of β-carotene content during the both of storage occurs due to the oxidation process that caused by the double bonds which contained in it is structure, and makes β-carotene is sensitive to oxidation [20,21], and a decreased of β-carotene content from carrot at vacuum cooling storage were higher (67.57%) than the carrot in the air cooling storage (56.42%) influenced by protein content and pH value. Low pH will be increasing a carotenoid oxidation by isomerization from trans into the cis [22] and protein molecules made a bonding with carotene molecules as a carotene bodies [2].

4. Conclusion

Air cooling storage was more effective storage techniques for extending the shelf life of carrot compared to the vacuum cooling storage which indicated from proximate content and β-carotene content. The decreased content of β-carotene, protein, and pH from carrot during vacuum cooling were higher than in the air cooling storage period, on the air cooling storage were 56.42%; 75.39%; 6.4-4.9 respectively and on the vacuum cooling storage were 67.57%; 95.17%; 6.4-3.5 respectively. The water content in the air cooling storage decreased by 2.67% while in the vacuum cooling storage decreased by 0.6% over the four weeks of storage and had increased 0.57% for twelve weeks of storage. The sugar content for air cooling storage increased up to eight weeks and decreased at the end of storage while the vacuum cooling storage decreased from the beginning of the storage period.

Acknowledgement
Authors thanks to Maintance Research Group (MRG) 2015 LPPM Sebelas Maret University for funding the research.

References

[1] Maiani G, *et al* 2009 *Mol. Nutr. Food Res.* ***53*** 194–218
[2] Fikselova M, Silhar S, Marecek J, Francakova H 2008 *Czech.J. Food Sci.* ***26*** 268–274
[3] Dang K T H, Singh Z, Swinny E E 2008 *J. Agric. Food Chem.* **56** 106667-10674
[4] Devanesan J N, Karuppiah A, Abirami C V K 2012 *J.Agrobiol.* **28** 119-128
[5] Wisudawaty P, Sugiarto, Pandji C 2002 *J. Agroindustri Indonesia.* **1** 25-30
[6] Leungwilai K, Beckles D M 2013 *J. Stored. Prod. Postharvest. Res.* **4** 35-43
[7] Fonseca S C, Oliveira F A R, Brecht J K 2002 *J.Food.Eng.* **52** 99-119
[8] Gibis D and Rieblinger K 2011 *Proc. Food Science.* **1** 229 –234
[9] Candra R M and Sucita D 2015 *J.CorelIT.* **1** 77–84
[10] Karnjanawipagul P, Nittayanuntawech W, Rojsanga P, Suntornsuk L 2010 *J. Phar. Sci.***37** 8-16
[11] Winarno F G 1995 *Kimia Pangan dan Gizi* (Jakarta: PT. Gramedia Pustaka Utama)
[12] Erickson M C 2010 *Compr Rev Food Sci Food Saf.* ***9*** 602–619
[13] Rashid N, Shimada Y, Ezaki S, Atomi H 2001 *Appl. Environ.* ***67*** 4064–4069
[14] Malik K and Garg F C 2011 *As. J. Food Ag-Ind.* ***4*** 359–364
[15] Barbara L, Tony C B P, Grahame W G 2000 *Microbiological Safety and Quality of Food. Volume 1* (Maryland: Aspen Publishers, Inc)
[16] Soliman T N, Farrag A F, Shendy A, Elsayed M M 2010 *J. Am. Sci.***6** 49-62
[17] Helmiyesi, Hastuti R B, Prihastanti E 2008 *Anatomi dan Fisiologi.* ***16*** 33–37
[18] Syafutri M I, Pratama F, Saputra D 2006 *J. Teknologi dan Industri Pangan.* **17** 1-11
[19] Careri M, Elviri L, Mangia A 1999 *J. Chromatogr A.* **854** 233-44
[20] Boon C S, Mcclements D J, Weiss J 2010 *Critical Reviews in Food Science and Nutrition.* **50** 515-532
[21] Penicaud C, Achir N, Mayer C D, Dornier M, Bohuon P 2011 *Fruits.* **66**
[22] Landrum J T 2009 *Carotenoids:Physical, Chemical, and Biological Functions and Properties,* (U.S : CRC Press)

IOP Conf. Series: Materials Science and Engineering **193** (2017) 012046 doi:10.1088/1757-899X/193/1/012046

Utilization of rice husk waste for Cd(II) adsorbent and its analysis using solid-phase spectrophotometry (sps)

S Saputro[1], M Masykuri[1], L Mahardiani[1], S Arini[1] and A H Ramelan[2]

[1] Chemistry Education, Faculty of Teacher Training and Education, Sebelas Maret University, Jl. Ir. Sutami 36A Surakarta, Central Java-57126, Indonesia
[2] Physics Department, Faculty of Mathematics and Natural Science, Sebelas Maret University, Jl. Ir. Sutami 36A Surakarta, Central Java-57126, Indonesia

E-mail: sulistyo68@yahoo.com

Abstract. This research aims to know the use of rice husk waste for metal ion adsorbent of Cd(II), the optimum mass which are needed, the sensitivity of SPS as a method to determine the reduction level of Cd(II) metal ions in the level of μg/L. This research was conducted by using an experimental method in the laboratory. The used adsorbent in this research is rice husk charcoal which activated using $ZnCl_2$ 10% solution. Determination of the optimum adsorbent mass was done with reacted 0.5; 1.0; 1.5; 2.0; and 2.5 g of adsorbent and 20 mL Cd(II) which have concentration 75.27 μg/L during 30 minutes. Analysis the levels of Cd(II) used SPS method and characterization of adsorbent used FTIR. The result showed that activated rice husk charcoal can be used as adsorbent for Cd(II) metal ions in the simulated liquid waste with the adsorption capacity of 0.58 μg/g, the optimum mass was 2.0 gram with percentage of the adsorbed Cd(II) of 77.87%, SPS is a sensitive method to determine the reduction levels of Cd(II) metal ions in the level of μg/L with the limit of detection (LOD) was 0.06 μg/L.

1. Introduction

Nowadays, the industry development in Indonesia is relatively rapid. It can be seen from many industries which producing various kinds of human needs such as paper industry, textile, skin tanner, etc. Along with those developments, many unneeded results will be produced which is the waste. One of the wastes is heavy metal which cause serious pollution to the environment if it exceeds the limit and has a very dangerous toxic substance and will cause serious disease to humans when it accumulates in the body [1]. One of the indicators which commonly use to detect water pollution is the contamination of heavy metals in it. It called as harmful heavy metals because it generally has a high mass density (5 g/cm^3) and a small concentration can be toxic and dangerous, and one of them is cadmium(II) [2].

Cadmium (Cd) is a heavy metal which is found as an environmental contaminant, both through natural occurrence and from industrial and agricultural sources. Foodstuff are the main source of cadmium exposure for the non-smoking general population. A number of studies have reviewed cadmium content in a range of different foods. In a collaborative effort under the SCOOP involving 13 EU Member States, cadmium concentrations in many foods ranged from non-detected to a high mean of 1.20 mg/kg for cephalopods. For rice, Japan was reported in having higher values than the general average, 0.061 compared to 0.017 mg/kg, respectively. The average concentration of cadmium in wheat was 0.054 mg/kg, vegetables ranged from 0.012 to 0.040 mg/kg, and in mollusk results varied from 0.20 mg/kg for cephalopods to 1.38 mg/kg for oysters [3].

Cadmium was monitored in 119 retail food categories within the framework of a national total diet study of the United Kingdom general population. Cadmium was presented at low concentrations in eleven of the food groups, and was below the levels of detection in meat, poultry, oils and fats, eggs, fresh fruits, beverages, milk and dairy products. Cadmium concentrations were highest in the offal (0.084 mg/kg) and nuts (0.065 mg/kg) groups. Food that was consumed in larger quantities made the greatest contribution to the dietary exposure; high contributors were potatoes (24%), miscellaneous cereals (21%) and bread (19%) [3].

Cadmium (Cd) is a heavy metal that is very dangerous. The use of Cd and their compounds are found in the battery industry, as a dye, photography, used in the manufacture of tetraethyl-lead, as a stabilizer and others. Additionally, it is widely used in light industries such as the processing of bread, fish processing, beverage processing, textile industry, and others. Thresholds Cd in drinking water both natural mineral water and bottled drinking water is 0.003 mg/L [4].

The determination of the concentration of cadmium in the water is very important even in very small concentrations, because if the cadmium-contaminated water consumed by humans, there will be cadmium ingestion in the body tissues. The impact of chronic cadmium poisoning is generally in the form of damage to the many physiological systems of the body. Body systems which are undermined by this Cd metal poisoning are the urinary system (kidney), respiratory system (respiratory/lung), circulatory system (blood) and heart. Beside that, the chronic poisoning also damage the reproduction glands, the olfactory system, and can lead to brittleness of the bones [5]. Thus, to determine the contamination of Cd (II) in the water at very small concentrations, it is necessary to use instrument with high sensitivity, namely solid-phase spectrophotometry (SPS).

SPS is one of the methods of analysis which is using UV-visible spectrophotometer and is based on spectrophotometric measurements directly from the solid phase that has absorbed the sample components. The advantages of this method of analysis compared to other methods is the simple operation, high sensitivity and accuracy, which reached the level μg/L, so it can detect the metal ions dissolved in very small concentrations [6]. This is evidenced by previous research on the determination of Cr(VI) in natural water by analysis of SPS obtained the limit of detection (LOD) of 0.014 g/dm^3 [7].

Several chemical and biological methods have been tried to remove heavy metals contained in sewage and surface waters, including adsorption, ion exchange, and separation with membranes. The adsorption process is more widely used in the industry because it has several advantages: more economical, it does not cause toxic side effects, and capable of removing material impurities.

The majority of Indonesia's population are farmers where most residents create rice as a staple food; so that the source of this staple is everywhere and particularly in the rural areas. Indonesia has an estimated number of 60,000 rice milling scattered throughout the area that produce waste such as rice husks 15 million tons per year. In large quantities, some rice milling machines can produce 10-20 tons of waste rice husks per day [8]. Various studies have reported that rice husk generally contain silica which is quite high ranging from 87-97%. Therefore, rice husk can be used as a source of silica in the silica-based material manufacture or adsorbent [9]. However, the use of rice husk is still very little, so that the hull remains as waste material that disturbs the environment. Thus, researchers interested in conducting research of the application of activated carbon from rice husk waste as adsorbent of Cd(II) with an analysis using SPS.

2. Method

2.1. Tools and materials

The used tools were a UV-visible spectrophotometer from K-MAC, FTIR spectrophotometer from Shimadzu, muffle furnace, oven, pan, analytical balance, volumetric flask, volume pipette, beaker glass, measuring cup, flask, erlenmeyer, watch glass, stirrer glass, drop pipette, aliquoting devices which are assembled using a syringe, blender, mortar and pestle, a 100 mesh sieve, porcelain bowls, stirrer bar, and a magnetic stirrer. Materials used are rice husk, $Cd(NO_3)_2$, $ZnCl_2$ 10%, resin AG Muromac 50W-X2, H^+ form (100-200 mesh), distilled water, chloroform, Whatman filter paper, blue litmus, and dithizone.

IOP Conf. Series: Materials Science and Engineering **193** (2017) 012046 doi:10.1088/1757-899X/193/1/012046

2.2. Research procedure

2.2.1. Production of adsorbent

Rice husk is cleaned and washed with distilled water, then it is dried at 105°C then charred in a muffle furnace at a temperature of 350°C for 1.5 hours. Results obtained are mashed and then sieved with a 100 mesh size.

2.2.2. Adsorbent activation

Charcoal soaks in a solution of $ZnCl_2$10% for 24 hours at room temperature. Filter and rinse the residue with distilled water until the filtrate become neutral, and then dried in an oven at 105°C for 24 hours. Adsorbent was tested by FTIR before and after activation.

2.2.3. Resin preparation

Muromac resin AG Muromac 50W-X2 H^+ form (100-200 mesh) dissolved in distilled water and silence a few moments until the resin becomes more fluffy.

2.2.4. Determination of Cd(II) calibration curve

A 20-mL of standard solution of Cd(II) 0 µg/L, 40 µg/L, 80 µg/L, and 160 µg/L respectively were added with 1 mL of 0.005% dithizone, and 0.06 mL resin. Stir it for 20 minutes and analyzed using UV-vis spectrophotometer at a wavelength of 485 nm and 615 nm and then absorbance difference of the two wavelengths determined, $\Delta A = A_{485nm}-A_{615nm}$, which ΔA will be made standard curve (ΔA vs concentration).

2.2.5. Determination of Cd(II) species in simulated liquid waste

Simulated liquid waste solution of 75 µg/L was taken 20 mL then added 1 mL of 0.005% dithizone, and 0.06 mL resin. Then, stir it for 20 minutes and analyzed using UV-Vis spectrophotometer at a wavelength of 485 nm and 615 nm. ΔA obtained will be substituted in equation Cd(II) calibration curve (ΔA vs concentration), so that the Cd(II) ions in the effluent can be known.

2.2.6. Determination of optimum adsorbent mass

Inserting a 20-mL of simulated liquid waste solution 75 µg/L in the 5 erlenmeyer. Insert each adsorbent with a variation of mass of 0.5; 1.0; 1.5; 2.0; 2.5 gram and stir it for 30 minutes. Each adsorbent was filtered with Whatman filter paper. The resulted filtrate was taken as much as 5 mL and then diluted with distilled water until 50 mL in volume. Furthermore, the filtrate diluted taken as much as 20 mL added with 1 mL of 0.005% dithizone, and 0.06 mL resin then stir it for 20 minutes and analyzed using a UV-vis spectrophotometer with a wavelength of 485 nm and 615 nm. ΔA obtained will be substituted in equation of Cd(II) calibration curve (ΔA vs concentration) so species of Cd(II) in a simulated liquid waste after the Cd(II) adsorption is known.

2.2.7. Determination of detection limit

Taking five blank solution 20 ml, then each added 1 mL of of 0.005% dithizone, and as much as 0.06 ml resin, then stir it for 20 minutes. Furthermore, it analyzed by UV-vis spectrophotometer at a wavelength of 485 nm and 615 nm. ΔA obtained will be substituted in the equation of the calibration curve.

3. Result and discussions

3.1. Production of activated charcoal from rice husk

Production of activated charcoal from rice husk consists of three stages: the stage of preparation, is the removal of impurities in rice husk, stage of carbonization, breakdown of organic matter into carbon by heating it at 350°C for 1.5 hours, and the activation stage, soaking with an activator solution which is $ZnCl_2$ 10% for 24 hours. $ZnCl_2$ 10% is used because it is a good activator for lignocellulose material like rice husk because lignocellulose material has high content of O_2 that can reacted with acid activating agent [10].

3.2. Fourier transform infra red spectra analysis

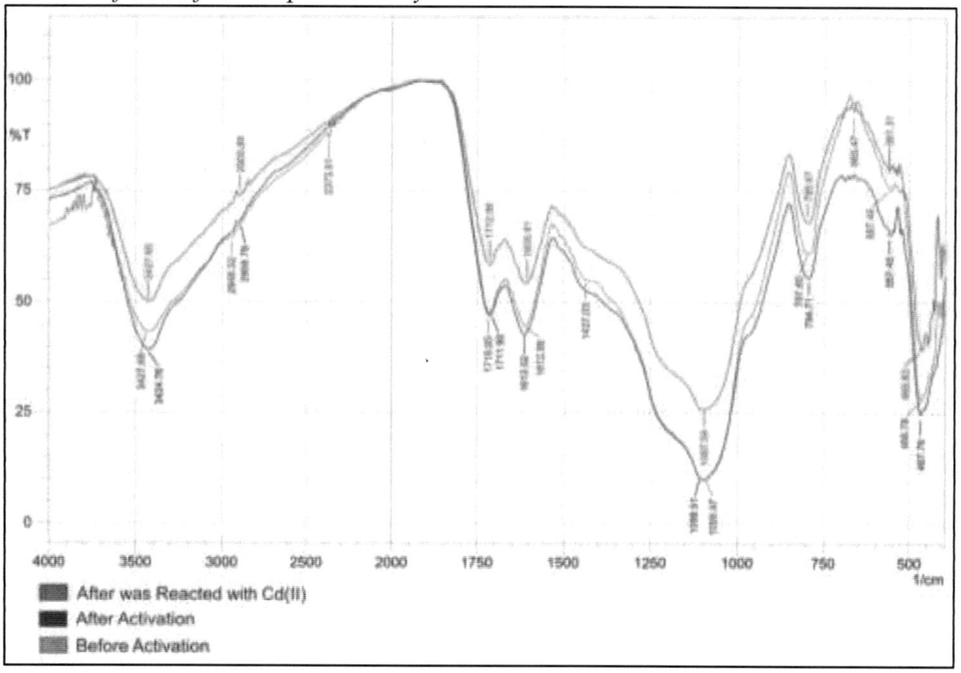

Figure 1. FTIR spectra of rice husk charcoal before activation, after Activation, and after reacted with Cd(II).

Based on figure 1, adsorbent before activation shows the presence of the –OH group from Si-OH (cyanol), Si-O from Si-OH, Si-O from Si-O-Si, -CH, $-CH_3$, C=C aromatic and C=O. In the adsorbent after activation and after contacting not shows any significant difference with the adsorbent before activation, it's just a slight shift in the wavenumber. The shift in the wavenumber of the groups indicates that such groups play a role in the process of adsorption of Cd(II) adsorbent activated charcoal rice husk. Based on changes in the infrared spectra can be seen that all the peaks there will be shorter or in other words the specific surface area decreases. This occurs because the adsorbent has interaction with metal ions which are absorbed so that the pores on the surface of the adsorbent decreases which is covered by Cd(II) adsorbed. It can be seen that there is no a new peak. This indicates that there is no a new group formed in the process of binding metal ions Cd(II) because the binding only disturb the vibration force that binds those ions Cd(II).

3.3. Determination of Cd(II) calibration curve

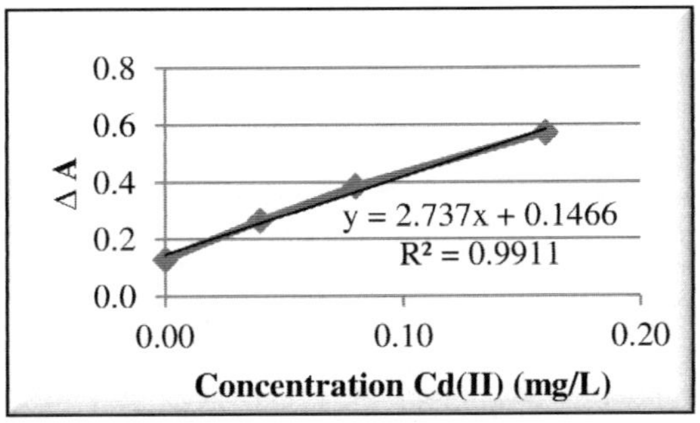

Figure 2. Calibration curve of Cd(II) solution.

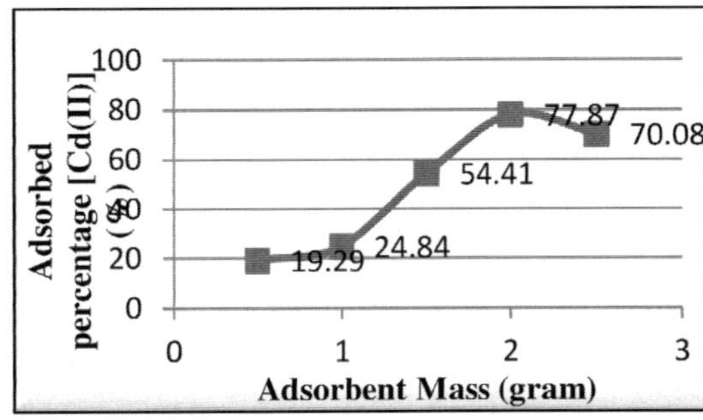

Figure 3. Determination of optimum adsorbent mass

Based on the graph in figure 2, y = 2.737x + 0.146 where y is ΔA and x is the concentration. The equation is used to determine the concentration of Cd(II) in simulated liquid waste.

3.4. Determination results of Cd(II) levels in simulated liquid waste

Table 1. Determination results of Cd(II) levels in simulated liquid waste

Sample	[Cd(II)] Calculation (μg/L)	ΔA	[Cd(II)] Actually (μg/L)
Sample Cd(II)	75	0.351	74.947
		0.353	75.595
Average			75.271

3.5. Determination of optimum mass adsorbent in a decrease levels of Cd(II)

Table 2. Determination of the optimum mass of adsorbent

Mass (gram)	Initial [Cd(II)] (μg/L)	ΔA	Final [Cd(II)] (μg/L)	Adsorbed [Cd(II)] (μg/L)	Adsorbed [Cd(II)] (%)
0.5	75.271	0.163	60.604	14.487	19.293
1.0	75.271	0.162	56.442	18.649	24.836
1.5	75.271	0.156	34.236	40.855	54.407
2.0	75.271	0.151	16.620	58.471	77.867
2.5	75.271	0.153	22.465	52.626	70.083

Based on data research in table 2 and the graph in figure 3, the optimum mass of adsorbent to adsorb Cd(II) metal ions in the simulated liquid waste was 2.0 gram with adsorbed concentration was 58.47 μ/L, percentage of adsorbed as 77.87% and adsorption capacity was 0.58 μg/g.

3.6. Determination of detecting limit

In this research, limit of detection (LOD) was 0.06 μg/L, so it can be said that the sample at a concentration of 0.06 μg/L can be read, so that the solid-phase spectrophotometry (SPS) is a sensitive and effective method to use in analysis the decreased levels of Cd(II) adsorption results of activated charcoal from rice husk in levels μg/L.

4. Conclusion

The result showed that activated rice husk charcoal can be used as adsorbent to adsorb Cd(II) metal ions in the simulated liquid waste with the adsorption capacity was 0.58 μg/g, the optimum mass of adsorbent to adsorb Cd(II) metal ions in the simulated liquid waste was 2.0 gram with percentage of adsorbed as 77.87%, solid-phase spectrophotometry (SPS) is a sensitive method to determine the reduction levels of

Cd(II) metal ions as an adsorption result of the activated rice husk charcoal in the level of μg/L with the limit of detection (LOD) was 0.059 μg/L.

Aknowledgement

Authors wishing to acknowledge assistance from Head of Chemistry Laboratory, Chemistry Study Program, Faculty of Teacher Training and Education, Sebelas Maret University, for using the facilities of the laboratory. This work was partially supported by the PUPT Grant for S.S (2015 and 2016) and by Competency Grant for S.S (2016) from the Ministry of Research, Technology and Higher Education, Indonesia.

References

[1] Danarto Y C 2007 *Jurnal Ekuilibrium* **6** 65
[2] Raikwar M K, Kumar P, Singh M and Singh A 2008 *Veterinary World* **1**(1) 28
[3] The European Commission on Cadmium in Food 2009 *The European Food Safety Authority J.* **980** 1
[4] Peraturan Menteri Kesehatan No.492/Menkes/Per/IV/2010 *Persyaratan Kualitas Air* Minum (Jakarta: Menteri Kesehatan)
[5] Palar H 2012 *Pencemaran dan Toksikologi Logam Berat* (Jakarta: Rineka Cipta)
[6] Sarenqiqige, Saputro S, Kai S, Satoda M, Matsuoka S and Yoshimura K 2013 *Anal Sci.* **29** 677
[7] Saputro S, Yoshimura K, Takehara K, Matsuoka S and Narsito 2009 *Anal Sci.* **25** 1445
[8] Affendi M *et al.* 2008 *Prosiding Seminar Nasional Sains dan Teknologi-II 2008* (Lampung: Universitas Lampung)
[9] Kalapathy U, Proctor A and Shultz J L 2000 *Bioresour Technol.* **73** 257
[10] Hsu LY and Teng H 2000 *Fuel Processing Technol.* **64** 155

IOP Conf. Series: Materials Science and Engineering **193** (2017) 012047 doi:10.1088/1757-899X/193/1/012047

The Effect of pH and Color Stability of Anthocyanin on Food Colorant

S Wahyuningsih[1], L Wulandari[1], M W Wartono[2], H Munawaroh[1], A H Ramelan[1]
[1]Inorganic Materials Research Group, Faculty of Mathematics and Natural Sciences, Sebelas Maret University, Jl. Ir. Sutami No. 36A Surakarta, INDONESIA
[2]Food Chemistry Research Group, Faculty of Mathematics and Natural Sciences, Sebelas Maret University, Jl. Ir. Sutami, Kentingan, Surakarta 57126, Central Java, Indonesia

E-mail: sayekti@mipa.uns.ac.id

Abstract. Anthocyanins are naturally occurring pigments of red and purple. Red anthocyanin pigments provide a strong and sharp and widely applied in various industries such as food coloring or drink. Anthocyanins isolated by maceration, extraction and thin layer chromatography (TLC). The extract has been obtained from the initial stages of maceration then separated into several fractions by chromatography to isolate fractions colored dark red. Identification of chemical compounds with TLC (Thin Layer Chromatography) is able to distinguish the fraction of anthocyanin produced. FTIR (Fourier Transform Infrared Spectroscopy) used to identification of the functional group of a compound. The UV-Vis absorption spectra have to produce maximum absorbance values that describe the intensity of anthocyanin spectra in different colors for different pH. Anthocyanins are more stable at low pH (acidic conditions) which gives a red pigment. Meanwhile, the higher the pH value of anthocyanin will provide color fading of the color blue. So as a food colorant, anthocyanin with a low pH or height pH has a significant effect on the food colorant.

1. Introduction

Anthocyanins are naturally occurring pigments of red and purple. Red anthocyanin pigments provide a strong and sharp and widely applied in various industries such as food coloring or drink. Anthocyanins are consumed as part of a normal diet and many times with higher levels than other flavonoid classes. The anthocyanin consumption in Europe are: 19.8 (Netherland) – 64.9 mg/d (Italy) (male), and 18.4 (Spain) – 44.1 mg/d (Italy) (female) [1].These compounds have been reported to exert health beneficial effects like cardiovascular protection and anti proliferative [2].

Anthocyanins are flavonoids of phenolic compounds. Whereas, the phenolic compounds are composed of a large group of organic substances, and flavonoids are an important subgroup [3].The flavonoid subgroup contains the anthocyanin's, one of the most broadly naturally source of colorants. On the anthocyanin structure (figure 1) there are 7 positions labeled R. R basically means that it can be occupied by almost any organic group like a methoxyl group, sugar, and the number of R that are occupied by specific substitutions would determine the color of the anthocyanin [4].

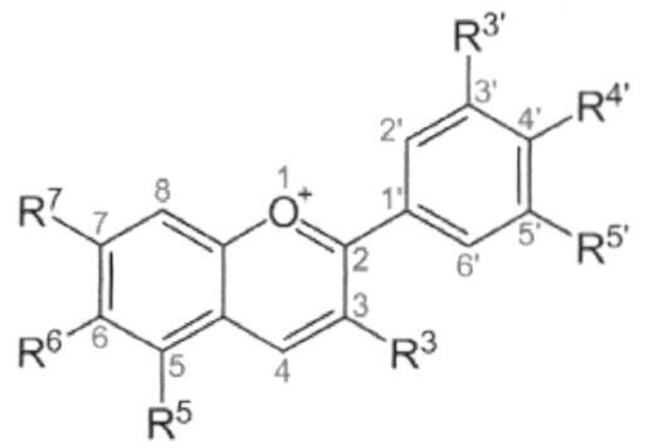

Figure 1. Structure of anthocyanin

Phenolic compounds are composed of a large group of organic substances, and flavonoids are an important subgroup. The flavonoid subgroup contains the anthocyanins, one of the most broadly naturally source of colorants including blue, purple, violet, red [5]. Anthocyanin comes from two words, "Antho" means flower and "cyanin: mean blue. Additionally, it is considered as a functional ingredient due to several anthocyanins has been shown to passes health-promoting properties [6]. The basic colors blue, purple, red and orange have a direct relation with the number of hydroxyl groups and indirect relation with the number of methoxyl groups. According to research the basic colors blue, purple, red and orange have a direct relation with the number of hydroxyl groups and indirect relation with the number of methoxyl groups [7]. Also, it has been showed that hydroxylation at position 4 changes the colors toward red tones. Anthocyanins are of great economic importance as pigments of fruit juices and wines. One of the main claims of the food industry nowadays is for natural colorants to replace synthetic red dyes, and anthocyanins are the principal candidates, however just anthocyanin and lees (sediment of the grape juice tanks) preparations are the only anthocyanin sources approved by FDA to be used for human food, while the main use is in the production of beverages and soft drinks [8]. However, anthocyanin instability limits their use, and different preparations have been evaluated to avoid anthocyanin degradation.

2. Experimental

2.1 Sample preparation

Rose flower was dried in an oven with temperature ± 50 °C. And then the dry sample was powdered using a blender.

2.2 Sample maceration and extraction

Extraction of rose flower powder used ethanol 96% and HCl 0.1 M as a solvent with ratio 4:1 in 24 hours. Residue and filtrate were separated by vacuum filtration. Then filtrate was evaporated to remove the solvent with a rotary evaporator, yielded concentrated extract.

2.3 Vacuum liquid chromatography (VLC)

Concentrated extract was fractionated by vacuum liquid chromatography to separate the polar and non-polar fraction. The eluent was used ethyl acetate and methanol. The eluate collected in a glass beaker and evaporated solvent with a rotary evaporator. Characterization eluate was conducted with Thin Layer Chromatography (TLC) and UV-vis spectrophotometry.

2.4 Gravity Chromatography

The chromatography used stationary phase Sephadex LH-20. 0.38 grams of sample eluted in the column of Sephadex. Samples eluted with solvent 0.1% HCl in methanol until the entire sample can be eluted out of the column. Each eluate evaporated solvent with a rotary evaporator. Characterization eluate was conducted with Thin Layer Chromatography (TLC), UV-vis spectrophotometry and Fourier Transform Infrared (FTIR).

IOP Conf. Series: Materials Science and Engineering **193** (2017) 012047 doi:10.1088/1757-899X/193/1/012047

2.5 Characterization of anthocyanin extracts in various pH

Measurement of the color intensity of anthocyanin extract was performed on six pH conditions (1, 3, 5, 7, 9, 11). Each solution was measured at the maximum wavelength with a spectrophotometer at range 400-800 nm.

3. Result and Discussion

3.1 Isolation, fractionation, and purification of compounds Roses Red

The maceration process of red roses carried out in acidic conditions with 0.1 M HCl (ratio 4:1). The addition of acid serves to denature cell membrane of plant, dissolve the pigment anthocyanin, flavonoid prevents oxidation and increases the polarity of the solvent that the dye is dissolved [9]. Figure 2 shows the results of maceration filtrate and concentrated extract of the red rose. Obtained concentrated extract as much as ± 422 g.

(a) (b)

Figure 2. (a) Maceration filtrate (b) Concentrated extract of red rose

The next separation process was fractionated with vacuum liquid chromatography (VLC) (figure 3). In the process of VLC performed used eluent EtOAc: Methanol, because before it was did TLC to determine the appropriate eluent. The results of separation with VLC obtained 12 fractions and identified with Thin Layer Chromatography (TLC). Figure 3 shows the results of identification by TLC and the fraction C, D, E, F, G, H, I, J have 1 spot.

(a) (b) (c)

Figure 3. (a) Fractionation process (b) Filtrate of VLC (c) TLC of fraction (eluent butanol: ethanol: water (6: 3: 2))

From figure 3 was known fraction C, D, E, and F showed the color of the fraction is red. The fraction C, D, E, F were identification by TLC using different polarity eluent to know the pattern of the separation of compounds. The results of TLC analysis of the fraction of C-F shown in figure 4.

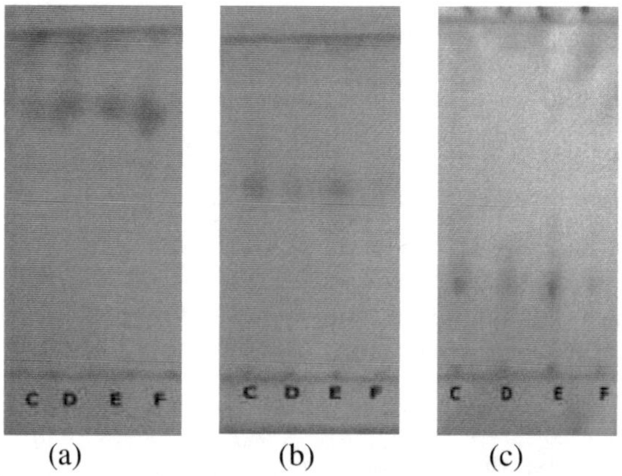

Figure 4. Identification of TLC with different eluent polarity (a) 2-propanol: acetic acid: water (7: 1: 2), (b) Butanol: Ethanol: water (5: 2: 3), (c) Butanol: acetic acid: water (4.1.5)

Figure 4 shown the fractions C, D, E and F have same separation profile, to determine the next phase separation did analysis with UV-Vis spectroscopy. The spectra of absorption UV-Vis is shown in figure 5 where the fraction C of the separation with VLC has a sharp peak in the 350-400 nm. The spectra indicated the fraction possibilities only one type, there are anthocyanin compounds. For further purification of fraction C was separated with chromatography gravity.

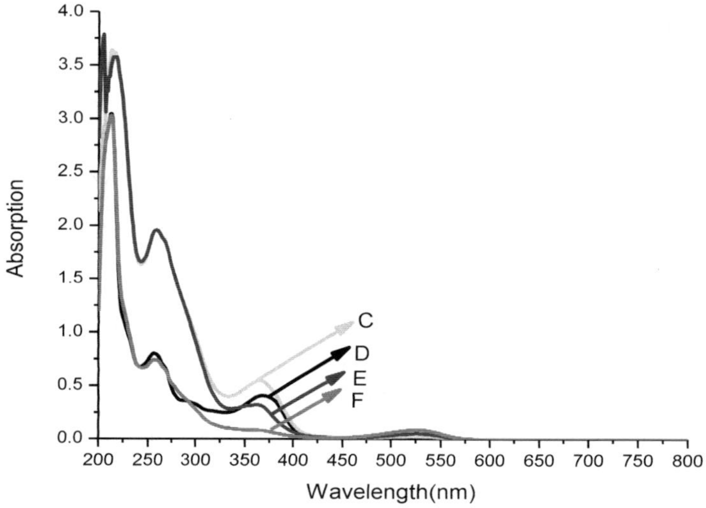

Figure 5. UV-Vis spectra of fractions C, D, E and F

The separation with gravity chromatography used stationary phase Sephadex LH-20 and eluent 0.1% HCl in methanol. Figure 6 shows the gravity separation process chromatography use Sephadex LH-20 column and the fraction result of separation. The figure show the color of the fraction is red, indicated anthocyanin compounds.

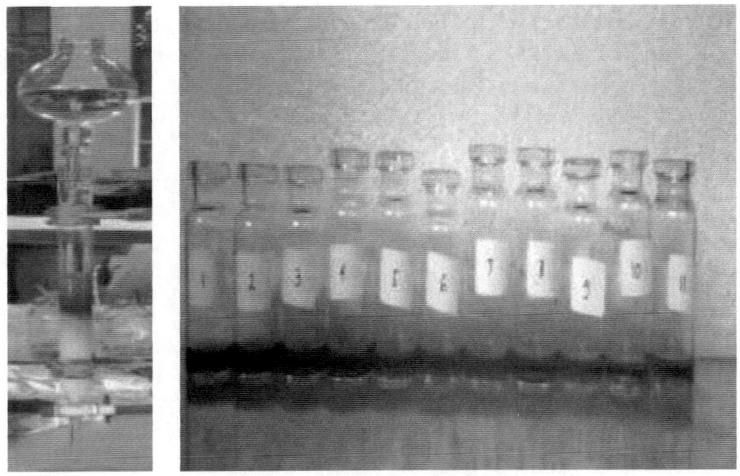

(a) (b)

Figure 6. (a) Process of separation by gravity column (b) Fraction of the result separation

The pattern of separation results gravity chromatographic can be determined by TLC identification with eluent butanol: ethanol: water (6: 2: 3) and shown in figure 7.

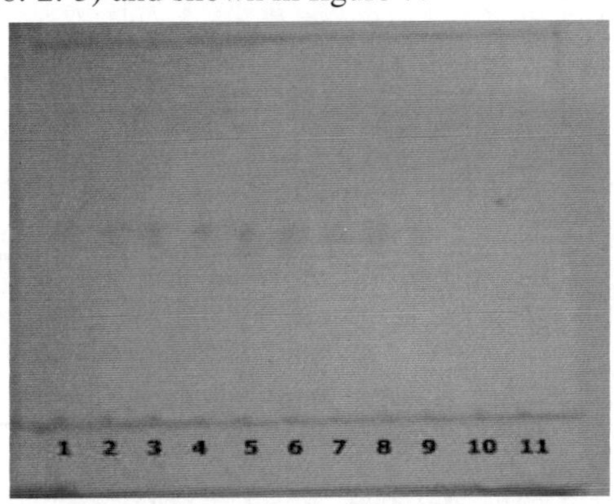

Figure 7. Results of TLC with eluent B: E: A = 6: 2: 3

From figure 7 it is known that the fraction of the number 7 is considered purer, that the purity test conducted using the eluent ethyl acetate: 2-propanol: water: formic acid (6: 2: 2: 1). Figure 8 shows the results of TLC provides one spot that it is possible to have been obtained anthocyanin pure compound. To ensure the purity of the compound then analysis of UV-Vis Spectroscopy and FTIR Spectroscopy.

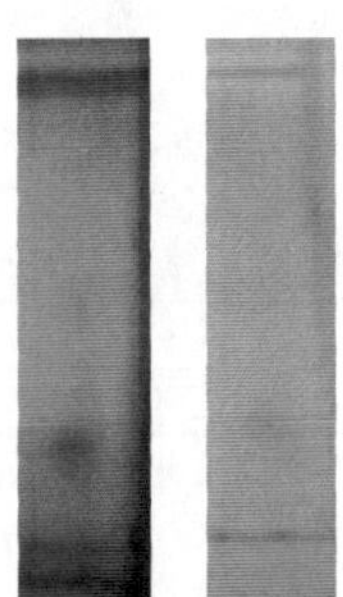

(a) (b)

Figure 8. TLC with eluent ethyl acetate: 2-propanol: water: formic acid (6: 2: 2: 1) (a) In terms of a UV lamp 254 nm, (b) after sprayed with Ce $(SO_4)_2$

The results of UV-Vis spectra of the rose extract is shown in igure 9. Rose extracts contained 3 peaks in the area 268 nm, 364 nm and 526 nm [10]. Isolated compounds only appeared two peaks, namely 288 nm and 526 nm. The maximum wavelength can be seen in table 3.

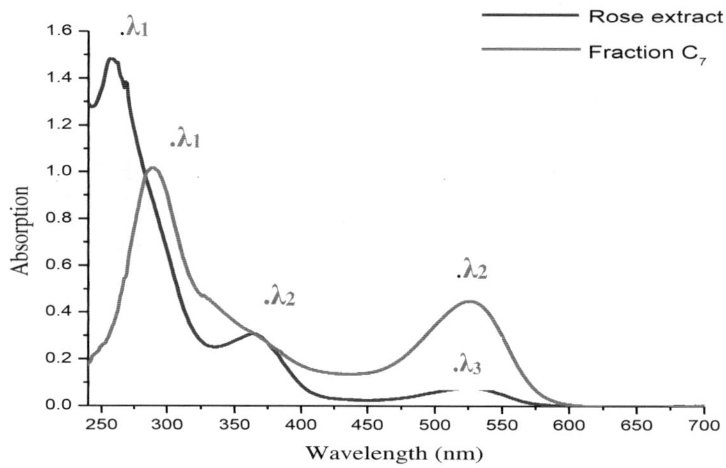

Figure 9. UV spectra extracts and isolates results (C7)

Table 1. The maximum wavelength of the compound isolates

λ	Extracts	Isolat	Wavelength (nm) Anthocyanin Cyanidin-3-sophoroside	Cyanidin-3,5-diglukoside
λ_1	256	288	283	270
λ_2	329	-	-	-
λ_3	526	526	526	526

Based on the results of spectra UV-Vis, C7 have aromatic groups in the area λ 288 nm. The aromatic group generally appears at λ 270-300 nm Beside it, in the area λ 526 nm originated from carbonyl groups conjugated (C=C). The wavelength of 526 nm is the wavelength anthocyanin cyanidin types [11].

From the IR analysis (table 2) can be seen that functional groups present in the compound C7. The existence of the aromatic group of spectra UV is amplified by the IR spectrum (figure 10) which shows the C = C aromatic absorption at wavenumber 1654 cm^{-1} and 677 cm^{-1} which indicate the presence of aromatic C-H bond. In addition, the aromatic group of IR also found in the C=C bonds alkenes appear in wavenumbers 1556 and 1516 cm^{-1}. C-O group detected at 1029 cm^{-1}, the hydroxyl group at wave number 3385 cm^{-1} and C-H absorption stretching appear at wave numbers 2943 and 2841 cm^{-1} which make up a group the sugar.

Table 2. Functional groups on the compound isolates

Functional group	Wavenumber (cm^{-1}) Isolate	Cyanidin
OH	3385	3370
C-H alifatic	2841	2840
C = C aromatic	1654	1660
C=C alkanes	1556 and 1516	1560
C-O	1029	1260
C-H aromatic	677	675

FTIR spectra C7 known that compounds have hydroxy groups, groups of alkanes and aromatic groups, which it can be concluded that the compound C7 isolate anthocyanin. In particular anthocyanins cyanidin compound has an absorption in the 3100-3400 cm^{-1} (OH);2900-2840 cm^{-1}(C-H aliphatic) 675-870 cm^{-1} (C-H aromatic) and 1660 cm^{-1} (C = C aromatic) [12].

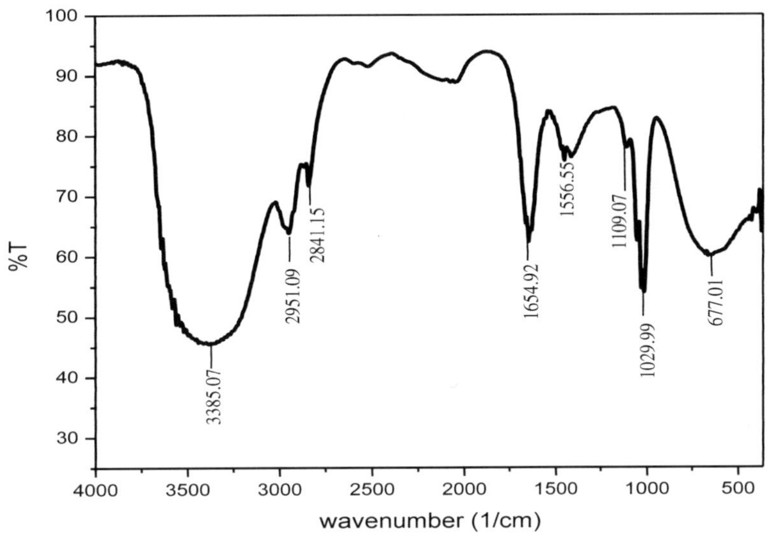

Figure 10. Infrared spectra of compound C7

3.2 Effect pH and stability color of anthocyanin

The stability of anthocyanin was influenced by several factors such as pH, temperature, light, and oxygen [13]. According to Clydesdale (1998) [14] and Markakis (1982) [15] Pigment anthocyanin (red, purple and blue) is an unstable molecule if there is a change in temperature, pH, oxygen, light, and sugar. Based on the analysis show that anthocyanins are red at low pH (acidic conditions), the higher the pH value of anthocyanins will provide color fading of colorless, yellow purple and blue.

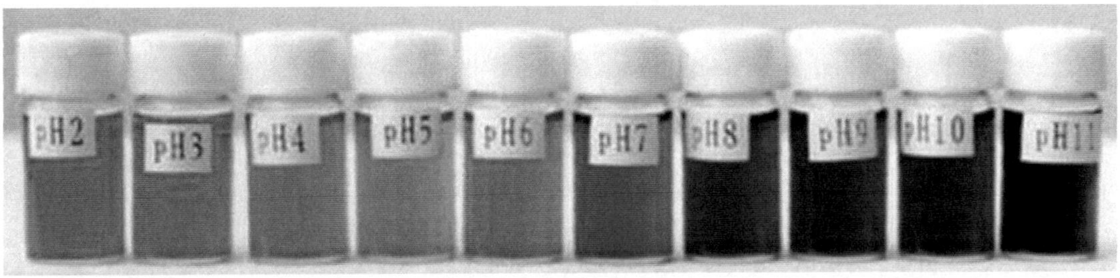

Figure 11. Color of anthocyanin in variation pH.

Figure 12 shows the absorption spectrum of anthocyanin shift to a greater wavelength in base condition or pH increase. Overall the spectrum shows a peak of anthocyanin at range 400-600 nm. In the pH 2 give spectra with high absorption.

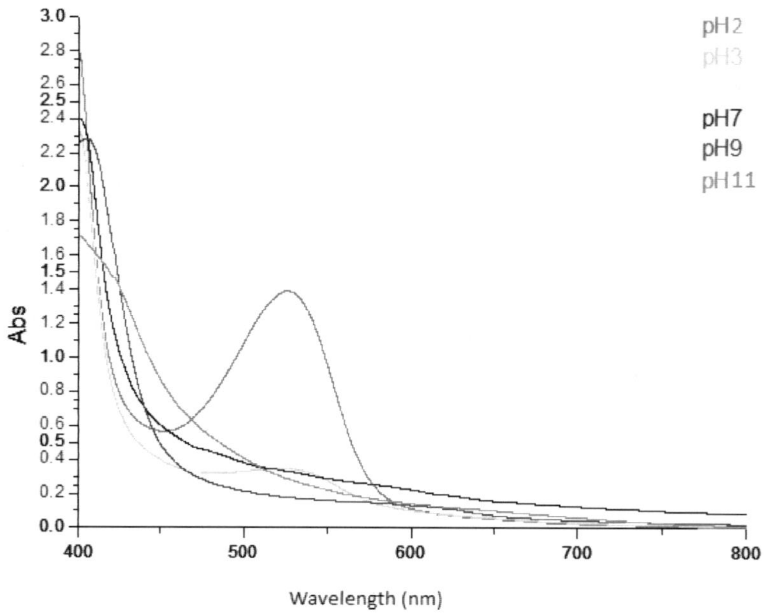

Figure 12. Effect of pH

Anthocyanins are stable at low pH. It becomes less stable when exposed to heat, causing a loss of color and browning. As a result, high temperature, increased sugar level, pH, and ascorbic acid can affect the rate of destruction. In solution, anthocyanins molecules are present in an equilibrium between the colored cationic form and the colorless pseudo base. This equilibrium is directly influenced by pH. pH is very important for the color of anthocyanins, some anthocyanins are red in acid solutions, violet or purple in neutral solutions, and blue in alkaline pH. The reason for this is right here this structure of the anthocyanins (red circle) is called flavylium cation the reason, at low pH the cyanidin molecule is protonated and forms a positive ion or cation, as the pH increases the molecules become deprotonated, at high pH the molecule forms a negative ion or anion [16, 17].This is the reason that most colorants containing anthocyanins can only be used at pH values below four. Additionally, anthocyanins can act as pH indicators.

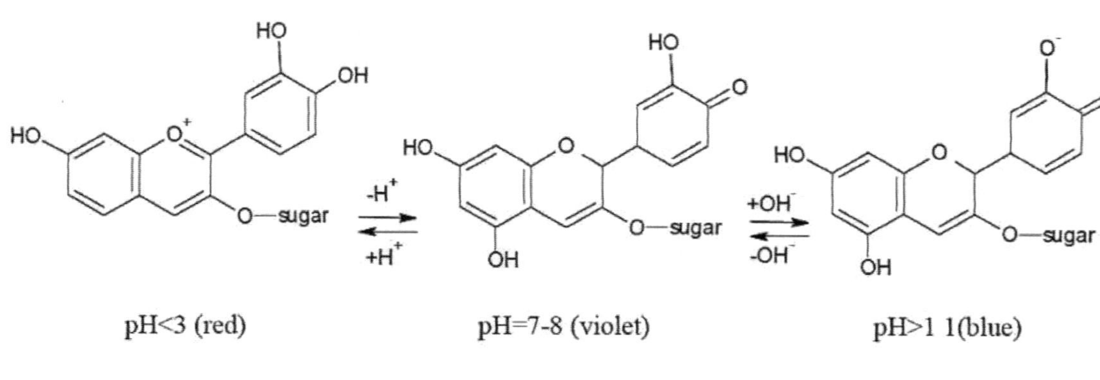

Figure 13. The equilibrium of anthocyanin

The anthocyanin belongs to the group of natural dyes responsible for several colors in the red–blue range, this spectrum depends on an abundance of anthocyanin in the natural source. As a result anthocyanin can be food colorant with variation pH.

4. Conclusion

The anthocyanin was successfully extracted from rose red flower by maceration method. The separation used vacuum liquid chromatography (VLC) and gravity chromatography. Characterization using TLC, UV-Vis, and FTIR shows the characteristic of anthocyanin compared by the standard of anthocyanin. Anthocyanin stable at pH low with red color. In the high pH, anthocyanin changed in to blue color. So, anthocyanin can be as food colorant.

References

[1] Zamora-Ros R, Knaze V, Lujan-Barroso L 2011 *Brit J Nut* **106** 1090–1099
[2] H Oliveira, N Wu, Q Zhang, J Wang, J Oliveira, V de Freitas, N Mateus and I Fernandes 2016 *Food & Function* **7** 2462-2468
[3] Yao LH, Jiang YM, Shi J F A, Tomas-Barber, Datta R N, Singanusong C, and Chen S S 2004 *Plant Foods for Human Nutrition* **59** 113–122
[4] Kong J M, Chia L S, Goh N K, Chia T.F, Brouillard R 2003 *Phytochem.* **64** 923
[5] Dai J and Mumper R J 2010 *Molecules* **15** 7313-7352
[6] A Mortensen 2006 *Pure Appl. Chem.* **78** (8) 1477–1491
[7] He F, Mu L, G-Liang Yan, Na-Na Liang, Qiu-Hong Pan, Wang J, Reeves M J and Chang-Qing Duan 2010 *Molecules* **15** 9057-9091
[8] Rymbai1 H, Sharma R R, Srivastav M 2011 *Int. J Pharm Tech Res CODEN (USA)* **3** (4) 2228-2244
[9] Arisasmita, Joek, Kuswardani I, dan Tjahjani L 1997 *Ekstraksi dan Karakterisasi Zat Warna dari Kulit Buah Manggis* (Surabaya: Universitas Katolik Widya Mandala)
[10] Saati E A, Simon B W, Yunianta and Aulanni'am 2011 *J Agr. Sci. Tech. A* **1** 1192-1195
[11] Harborne J B 1967 *Comparative Biochemistry of the Flavonoids* (London: Academic Press)
[12] Hurst W J 2002 *Functional Foods & Nutraceuticals Series (Book 4)* (New York: CRC Press)
[13] Basuki N, Harijono, Kuswanto, & Damanhuri 2005 *Agravita* **27** (1) 63-68
[14] Clydesdale F M 1998 *Color: origin, stability, measurement and quality in Storage Stability. Taub, I.A. & Singh, R. P (Ed)* (New York: CRC Press LCC)
[15] Markakis P 1982 *Anthocyanins as food additives: Anthocyanins as food colours* (New York: Academic Press)
[16] Brouillard R & Dangles O 1994 *Food Chem.* **51** 365–371
[17] Jackman R L & Smith J L 1992 *Anthocyanins and betalains In Hendry G A F & Houghton J D (Eds.), Natural food colorants* 182–241 (Glasgo: Blackie)

IOP Conf. Series: Materials Science and Engineering **193** (2017) 012048 doi:10.1088/1757-899X/193/1/012048

Indigo Dye Derived from *Indigofera Tinctoria* as Natural Food Colorant

S Wahyuningsih[1*], A H Ramelan, D K Wardani, F N Aini, P L Sari, B P N Tamtama, Y R Kristiawan

[1]Inorganic Material Research Group, Faculty of Mathematics and Natural Sciences, Sebelas Maret University, Ir. Sutami Street No. 36 A, Surakarta, Central Java, 57125, Indonesia.

*E-mail: sayekti@mipa.uns.ac.id

Abstract. Recently, the uses of dyes are increasingly widespread especially in foods and beverages as food colors to attract the consumers. The dye that currently attracts is indigo. Indigo is a group of carbonyl compounds, one of the oldest known dyes in terms of natural blue dye derived from the *Indigofera tinctoria* plant. The use of indigo as a natural food dye intended to reduce the use of synthetic dyes are carcinogenic impact. The method used in this study includes the analysis of indigo using UV-Vis spectrophotometry and FTIR analysis. Based on the UV-Vis Spectrophotometer analysis results with the various concentrations of 0.002 mg/mL; 0.004 mg/mL; 0.006 mg/mL and 0.008 mg/mL were obtained maximum absorption peak at wavelength of 550-700 nm. The indigo dyes in various concentrations produce a stable pH at an average pH 9, therefore it can make the colors not easily fade (strong staining). Based on infrared spectrophotometer measurement were obtained absorption spectrum at 3100-3500 cm^{-1} as primary N-H and secondary amine, 1600 cm^{-1} as aromatic C=C, 1000-1350 cm^{-1} as aromatic C-N, 690-900 cm^{-1} as aromatic C-H.

1. Introduction

In recent years, there has been a revival of the use of dyes and colors of natural origin for coloring food, pharmaceutical, cosmetic and textile products. Among the natural dyes which are becoming widely recognized throughout the world, indigo which is one of the oldest known natural dyes [1], is a derivative of the colorless glucosides of the enol form of indoxyl, e.g. indican (indoxyl-D-glucoside). Indigo, after which this group of carbonyl dyes is named, is one of the oldest known natural dyes. It is a derivative of the colorless glucosides of the enol form of indoxyl, e.g.indican [2,3].

Extraction of indigo from the leaves of *Indigofera tinctoria Linn.* investigated chemical constituents including the chemical structure of major components in the water extract from both kinds of plant to develop the extraction method for preparation of ready-to-use natural dyes [4]. Coloring matter (Indigotin) from *Indigofera tinctoria* plant is usually present abundantly in flowers which give blue color mainly used to dye linen and hairs. Flavonoids, terpenoids, alkaloids, glycosides, Indigotine, Indirubin, rotenoids are the other related compounds abundantly present in the plant. These compounds were found to be

IOP Conf. Series: Materials Science and Engineering **193** (2017) 012048 doi:10.1088/1757-899X/193/1/012048

responsible for many pharmacological activities such as antihyperglycemic activity, antioxidant, anti-inflammatory activity, antibacterial, antihepatoprotective activity, antidiabetic activity and anticonvulsive agent [5].

Classification of scientifically conducted on alkaloid content: indicant glycosides, non-protein amino acids, nitro-propionic acid esters, cyanogenic glycosides, guanidine alkaloids, phenolic acids, phenolic glycosides flavonoids, isoflavonoids. Types of *indigo* which containing alkaloid compounds as above indicated as a dye [6]. The use of indigo as a dye should be depend on the color. The dye stability will be influence by pH and their chromophore certain in dye.

2. Materials and Methods

2.1 Materials

The instruments were used for analysis are UV-Vis spectrophotometer Lambda 25 (Pelkin Elmer), and Fourier Transform Infra Red (FTIR) Prestige 21 (Shimadzu). The Indigofera dyes extract was from *Indigofera Tinctoria* leaf as commercial product in Parakan, Temanggung, Central Java Indonesia.

2.2 Preparation

Indigo dyes of 0.002; 0.004; 0.006; and 0.008 mg, were dissolved in 100 mL of distilled water, respectively to obtain the mother liquor indigo. Then, the solutions were measured its pH. Influence of concentration on pH were studied. The homoogen solution of indogo dyes were analyzed by UV-Vis spectrophotometer to determine Indigo dye absorption spectra in various concentration samples. UV-Vis spectrophotometer analysis also aims to determine the wavelength of maximum absorption of the indigo. The chromophore groups of indigo was identify by FTIR.

3. Results and Discussion

Indigo dye (figure 1) ussually has a contained of flavonoids, terpenoids, alkaloids, glycosides, indigotine, indirubin, rotenoids. These compounds were found to be responsible for many pharmacological activities such as antihyperglycemic activity, antioxidant, anti-inflammatory activity, antibacterial, antihepatoprotective activity, antidiabetic activity and anticonvulsive agent [7]. Because of their functions, indigo dyes can be use as a functional food dye. Based on indigo structural above, the different structure of indigo dye will be occur in acid and base conditions, which shown in figure 2.

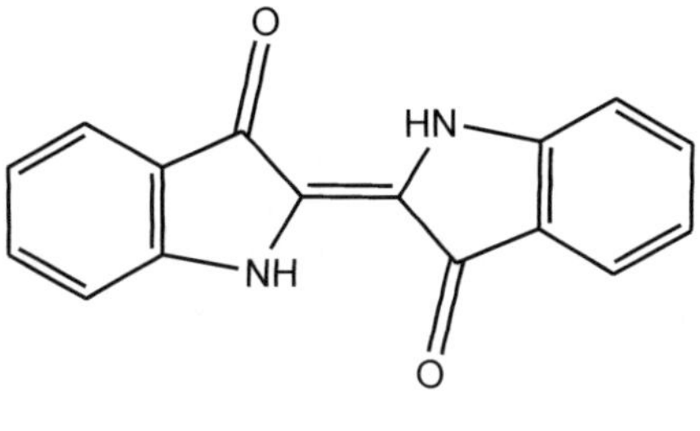

Figure 1. Structure of indigo [8-9]

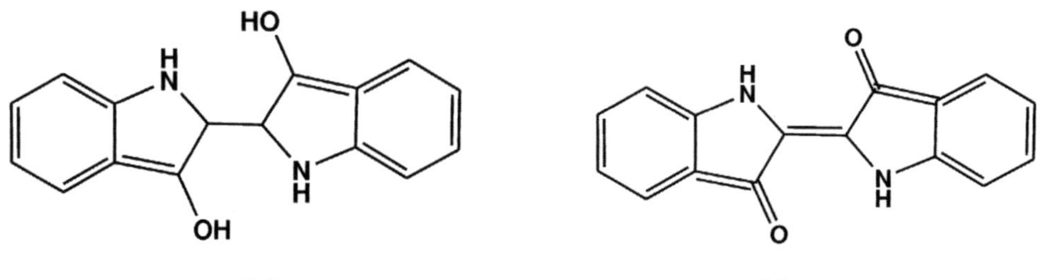

Figure 2. The indigo dyes structure in (a) acid condition and (b) base condition [4]

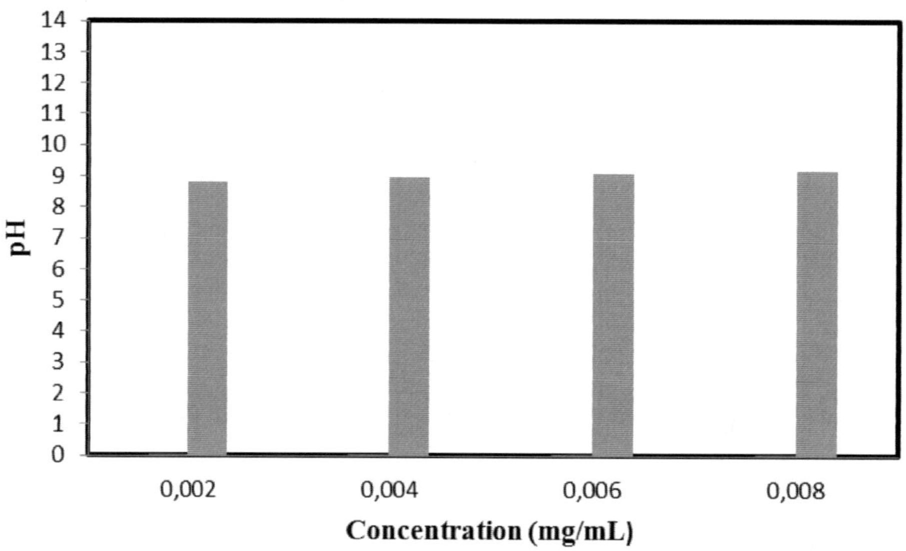

Figure 3. pH of indigo in various concentration (mg/mL)

The pH measurement in various consentration indigo dyes shown that the higher the indigo concentration, the higher the pH value. In the range pH of 8,8 – 9,2 is a stable of indigo dyes condition. Therefore, those conditions produce the color do not easily fade (strong staining). The absorption spectrum measurement of indigo dyes performed at a wavelength in range 200-800 nm. Indigo absorption spectrum was shown in figure 4.

IOP Conf. Series: Materials Science and Engineering **193** (2017) 012048 doi:10.1088/1757-899X/193/1/012048

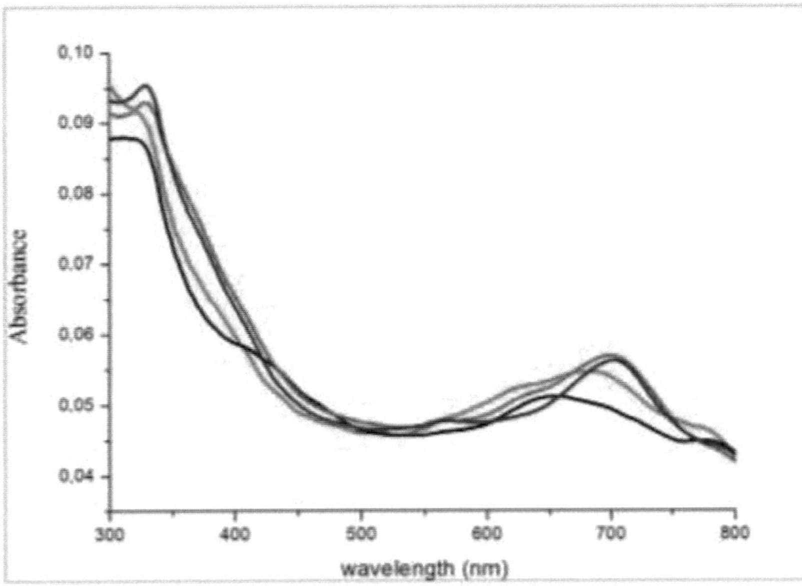

Figure 4. The UV-Vis absorption of indigo dyes in various concentration of 0.02 mg/mL (▬) ; 0.04 mg/mL (▬); 0.06 mg/mL (▬); and 0.08 mg/mL (▬)

The UV-Vis absorption of indigo dye by using UV-Vis spectrophotometer showed that the maximum absorption peak is at a wavelength in the visible range of 550-700 nm, where the absorbance maximum reached at a concentration of 0.008 mg/mL. In general, the concentration of dyes affected the color intensity. The strong spectrum in the whole visible range have gained the blue color of indigo dye. The blue color of indigo dye cause of the chromophore group in chemical structure. Electronic transtitions of indigo chromophore group from both $\pi \rightarrow \pi^*$ and $n \rightarrow \pi^*$ were in visible range.

The chromophore groups of dye was also have been identify by FTIR spectra. The FTIR absorption spectra of indigo dyes was shown in figure 5. Based on infrared spectrophotometer measurement were obtained absorption spectrum at 3100-3500 cm^{-1} as primary N-H and secondary amine, 1600 cm^{-1} as aromatic C=C, 1000-1350 cm^{-1} as aromatic C-N, 690-900 cm^{-1} as aromatic C-H (table 1).

Table 1. The functional groups of Indigo from IR spectra

Wavenumber (cm^{-1})	Functional Group
3100-3500	N-H Primary and Secondary Amines
1600	C=C Aromatic
1350-1000	C-N Amines
900-690	C-H Aromatic

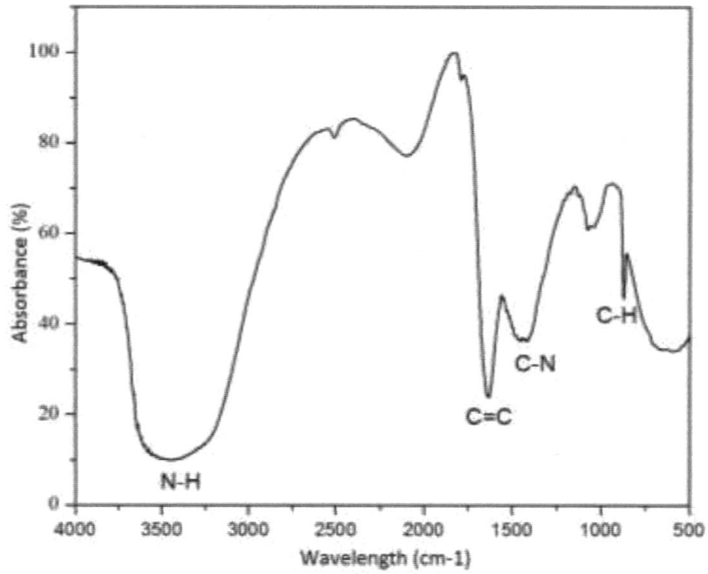

Figure 5. FTIR spectrum of Indigo

4. Conclusion

Indigo dyes derived from *Indigofera tinctoria* plant was potentially as a natural blue dye for food colorant. The stable blue color of indigo dyes produce by strongly UV-Vis absorption in the broad range of 550-700 nm is base condition (8.8 – 9.2). The blue color on Indigo dye was generated from an electronic transition of the chromophore group in indigo dye structure as N-H primary and secondary amine, C=C aromatic, C-N amine, and C-H aromatic.

References

[1] Ensley B D, Ratzkin B J, Osslund T D, Simon M J, Wackett L P, Gibson D T 1983 *Science* **222** 167-169
[2] Minami Y , Kanafuj T, Miura K 1996 *Biosci.Biotech.Biochem.* **60** 147-149
[3] Minami Y, Takao H, Kanafuji T, Miura K, Kondo M, Hara-Nishimura I, Nishimura M, Matsubara H 1997 *Plant Cell Physiol* **38** 1069-1074
[4] Chanayath N, Lhieochaiphant S, Phutrakul S 2002 *CMU. J.* **1** 149-160
[5] Kimberly Hamblin Hart, Paulalan Cox 2002 *J. Ethnobiol.* **20** 303
[6] Muzayyinah M 2014 *BIOEDUKASI* **7** 23-26
[7] Chaitanya L G 2014 *Res. J. Chem. Sci.* **4** 87-96
[8] Komboonchoo S, Bechtold T 2009 *J. Clean. Prod.* **17** 1487-1493
[9] Ghani M A, Stern B, Edwards H G M, Janaway R 2012 *Vib. Spectrosc.* **62** 98-109

IOP Conf. Series: Materials Science and Engineering **193** (2017) 012049 doi:10.1088/1757-899X/193/1/012049

Spectroscopic Study of Green Tea (*Camellia sinensis)* Leaves Extraction

A Marzuki[1*], V Suryanti[2], A Virgynia[1]

[1]Department of Physics, Sebelas Maret University, Surakarta 57126, Indonesia
[2]Department of Chemistry, Sebelas Maret University, Surakarta 57126, Indonesia

* E-mail: amarzuki@mipa.uns.ac.id

Abstract. This paper reports the analysis of UV-VIS-NIR absorption spectra of different concentrations of green tea (Camellia sinensis) leaf extract in two different solvent systems (chloroform and ethyl acetate). In those solvents, two different peaks characterizing green tea are observed at different wavelengths, namely 296 nm and 329 nm (extracted in chloroform) and 391 nm and 534 nm (extracted in ethyl acetate). We then investigated the absorption spectra change as function of green tea concentration in both solvents. We found that light absorption increases linearly with the increase of green tea concentration. Different wavelengths, however, respond this change differently. However, the way it changes is wavelength dependence.

1. Introduction

Tea is one of widely consumed traditional beverages. Different countries have different way on looking at tea and thus forming a different tea culture. This unique culture includes the way tea is produced and consumed as well as the aesthetics of tea drinking. British tea culture and Chinese tea culture are two important examples not only on how tea is served but also on tea preparation. In part of China, the tea beverage is prepared by brewing with salt and butter.

In addition to being a fresh drink, tea has also been used as traditional herb which has many benefits for human health such as improved brain function, fat loss and a lower risk of cancer. For whatever reason the tea is consumed, maintaining the tea quality is very important. For a given tea drink quantity, a quality standard to meet to the general people's flavour has to be taken. This means that density of a chemical substance characterizing the tea flavour for a given quantity has to be relatively constant.

Up to now, laboratory based analysis is still considered as the most accurate and reliable methods used to determine the quality as well as the variety of green tea. This outstanding method, however, are time consuming and expensive. To overcome this problem, many efforts have been devoted in order to develop both method and an instrument for analysing green tea simply, cheaply and accurately. Among them are methods and instrument used to determine the content of green tea [1-5] as well as to identify the green tea varieties [6].

In the last few decades, fibre sensors have received increasing attention due to their unique physical properties such as small size, flexible, and immunity to electromagnetic interference. Nowadays, fibre sensor haven been used in almost all areas where conventional sensors have long been regarded as the prime sensors [7 – 9]. Optical fibre liquid and gas sensors have also been researched over years [10-11] and many of them are already available in the market. This paper presents an early study aims at determining the absorption peak characterizing green tea as function of its concentration using fibre sensor works based on evanescence wave absorption.

IOP Conf. Series: Materials Science and Engineering **193** (2017) 012049 doi:10.1088/1757-899X/193/1/012049

2. Principle of operation and Experiment

Evanescence field based Fibre sensor reported in this paper works in the same principle as UV-VIS-NIR spectrophotometer. As polychromatic light passes through a transparent material, light at a wavelength λ corresponding to energy difference between ground state and an exited state of a substance composing the material will be absorbed. The amount of light absorbed is determined by the absorbing substance concentration.

Whereas absorption process by a substance in UV-VIS_NIR spectrophotometer occurs along the transmission line, the process in evanescence fibre sensor occurs at certain area at which electromagnetic field of the light transmitting through optical fibre penetrates out of the fibre area and interacts with the surrounding substances. In this type of fibre sensor, the absorbed light depends not only on the concentration of the absorbing substance but also on the electromagnetic field depth penetration (d_p) as expressed by:

$$d_p = \frac{\lambda}{\frac{2\pi}{n}\sqrt{sin^2\theta - n^2}} \quad (1)$$

Where λ, θ, and n are light wavelength, incident angle at core-peeled cladding interface, and core refractive index respectively.

Evanescence fibre sensor used in this experiment was made by removing part of the POF. This was done by polishing POF which was mounted on the surface of the resin bar. Polishing this surface gently using 5000 grit of abrasive polishing paper will unclad part of the POF.

Figure 1 is the experimental setup used to measure the absorption spectra of samples using evanescence fibre sensor. One end of the POF was coupled to a white (polychromatic) light source and the other end to a spectrophotometer. The sensing element of the sensor was placed in liquid cell. As the white light was launched from one end of the fibre, light will propagate in the fibre core with multiple total internal reflections (TIR). Light is propagating through fibre optic by mean of multiple internal reflections. Since the clad of the sensing element was removed, the evanescence wave within this area leaks into the material surrounding the unclad fibre. If the material contains a substance with electronic energy different between ground state and an exited state correspond to any wavelength within the range of the launched light, part of the leakage light (evanescence wave) is absorbed. Therefore, the intensity of light at a particular wavelength coming out at the end of the fibre sensor decreases. In this experiment, the absorption strength will then be detected by spectrometer and displayed by personal computer (PC).

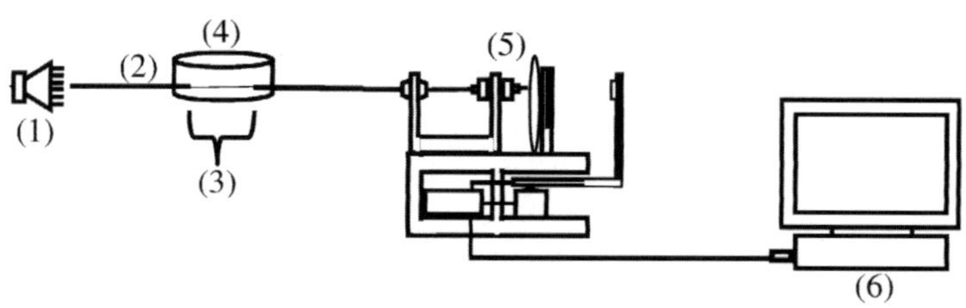

Figure 1. The experimental setup used to measure absorption spectra of green tea. (1) light source, (2) Polymer optical fibre, (3) unclad *fibre sensor*, (4) liquid cell, (5) spectrometer (6) PC

3. Results and Discussion

Figure 2 shows absorption spectra of green tea recorded at room temperature using UV-VIS spectrophotometer both for those extracted in (a) chloroform and (b) ethyl acetate. It can be seen from figure 2 that absorption peaks characterizing green tea extraction in both solutions are different. Green tea extract in chloroform is characterized by two main absorption peaks located at 296 nm and 329 nm, while for that in ethyl acetate their unique absorption peaks are 391 nm and 666 nm. It is also clearly seen that the intensity of light within measured wavelength range depends on the tea concentrations. In all wavelengths, the absorbance increases with increasing green tea concentration. In fact, we have used this finding to presume that green tea concentration in a single green tea solution can be determined by looking at light absorbance at any particular wavelength using equation:

$$A = -logT = \varepsilon\, c\, l \qquad (2)$$

With A, T, ε, c and l are absorbance, transmittance, extinction coefficient, speed of light, and sample path length, respectively.

Evanescent wave fibre sensor reported in this paper was developed based on the above finding. Absorption process in UV-VIS-NIR Spectrophotometer is different from that in evanescent fibre sensor. The measured material in UV-VIS-NIR spectrophotometer absorbs relatively intense light passing through it, whereas in evanescent fibre sensor the measured substance absorbs relatively weak light penetrating out of the fibre core within the unclad fibre area with depth penetration *d* following equation 1.

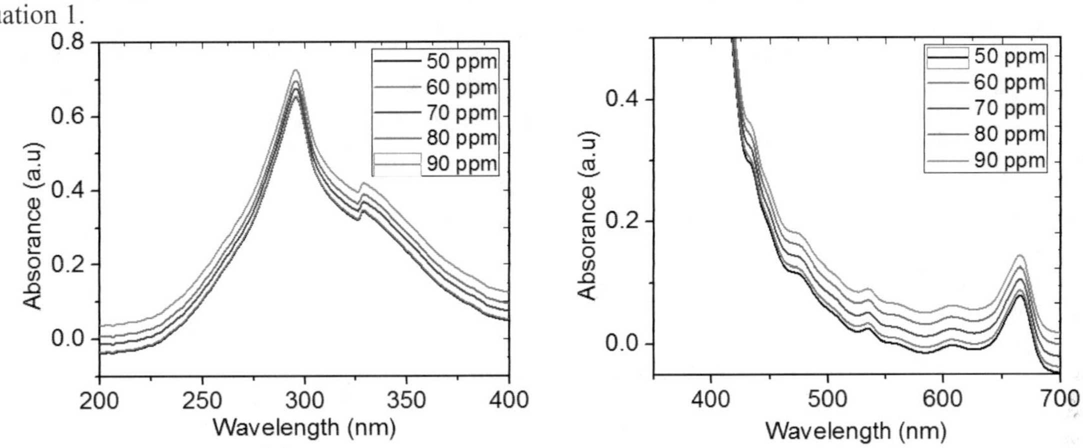

(a) (b)

Figure 2. Absorption spectra recorded at room temperature using UV-VIS Spectrophotometer for different concentrations of green tea extracted in (a) Chloroform and (b) ethyl acetate

Figure 3 shows the absorbance recorded using evanescent fiber sensor of green tea extracted in two different solutions: (a) chloroform and (b) ethyl acetate. These results are in a good agreement with those recorded using UV-VIS-NIR Spectrophotometer in that light absorbance increase with the increase of green tea concentration (equation 2). Following equation 1, increasing green tea concentration also means the index of refraction increases. This results in increasing depth penetration. As absorption is a process of material-light interaction treating a beam of light as a stream of particles, deeper light penetration into the sensed material means that more light particles will be absorbed by the electrons composing the sensed materials.

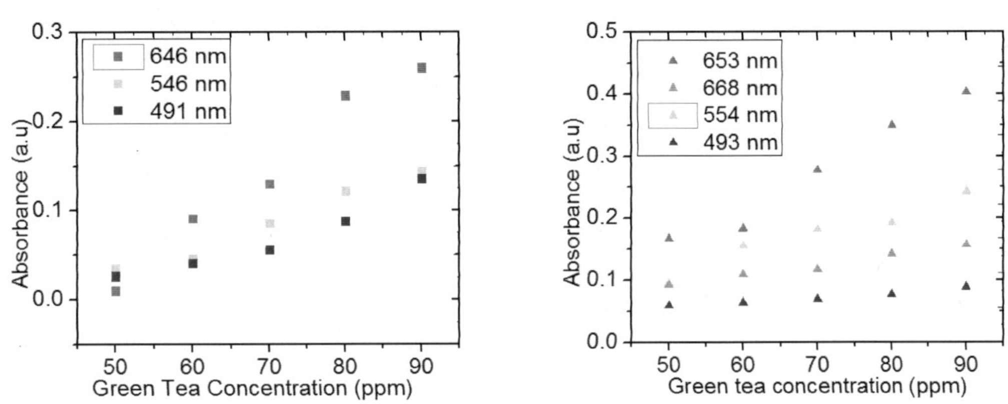

(a) (b)

Figure 3. Concentration dependence of evanescent wave absorbed by green tea extracted in (a) chloroform and (b) ethyl acetate.

Further from figure 3, it can be seen that light absorbed by the material increases with the increase of wavelength (λ). This finding is in a good agreement to that expressed in equation (1) at which depth penetration (d) is linearly related to wavelength. Increasing the sloop as the wavelength increases is attributed by both increasing wavelength and concentration simultaneously.

4. Conclusions

Spectroscopic properties of green tea extracted both in chloroform and ethyl acetate have been studied. Their absorption spectra within the wavelength ranging from 200 – to 700 nm have been measured as function of green tea concentration using UV-VIS-NIR spectrophotometer. We have found that the absorbance increases with the increase of green tea concentration. Based on this result, we have used it to examine an evanescence fiber sensor designed for measuring single component solution of green tea. We have shown that light absorbed by mechanism of evanescence increase linearly with the increase of green tea concentration. This result suggests that the evanescent fiber sensor developed in this research might be potentially used to monitor the quality of green tea drinks.

Acknowledgement

This work has been financially supported by the PNBP Sebelas Maret University, Surakarta Indonesia.

References

[1] Chen Q, Zhao J, Guo Z, Wang X, 2010, J. Food Comp. Anal 23 (4) 353-358
[2] Qi S, Ouyang Q, Chen Q, Zhao J 2014 J Pharm Biomed Anal 97 116-122
[3] Kumar A S, Shanmugam R, Nellaiappan S, Thangaraj R 2016 Sens Actuators B Chem 227 352-361
[4] Chen Q, Zhang D, Pan W, Ouyang Q, Li H, Urmila K, Zhao J 2015 Trends Food Sci Tech 43 63-82
[5] Liu L, Fan Y, Fu H, Chen F, Ni C, Wang J, Yin Q, Mu Q, Yang T, She Y 2017 Anal. Chim. Acta 963 119-128
[6] Chen Q, Liu A, Zhao J, Ouyang Q 2013 J Pharm Biomed Anal 84 77-83
[7] T Bremer T, Reinsch K, Leen G, Roth B, Lochmann S, Lewis E 2017 Sensor Actuat A-Phys 256 84-88
[8] Marzuki A, Heriyanto M, Setiyadi I D, Koesuma S 2015 J Phys Conf Ser 622 012059
[9] Nishiyama M, Watanabe K 2014 Sensor Actuat A-Phys 209 154-160
[10] Matjasec Z, Donlagic D 2017 Sens Actuat B Chem 242 577-585
[11] Kitture R, Pawar D, Rao Ch N, Choubey R K, Kale S N 2017 J alloy Comp 695 2091-2096

IOP Conf. Series: Materials Science and Engineering **193** (2017) 012050 doi:10.1088/1757-899X/193/1/012050

Chemical Constituents of *Luffa acutangula (L.)* Roxb Fruit

V Suryanti*, S D Marliyana and I Y Astuti
Department of Chemistry, Faculty of Mathematics and Natural Sciences, Sebelas Maret University, Jl. Ir. Sutami 36A Surakarta, Central Java 57126 Indonesia

*Email: venty@mipa.uns.ac.id

Abstract. The phytochemical screening conducted on ethanol extract of *Luffa acutangula (L.)* Roxb's fruit revealed the presence of alkaloids, saponins, carotenoids and terpenoids and the absence of flavonoids, tannins and anthraquinones. The GC-MS of the analysis *L. acutangula (L.) Roxb*'s fraction resulted in the identification of six compounds. The compounds that could be identified were 2,3-dihydro,3,5-dihydroxy-6-methyl-(4H)-pyran-4-one; 3,7,11,15-tetramethyl-2-hexadecen-1-ol; (3β, 20R)-cholest-5-en-3-ol; *n*-hexadecanoic acid; 9, 12, 15-octadecatrienoic acid methyl ester and citronellyl tiglate. The present study provides evidence that *L. acutangula*'s fruit contains medicinally important bioactive compounds and this justifies the possibly use of these fruits as traditional medicine for treatment of various diseases.

1. Introduction
Phytochemical compounds are known to have beneficial importance in industrial and medicinal sciences. Although phytochemical tests are helpful in discovering bioactive compounds for therapeutic agents, food additives, agrochemicals and industrial chemicals, only a small number of plant has been investigated phytochemically. Plant based natural constituents can be derived from different parts of the plant like bark, leaves, flowers, roots, fruits and seeds [1-5].

The phytochemical screening have been conducted in Cucurbitaceae family members, such as *Trichosanthes anguina L.* and *Sechium edule Jacq. Swartz* [6-7]. Alkaloids, tannins, polifenol, saponin, cardenolin/bufadienol and flavonoid have been found to be present in the ethanol extract of *T. anguina L.*'s fruits, while anthraquinones have been found to be absent [6]. The phytochemical analysis of ethanol extract of *S. edule Jacq Swartz*'s fruits revealed the presence of alkaloids, saponin, cardenolin/bufadienol and flavonoid and the absence of tannin, phenolics and anthraquinones [7].

Luffa acutangula (L.) Roxb (ridge gourd) belongs to Cucurbitaceae family. It is widely growing vegetative climber and found throughout Indonesia. The fruits usually are taken as vegetables. The plant has been reported to have various medicinal properties such as treatment of jaundice, splenic enlargement and laxative. It is also proved as CNS depressant used traditionally in insect bites [8]. The plant also possesses potent α-glucosidase inhibitory effect [9]. The present study was designed to investigate the presence of various phytochemicals constituents in *L. acutangula (L.) Roxb*'s fruits.

2. Experimental
2.1. Materials and Instruments
Fresh fruits of *L. acutangula (L.) Roxb* were purchased from local market in Surakarta, Indonesia. All the chemicals used were of analytical grade, purchased from Merck Chemical Company (Merck, Germany).

The chemical compositions of the sample was studied by GC Hewlett Pacard 5890 series II and GC-MS Shimadzu QP-5000. Components were separated on CD sil 5 CB non polar column. The oven temperature was programmed at 120-300°C at a rate of 10°C/min. Sample injection volume was 0.3 μL with a split ratio 1:10 and pressure at the column inlet 10 kPa with helium carrier gas at 1 ml/min flowrate. Compounds were identified by comparison of mass spectra with those in the Wiley and NIST Libraries.

2.2. Preparation of Fruit Extract
The fruit samples (2.5 kg) were cleaned, dried in the oven at 80°C for 4 h and ground. The powdered material (42 g) was transferred into a Soxhlet apparatus and extracted in the Soxhlet extractor using 350 mL petroleum ether for 6 h. The residue was sequentially macerated with ethanol and allowed to stand for 24 h and then filtered. The extract then was concentrated using a rotary evaporator and dried under vacuum dessicator to yield 30 g (32 mL) of brownish black liquid. This ethanol extract was used for phytochemical test.

2.3. Screening of Phytochemical Components for the Ethanol Extract
The plant extract was subjected to qualitative tests adopting standard procedure for the identification of the phyto constituents [10-15].

2.3.1. Test for Alkaloids
Ethanol extract (0.25 g) was stirred with 3 mL of 1% hydrochloric acid on a steam bath. The filtrate (1 ml) was treated with few drops of Draggendoff's reagent, after which it was observed whether the alkaloids were present or absent. Alkaloids, if present in the solution of sample, will react with Dragendorff's reagent and produce an orange or orange red precipitate formation.

2.3.2. Test for Flavonoids
Test solution when treated with sodium hydroxide solution, shows increase in the intensity of yellow color which would become colorless on addition of few drops of dilute hydrochloric acid, indicates the presence of flavonoids.

2.3.3. Test for Tannins
Ethanol extract (0.25 g) was dissolved in 10 ml distilled water and filtered. A few drops of 0.1% ferric chloride were added to the filtrate. The appearance of intense brownish-green or a blue-black colour indicated the presence of tannins in the test samples.

2.3.4. Test for Terpenoids
Ethanol extract (0.25 g) was mixed with 1.5 ml of chloroform ($CHCl_3$) and concentrated H_2SO_4 (1.5 ml) was carefully added to form a layer. A reddish brown colouration of the interface was formed to indicate positive results for the presence of terpenoids.

2.3.5. Test for Saponins
Ethanol extract (0.25 g) was boiled together with 10 mL of distilled water in a water bath and filtered. Ten mL of the filtered sample is mixed with 1 mL of distilled water in a test tube and shaken vigorously to obtain a stable persistent froth. The frothing is then mixed with 3 drops of olive oil and observed for the formation of emulsion, which indicates the presence of saponins.

2.3.6. Test for Anthraquinones
Ethanol extract (0.25 g) was taken into a dry test tube and 3 ml of chloroform was added and shaken for 5 minutes. The extract was filtered and the filtrate was shaken with equal volume of 10% ammonia solution. A pink violet or red colour in the lower layer indicates the presence of anthraquinone.

2.3.7. Test for Carotenoids
Ethanol extract (0.25 g) was extracted with 10 ml of chloroform in a test tube with vigorous shaking. The resulting mixture was filtered and 85 % sulphuric acid was added. A blue colour at the interface showed the presence of carotenoids.

2.4. Column Chromatography
Modification of the method of Selowa *et al.* [16] was used to elute the fractions during column chromatography. A suspension of silica gel 60 and n-hexane was poured into a glass column (50 cm long × 1.5 cm diameter) up to a height of 30 cm being careful to prevent formation of gaps and bubbles and equilibrated with 100% n-hexane. A mixture of 2 g ethanol extract and 2 g of the silica gel 60 in 100% n-hexane was loaded onto the column and eluted with 100% n-hexane first and then with the solvent combination of n-hexane and ethanol with increasing volume of ethanol (polarity). The fractions of 2 mL of eluents were collected on the basis of solvent polarity and/or colour separation. Thin layer chromatography (TLC) was carried out and the fractions with similar profiles could be pooled together. The fractions were dried in a rotary evaporator after which they were dried to a constant weight in a fume cupboard.

3. Results and Discussion

3.1. Phytochemical Screening of the Ethanol Extract
The phytochemical active compounds of *L. acutangula (L.) Roxb*'s fruits were qualitatively analyzed and the results are presented in table 1. Based on the presence or absence of colour change indicate positive and negative results are indicative. In the screening process of *L. acutangula (L.) Roxb's* fruit extract, alkaloids, terpenoids, carotenoids and saponins gave positive results whereas flavonoids, tannins and anthraquinones gave negative results. Tupe *et al.* (2013) reported that the phytochemical analysis of aqueous leaf extract of *L. acutangula (L.) Roxb* showed the presence of alkaloids, tannins, carbohydrates, saponin and terpenoids and the absence of flavonoids, anthocyanin and steroids [17]. In the present study, we have found that the biologically active phytochemicals were present in the ethanol extract of the *L. acutangula (L.) Roxb's* fruits. The medicinal properties of *L. acutangula (L.) Roxb*'s fruits may be due to the presence of above mentioned phytochemicals.

Table 1. Phytochemical analysis of the ethanol extract of *L. acutangula (L.) Roxb*'s fruits.

Phytochemical Components	Test	Results
Alkaloids	Draggendoff Test	+
Flavonoids	Alkaline Reagent Test	-
Tannins	Ferric Chloride Test	-
Terpenoids	Salkowski Test	+
Saponins	Foam Test	+
Anthraquinones	Borntragers test	-
Carotenoids	Sulphuric Acid test	+

+ = Presence; - = Absence

3.2. GC-MS Analysis
The GC-MS analysis of *L. acutangula (L.) Roxb* was carried out for three fractions, however only one fraction could detect the peaks of the compounds. Nineteen (19) of bioactive phytochemical compounds were shown in the chromatogram (figure 1). Tables 2 show the compounds found in the *L. acutangula (L.) Roxb* fraction. The compounds that could be identified were 2,3-dihydro,3,5-dihydroxy-6-methyl-(4H)-pyran-4-one; 3,7,11,15-tetramethyl-2-hexadecen-1-ol; (3β, 20R)-cholest-5-en-3-ol; *n*-hexadecanoic acid; 9, 12, 15-octadecatrienoic acid methyl ester and citronellyl tiglate. Some compounds

remained unidentified because of a lack of library spectra of the corresponding compounds. Cecheovska *et al. (*2011) reported that compound 2,3-dihydro,3,5-dihydroxy-6-methyl-(4H)-pyran-4-one found in prunes has antioxidant activity [18].

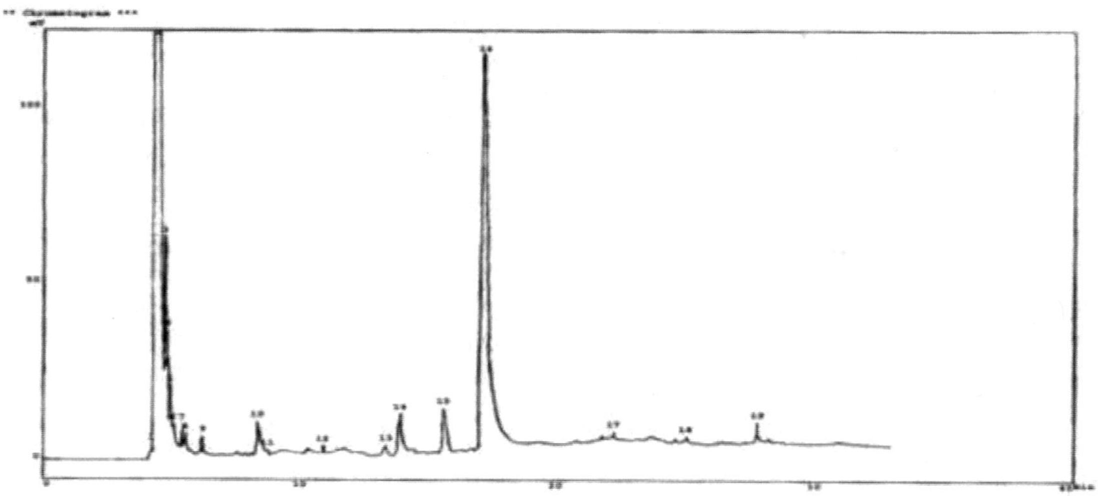

Figure 1. GC-MS chromatogram of the fraction of *L. acutangula (L.) Roxb*'s fruits.

Table 2. Compounds that could be identified in the fraction of *L. acutangula (L.) Roxb*'s fruits.

No.	Name	Molecular Formula	Molecular Weight
1.	2,3-dihydro,3,5-dihydroxy-6-methyl-(4H)-pyran-4-one	$C_6H_8O_4$	144
2.	3,7,11,15-tetramethyl-2-hexadecen-1-ol	$C_{20}H_{40}O$	296
3.	(3β, 20R)-cholest-5-en-3-ol	$C_{27}H_{46}O$	386
4.	*n*-hexadecanoic acid	$C_{16}H_{32}O_2$	256
5.	9, 12, 15-octadecatrienoic acid methyl ester	$C_{19}H_{32}O_2$	292
6.	citronellyl tiglate	$C_{15}H_{26}O_2$	238

4. Conclusion

L. acutangula (L.) Roxb's fruits have potential to act as a functional food and a source of useful drugs because of the presence of various phytochemical components. The ethanol extract of *L. acutangula* (L.) Roxb's fruit shows the presence of alkaloids, saponins, carotenoids, and terpenoids and the absence of flavonoids, tannins and anthraquinones. The compounds that could be identified by GC-MS analysis were 2,3-dihydro,3,5-dihydroxy-6-methyl-(4H)-pyran-4-one; 3,7,11,15-tetramethyl-2-hexadecen-1-ol; (3β, 20R)-cholest-5-en-3-ol; *n*-hexadecanoic acid; 9, 12, 15-octadecatrienoic acid methyl ester and citronellyl tiglate.

References

[1] Borris R P 1996 *J. Nat. Prod. Res.* **51** 29-38
[2] Habila J D, Bello I A, Dzikwe A A, Ladan Z and Sabiu M 2011 *Aust. J. Basic Appl. Sci.* 5 537-543.
[3] Rahman S M M, Pervin S, Quader M A and Hossain M A 2009 *Indo. J. Chem.* **9(3)** 470-473.
[4] Rahman, S M M, Munira S and Hossain M A 2008 *Indo. J. Chem.* **8(3)** 459-462.
[5] Rahman S M M, Shabnom S, Quader M A and Hossain M A 2008 *Indo. J. Chem.* **8(2)** 242-244.
[6] Suryanti V, Marliyana S D and Kristinawati D 2005 *J. Alchemy* **4(2)** 28-34.

[7] Marliyana S D, Suryanti V and Suyono 2005 *Biofarmasi* **3(1)** 26-31.
[8] Misar A V and Upadhye A S 2004 *Indian J. of Pharm. Sci.* **66(4)** 463-465.
[9] Andrade-Cetto A, Becerra-Jimenez J and Cardenas-Vazquez- R 2008 *J. Ethnopharcol.* **116** 27-32.
[10] Harborne J B 1973 *Phytochemical Methods: A Guide to Modern Techniques of Plant Analysis.* (London : Chapman and Hall)
[11] Trease G E and Evans W C 1983 *Orders and Families of Plant in Pharmacognosy.* (Oxford : Oxford University Press)
[12] Sofowora A 1993 *J. Ethnopharmacol.* **38(2-3)** 209-214.
[13] Sofowora A 1996 *J. Altern. Complement. Med.* **(2)** 365-372.
[14] Krishnaiah D, Devi T, Bono A and Sarbatly R 2009 *J. Med. Plants Res.* **3** 67-72.
[15] Zohra S F, Meriem B, Samira S and Muneer M S A 2012 *J. Nat. Prod. Plant Resour.* **2(4)** 512-516.
[16] Selowa S C, Shai L J, Masoko P, Mokgotho M P and Magano R 2010 *Afr. J. Tradit. Complement. Altern. Med.* **7(2)** 98-103.
[16] Tupe S B, Patil P D, Thoke R B and Aparadh V T 2013 *Int. Res. J. of Pharm. App. Sci.* **3(1)** 49-51.
[17] Cecheovska L, Cejpek K and Konecny M 2011 *Velisek J* **(233)** 367-376.

IOP Conf. Series: Materials Science and Engineering **193** (2017) 012051 doi:10.1088/1757-899X/193/1/012051

Barns Society as a Buffer Stock (Case in Nglaris Village Bener District Purworejo Regency)

E W Riptanti[1], A Qonita[1], Suprapti[1]
[1]Agribusiness Department, Faculty of Agriculture, Sebelas Maret University

E-mail: erlyn4@yahoo.com

Abstract. Nglaris village is a mountainous area with most of its agricultural lands are used for growing annual crops, while only small agricultural areas are for growing rice. However, the post-harvest storage, in the form of barn society, can supply the people's needs of rice. Article describes the function of barn society as a buffer for foods supply. Research design was exploratory research, aims to generate in-depth qualitative and contextual data. The technique applied in this research was a case study. The cost for barn society maintenance was collected from the joint venture between farmers united in the barn society group. The barn society has 20 tons capacity for unhulled rice. The farmers obtained the income from working during the harvest time and become a labour harvester to outside Nglaris or called "boro". The farmer (field owner) gave natura wages ("bawon"). Bawon calculated based on the result of total harvested unhulled. Supply for buffer of foods in the dry season was at least 50% of the capacity barn society's, so that it can secure the supplies of staple foods. The barn society is specifically advantageous during the dry season because each farmer could lend and should be returned within one year.

1. Introduction

Foods insecurity is commonly occurred and become the defining characteristic at many developing countries and dry land ecosystems. People living in dry land region, continue to struggle with hunger, frequent famines and related consequences such as loss of lives, savings and wealth, disease and despair. Furthermore, the prevalent foods insecurity mitigation measures adopted by the inhabitants coupled with poor agricultural practises especially in soil and water conservation have also considerably impacted on the prevailing foods security situation in the area [5]. The main cause of foods insecurity resulting the low income and poverty experienced by smallholder farmers is low crop and animal productivity [2].

One of the indicator to measure foods security in a region is by reviewing the supply of foods crops production (rice and secondary/ "palawija" crops). In the middle of agricultural land constraints in Purworejo regency, the rice production increases for 7.89% in 2012 and 2013, with harvest productivity as many as 5.64 ton/ha [18]. Nglaris village is one of the mountainous area in Bener district, Purworejo regency where most of its residents are doing farming for living. The rice production is not enough for the residents' needs. The village only has two hectares of rice field area. Meanwhile, the regency's population is recorded more than 1,200 people [5]. The plantation crops grown in the village is coconut, while for fruits are thorny palm, duku and jackfruit. The crops harvest from the commodities are not enough for the staple food needs [9].

The community's responds to adopt different adaptive strategy according to the available choices which are provided for them, such as by seeking for additional income from both on-farm and off-farm works [14]. People's migration to urban centre area to seek employment wages [16]. Migration of work

IOP Conf. Series: Materials Science and Engineering **193** (2017) 012051 doi:10.1088/1757-899X/193/1/012051

force for villagers, mostly affected by some factors, such as family's poor economy condition, marginal condition of natural resources, and attraction or driving factor which lead them for migrating [3].

The drought season, where the annual crops are not harvested, most of the residents in Nglaris village migrate to the rice-producing region, as a hodge. They are paid with unhulledrice which is brought home for main consumption, while some of the unhulled rice are kept in the community's barn society. The collective management of foods supply by community, in the form of barn society has become Indonesians tradition for a long time, including in Nglaris village. The existence of barn society has ups and downs, while its role is adjusted with recent dynamics. The foods management is related to the issue of how to manage the foods supply. This means that the foods supply management is one of the aspect which commonly, has not been assessed well. The purpose of this research is to assess the barn society function as the buffer of foods.

2. Research Method

This research method used an exploratory research to generate an in-depth and contextual of quantitative and qualitative data. A case study was applied for this research technic. This research was conducted for "Subur Makmur" barn society in Ngaris village, Bener district in 2016. The key informant involves the members and committees of the community barn society, village officers, rice mills owners and rice traders. The data collected through in-depth interview, observation, and recording. The data were analyzed using analytic descriptive method.

3. Result

Most of the Nglaris village residents did the farming for living, but the income from this work sector was not enough to meet the staple food needs. They were migrating in groups to meet their staple food needs by working as a hodge in the agricultural area during the drought season for fruits. Each group migrate for three to four times as a hodge in a year. these were the family respond or way to face the foods supply issue, and another strategy according to Maxwell and Caldwell [13], Ziaei, et al [22] and Gupta, et.al [10] in table 1.

Table 1. Family Strategy to Access The Foods

No.	Strategy	Family Habit
1.	Changing foods pattern	a. Relying on cheap and available foods
2.	Increasing the family's short-term foods supply	a. Borrowing raw foods materials from friends or relatives
		b. Paying foods by instalments
		c. Collecting raw foods materials and crops
		d. Recycling the leftover raw foods materials for seeds during the next year growing season
3.	Decreasing the amount of foods consumption	a. Sending kids to eat with the neighbours
		b. Sending family members to work outside
4.	Foods portion strategy	a. Controlling the foods portion for adults, so that kids are able to eat more foods
		b. Mobilizing all family members
		c. Reducing foods portion consumed in a day
		d. Fasting or committing a day without foods

Source: Maxwell and Caldwell [13], Ziaei, et al [22] and Gupta, et.al [10].

Each group migrates with five to eight adult members both men and women, which commonly has family relationship. The group members work as a hodge to the farmers or wholesaler. The farmers (rice field owner) will pay them with natura wages ("bawon"). Bawon calculated based on the result of total harvested dry grain or unhulled rice, with comparison 1 : 8 or 1 : 7 from the total harvested unhulled rice. It means one part for them and eight (or seven) part for the farmers. Meanwhile, the wholesaler will pay them with comparison 1 : 10. from total harvested unhulled rice. The wholesaler give smaller wages but routine work. The unhulled rice collected to pay living expenses during the migration period and to pay staple food needs. Meanwhile, the remains supply of unhulled rice is stored inside the barn society. This is according to Rahman, et. all [17] who said that the importance to develop community tradition in an individual way to practise for foods supply. Each group brings at least four to five tons of unhulled rice in once migration. Unhulled rice wages of working in the agricultural center which contrasted with the results of other researchers in cropping strategy.

The effort to increase foods supply could be done with community's foods supply development [12]. A program of raw foods materials storage is essential to support for ourselves and our family members in an emergency situation. This may be resulted in a situation where financial resources to purchase foods would not be available or may be appreciably decreased [7,8]. The good management of foods supply in the family level, community, regional government and central government, is one of the key factor to achieve the foods security and to minimalize the risk of foods insecurity [17]. Foods supply in the community level, is responded by the government with the issuance of minister regulation No. 30 in 2008 about the village foods supply.

The barn society socially functioned in collecting unhulled rice from the migrated farmers and to overcome the foods insecurity during drought season. The barn society has been proven in overcoming villagers problems during the drought season or foods insecurity period, according to Witoro, et.al, [21] and Ahmad, et.al, [1]. Furthermore, the barn society functioned as the foods price controller. In this case, the farmers will store the harvested crops (due to overproduction or during harvest time) in several months, then sell them in the period when the rice hit its most profitable price [1,4].

The case occurred at "Subur Makmur" barn society in Nglaris village, is that the barn society is filled with the remains amount of unhulled rice after it is reduced for basic foods needs and household foods supply. The barn society's active members are as many as 148 people. The ability of smallholder farmers, individually or in groups to effectively and efficiently store a several part of their harvest at farm level, can be a major contributing factor for attaining foods sufficiency and eliminate wide spread of hunger in the region [2].

The farmers who are also the community barn society members, sell between 25 kg and 50 kg of unhulled rice with 12% of water content to the barn society's committee. This is the form of social responsibility in maintaining the barn society. The 20 tons of barn society capacity is fullfilled when the migrated farmers back to their village while bringing the unhulled rice. If the barn society is overcapacity, then the committee will sell the unhulled rice to the wholesaler or rice mill owners who are able to pay the highest price.

During the drought season, the barn society's committee sell the unhulled rice to the community members, with at least 10% to 20% below the average price. In addition, the barn society community also gives a lend to the members with maximal 50 kg of unhulled rice. The unhulled rice should be returned within a year, by giving additional 10% from the amount of borrowed unhulled rice. This activity has reduced seasonal gap in foods consumption [11]. Besides that, barn society also has social function, which is to give maximal 40 kg of rice for social aids to the disadvantaged families, such as experiencing death of family member, sickness and unable to work due to several conditions, or any other disadvantages condition.

The supply of unhulled rice in the barn society during the drought season is minimal 50% of the barn society capacity. This is to anticipate the emergency needs. The committee of barn society should be responsive toward the unhulled rice supply availability and should be able to predict the members needs, so that the unhulled supply in the barn society is always available. The rice mills owners/ wholesaler is prohibited to conduct a speculation act against the market price, because the barn society function as a buffer of supply during the drought season, according to the research conducted by Wahyudi, et.al [19] and Wahyudi, et.al, [20]. The impact is that farmers are easy to access unhulled rice in lower price than the market price. Thus, it needs serious steps in securing the buffer of supply in the barn society, in order to avoid the market from speculation act [6].

Local wisdom is shown by the migrated farmers who bring home the unhulled rice. The customs should be maintained due to its importance for the foods security at the family and village level. The community empowerment is developed to strengthen the local wisdom above, from the power of management side, where the barn society functions for buffer of supply. The form of community empowerment goals is to enhance the farmers' role in managing the barn society management by storing the harvested crops.

4. Conclusion

Nglaris village has a local wisdom where the migrated farmers has their customs to bring household their wage in the form of unhulled rice. The remains unhulled rice after it is reduced for the family's consumption and foods supply, is sold to to the community barn society to fill the barn society's supply. The community barn society functions as the buffer of supply during the drought season and is proved to maintain the price stability in the Nglaris village.

Acknowledgements

The author would like to thank the Ministry of Research, Technology and Higher Education for financing this research in the form of grant scheme for University's Qualified Research in the first year in 2016. The author would also like to thank to every party who has given the access for data and information.

References

[1] Ahmad M dan Imron R 2015 *Model Revitalisasi Fungsi dan Peran Lumbung Pangan Desa untuk Meningkatkan Ketahanan Pangan* University Research Colloquium
[2] Albinus M M 2012 *ARPN J. Sci. Tech.* **2** 814-821
[3] Bambang Winarso 2014 *J. Penelitian Pertanian Terapan* **14** 1-14
[4] Bhartendu KC and Lieutenant CAR 2015 *J. Bus. Manage.* **17** 37-43
[5] Boniface NW, Kennedy JO, and Telesia MM 2014 *AJAFS* **2** 52-59
[6] *Brian Wright* 2009 *Speculators, Storage, And The Price Of Rice* (California: Giannini Foundation of Agricultural Economics University of California)
[7] Charlotte PB and Deloy GH 1995 *Foods Storage In The Home* Cooperative Extension (Utah : Utah State University)
[8] Elizabeth LA and Judy AH 2011 *Foods Storage for Safety and Quality* Cooperative Extension. (Georgia: Georgia University)
[9] Erlyna WR, Suprapti S, Aulia Q 2016 *Revitalisasi Lumbung Pangan Lokal dalam Meningkatkan Ketahanan Pangan Pada Rumah Tangga Miskin di Jawa Tengah* Laporan Penelitian Lembaga Penelitian dan Pengabdian kepada Masyarakat Universitas Sebelas Maret.
[10] Gupta P, Kalyani S, Veenu S, Sidharth A, Pulkit M 2015 *J. Foods Sec.* **3** 6-10
[11] Karna B and Maisy W 2015 *J. De. Econ.* **115** 200-216
[12] Nuhfil H 2012 *Agr. Econ. Elect. J.* **1** 1-10
[13] Maxwell, Daniel, Caldwell, Richard 2008 *The Coping Strategies Index : Field Methods Manual* Second Edition Cooperative for Assistance and Relief Everywhere, Inc. (CARE).
[14] Mohamednour AG, Elnour AE, Abdel SO, Sid Ahmed 2015 *Int. J. Agr., Forestry and Fisheries* **3** 249-254
[15] Monografi Desa 2015 *Monografi Desa Nglaris 2015* Pemerintah Desa Nglaris Kecamatan Bener Kabupaten Purworejo
[16] Mwang'ombe AW, Ekonya WN, Muiru WM, Wasongo VO, Mnene WM, Mongare PN, Chege SW 2011 *J. Environ. Sci. Tech.* **4** 403-410
[17] Rahman HPS, Purwoto A, Hardono GS 2005 *J. Forum Penelitian Agroekonomi* **23** 73-83
[18] Statistics of Purworejo Regency 2014 *Purworejo Regency in Figure* Statistics of Purworejo Regency
[19] Wahyudi S, Senator NB, Andi C, TMA Ari S 2010 *Proceedings of the World Congress on Engineering* **3** 2129-2134

[20] _______ 2012 *J. Eng. Sci.* **44** 101.1- 101.10
[21] Witoro, Yusup N, Martua S 2006 *Lumbung Pangan Jalan Menuju Keserdiaan Pangan* (Bogor: Koalisi Rakyat untuk Kedaulatan Pangan)
[22] Ziaei M, Farhad SB, Farshid E, Ali K 2013 *Int. J. of Agr. Crop* **6** 225-230

IOP Conf. Series: Materials Science and Engineering **193** (2017) 012052 doi:10.1088/1757-899X/193/1/012052

Community food pattern and nutrition among toddlers in district of Pacitan, East Java, Indonesia

A Lestari[1], D Hanim[2], E Listyaningsih[3], Supriyadi[4]
[1]Doctoral Student, Focus in Health Promotion Study Program of Extention Development /Community Empowerment, Post Graduate UNS, Faculty of Medicine UNS, and P4GKM LPPM UNS, Surakarta, Indonesia
[2] Head of Nutrition Science Study Program, Post Graduate UNS, Faculty of Medicine, and P4GKM LPPM UNS, Surakarta, Indonesia
[3] Faculty of Medicine UNS, and P4GKM LPPM UNS, Surakarta, Indonesia
[4] Faculty of Agriculture UNS, LPPM UNS, Surakarta, Indonesia

E-mail: leslietarinick@yahoo.com

Abstract. Aim of this study was to asses and formulate food consumption patterns among under five children in accordance with the desirable dietary pattern score and Recommended Dietary Allowances to the target of East Java's desirable dietary pattern. Toddlers who have nutritional status less than normal are 29.44%. Toddlers with good nutritional status are 59.39% and obesity toddlers are 11.16 %. Families of respondents were able to provide the food of the crop and compound ($p = 0.81$; $r = 0.11$). The ability of the respondent's family to buy vegetables with $p = 0.06$; $r = 0.19$. Animal side dish food spending patterns showed that the family of respondents who consume meat poultry, including eggs was 92.1%. Families fruits spending patterns of respondents was 70.8% had a habit of eating fresh fruit so that every day can be provided. The pattern of food consumption in Pacitan already reflected the pattern of nutritionally balanced food, especially in the fulfillment of carbohydrates, vegetable-sourced protein, animal-sourced protein. Nutritional status of toddlers was ideal. The main driving factor for the improvement of nutritional status of toddlers was nutrition awareness against food diversification.

1. Introduction

Recently, there are a lot of researchers who determine that the relationship among starvation, nutritional imbalance and prevalence of both infection and degenerative ilnesses cannot be separated from the problem of food security at family level and Desirable Dietary Pattern at community level, particularly in Pacitan District, East Java [1].

The Desirable Dietary Pattern is defined as a composition of a variety of food or food groups based on their contribution of energy, both in absolute and relative contribution to the total energy in terms of the availability and the consumption of food, which is able to meet the needs of nutrition to consider various aspects including social, economic, cultural and religious aspects as well as the taste [2].

In Pacitan District, the highest nutritional problems include malnutrition,imbalance composition of the dishes and insufficiency of the overall consumption to support the body. Toddlers (1-5 years old) are the age group that most often suffer from Protein Energy Deficiency.

Based on the results of Nutritional Status Monitoring survey in 2001,the energy consumption in Pacitan was to 2,279 cal / person / day. The amount of energy, wich was about 190,10 cal or about 8, 34% of the energy contribution were from tubers [3]. Cassava and various processed products were still widely consumed to substitute rice [4]. Thus, Pacitan has the potential diversification of staple food items. Diversification of food sources of other nutrients needs to be studied further [5]. Therefore, this study aimed to examine the patterns of food and the nutritional status of children under five years old in 12 villages in 12 subdistricts in Pacitan.

2. Experimental

This study was observational analytic with cross-sectional design. Retrieval and analysis of data was done in March-July 2014. Selection of research subjects was determined by purposive sampling. A number of 197 todlers under five years old in Pacitan District were includid in this study. Data / information was collected by observation, interviews, questionnaires, and focus group discussions. Toddler's food pattern data per Subdistrict was measured according to the Recommended Dietary Allowance (RDA). Data were analyzed using Chi-Square. The concluding result will be used to formulate recommendations for the Office of Food Security and Public Health Service in Pacitan.

3. Result and Discussion

3. 1. Result

The choice of location was done with consideration of their village community nutritional problem such as toddlers malnutrition. The average education level was high school with employment status in private sector (trader, farmers, fishermen) with a value of $p = 0.175$; $r = 0.74$. Families of respondents were able to provide the food of the crop and compound ($p = 0.81$; $r = 0.11$). The ability of the respondent's family to buy vegetables (vegetable side dishes) with $p = 0.06$; $r = 0.19$. Animal side dish food spending patterns showed that the family of respondents who consume meat poultry, including eggs was 92.1% ($p = 0.00$; $r = 0.921$). Families fruits spending patterns of respondents was 70.8% had a habit of eating fresh fruit so that every day can be provided papaya, banana, water guava which were bought from fruit sellers around each village ($p = 0.02$; $r = 0.708$).

Table one shows the result study of respondent's characteristics in 12 villages, includid number of respondents, average family member, average of mother's education level and average of household's salary per month.

Table 1. Analysis of Respondent's characteristics per village (12 villages)

Name of villages	Number of Respondents (person)	Number of family	Level of Education	Family income (Rupias)
Arjosari	25	3-4	SMP-SMA	220000±2000000
Belah, Donorejo	40	3-4	SMP-SMA	137500±2000000
Dersono, Pringkuku	13	3-4	SMP-SMA	423100±2000000
Jatigunung Tulakan	8	3-4	SMP-SMA	687500±2000000
Jeruk Bandar	13	3-4	SMP-SMA	423100±2000000
Kebonagung	11	3-4	SMP-SMA	500000±2000000
Punung	15	3-4	SMP-SMA	370000±2000000
Pagerejo, Ngadirojo	24	3-4	SMP-SMA	830000±2000000
Pakisbaru Nawangan	16	3-4	SMP-SMA	347000±2000000
Sirnoboyo, Pacitan	6	3-4	SMP-SMA	333000±2000000
Sudimoro	12	3-4	SMP-SMA	458000±2000000
Tegalombo	16	3-4	SMP-SMA	347000±2000000
Statistic test	p=0.00; r=0.99	p=0.459;r=0.222	p=0.367;r=0.187	p=0.474;r=0.430

Source: Primary data processed in 2014.

According to table 1 it can be seen that the average number of family members as much as 3-4 people per household (p = 0459; r = 0222). This means that statistically there is not real difference between the number of families in the rural area with a village in the central district, that every family have three to four members.
Table 2 shows analysis of nutritional status and feeding frequency among toddlers at posyandu in District of Pacitan, East Java.

Table 2. Analysis of nutritional status dan feeding frequency among toddlers (n=197)

Nutritional status of toddlers	Feeding frequency (per day)			
	>3x per day	3x per day	<3x per day	TOTAL (%)
Thin	6	51	1	58(29.44%)
Good (normal)	4	110	3	117(59.39%)
Obesity	2	20	0	22(11.16%)
Total	12(6.09%)	181(91.87%)	4(2.03%)	197(100%)
Value p=0.079; r=527				

Source: Primary data processed in 2014

According to table 2, there were found some toddlers who eat fewer than three times per day. There were four toddlers (2.03%) who had nutritional inadequacy which was not correspond to the recommended dietary allowance. However, when viewed from the statistical analysis was not significant (p = 0.079; r = 527). This is in accordance with the WHO agreement stating malnutrition in the community will be a public health problem if it happens more than 5%.

Table 3. Analysis of the nutritional status and pattern of milk drinking among toddlers

Nutritional status of toddlers	Milk drinking frequency per day				TOTAL (%)
	never	Rarely	often	every day	
Thin	0	22	22	16	60(30.15%)
Good (normal)	4	11	40	62	117(58.79%)
Obesity	0	2	6	14	22(11.06%)
Total	4(2.01%)	35(17.59%)	68(34.17%)	92(46.23%)	199 (100%)
value p=0.000; r=0.277					

Source: Primary data processed in 2014

Based on table 3 it can be seen that there were four toddlers (2.03%) who were not consume the milk. It was found that 35 toddlers (17.17%) who rarely drank milk.This shows the habit of milk consumption among toddlers is still not appropriate with Recommended Dietary Allowances (RDA). However, when viewed from the statistical analysis there was a significant relationship (p = 0.000; r = 0277) between nutritional status and milk consumption habits. Furthermore, there were 27.7% nutritional status of toddlers who were not affected by the consumption habits of milk.

Here is a description of the consumption of Pacitan District residents during the study period in the energy, the percentage of Nutrition Adequacy Score, and scores Desirable Dietary Pattern wich have been recorded from March to May 2014.The food group in the Desirable Dietary Pattern are nine items (Hardinsyah and Martianto, 1992) :
a. Rice - the food grains from cereals commonly as a staple food.
b. Bulbs - tubers which is the food of roots or tubers are edible.
c. Animal food is a food of animal and preparations.
d. Oils and fats are of vegetable foodstuffs that have fat and oily and fat from the animal.

IOP Conf. Series: Materials Science and Engineering **193** (2017) 012052 doi:10.1088/1757-899X/193/1/012052

e. Oily fruit and seeds are good food containing oils from fruits and seeds.
f. Nuts - Nuts are seeds that contain high levels of fat.
g. Sugar and the processed products.
h. Vegetables and fruits.
i. Others which are condiments.

Table 4. Calculation results of Desirable Dietary Pattern (DDP) residents in Pacitan

number	Food group	Energy (Kal) standart	Energy (Kal) Pacitan	% DDP standart	% DDP Pacitan	Quality	DDP score
1	Rice	1100	1000	50	45.45	0.5	22.725
2	Bulbs	132	172	6	7.82	0.5	3.91
3	Animal food	264	214	12	9.72	2	19.44
4	Oils and fats	220	185	10	8.41	0.5	4.205
5	Oily fruits and seeds	66	56	3	2.54	0.5	1.27
6	Nuts	110	100	5	4.54	0.5	2.27
7	Sugar and the processed products	110	85	5	3.86	0.5	1.93
8	Vegetables and fruits	132	105	6	4.77	5	23.85
9	Others which are condiments	66	55	3	2.5	0	0
	Total	2200	1972	100	89.61		79.6

Source: Primary data processed in 2014.

Based on table 4 it can be seen the level of nutrient consumption of energy intake on the respondent (mother) 1972 Kal per day with desirable dietary pattern amounted to 89.61%. When compared with the value of Recommended Dietary Allowance (RDA), it can be assessed the adequacy of each nutrient. Energy and protein sufficiency level most of the respondents were already quite good (> 80% RDA). There are about 181 of the 199 respondents (90.95%) have met the needs of energy and protein.

IOP Conf. Series: Materials Science and Engineering **193** (2017) 012052 doi:10.1088/1757-899X/193/1/012052

Table 5. Analysis of food diversification effort at household level (n=190)

Name of Village	Effort of Diversification						Total
	No effort	Meal 3x per day	Meal with veg	Meal with fruits	4 sehat 5 sempurna	Balance nutrition	
Arjosari	15	1	0	0	5	0	21(11.05%)
Belah, Donorejo	25	3	0	0	5	4	37(19.47%)
Dersono, Pringkuku	4	2	1	1	4	1	13(6.84%)
Jatigunung Tulakan	4	0	0	0	0	4	8(4.21%)
Jeruk Bandar	7	1	0	0	2	2	12(6.36%)
Kebonagung	9	0	0	0	2	0	11(57.89%)
Punung	6	3	0	0	5	0	14(7.37%)
Pagerejo, Ngadirojo	11	4	0	2	4	3	24(12.63%)
Pakisbaru Nawangan	8	0	0	0	5	3	16(8.42%)
Sirnoboyo, Pacitan	3	0	1	0	1	1	6(3.16%)
Sudimoro	6	0	1	0	4	1	12(6.32%)
Tegalombo	10	0	1	0	4	1	16(8.42%)
Total	108 (56.84%)	14 (7.37%)	4 (2.11%)	3 (1.57%)	41 (21.58%)	20 (10.52%)	190 (100%)
Value of p=0.68 ; r=0.94							

Source: Primary data processed in 2014.

Analysis of food diversification effort at the household level which are an effort to be able to provide meal 3 times per day, eat vegetables, eat fruits after the main meal, eat with menu 4 healthy 5 perfect, and efforts to meet the nutritional balanced statistically showed no significant relationship with consumption patterns every day ($p = 0.68$; $r = 0.94$). Thus there were 94% of families that their nutritional adequacy did not depend on diversification efforts at the household level.

3. 2. Discussion

The choice of location was done with consideration of their village community nutritional problem such as anemia in pregnant mothers and todlers malnutrition. The average degree of anemia in pregnant mothers were mild (hemoglobin levels of more than 10 g / dl). The average education level was high school with employment status in private sector (trader, farmers, fishermen) with a value of $p = 0.175$; $r = 0.74$. Families of respondents were able to provide the food of the crop and compound ($p = 0.81$; $r = 0.11$). The ability of the respondent's family to buy vegetables (vegetable side dishes) with $p = 0.06$; $r = 0.19$. Animal side dish food spending patterns showed that the family of respondents who consume meat poultry, including eggs was 92.1% ($p = 0.00$; $r = 0.921$). Families fruits spending patterns of respondents was 70.8% had a habit of eating fresh fruit so that every day can be provided papaya, banana, water guava which were bought from fruit sellers around each village ($p = 0.02$; $r = 0.708$).

IOP Conf. Series: Materials Science and Engineering **193** (2017) 012052 doi:10.1088/1757-899X/193/1/012052

4. Conclusion

The pattern of food consumption in Pacitan already reflected the pattern of nutritionally balanced food, especially in fulfillment of carbohydrates, vegetable protein, animal protein. Nutritional status of todler was ideal, but there were still two villages with under nutrition. Posyandu activities were quite active in ensuring monitoring the nutritional status of todler and pregnant women. The driving factor of nutritional status improvement of todler was nutrition awareness against food diversification and driving factor of improving the nutritional status of pregnant women was the service of antenatal care (ANC). The level of nutrient consumption of energy intake on respondent (pregnant mother) was 1972 Kal per day with Desired Food Pattern percentage of 89.61. Level of energy and protein sufficiency among respondents were quite good (more than 80% RDA), which was about 90.95%.

Reference

[1] Departemen Kesehatan 2001 *Riset Kesehatan Dasar (RISKESDAS)* (Jakarta)
[2] Hardinsyah, Baliwati Y F, Martianto D, Rachman H S, Widodo A, and Subiyakto 2001 *Pengembangan Konsumsi Pangan dengan Pendekatan Pola Pangan Harapan* Kerjasama antara Pusat Studi Kebijakan Pangan dan Gizi Lembaga Penelitian Institut Pertanian Bogor dengan Pusat Pengembangan Konsumsi Pangan Badan Bimas Ketahanan Pangan Departemen Pertanian (Jakarta)
[3] Dinas Kesehatan Kabupaten Pacitan 2001 *Profil Kesehatan Kabupaten Pacitan* (Pacitan : District Health Office of Pacitan)
[4] Hanafie, R 2010 *Jurnal Sosial Ekonomi Pertanian (J-SEP)* **4** 2
[5] Hanafie. R, 2006. *Jurnal Widya Agrika* **4** 3
[6] FAO-RAPA 1989 *Report of The Regional Expert Consultation of The Asian Network for Food and Nutrition and Urbanization* (Bangkok)
[7] Hardinsyah dan Martianto, D 1992 *Gizi Terapan*. (Bogor: Institut Pertanian Bogor)
[8] ILSI dan FAO 2003 *SEA-RDA Harmonization* (Geneva: Drafting)
[9] Nutrition and Health Surveillance System (NSS). Helen Keller Indonesia 2004 *Indonesia Crisis Bulletin*. **5** 1
[10] Sacchet, D.L, Brian, R, Gordon, H and, Tugwel, P 1991. *Clinical Epidemiology: Basic Science for Clinical Medicine* Second Edition (Washington : Amazon)
[11] Sukirman, 1996. *Ketahanan Pangan: Konsep, Kebijakan dan pelaksanaannya*. (Makalah) (Jakarta: Departemen Pertanian Republik Indonesia)

IOP Conf. Series: Materials Science and Engineering **193** (2017) 012053 doi:10.1088/1757-899X/193/1/012053

Sustainable Food Security Measurement: A Systemic Methodology

W Findiastuti[12], M L Singgih[1] and M Anityasari[1]

[1] Industrial Engineering, University of Trunojoyo, Indonesia
[2] Industrial Engineering, Institute Technology of Sepuluh Nopember, Indonesia

Email: weny123@gmail.com

Abstract. Sustainable food security measures how a region provides food for its people without endangered the environment. In Indonesia, it was legally measured in Food Security and Vulnerability (FSVA). However, regard to sustainable food security policy, the measurement has not encompassed the environmental aspect. This will lead to lack of environmental aspect information for adjusting the next strategy. This study aimed to assess Sustainable Food security by encompassing both food security and environment aspect using systemic eco-efficiency. Given existing indicator of cereal production level, total emission as environment indicator was generated by constructing Causal Loop Diagram (CLD). Then, a stock-flow diagram was used to develop systemic simulation model. This model was demonstrated for Indonesian five provinces. The result showed there was difference between food security order with and without environmental aspect assessment.

1. Introduction

Sustainable food security was discussed in 1993 by [1]. It was mentioned that sustainable development should encompassed aspects of food, agriculture and people. Research done by [2, 3] then discussed about the importance of nature stability in agricultural to maintain the food security. Discussions about measurement of food security and indicator development were done by [4] which used sustainable land resources as indicator of sustainable food security. In 2004, Atlas of Sustainable Food Security was published by [5] Indicators were reconstructed from basic food security indicators of previous publications. However, environmental indicators were built based on ecological footprint method. It assessed deforestation, remaining land and water resources that could be supplied by nature for future demand. Although it can give information resources supplied by nature for future demand, rising population and consumption of natural resources in recent years would generate problem in assuming time remaining for the natural supplies. For example: it is assumed that fresh water supplies will sufficient for next five years, 'doing business as usual' in rising population and consumption would bias those of information. Unsustainable process of food fulfillment would degrade environment rapidly that we could not restrain. The assumption of five years might turn to one year. This paper tried to reconstruct indicators of sustainable food security, by applying eco-efficiency as environment assessment for the whole indicators.Known as widespread use of sustainable performance assessment tool in industries, eco-efficiency has extended into broader scope of applications. Economic consideration of environmental assessment in integrated index has simplified the next step of decision making at higher management. Emerged as business link to sustainable development, it results win-win solutions for economic and environment performances. Industrialists can maintain the environment while enhance their revenues [6, 7]. This has made it applied in broader scope of studies. As regional sustainable assessment, eco-efficiency was applied excellently in European Union by [8] and Finnish region of Kymenlaakso by [9]. In government sector, it helped sustainable development policy makings in agricultures [10-12], transportations [13], and development of environmental regulation of industry areas [14, 15]. The difference between micro and macro level economics is a higher concern of social aspect in macro level. In micro level or industrial level, social aspect could be achieved separately with eco-efficiency assessment, such as programs of corporate social responsibility (CSR), otherwise in macro level or governmental level, social aspect is significant in every decision made. This paper proposed complete the triple bottom line of sustainable: economic, environment and social in eco-efficiency performance measurement especially for macro level applications supported with systemic approach to construct indicators and relationships between them. This systemic approach was an extended idea of system dynamic supported performance measurement developed by [16]. Similar research done by [8, 9, 17], used social indicators in separated part of eco-efficiency ratio and were not supported by systemic view of indicators relationships. That caused lack of information of what effects, feedbacks and trade-offs between them, even though social indicators related to economic and environment indicators have been identified. The systemic view of indicators relationships helps policy maker in trade-offs considerations for every indicator.

However, some researchers argued that being eco-efficient doesn't mean completely sustainable. Being eco-efficient is minimizing resource and energy used, so cost of raw material resource and risk of waste generated could be minimized as well. Eco-efficiency has not required yet for the social indicators of sustainable development, since it is only incorporate indicators of economic and environment. Besides those criticisms, eco-efficiency helps units deciding the beneficially way tobe sustainable. The more eco-efficient activities, the more activities with environmental considerations that would accelerate goal achievement of sustainable development [6, 9].

This study was done for indicators identification of eco-efficiency on sustainable food security. Using a systemic approach by constructing Causal Loop Diagram (CLD), complete economic, environment and social indicators of eco-efficiency were identified. A 'new' eco-efficiency indicators come up from constructing process of relationship between existing indicators of food security in CLD. Indicators were identified according to eco-efficiency ratio as mentioned in Eqn. (2). Existing indicators were indicators of food security in Food Security and Vulnerability Atlas of Indonesia 2009 developed by [18].

2. Eco-Efficiency in Micro And Macro Levels

Concept of eco-efficiency was initiated in 1992 by World Business Council of Sustainable Development (WBCSD) at Earth Summit Rio de Janeiro. It is defined as: 'Eco-efficiency is achieved by the delivery of competitively-priced goods and services that satisfy human needs and bring quality of life, while progressively reducing ecological impacts and resource intensity throughout the life cycle to a level at least in line with the earths estimated carrying capacity'. In quantitative term, it is stated as ratio of product or service value (added) to its environmental impact and written as [19, 20]:

$$\text{Eco-efficiency} = \frac{\text{Product or service value (added)}}{\text{Environmental impact}} \quad (1)$$

Eqn. (1) is a general form of eco-efficiency and could have differed interpretations. The basic ratio of eco-efficiency is a modified output per input of basic ratio of efficiency which maximizing output for minimizing input would be better efficiency result. Product or service value (added) as the numerator explains the economic performance of unit measured, what can unit gain from every activity done. Environmental impact, the denominator is related to resource, energy used or waste resulted for the environment. This form of ratio is commonly used in micro level or industrial level studies.

There is no consensus for exact form of eco-efficiency ratio used in any activities. It depends on research purpose, scope, and data available. Research done by [11] used ratio of 'served function to its environmental impact' for measuring number of land space could serve by crop protection product. For environmental-impact emphasized, [21] compared eco-efficiency by applying 'environmental impact points per 1000 euro'. In macro or government level of applications, European Environment Agency (EEA) quantified eco-efficiency as [19]:

$$\text{Eco-efficiency} = \frac{\text{more welfare}}{\text{less nature}} \quad (2)$$

Eqn. (2) interpreted as economic development gained for less nature used and disruption. [8] and [15] referred 'more welfare' to Gross Domestic Product (GDP) which is an economic performance in macro level.

IOP Conf. Series: Materials Science and Engineering **193** (2017) 012053 doi:10.1088/1757-899X/193/1/012053

Nevertheless, the term 'more welfare' should be referred to achievement of meeting the human needs or quality of life that not always measured with money. Reduced number of poverty and hunger, increased number of higher education or reduced number of unemployment could be one of interpretations of 'more welfare'. So it is more than prosperity of the people in a region.

3. Case of Sustainable Food Security

Food security is a performance measurement of how well a region be able to fulfill its people with sufficient food at all times. It was defined by FAO in World Food Summit 1996 as: 'Food security exists when all people, at all times, have physical and economic access to sufficient, safe and nutritious food that meets their dietary needs and food preferences for an active and healthy life'. This definition is detailed into 4 dimensions of, food availability, food access, food utility and food stability [22]. Legal food security assessments were done by World Food Program (WFP) of FAO and published as food security map, for example: Food and Vulnerability Atlas (FSVA), The Food Security Atlas, Food and Nutrition Security Atlas, Vulnerability Analysis and Mapping, and The Food Insecurity. Since arising concern of environmental degradation, those of assessments were accomplished with environmental considerations such as: climate change, deforestation, and rainfall fluctuation.

Eco-efficiency is a sustainability performance measurement by using ratio of economic performance to its ecology performance. Proposed by World Business Council of Sustainable Development (WBCSD) at Earth Summit Rio de Janeiro 1992 as accountability of industrialist to environment degradation, eco-efficiency has appeared to be a win-win solution for realization of sustainable development. With that of principle, eco-efficiency become wide spread accepted by companies for sustainable development assessment and extended to broader level of economic activities, such as in regional or national level [23, 24].

As rising population growth, focus of environmental researches are widen to consumption side under the concept of Sustainable Consumption, whereas consumption side are blamed on the negation of resource efficiency achievement through their increasing demand namely rebound effect [25-27]. The challenge is more than just doing cleaner production, saving energy and water. At the same time, consuming products efficiently and do not speed up more productions has to be integrated fashion with the sustainable production. We need production and consumption as a whole system concerned [26]. However, dealing with environmental system of production-consumption means dealing with complex, time dependent and dynamic system. Strategy implemented on the production-consumption system based on conventional performance measurement sometime doesn't appropriate the objectives decided. Conventional performance measurement results are mostly used for controlling tools rather than improvement tools. An explicit description of system response to every changing condition is needed to support goal achievement effectively [28, 29].

This paper objective is developing dynamic eco-efficiency production-consumption conceptual framework to overcome the challenge of environmental complex system problem. Dynamic eco-efficiency of production-consumption is eco-efficiency measurement combined with system dynamic principles approach for production-consumption system. The differences with previous concepts built by [30, 31] are explicit description and inclusion of time dependence relationship within production and consumption in development of eco-efficiency measurement.

In this paper, the conceptual framework would be built on the implementation to measurement of sustainable food security. Eco-efficiency assessment of sustainable food security conducted based on food system that consists of food production, distribution and consumption and relates to food security measurement dimensions.

4. Eco-Efficiency of Sustainable Food Security

Officially, the term eco-efficiency is then defined by the World Business Council of Sustainable Development (WBCSD) in Rio de Janeiro Earth Summit in 1992 as [32]:

"Eco-efficiency is achieved by the delivery of competitively-priced goods and services that satisfy human needs and bring quality of life, while progressively reducing ecological impacts and resource intensity throughout the life-cycle to a level at least in line with the earth's estimated carrying capacity".

The essence of this definition is "satisfy human needs and bring quality of life, while progressively reducing ecological impacts and resource intensity". While protecting environment is considered as the extended cost, this term tries to clarify that protecting environment could be a profit for the business.

Derived from simple efficiency formula, output/input, (1) illustrates economic and environmental performance represented by numerator and denominator above. Product or service value as output representing economic performance refers to the value of the product and service produced by a company, the sector or the overall economic activity. While the denominator environmental influence as the input is the effect on the environment per unit of the product or service that is produced. Each numerator and denominator are measured based on indicators which is determined correspond to unit economic activity measured.

There is no standard of eco-efficiency ratio or indicator used for any economic activity. The ratio and indicator could be adapted to unit activity measured that assesses economic performance to its environmental impact. Nevertheless, for macro or meso level of economic activity, European Environment Agency (EEA) suggested using ratio of more welfare to less nature [33].

For sustainable food security, this paper uses ratio of food security achievement to its environmental influence as represented as:

$$\text{Eco-efficiency of food security} = \frac{\text{food security}}{\text{Environmental influence}} \quad (3)$$

Equation (2) above describes: numerator food security refers to food security achievement which is measured by four indicators of Food Availability, Food Access, Food Utility and Food Stability. Denominator environmental influence refers to environmental impact of food security achievement with production-consumption point-of-view.

5. Eco-efficiency Based on System Dynamic

System dynamic approach on environmental issues was first performed in 1972 in the book Limits to Growth to describe the simulated world began in 1900-2100 with 12 scenarios and analyse its response to environmental conditions [34]. In connection with the determination of the policy environment, [35] identified that the system dynamic used in the environmental impact assessment, solid waste management, analysis of greenhouse gas emissions and global warming, water resources planning, environmental planning and management, environmental sustainability, ecological modelling and so on. According to its function, in the field of environmental system dynamic approach is widely used in analysing the effects of policies in the field of environment in the span of the next few years [36], then began to be integrated with environmental performance measurement systems such as the study by [37, 38] with the consideration that the performance measurement system is the basis for policy-making in determining the next strategy. In this study, system dynamic will be used in systemic approach to eco-efficiency performance measure. Eco-efficiency is one of the performance measurement tool used to measure the level of sustainability or the sustainability of an activity. This measurement uses a process approach to activities taking into account the economic and ecological performance in the form of a ratio of value added to the economy gained as a result of environmental pressures. Increasing complexity of the systems involved in an activity, the measurement of eco-efficiency by the narrow limits of a system cannot represent the actual system sustainability. Here a holistic approach or whole on a system is needed to be able to measure the level of sustainability of actual system.

Eco-efficiency measurements with system dynamic approach is based on a review that the system dynamic approach is needed to be able to complete the performance measures with an explanation of the relationship between the factors that affect performance, trade-offs that occur between these factors and the behaviour of the system using causal loop diagrams and stock and flow diagram in response to policies that have been taken at the time when the next performance monitoring policy implementation strategies. System dynamic approach would also explain the non-linear interactions, delay, feedback loops and other elements that refer to the dynamic complexity of using quantitative simulation.

5. Causal Loop Diagram (CLD) development

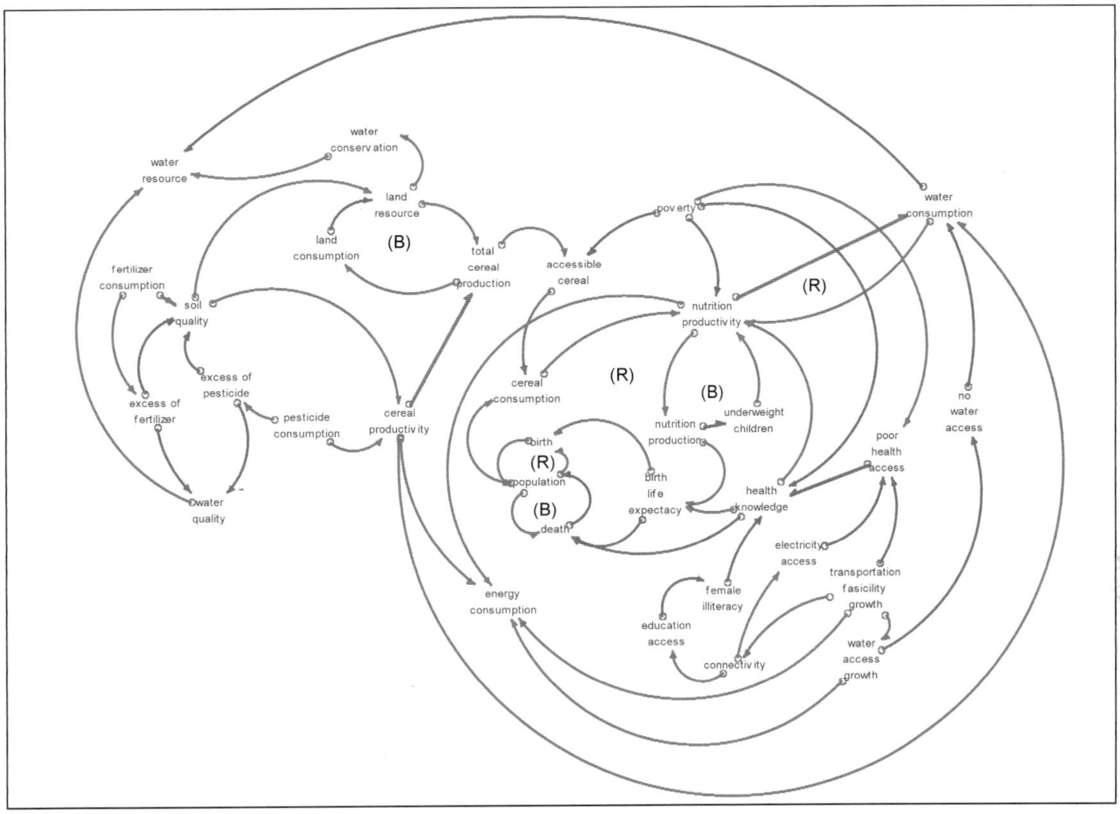

Figure 1. Causal Loop Diagram of Food Security

Figure 1 describes causal loop of eco-efficiency of sustainable food security. Conducting the overall measurement of sustainable food security eco-efficiency, indicators of food security achievement and environmental influence are defined by performing the influence relationship within subsystem of food system and subsystem of food security dimensions. This description of relationship supports the identification of intervention actors of the system that can be used as the indicator.

Following measurement framework designed by [29], relationships between sub systems then support the indicators of eco-efficiency identification by deploying as figure 3. As Equation (2), eco-efficiency of food security consists of numerator and denominator indicators. Figure 3 shows the numerator as food security and Denominator as Environmental Influence indicators deployment refers to food security and relationship with Food Production. Environmental influence is measured by food production influence to environment defined by its irrigation, fertilizer and seeds. Figure 4 describe eco-efficiency area after the simulation, and there is change in sown area order of five provinces of Indonesia.

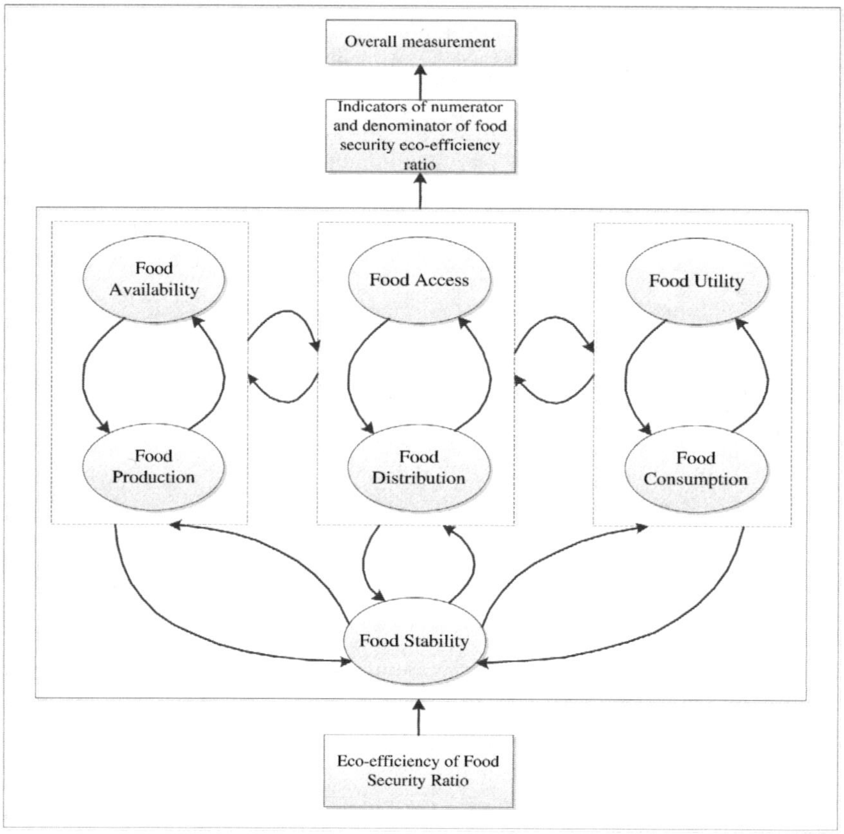

Figure 2. Framework Of Dynamic Eco-Efficiency for Sustainable Food Security Measurement

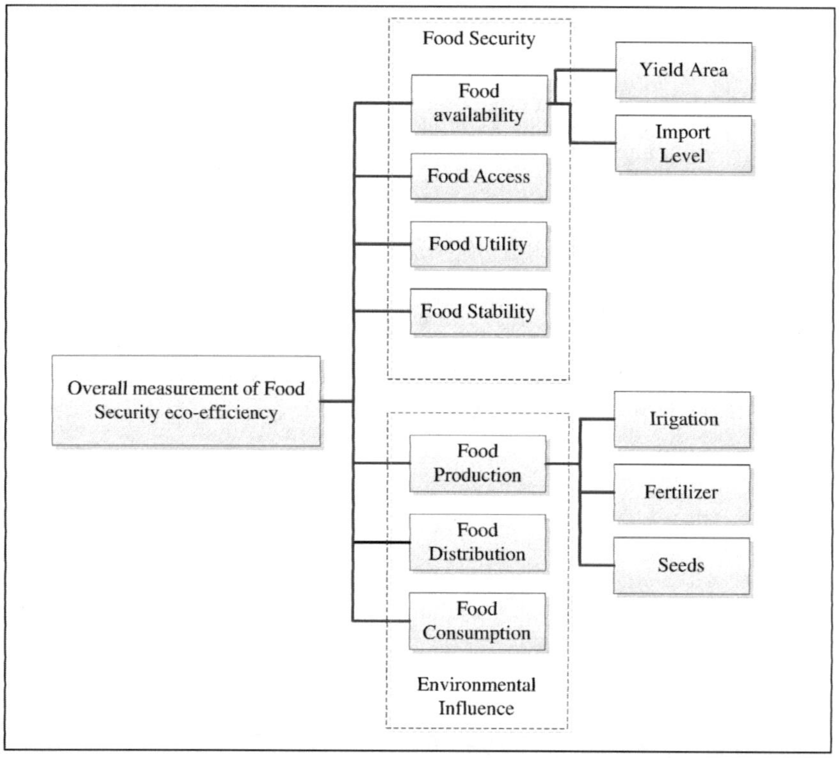

Figure 3. Food Security Eco-efficiency Deployment

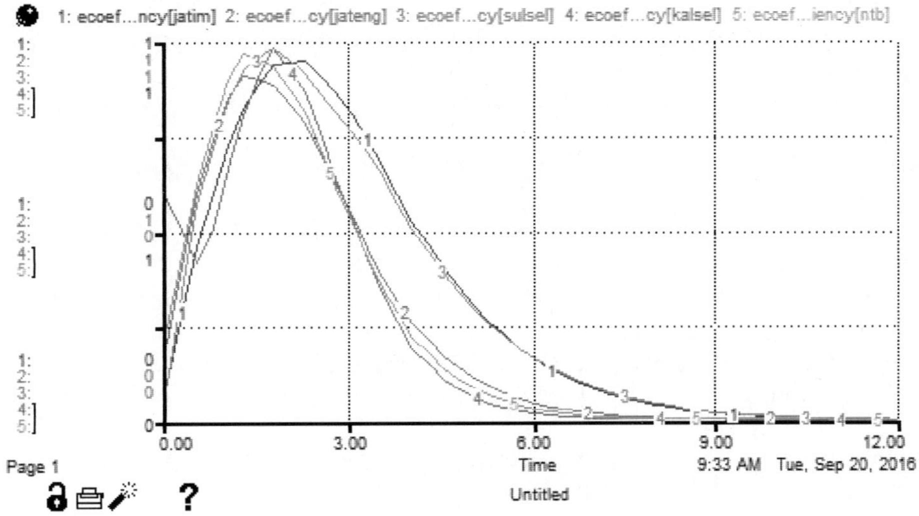

Figure 4. Eco-efficiency after simulation

6. Conclusion

This paper objective is to develop a eco-efficiency for sustainable food security measurement. Eco-efficiency of sustainable food security needs system dynamic combining to cope with complexity, time dependent and dynamic system that can't be cope by ordinary performance measurement. The use of system dynamic supports identification of intervention actors of system that can be used to be the indicator of measurements beside the standard indicators

7. References

[1] Speth J G 1994 Towards Sustainable Food Security. In: *Sir John Crawford Memorial Lecture,* (Washington DC: Consultative Group on International Agricultural Research (CGIAR))
[2] Cassman K G and Harwood R R 1995 *Food Policy* **20** 439-54
[3] Pinstrup-andersen P and Pandya-lorch R 1998 *Ecological Economics* **26** 1-10
[4] Van Keulen H, Kuyvenhoven A and Ruben R 1998 *Agric Sys* **58**
[5] MSSRF and WFP 2004 Atlas of the Sustainability of Food Security in India. (India: M. S. Swaminathan Research Foundation)
[6] Huppes G and Ishikawa M 2005 *J. Ind. Ecol* **9** 25-41
[7] PicazoTadeo A J, Gómez-Limón J A and Reig-Martínez E 2011 *J. Environ. Manage.* **92** 1154-64
[8] Hinterberger F and Scneider F 2001 Eco-efficiency regions: Toward reducing Total Material Input. In: *7th European Roundtable on Cleaner Production,* ed S E R Institute (Lund, 2-4 May 2001
[9] Mickwitz P, Melanen M, Rosenstrom U and Seppala J 2006 *J Clean Prod* **14** 1603 - 11
[10] Reith C C and Guidry M J 2003 *J. Environ. Manage.* **68** 219-29
[11] Jonge A M d 2004 *Crop Prot.* **23** 1177-86
[12] Basset-Mens C, Ledgard S and Boyes M 2009 *Ecol. Econom.* **68** 1615–25
[13] D'Agosto M and Ribeiro S K 2004 *Transp. Res. Part D* **9** 497-511
[14] Wang Y, Liu J, Hansson L, Zhang K and Wang R 2010 *J. Clean. Prod.*1-8
[15] Zhang B, Bi J, Fan Z, Yuan Z and Ge J 2008 *Ecol. Econom.* **68** 306-16
[16] Santos S P, Belton V and Howick S 2002 *Int. J. Oper. Prod. Manage.* **22** 1246-72
[17] Seppäläa J, Melanen M, Mäenpää I, Koskela S, Tenhunen J and Hiltunen M-R 2005 *J. Ind. Ecol.* **9** 117-30

[18] WFP, DKPRI and DEPTANRI 2009 A Food Security and Vulnerability Atlas of Indonesia. In: *Food Security Atlas,* (Indonesia: FAO)
[19] WBCSD 2000 Eco-efficiency: Creating more value with less impact. In: *Eco-efficiency*: WBCSD)
[20] DeSimone L D and Poppof F 2000 *Eco-efficiency: The Business Link to Sustainable Development* (Massachisett: MIT Press)
[21] Wursthorn S, Poganietz W-R and Schebek L 2011 *Ecological Economics* **70** 487-96
[22] FAO 2006 Food Security Policy Brief
[23] DeSimone L D and Popoff F 1997 *Eco-efficiency: The business link to sustainable development* (London: The MIT press)
[24] Hinterberger F and Scneider F 2001 Eco-efficiency of Regions: Toward reducing total material input. In: *7th European Roundtable on Cleaner Production,* (Lund: Sustainable Europe Research Institute)
[25] Huppes G and Ishikawa M 2005 *Journal of Industrial Ecology* **9**
[26] Clark G 2007 *Journal of Cleaner Production* **15** 492-8
[27] Mont O and Plepys A 2008 *Journal of Cleaner Production* **16** 531-7
[28] Ghalayini A M, Noble J S and Crowe T J 1997 *International Journal of Production Economics* **48** 207-25
[29] Santos S, Belton V and Howick S 2002 *International Journal of Operation and Production Management* **22** 1246-72
[30] Ericksen P J 2008 *Global Environmental Change* **18** 234-45
[31] Ericksen P J, Ingram J S I and Liverman D M 2009 *Environmental Science and Policy* **12** 373-7
[32] Verfaillie H A and Bidwell R 2000 *Measuring eco-efficiency: A guide to reporting company performance* (WBCSD)
[33] WBCSD 2000 Eco-efficiency: Creating more Value with less Impact. World Business Council for Sustainable Development (WBCSD)
[34] Meadows D, Randers J and Meadows D 2004 Limits to Growth: The 30-Year Update. (Chelsea: Green Publishing Company)
[35] Anand S, Vrat P and Dahiya R P 2006 *Journal of Environmental Management* **79** 383-98
[36] Shen Q, Chen Q, Tang B-s, Yeung S, Hu Y and Cheung G 2009 *Habitat International* **33** 15-25
[37] Fagan J E, Reutera M A and Langford K J 2010 *Resources, Conservation and Recycling* **54** 719-36
[38] Jin W, Xu L and Yang Z 2009 *Ecological Economics* **68** 2938-49

IOP Conf. Series: Materials Science and Engineering **193** (2017) 012054 doi:10.1088/1757-899X/193/1/012054

Soil Fertility Status, Nutrient Uptake, and Maize (*Zea mays* L.) Yield Following Organic Matters and P Fertilizer Application on Andisol

S Minardi [1], S Harieni [2], A Anasrullah [1], H Purwanto [1]
[1] Department of Soil Science, Faculty of Agriculture, Sebelas Maret University, Surakarta Jl. Ir. Sutami 36 A Kentingan 57126, Surakarta, Indonesia
[2] Department of Agrotechnology, Faculty of Agriculture, Tunas Pembangunan University, Surakarta, Indonesia

E-mail: minardi_sam@yahoo.com

Abstract. Objective of this study were to elucidate effects of organic matters and P fertilizer application on soil fertility status, nutrient uptake and maize yield in the Andisol. This experiment consisted of two factors. The first factor comprised of four levels of organic matters input (without organic matter, manure, rice straw, and *Gliricidia sepium* leaves), with the application dosage 10 $t.ha^{-1}$ and the second factor comprised of three levels of P fertilizer application (without P addition (control), 50 kg P_2O_5 ha^{-1}, 100 kg P_2O_5 ha^{-1}). Results of this study showed that organic matters and P fertilizer application improved soil fertility status, especially pH, soil organic C, cation exchange capacity (CEC), available P which resulted in an increase in P uptake that improve yield of maize. The highest yield of maize (corn cob) was obtained through application *Gliricida sepium* (8.40 $t.ha^{-1}$), followed by manure (6.02 $t.ha^{-1}$) and rice straw (5.87 $t.ha^{-1}$). Application of 50 kg P_2O_5 Ha^{-1} yield was (5.76 $t.ha^{-1}$) and application of 100 Kg P_2O_5 Ha^{-1} yield was (6.12 $t.ha^{-1}$).

1. Introduction

The one type of land troubled by the presence of phosphate is Andisol [1][2]. Andisol actually have a relatively high content of total P 160-500 mg.100 g^{-1} soil, but the number of P available to plants is very low, only about 1% of the total P contained in the soil [2][3]. As the third after water and nitrogen, the element P is estimated to be limiting growth and crop production [3]. P inorganic fertilizer application to increase the availability of P that can be utilized by plants, it is less efficient, since only 8-13% are available and can be absorbed by plants, the rest is converted into a form insoluble phosphate compounds and unavailable to plants [4][5].

Organic matter, is one effort that can be done because it can increase the availability of soil P [6][7][8][9]. The use of organic matter (manure, mulch and compost) has been shown to increase soil fertility [10]. Some research indicates the real role of organic matter in improving the availability of P in Andisol [11]. Susilowati [12] reported that organic matter in the form of *Gliricida sepium*, *Leucaena glauca* and *Zea mays* in Andisol very markedly increase the availability of P after incubation of 30 days and 60 days. Research conducted Safitri [13] showed that the use of organic materials such as prunings Tithonia and Tephrosia in Andisol able to increase the availability of P. Increased availability of P as a result of organic matter can occur because during the process of decomposition of organic materials produce dissolved organic acids such as citric acid, malic acid, and acetic acid which plays an important role in the binding of Al and Fe that P became available [14][15][16]. Okajima [17] and Bell and Beessho [18] said that the use of organic materials is based on the assumption of the role of humic acid and fulvic relation to its ability to reduce the occurrence of bond between phosphorus and aluminum or iron in the soil. Andisol in the area of Tawangmangu, Central Java, Indonesia (study site), cow manure, rice straw and crop plants Gliricidia *(Gliricidia sepium)* is common and has been used by farmers since availability is quite a lot. This research aims to exploit the potential of organic matter (manure, rice straw and *Gliricidia sepium)* and P fertilizer to improve soil fertility in providing Andisol P element on the corn crop.

2. Experimental

A field experiment was conducted in experimental station Seed Food Crops and Horticulture, Tawangmangu, Karanganyar, Central Java, Indonesia. The soil type Andisol with loam texture (45% sand, clay 16% and silt 39%), pH H_2O (5.6), high soil organic matter (4.96%) and nutrient P-available is very low (3.49 mg.kg^{-1}) and a high CEC (32.15 mg.kg^{-1}) The organic material used is manure, rice straw and *Gliricidia sepium* and seed corn (*Zea mays* L.). In addition, to use of urea and KCl fertilizers, also used pesticides to control pests and diseases that arise during the course of the study in the field.

This study was conducted using factorial trial with two factors conducted on the experimental plot size of 3 x 1.5 m. The first factor was the kind of organic material (10 t.ha^{-1}), which consists of four levels, ie, without organic matter (B0), manure (B1), the provision of rice straw (B2) and *Gliricidia sepium* (B3) , The second factor was the provision SP36 which consists of three levels, namely without fertilizer P (P0), 50 kg P_2O_5 ha^{-1} (P1) and P_2O_5 100 kg.ha^{-1} (P2). Twelve treatments combination were set out in a randomized block design with three replications. At each treatment was given basic fertilizer urea (300 kg.ha^{-1}) and KCl (50 kg.ha^{-1}) as a basal fertilizer.

Variable corn growth observed absorption P (calculated by multiplying the dry biomass of the top with a concentration of P-plant tissue) and corn yields are expressed in weight of corn cobs per plot at the end of the experiment. For soil analysis conducted on P-available (extracted by the method of Bray-II) and P-total (extract H_2SO_4 1M), soil pH (in 1: 2.5 ratio of soil: water), soil organic matter (method of Walkley and Black), CEC (using ammonium acetate at pH 7.0), carried out the beginning and end of the experiment. The data were then analyzed statistic. The results of the analysis are presented in Table ANOVA on a real level (5%) and highly significant (1%). To distinguish between the mean treatments used multiple range test Duncan Multiple Range Test (DMRT).

3. Results and Discussion

From an initial analysis of the soil chemical properties Andisol with criteria according [4], especially slightly acid pH H_2O (5.6), NaF pH (11.1), soil organic matter (4.96%), P-available nutrients is very low (3.49 mg.Kg^{-1}) and a high CEC (32.15 mg.Kg^{-1}). Referring to the key guidelines for the estimation of the suitability of land especially soil fertility (pH, CEC, BS, SOM and available P) the fertility of the soil include the category of low to moderate [4]. The low nutrients available P can be a major limiting nutrient in plant growth that is less favorable for crops [19][3]. Organic matter is an effort to increase the availability of P that can be utilized by plants [6]. Characteristics of organic materials used in this study, are presented in Table 1,

Table 1. Characteristics of organic materials

Characteristics of organic materials	Manure	Rice straw	*Gliricidia sepium*
Organic C (%)	29.90	43.44	46.11
Total N(%)	0.67	0.85	2.28
Total P (%)	0.22	0.36	0.61
Ratio C/N	43.97	50.49	20.12
Ratio C/P	130.03	117.42	74.37
Total organic acid (%)	21.56	26.91	35.53
Humic acid (%)	16.71	15.92	22.17
Fulvic acid (%)	3.22	7.70	8.10

IOP Conf. Series: Materials Science and Engineering **193** (2017) 012054 doi:10.1088/1757-899X/193/1/012054

The use of organic materials was able to improve soil fertility all parameters (pH, CEC, SOM and P), as presented in (Table 2) and (Figure, 1,2 and 3). Increased pH (H_2O) as a result of the provision of the three kinds of organic materials, the highest row in the treatment of *Gliricidia sepium*, manure and rice straw. P fertilizer treatment, achieved at the highest dose of 100 kg ha $^{-1}$ $P_2 0_5$ followed by 50 kg ha^{-1}.

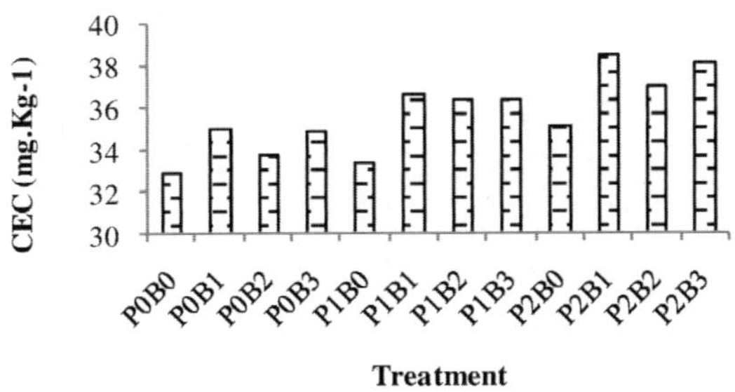

Figure 1. Effect of treatment on CEC

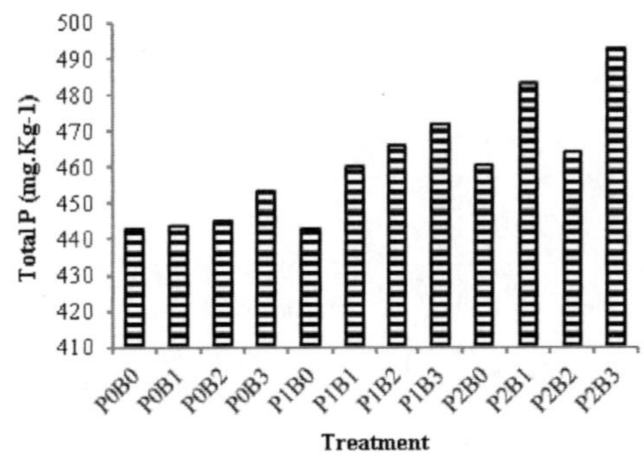

Figure 2. Effect of treatment on Total P

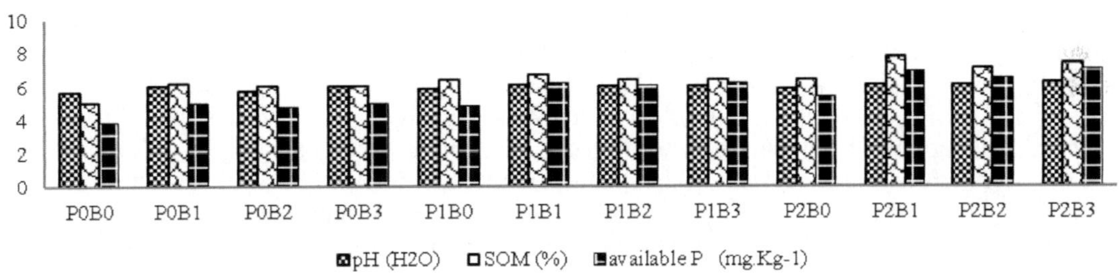

Figure 3. Effect of treatment on pH, soil organis matter (SOM), available P

Description:
P0 = without P fertilizer, P1 = P fertilizer dose of 50 Kg.Ha^{-1}, P2 = P fertilizer dose of 100 Kg.Ha^{-1}.
B0 = without OM, B1 = manure, B2 = rice straw and B3 = *Gliricidia sepium*

Increased pH (H_2O) in the combination treatment achieved the highest *Gliricidia sepium* and P fertilizer dose of 100 kg ha^{-1} (P2B3) of 5.66 to 6.25, an increase of approximately 10%. The rising trend pH (H_2O) very likely caused by the ligand exchange reactions between organic anions decomposition of organic matter, especially humic and fulvic acids to the free -OH on local exchanges, so the effect on improvement of OH ions in the soil solution [20][15][21]. Bell and Beessho [18] says that the role of organic acids such as humic acid and fulvic, in reducing the bond between phosphorus and aluminum or iron in the soil. The humic acid and fulvic has high capability in forming a chelate with several cations, especially aluminum and iron was decreased activeness in adsorb P [9], this situation will affect the availability of P [22][23][10].

Increased P-provided may occur due to the release of P from organic materials are added, as well as the indirect influence of the organic material present in the P sorption complex of soil. The organic material is known to reduce the P sorption by iron oxides and Al, and also colloidal clay contained in the soil [14][15][16][18]. Agree with [18] and [21], humic acid content (22.17%) and fulvic acid (8.10%) higher than *Gliricidia sepium* than organic materials others (Table 1), are believed to play a role in increasing the pH (H_2O) which in turn will positively affect the availability of P in the soil. The provision of real-P and organic materials potential in improving P and P available, which is a growth limiting nutrient corn plant other than N, K and soail organic matter [23]. Increased P-provided by a third use of organic materials in this study, in order are *Gliricida sepium* > manure > rice straw, which is thought to be caused by the activity of organic acids, especially humic and fulvic acids. Increased P provided the highest achieved in the combination treatment of *Gliricidia sepium* and P fertilizer dose of 100 kg.ha^{-1} (P2B3) from 3.83 mg.kg^{1} to 7.06 mg.kg^{1} or an increase of approximately 45%. (Table 2).

Table 2. Characteristics of Andisol after being treated with the organic material and P fertilizer.

No.	Treatment	pH (H2O)	pH (NaF)	SOM (%)	CEC (mg.Kg^{-1})	available P (mg.Kg^{1})	Total P (mg.Kg^{1})
1	POBO	5.66	10.39	5.05	32.88	3.83	443.14
2	POB1	5.98	10.16	6.15	35.11	4.98	444.01
3	POB2	5.68	10.34	6.04	33.78	4.72	445.17
4	POB3	6.01	10.30	6.05	34.88	5.02	453.31
5	P1BO	5.85	10.40	6.38	33.38	4.85	442.96
6	P1B1	6.10	10.41	6.71	36.67	6.19	460.33
7	P1B2	6.05	10.35	6.39	36.44	6.02	466.12
8	P1B3	6.00	10.42	6.40	36.49	6.19	472.08
9	P2BO	5.90	10.49	6.38	35.22	5.40	460.48
10	P2B1	6.12	10.59	7.74	38.60	6.86	483.31
11	P2B2	6.08	10.38	7.07	37.05	6.53	464.36
12	P2B3	6.25	10.43	7.39	38.16	7.06	492.59

From the analysis of the soil at the end of the experiment (Table 2), suggesting the provision of three kinds of organic matter and fertilizer P, affecting the changes in soil properties associated with the availability of P in the soil, especially pH (H_2O), CEC and SOM (Figure 4). Correlation test results showed that the pH (H_2O), the Commission and SOM positively correlated with P-available (r = 0.821, r = 0.967 and r = 0.915), which gives the sense that an increase in pH (H_2O), CEC, and OM will be followed by an increase P-available. The changes P increased availability will result in an increase in the value of P uptake and P concentration in plant tissue. The results showed that the use of three kinds of organic materials (B1, B2 and B3) and fertilizer P (P1 and P2). Overall also been proven to increase the P concentration in plant tissue were significantly ($p < 0.001$).

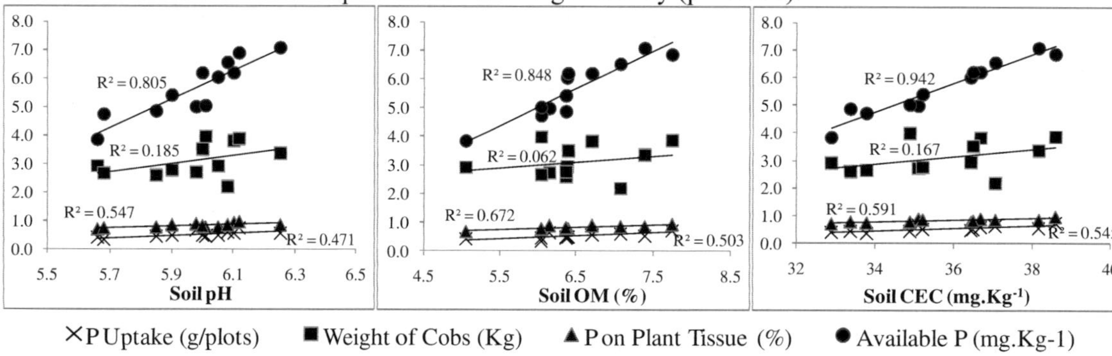

Figure 4. The effect of soil pH, soil organis matters, and soil CEC on P uptake, weight of Obs (maize yield), P on plant tissue, P available improvement.

The increase in P uptake of plants is determined by the concentration of P in the soil as well as the ability of plants to absorb elements of P in the soil. Where organic matter serves to meet the needs of plant nutrients that allow the plant to grow and develop properly. P is required for root development. Rooting more developed will allow for more nutrient absorption. Clarified by Foth [20], P uptake of plants is largely determined by the contact with the roots of P, P concentration in soil solution and the ability of the plant. As reported Darman [10] that the availability of P will increase P uptake which in turn affects the increase in plant dry weight and yield of corn (Table 3) and (Figure 5 and 6). P fertilizer is needed in a corn plant growth and yield [24] [25] [26].

Table 3. Effect of organic matter and fertilizer P to shoot dry weights (g), of P plant tissue (%), P uptake (g) and the weight of cobs per plot (Kg)

No	Treatment	Shoot dry weights (g)	P on plant tissue (%)	P Uptake (g)	Weight of cobs (Kg)
1	POBO	58.23	0.68	0.40	2.91
2	POB1	72.77	0.88	0.64	2.71
3	POB2	47.00	0.74	0.34	2.64
4	POB3	56.54	0.76	0.43	3.96
5	P1BO	55.99	0.78	0.44	2.59
6	P1B1	61.88	0.90	0.56	3.82
7	P1B2	60.95	0.78	0.48	2.93
8	P1B3	61.52	0.79	0.49	3.51
9	P2BO	58.81	0.83	0.49	2.77
10	P2B1	76.62	0.94	0.72	3.87
11	P2B2	71.01	0.85	0.60	2.18
12	P2B3	62.08	0.85	0.53	3.35

Description:
P0 = without P fertilizer, P1 = P fertilizer dose of 50 Kg.Ha^{-1}, P2 = P fertilizer dose of 100 Kg.Ha^{-1}.
B0 = without OM, B1 = manure, B2 = rice straw and B3 = *Gliricidia sepium*

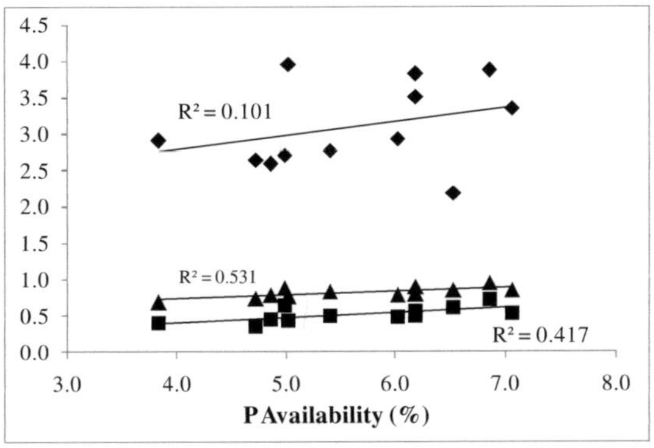

Figure 5. The effect of P availablility on weight of cobs, P uptake, P on plant tissue

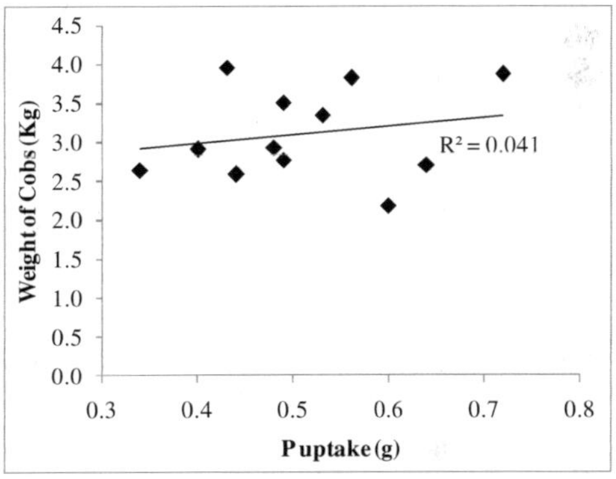

Figure 6. The effect of P uptake on weights of the cobs .

4. Conclusion

Overall, this study proves that the organic material and P fertilizers can improve soil fertility and yield of corn in Andisol. Improving the status of soil fertility and yield of corn, mainly due to the effect of treatment than the use of organic materials with the addition of P fertilizer treatment, while their interaction is limited to the results of this trial was not yet able to provide increased yield of corn. Successive loss results obtained on providing the highest cob *Gliricida sepium* (8.40 t.ha^{-1}) > Manure (6.02 t.ha^{-1}) > Rice straw (5.87 t.ha^{-1}). Application of 50 kg P_2O_5 ha^{-1} yield was (5.76 t.ha^{-1}) and application of 100 kg P_2O_5 ha^{-1} yield was (6.12 t.ha^{-1}).

5. Acknowledgements

The authors wish to express their gratitude for Dr. Danar Praseptiangga for his valuable discussions and Mr. Sidik Pramono for his technical supports.

IOP Conf. Series: Materials Science and Engineering **193** (2017) 012054 doi:10.1088/1757-899X/193/1/012054

6. References

[1] Tan K 1987 *The Andisols in Indonesia* (New York: Van Norstrand Reinhold Company)

[2] Darmawijaya M 1990 *Klasifikasi Tanah* (Bandung: Balai Penelitian Teh dan Kina)

[3] Sanchez P 1992 *Properties and Management of Soils in the Tropics.* (New York: John Willey & Sons)

[4] Roesmarkam A and Yuwono N *Ilmu Kesuburan Tanah* (Yogyakarta: Penerbit Kanisius)

[5] Tisdale S, Nelson W and Beaton J 1985 *Soil fertility and fertilizers.* (New York: Mc Macmillan Co)

[6] Schnitzer M 1991 *Soil Sci* **151** 41–58

[7] Schnitzer M 1997 *Pengikatan Bahan Humat oleh Koloid Mineral Tanah. dalam: Interaksi Mineral Tanah Dengan Bahan Organik dan Mikrobia* ed P M Huang and M Schnitzer (Yogyakarta: Gadjah Mada University Press)

[8] Stevenson F J and Fitch A Kimia Pengkomplekan Ion Logam dengan Organik Larutan Tanah *Interaksi Mineral Tanah Dengan Bahan Organik dan Mikrobia* (Yogyakarta: Gadjah Mada University Press) pp 333–76

[9] Suntoro, Syekhfani, Handayanto E and Sumarno 2001 *Agrivita* **23**

[10] Darman S 2008 *J. Agrol.* **15**

[11] Minardi S 2002 P *Sains Tanah. J. Penelit. Ilmu Tanah dan Agroklimatologi.* **1** 16–23

[12] Susilowati D 1997 *Pengaruh Pemberian Beberapa Bahan Organik Terhadap Kurva Jerapan P dan Beberapa Sifat Kimia Lainnya Pada Andisol, Bendosari Malang dan Ultisol, Karta Lampung.* (Malang: Fakultas Pertanian Universitas Brawijaya.)

[13] Safitri E 2001 *Pengaruh Penggunaan Bahan Organik Asal Tithonia diversifolia dan Tephosia candida Untuk menurunkan Jerapan P Pada Andisol.* (Malang: Fakultas Pertanian Universitas Brawijaya)

[14] Hidayah Utami S N, Handayani S and Maas A 2000 *J. Ilmu Tanah dan Lingkung.* **2** 7–12

[15] Hue N V, Craddock G R and Adams F 1986 *Soil Sci. Soc. Am.* **50** 28–34

[16] Minardi S 2008 *Agrineca* **8**

[17] Okajima H Reclamation and improvement of Soils Derived from Volcanic Ash. *Soil Derived From Volcanic Ash in Japan* ed Y Inshizuke and C A Black pp 64–83

[18] Bell I C and Beessho T 1993 Assesement of Al detoxification and plant response *In Soil Organic Matter Dynamic and Sustainability of Tropical Agiculture* ed K Mulongoy and R Merck (New York: John Wiley & Sons)

[19] Parfitt R and Kimble J 1989 *Soil Sci. Soc. Am.* 971 – 977

[20] Foth H D 1994 *Fundamentals of soil science. Terjemahan E.D. Purbayanti, D.R., Lukiwati, dan R. Trimulatsih. Dasar-dasar Ilmu Tanah.* (Yogyakarta: Gadjah Mada University Press)

[21] Kononova M M 1966 *Soil Organic Matter. Its Nature, its Rule in Soil Formation and Soil Fertility. 2nd edition.* ed B Cinnati. (Toronto. London. Melbourne)

[22] Syekhfani 1997 *Hara, Air, Tanah dan Tanaman.* (Malang: Jurusan Tanah, Fakultas Pertanian. Universitas Brawijaya)

[23] Nursyamsi D, Husnain, Kasno A and Setyorini D 2005 *J. Tanah Iklim* **23** 13–23

[24] Permadi 2005 *J. Agrivigor* **5** 9–15

[25] Polakitan A, Kaunang R, Polakitan D and Taulu L 2004 *Respon tanaman jagung (Zea mays L.) terhadap pemupukan P pada Tanah Podzolik Merah Kuning. Dalam Prosiding Seminar Nasional Klinik Teknologi Pertanian sebagai Basis Pertumbuhan Usaha Agribisnis menuju Petani Nelayan Mandiri* (Bogor: PusatPenelitian dan Pengembangan Sosial Ekonomi Pertanian)

[26] Purnomo J 2007 Respon tanaman jagung terhadap pemberian pupuk fosfat pada tanah Inceptisol dari Bogor *Pros. Lokakarya Nasional Inovasi Teknologi Pertanian Mendukung Hari Pangan Sedunia 2007* ed D Subardja, R R. Saraswati, H S Mamat, P Setyanto, D Setyorini, Wahyunto, M Noor and Irawan (Bandar Lampung)

IOP Conf. Series: Materials Science and Engineering **193** (2017) 012055 doi:10.1088/1757-899X/193/1/012055

Law Enforcement of Consumer Protection for Safe Food Packaging in The Decisions of Criminal Justice

F Wiryani[1], Herwastoeti[1], M Najih[1], A Haris[2]
[1]Faculty of Law, University of Muhammadiyah Malang, Malang, Indonesia
[2]Faculty of Islamic Study, University of Muhammadiyah Malang, Malang, Indonesia

E-mail: fifik@umm.ac.id

Abstract. The right to a safe food is a human rights protected by the 1945 Constitution and legislation, including the Health Act, the Consumer Protection Act and Food Act. The law governing the rights and obligations of consumers; rights, obligations and responsibilities of businesses, as well as prohibitions and sanctions for businesses that violate. Food consumers aggrieved can file a non-litigation legal action and / or litigation. Non-litigation legal efforts made through negotiation or mediation or through Consumer Dispute Resolution Body (BPSK). The litigation efforts made by filing a lawsuit for damages to the court and / or reporting the case to the criminal law enforcement. This study specifically examines the enforcement of criminal law in the judgment as a safeguard against food consumers. Sanctions provisions setting a strategic role in an effort to make the protection of consumers of food. Patterns general formulation of the maximum penalty in the third Act is not appropriate because it too gives flexibility for the judge to make a decision as low to the Defendant. Facts on society, business agent has a dominant and strong position compared with consumers of food. These favorable conditions business agent position and vice versa less give legal protection to the Consumer Food. Preferably the pattern formulation penalty of criminal acts in the field of food using a specific minimum and maximum public.

1. Introduction

Food is the most substantial and essential basic need in the human's life [1]. Its fulfillment is a part of human rights [2] listed in the Universal Declaration of Human Rights (UDHR), the 1945 Constitution of Indonesia, the Law on Human Rights, the Law on Health, the Law on Consumer Protection, and the Law on Food as well as in their regulations of implementation.

Food is also the basic component to materialize the qualified human resources. Accordingly, food supply is to constantly be in sufficient availability accompanied by food safety, quality, nutrients, and varieties. However, society's purchasing power must also be taken into account and this food supplying is not to go against the principles of religions, faith, and society's culture sustainably [3]. Food safety refers to the condition and effort required to prevent food from biological and chemical harm as well as any other defective substances. As it is to be conducted in line with the religious and cultural values, the food safety effort is to ensure that the food supply would be safely consumable [4] and promote no defects to the people's health. The enforcement of food safety should be conducted through the provisions of food sanitation, food additives regulation, genetically modified food products and food irradiation regulation, food packaging standard, food safety and quality warranty, and required halal product certification [5].

Every packaged food circulated in the community is to be labeled. Label refers to any form of information related to food presented in the form of photograph, text, photo-text combination, or any other forms included in or printed out on the food packaging [6].

The labeling is regulated in the various legislations such as Act No. 18 Year 2012 on Food, Act No. 8 of 1999 on Consumer Protection, Act No. 36 of 2009 on Health, PP 69 of 1999 on Food Label and Advertisement, PP 102 of 2000 on National Standardization, Perka BPOM No. HK.03.1.23.11.11.09909 of 2011 on Supervision Claims In Processed Food Labelling and Advertising, Decree No. 033 of 2012 on Food Additives and Jo Decree No. 034 of 2012 on the Maximum Limit Melamine In Food, as well as the Director General of POM Decree No. 02 591 / B / SK / VIII / 91 which amend Regulation of the Minister of Health No. 180 / Menkes / Per / IV / 1985 on food expiration.

The administration of label certification on the packaged food product by Food and Drug Monitoring Agency takes hold of a strategic role as an attempt to ensure the safety of food products. Each packaged and marketed food product in Indonesia is required to meet the national standards and add an information label in Indonesian Language (Bahasa Indonesia) explaining the product name, compositition, size net weight/volume, name and address of the company, production dates, dates of expiration, and/or halal certification [7].

However, the existed statutory laws and institutional regulations authorized to perform and control the safe packaged food labeling and halal certification have not fully safety guaranteed.

Based on the research of the National Consumer Protection Agency (BPKN) there are four major problems in food safety, namely: 1) Food poisoning as a result of damaged and contaminated food, or mixed with any adverse substances; 2) The use of restricted food additives; 3) The incompatibility of food label with the standard provisions; and 4) Food expiration [8].

According to BKN label problems receive less attention from both the consumer and manufacturers, only 6.7% of the consumers pay attention to the whole information on the label. In Indonesia, we can easily find food products which do not have complete label information on their packaging [9]. Lina Apriana Sari et al. found that many food manufacturers knowingly re-distribute and sell their expired food products [10]. Aisyah Herlita Setyaningrum also found the circulation of imported packaged food products which do not meet the safety requirements of the national standard provisions [11]. Based on the annual report of POM 2015, was found violations of the labeling of food product packaging, including the halal label, as well as tables 1 and 2 below [12]:

Table 1. Violations of the labeling of food product packaging

Type of Product	comply the regulation	Not comply	Total	Percentage
MD	10.175	321	10.496	3%
ML	795	151	946	16%
PIRT	273	221	494	45%
TOTAL	**11.243**	**693**	**11.936**	**6%**

MD: Licensed Domestic Product
ML: Licensed Foreign Product
PIRT : Local Licensed micro enterprises

Table 2. Violations of the labeling of food product packaging

	Not comply	PIRT Label
not mention the name and address of manufacturer / importer	33 offense	96 offense
not mention the production code	678 offense	847 offense
not mention the expired date	62 offense	283 offense
composition incomplete / not appropriate	68 offense	226 offense
Netto	42 offense	42 offense
Not in Indonesian	1 offense	0 offense
misleading claims	2 offense	3 offense

The impact of unsafe food consumption is the occurrence of food-borne illness and death cases. Based on the reports of the Food and Drug Monitoring Agency in 2005, from the 152 exceptional food poisoning incidents during 2004, 7295 people suffered from food poisoning and 45 lives were claimed [13]. In 2015, there were reported remarkable incident of food poisoning, as many as 61 cases, exposure to 8263 people, as many as 2,251 people ill and died as many as three people [14].

The World Health Organization (WHO) estimates that the ratios between the reported food poisoning incidences and the actual incidences are 1:10 in the developed countries, and 1:25 in the developing countries. Based on this estimation, in 2015 the possibility of actual incidences in Indonesia at that time was approximately 206 575 people were food-poisoned, 56 275 sick and 75 people died.

Food entrepreneurs who do not comply with the provisions of safe packaged food labeling can be sentenced with administrative and/or criminal sanctions. The victim consumers may report the provision violation to the local Food and Drug Monitoring Agency and/or to the Police Offices. They are also allowed to sue for damages to the violating entrepreneurs. This paper specifically examines the application of criminal sanctions in the judgment.

2. Experimental

2.1 Problem Statement

This study discusses two issues: 1) How is the analysis of the regulations of consumer legal protection on safe packaged food; and 2) How is the analysis of the implementation of consumer legal protection in the judge's decision related to the criminal acts on the packaged food.

2.2 Research Methods

This study used juridical doctrinal approach. The type of data was secondary data obtained from primary and secondary sources in law. The data from the primary legal material were in the form of statutory laws regulating food and consumer protection as well as the judge's decision related to the criminal acts on the packaged food. As for the data from the secondary legal materials, they were taken from journals, reference books, academic papers, articles, studies, and printed or electronically- presented news related to the study. The data were collected by using literature and document searches both physically and electronically. The collected data were analyzed by employing content analysis combined with the principles of statutory laws as well as prescriptive analysis [15].

3. Result and Discussion

3.1 Analysis of the Consumer Legal Protection on Safe Packaged Food

This study on the food consumer legal protection would be viewed from the aspect of the regulations related to the rights and obligations of the consumers, the rights and obligations of the entrepreneurs, and the prohibition and sanction imposed on the entrepreneurs.

3.1.1 Rights of the Consumers and Obligations of the Entrepreneurs

The rigths of the consumers stipulated in Article 4 of the Law on Consumer Protection are*:1)* to obtain comfort, security and safety in using product; 2) to choose and obtain Point 2) to obtain correct, clear and honest information on the condition and warranty of product; 3) to obtain itmpensation, redress and/or substitution, if the product received are not in accord with the agreement or not received as requested.

The obligations of the entrepreneurs: 1) to act in good faith; 2) to provide correct, clear and honest information with regard to the condition and warranty; 3) to guarantee the product based on the prevailing quality standard provisions; 4) to provide compensation, redress and/or substitution for the damages caused by the use, consumption and application of the product; 5) to guarantee the food products produced and/or traded complied with the food safety and quality standards; 6) to have a legal license for distributing and marketing; 7) to use safe and food-grade packaging materials.

3.1.2 Obligations of the Consumers and Rights of the Entrepreneurs

The obligations of the customers [16]· are: 1) to read or follow the information instructions and application or usage procedures of the product; 2) to act in good faith; 3) to pay for the price in accordance with the agreed.

IOP Conf. Series: Materials Science and Engineering **193** (2017) 012055 doi:10.1088/1757-899X/193/1/012055

The rights of the entrepreneurs [17] *are: 1)* to receive the payment in accordance with the sales agreement; 2) to obtain legal protection from the consumer's acts of bad faith; 3) to conduct proper self-defense in the legal settlement; 4) to be exempted from compensation liability to the customers, if: a) the customers' losses and damages caused by the consumption of the goods purchase are due to the customers' faulty acts; b) the goods are evidently not to be circulated or not intended to be distributed; c) the goods defect appears at a later date after purchasing; d) the negligences are due to the customers themselves; e) the customers' claims are overdue more than 4 years.

3.1.3 Prohibition and Sanction Imposed on the Entrepreneurs

As the effort to protect food consumers, a set of prohibiton imposed on the entrepreneurs of packaged food product was promulgated. The entrepreneurs are not allowed: 1) do not use the label to the specified standards on their product; 2) to not clearly and explicitly mention the infomation in the label; 3) to use harmful or non-foodgrade food packaging; 4) to remove, cover, replace the label, relabel and/or change the dates of expiration.

Entrepreneurs violating the above-mentioned prohibition are threatened with administrative and criminal sanctions. The imposed administrative sanctions do not repeal their criminal sanctions. The type of administrative sanctions and criminal sanctions respective regulations are:

Table 3. The Regulation of Administrative Sanctions

The Types of Administrative Sanction	**ACT**		
	Consumer Protection Protection	**Food**	**Health**
Fine	√	√ (50 M ≤)	–
Compensation stipulation	√ (200 M ≤)	√	–
Termination offers, promotions and advertising Temporarily	√	√	√
Temporary product distribution disallowance	√	√	√
The obligation to pull out goods from circulation by Manufacturers	√	√	√
Confiscation and destruction	–	–	√
Temporary production termination	√	√	√
Revocation of product distribution permit;7	–	√	√
Revocation of production permit	–	√	–
Revocation of business permit.	–	√	–

Sources: Act No. 8 on 1999 [18]; Act No. 18 on 2012 [19]; and Act No. 36 on 2009 [20].

The administrative sanctions provided by BPOM or local governments as providers of food packaging distribution license. Application of criminal sanctions based on a court decision made through the mechanism of criminal justice. Criminal sanctions is ultimum remedium, which attempts to latest sanctions If the attempt is given administrative sanctions not effective. Implementation of administrative given, does not remove the criminal sanction.

Types of criminal sanctions based on the regulations are: Principal Penalties and additional penalties. Principal Penalties, consists of:

Table 4. Comparison Regulations of Principal Penalties for Food Crime Sanction

Type of Sanction	Criminal Sanction		
	Consumer Protection Act	Food Act 7/1996	Food Act 18/2012
Imprisonment	≤ 5 years	1 – 5 years	1 – 10 years
Fine	IDR ≤ 2 Milyars	IDR 120 – 600 Milyars	IDR 2 – 10 Billions

Sources: Consumer Protection Act [21]; Food Act 7/1996 [22], Food Act 18/2012 [23]

Additional Penalties, which consists of: 1) Confiscation of certain goods; 2) Announcement of judge's decision; 3) Payment for damages; 4) Injunction to stop certain activities that cause damages to the consumers; 5) The obligation to pull out goods from circulation; 6) Revocation of business permit; or 7) Revocation of certain rights [24].

Entrepreneurs violating the provisions above may be defined as conducting criminal acts which may be the subject to administrative sanctions and/or criminal sanctions.While for the Law of Health, there are no criminal provisions related to this study. However, this law imposes administrative sanctions as well.

The pattern of punishment in the Law is assigned by the law makers with the severity of the crime or violation as the consideration. Whereas, the stipulation of serious or minor criminal threats is determined based on the eminence or the level of the crime and the damage it causes. This can be seen in the maximum penalties which are dependent on the type of crime committed. From the same type of crime, there are several different maximum penalties in different Law such as in Law on Consumer Protection, Law on Food of 1996, and Law on Food of 2012.

3.2 An Analysis of the Implementation of Penalties on the Violation of Labeling and The Circulation of Packaged Food in Judicial Verdict Provision

Legislative regulation is one role of the state in the effort to achieve the welfare of the citizens. It is also a preventive endeavor to protect and guarantee the rights of the citizens. As for the Judicial Decree, it was the last bastion for the justice-seekers who have been denied their right.

Based on a random investigation, it was found that several samples of court judgments that adjudicated the perpetrators for violating the legislation in the field of packaged food. The court judgment is one of law enforcement portraits in food consumer protection. From the 33 (thirty three) court judgments which were randomly taken, there was a wide gap between public prosecutors' demands and the court judgment with its maximum criminal charge. Based on 33 (thirty three) Judicial Decree which were randomly procured, it reveals:

Table 5. Judicial decision Crime Criminal About Food

The Article'crime	The Criminal Sanktions	Prosecutions	court decisions
Food additives Crime **Food act 7/ 1996:** **Pasal 10 (1)** **Consumer Protection Act 8/ 1999:** **Article 55 b & 8(1) a**	Imprisonment: ≤ 5 years &/ Fine: Act 7/1996: 120-600 Milyars Act 8/1999: 2 Billions	Imprisonment:5-12 months; and Fine: IDR 2,5 – 50 Milyar; Evidence seized and destroyed	Imprisonment 4-12 months & Evidence seized and destroyed. [25]
unlawful importation and distribution of food **Act 7/ 1996:** **Article 36 (2) & 58 k** **Act 8/ 1999: article 8 (1) a**	Imprisonment:≤3years &/ Fine: Act 7/1996: IDR 120-600 M. Act 8/1999: IDR 2 B	Imprisonment: 6-10 months + Fine IDR. 4-10 Millions + Evidence seized and destroyed. Imprisonment:1-10 months +Evidence seized & destroyed	Fine IDR. 5 Millions + Evidence seized and destroyed [26]
No food distribution license **Sct 18/ 2012:** **Article 91** **Article 142**	Imprisonment:: ≤ 2 years &/ Fine: ≤ IDR 4 Billions Act 7/1996: IDR 120-600 M. Act 8/1999: IDR 2 B.	Fine: IDR. 5–15 Millions + Evidence seized and destroyed	Imprisonment: 7 months [27]
Label Unlawful **Act 7/1996** **Article 30 (3) & 58 h**	Imprisonment: ≤ 3 years & /or Fine: ≤ IDR 360 M	Imprisonment: 6 months + Fine IDR. 5 Milyars + Evidence seized & destroyed	penjara 6 bulan + Evidence seized and destroyed [28]
Nofooddistribution license **Act 18/2012** **Article 141 & Article 89**	Imprisonment: ≤ 2 years &/ Fine: ≤ IDR 4 Billions	Imprisonment: 1-8 months + Evidence seized and destroyed.	Imprisonment: 8 Months [29]
advertising and labeling violations **Act 8/1999: Article 62(1)** **the ministerial decree of religion 518/ 2001: Article 8(1) a**	Penjara: ≤ 2 years &/ Fine: ≤ IDR 4 Billions	Imprisonment: 4 years months + Evidence seized and destroyed.	Imprisonment: 23 days + Evidence seized and destroyed [30]

IOP Conf. Series: Materials Science and Engineering **193** (2017) 012055 doi:10.1088/1757-899X/193/1/012055

The table 5 describes that: 1) All judges decided to punish the defendant with a penalty that was equal or lower than the demand submitted by the Public Prosecutors; 2) There were two verdicts that had the widest gap between the court judgment and the maximum penalty, they were: a) The court judgment of Nganjuk District Court Number 254/Pid.B/2014/PN.Njk. where the maximum penalty is **2 years imprisonment** or Rp. 4,000,000,000.00 fine; and b) the Judges only imposed **15 days imprisonment** and evidences to be deprived or destroyed. The court judgment of Singkawang District Court Number 36/Pid.SUS/2013/PN.SKW where the maximum penalty according Law Number 7 of 1999: **5 years imprisonment** or Rp. 2,000,000,000.00 fine; and according Law Number 7 of 1996: **3 years imprisonment** or Rp. 360,000,000.00 fine, whereas the Judges only gave **23 days imprisonment** and evidences to be deprived and destroyed.

The low of the Public Prosecutors' prosecution against the defendants who committed a criminal offense in food sector compared to the penalty could be interpreted that there is a tendency that the Law Enforcers still did not pay serious attention to the importance of law enforcement for food business operators. This verdict also indicated that the law enforcers were still not sensitive to the impacts or the potential impacts on consumers' health or even their lives in general. This also resulted in other (food) business operators being undeterred in committing similar violations to gain more profits.

4. Conclusion

The laws that regulate food consumers' protection are Law Number 36 of 2009 about Health, Law Number 8 of 1999 about Consumer Protection, Law Number 18 of 2012 about Food which substitutes Law Number 36 of 2009 about Health that did not regulate the provision of the sanction.

Law on food consumers' protection implemented general maximum punishment system. The system made use of the highest penalty in the criminal offense in food sector. In this system, the Public Prosecutors could freely lay a demand and the Judge could give the court judgment in the matter of the length or the amount of the penalty for the defendant. However the court judgment should not exceed the maximum penalty.

Meanwhile, as seen from the pattern of sanction implementation, Law Number 8 of 1999 and Law Number 18 of 2012 used alternative pattern. With this pattern, the Public Prosecutors could enact the demand and the Judges decided upon the type of the penalty alternatively that is imprisonment or a fine.

In the studied court judgments, the Public Prosecutors' demands were always far below the maximum penalty and there was no judge who gave penalty above the Public Prosecutors' demand. In the pattern of sanction implementation, 4 (four) variations were used, they were: 1) implementing consolidated punishment of imprisonment, fine, and additional punishment; 2) implementing of consolidated punishment of imprisonment and fines; 3) implementing only one type of punishment either imprisonment or fines with additional punishment; 4) implementing imprisonment or fines only.

To minimize the rate of crime in food sector and to increase food consumers' protection, thus it is suggested to: The Government to make changes in the regulation of specific minimum and general maximum penalty and the pattern of the implementation of cumulative and/or alternative sanction.

References

[1]. Nurhayati and Irna 2009 *Jurnal Mimbar Hukum* **21** (2) 203
[2]. Hamzah 2011 *Jurnal Media Hukum* **18** (1) 17-34
[3]. Act Number 18 on 2012 about Food, Preamble's considering Point a and b, and the General Provisions I.
[4]. Act Number 18 on 2012 about Food, Article 1 (5); juncto Article 41 a; juncto Article 59 b; juncto Article 60 (1), and Article 67.

[5]. SEAMEO COOPERATIVE Program (2013, March 26) *Keamanan Pangan, Gizi Buruk serta DampakSosio-Ekonominya*http://gizi.depkes.go.id/wp-content/uploads/2012/05/Food_Safety_Dadi.pdf
[6]. Act Number 18 on 2012 about Food, See Article 69; juncto the General Provisions of the Explanation
[7]. SEAMEO COOPERATIVE Program (2013, March 26) *Keamanan Pangan, Gizi Buruk serta DampakSosio-Ekonominya*http://gizi.depkes.go.id/wp-content/uploads/2012/05/Food_Safety_Dadi.pdf
[8]. Anonimous (2013, August 15) http://www.konsumencerdas.co.cc
[9]. Anak Agung, Diah A (2013, March 11) *Perlindungan Hukum Konsumen Dalam Pelabelan Produk Pangan* http://www.pps.unud.ac.id/thesis/pdf_thesis/unud-376-95861053-final%20tesis.pdf
[10]. Lira A S N, Sidik M, Harianto D (2013, March) *Tinjauan Yuridis Terhadap Perlindungan Konsumen Atas Beredarnya Makanan Kadaluwarsa* http://repository.usu.ac.id/bitstream/123456789/26811/5/Abstract.pdf
[11]. Herlita A (2013, March 12) *Perlindungan Hukum Bagi Konsumen Makanan Kemasan Impor di Yogyakarta* http://simpus.uii.ac.id/search_adv/?n=000148&1=100&b=1&j=SK
[12]. BPOM (2016, April 2016) *Laporan Tahunan Badan POM Tahun 2015* http://www.pom.go.id/new/index.php/browse/laporan_tahunan/19-02-2007/19-02-2017/1
[13]. ICD/ SEAMEO COOPERATIVE Program (26 Maret 2013) SEAMEO-Tropmed Regional Center for Community Nutrition, University of Indonesia: *Keamanan Pangan, Gizi Buruk serta Dampak Sosio-Ekonominya* (Food Safety, Malnutrition and Its Socio-Economic Impact), on http://gizi.depkes.go.id/wp-content/uploads/2012/05/Food_Safcty_Dadi.pdf
[14]. BPOM (10 April 2016) Laporan Tahunan Badan POM Tahun 2015, on http://www.pom.go.id/new/index.php/browse/laporan_tahunan/19-02-2007/19-02-2017/1,.
[15]. Mahmud P M 2005 Penelitian Hukum (Prenada Media Group)
[16]. Act Number 8 on 1999 about Consumer Protection, Article 5.
[17]. Act Number 8 on 1999 about Consumer Protection Articles: 6, 19 (5), 24 (3), and 27.
[18]. Act No. 8 on 1999 about Consumer Protection, Article 60.
[19]. Act No. 18 on 2012 about Food, Article 106.
[20]. Act No. 36 on 2009 about Health, Article 201 (2).
[21]. Act No. 8 on 1999 about Consumer Protection, Articles 61-63.
[22]. Act No. 7 on 1996 about Food, Articles 55-59.
[23]. Act No. 18 on 2012 about Food, Articles 133-148.
[24]. Act No. 8 on 1999 about Consumer Protection, Article 63; juncto Act No. 18 on 2012 about Food, Article 148.
[25]. The court judgment of Depok District Court Number 638/Pid.SUS/2010/PN.Dpk.; The court judgment of Jakarta Selatan District Court Number 222/Pid.SUS/2013/PN.Jkt.Sel.; The court judgment of Singkawang District Court Number 19/Pid.SUS/2013/PN.SKW; The court judgment of Pekalongan District Court Number 174/Pid.B/2013/PN.Pkl.
[26]. The court judgment of Tuban District Court Number 96/Pid.Sus/2015/PN.Tbn.
[27]. The court judgment of Ungaran District Court Number 5/Pid.Sus/2013/PN.Ung.; The court judgment of Tarakan District Court Number 256/Pid.B/2013/PN.Trk.; The court judgment of Surabaya District Court Number 2825/Pid.B /2014/PN.; The court judgment of Payakumbuh District Court Number 97/Pid.Sus/2014/PN.Pyh.
[28]. The court judgment of Nganjuk District Court Number 254/Pid.B/2014/PN.Njk.; The court judgment of Palangkaraya District Court Number 962/Pid.Sus/2014/PN.Plg.; and The court judgment of Palangkaraya District Court Number 963/Pid.Sus/2014/PN.Plg.
[29]. The court judgment of Ketapang District Court Number 159/Pid.B/2010/PN.KTP.; and The court judgment of Ketapang District Court Number 108/Pid.Sus/2012/PN.KTP.

[30]. The court judgment of Pekalongan District Court Number 341/Pid.B/2013/PN.Pkl.; The court judgment of Pekalongan District Court Number 279/Pid.B/2014/PN.Pkl.; The court judgment of Bandung Court Number 126/Pid./2014/PT.BDG.

IOP Conf. Series: Materials Science and Engineering **193** (2017) 012056 doi:10.1088/1757-899X/193/1/012056

Staple Food Self-Sufficiency of Farmers Household Level in The Great Solo

Darsono[1]
Agribusiness Study Program, Faculty of Agriculture, Universitas Sebelas Maret
Jl. Ir. Sutami No. 36 Surakarta, Central Java Indonesia 57126.

E-mail: darsono.uns.solo@gmail.com

Abstract. Analysis of food security level of household is a novelty of measurement standards which usually includes regional and national levels. With household approach is expected to provide the basis of sharp food policy formulation. The purpose of this study are to identify the condition of self-sufficiency in staple foods, and to find the main factors affecting the dynamics of self-sufficiency in staple foods on farm household level in Great Solo. Using primary data from 50 farmers in the sample and secondary data in Great Solo (Surakarta city, Boyolali, Sukoharjo, Karanganyar, Wonogiri, Sragen and Klaten). Compiled panel data were analyzed with linear probability regression models to produce a good model. The results showed that farm households in Great Solo has a surplus of staple food (rice) with an average consumption rate of 96.8 kg/capita/year. This number is lower than the national rate of 136.7 kg/capita/year. The main factors affecting the level of food self-sufficiency in the farmer household level are: rice production, rice consumption, land tenure, and number of family members. Key recommendations from this study are; improvement scale of the land cultivation for rice farming and non-rice diversification consumption.

1. Introduction

Food crisis continues to arise when the macro economic factors tend to shock. On the other hand, Indonesia is the largest on the rice as staple food consumption in Asia with the 169 kg/capita/year, average on Asia is 100kg/capita/year [1]. National program to keep stability staple food with the challenges of food insecurity staple sourced from many aspects, such as population control, reduced productivity of staple food commodities, the use of food for renewable energy, even political issues [2]. Programs to achieve food security principal indicated stuck on the macro policy aggregation that are so often ineffective to achieve sustainable food security principal. There are necessary to find the most upstream source of problems in formulating policy.

Analysis of household food self-sufficiency rate is new challenges in the self-sufficiency on the longstanding term [3]. Achievement of food self-sufficiency is generally an increase in the availability of food to the scope of the region, or nationwide. Its main target is the food commodities from agricultural products such as rice, corn, soybeans, peanuts, green beans, cassava, sweet potato. The strategy for food self-sufficiency is to substitute imports with expected target is to increase food production by farmers. Results of food availability met by domestic products (not imported) [4]. The concept of self-sufficiency has been established, then develop a new conception of the self-sufficiency and food sovereignty. But the last major debate is still going on in the world.

Concentration of staple food insecurity and poverty in Indonesia are mostly located in the rural households who are farmers (60%) of the total national poverty rate (11.3 %) [2,5]. Formula-based food policy tends to macro data that are often not able to uncover specific issues existing household level, where the level is the main problem. These were based on the fact that food insecurity situation occurs where food is available but can not be accessed due to the limited resources of households owned economic (income, employment, economic resources). This is consistent with the opinion of Saliem et. al. [6], and Litle et. al. [7], that food production is not the sole determinant of food security, it is only one factor. To date in Indonesia, among many practical and bureaucrats do not understand the sense of self-sufficiency to food security. As a result of these circumstances the concept of food security is often identified with increased production or supply of sufficient food [8]. Darsono Studies [3] find factors that affect the level of household food security in the dryland farmers Wonogiri are: family size, consumption of rice, the total income of farm households, and the exchange rate farmers. Maintain household food security in the drylands Wonogiri district by promoting policies staple and secondary food crops, increased participation of the workforce is employed in agriculture crops and improve farmers' exchange value.

Deepening analysis food security at farmers household level is important to provide a measure of the dynamics on the local staple food insecurity will contribute staple food security policy formulation. More specifically is, farmers as producers of staple food commodities is also indicated on the position of at risk of food shortages. A phenomenon that should be explored in order to find the main factors that cause it. This study is expected to answer the paradoxical phenomenon. The use of object-scale survey in Great Solo region will describe the dynamics of self-sufficiency in staple food on the rice farm household level because this area is one of the national rice granary. The purpose of this study are: (1) identify the condition of self-sufficiency in staple food farming households in Great Solo, (2) analyze the adequacy of staple food availability, in Great Solo, (3) found the main factors affecting the dynamics of self-sufficiency in staple foods farm household level in Great Solo.

2. Experimental

This study is an extension of the previous regional scale studies used only one district (Wonogiri) into all the termed former residency of Surakarta (Surakarta city; and district of Boyolali, Sukoharjo, Karanganyar, Wonogiri, Sragen, Klaten) termed Great Solo or Surakarta Kingdom. The basic method is a descriptive study, with in-depth analysis [10]. Food commodities analyzed were rice as a main food and micro strategic base on World Food Programme Guidelines [11]. Research sites in Surakarta (for secondary data); Boyolali, Sukoharjo; Karanganyar; Wonogiri (based on previous year data); Sragen, and Klaten.

2.1. Food Security Analysis

To measure the level of household food security levels used rice balance formula: SPSE = (Number of monthly rice production/S) - (consumption of all households for one month/D). Balance between supply (S) and Demand (D = private consumption needs) there are 3 levels, that is surplus/food secure (S greater than D, the value of SPSE> 1), balanced (S = D, SPSE value = 1), and deficit/not food secure (S less than D, the value of SPSE <1).

To determine the level of self-sufficiency in staple food farming households developed semi-log regression equation as follows:

$$SPSE = a+b_1\log(PRODUCTION)+b_2\log(CONSUMPTION)+b_3\log(NTP)+b_4\log(LPST)+ b_5\log(PDKK)+b_6\log(JAK)+b_7\log(PTRT)+ b_8\log(USIA)+ei \quad (1)$$

Description: SPSE: Balance of rice farming households, namely the balance of production and consumption of households is the difference between production and consumption is approached in the last one month (kilograms). PRODUCTION: Total production of rice in the farm for 1 (one) month in the last growing season (kilograms). CONSUMPTION: Total consumption of rice whole family for 1 (one) month (kilograms). NTP: Farmers Exchange value, is the ratio between the index of prices received by farmers (It) with the index of prices paid by farmers (Ib) expressed as a percentage. NTP is conceptually measuring the ability of the exchange of goods of agricultural products produced by farmers with goods or services consumed by households of farmers and their purpose in producing agricultural products. Formula for the calculation of the Farmers Exchange:

IOP Conf. Series: Materials Science and Engineering **193** (2017) 012056 doi:10.1088/1757-899X/193/1/012056

$$NTP = \frac{It}{Ib} \times 100 \qquad (2)$$

Description: NTP = Farmers Exchange, It = index price received by farmers, Ib = Index price paid by farmers.
LPST: total rice cultivation area in rice farming (hectares). PDKK: Education of household head (years). JAK: Number of family members (people). PTRT: total household income (Rupiah). AGE: age of the household head farmer (years). ei: Error (error) arising on observations to i assumed as random variables independently distributed with mean equal to zero (Ordinary least square /OLS method). a: intercept, and b: regression coefficient.

Variable coefficient estimation was conducted using OLS (Ordinary Least Squares) then the error term (e) is minimized. Panel data is used as a combination of spatial data (from analysis LQ) and the moment of data (cross section) of the farm household level primary data.

2.2. Testing the model
To determine how much influence estimator variables on the level of household food security of farmers, as well as test the accuracy of the model used coefficient of determination (R^2). Value (R^2) ranging from 0 to 1, the greater the value (R^2), the greater the influence of variables estimators of the amount bid. To determine whether the variables used jointly affect household food security levels farmers used the F test with a confidence level of used 90%, 95%, and 99%.To determine the effect of each predictor variable on the level of food self-sufficiency farming households t test was used with a level of confidence that is used 90%, 95%, and 99%. To determine which variables are most influential on food self-sufficiency standard used value of the partial regression coefficient t-test analysis. Estimation of variable coefficients using SPPSS software version 19.

2.3. Classical Assumptions Testing
Once the model is obtained then have to test the model includes BLUE (Best Linear Unbiased Estimator) or not. The model is said to BLUE if it fulfills the following requirements; non multicollinearity; avoid multicollinearity can be detected by looking at the value of the matrix Pearson Correlation (PC) with a value of less than 0.8 [12], [13]. Not a case of Heteroskidastity; heteroskedatisitas test conducted by park test and scatterplot charts are spread randomly. Not a case of autocorrelation; performed using statistical test of Durbin Watson d with the following criteria: (1) 1.65 $<$DW $<$2.35, which means no autocorrelation, (2) 1.21 $<$DW $<$1.65 or 2, 35 $<$DW $<$2.79 which means it can not be inferred, and (3) DW $<$1.21 or DW$>$ 2.79 which means there is autocorrelation.

2.4. Data and Data Sources
Primary data include household farmers who represent districts with a base of rice (25 farmers) and non-rice basis (25 farmers), total data are 50 farmers. The selection of samples was done by the farmers proporsive with random sampling based on considerations such as the representation of research surplus and deficit dynamics rice household level, education level, family size, and total household income level.

3. Results and Discussion
3.1. Analysis Farmer Household Staple Food Conditions
In Figure 1, note that the balance of rice farming households in Great Solo surplus situation for an average of 2750.2 kg/capita/year. If the number is a stock that is actually stored by the farmers, in the farm household level in Great Solo reached sufficiency/food staple rice self-sufficiency. Unfortunately sizable surplus was "forced" to be sold to meet non-food needs of families whose number is much higher than the basic food needs. This was compounded by a low on farmers exchange value. The level of per capita consumption of rice farming households Great Solo average of 96.8 kg/capita/year. This number is much lower than the national rate of 136.7 kg/capita/year [14], [15], [16].

3.2. Key factors Farmers Household Staple Food Self-Sufficiency
Test results merit level models Great Solo region indicated by the R^2 value of 0.266 manifest until the confidence level up to 99 % (α = 1 %) is indicated by the F test Because less powerful models can explain the dynamics of food security at the regional level, then the next (in the next section) will be analyzed at the micro household level. Model at the household level, the real value of R^2 is 0.997 confidence level up to 99 % (α = 1 %). It means that all the independent variables were entered

(specified) in the model is able to better explain the dynamics of real influencing on household food security of farmers in Great Solo at 99%. The factors outside the model that describes the dynamics of food security is only about 1 %.

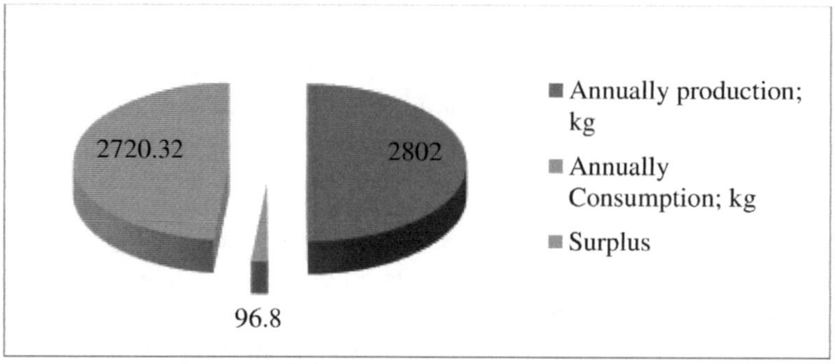

Sources: Primary Data Analysis (2013).

Figure 1. Household Staple Food Balance Sheet Regional Farmers Great Solo

Testing for the classical regression model assumptions used include (1) multicollinearity test based on the value of Pearson Correlation Matrix is known that the correlation between the independent variables are worth no greater than 0.8. It can be concluded that there is no multicollinearity among the independent variables affecting food security in the region and the Great Solo farm household level, (2) autocorrelation test to detect the presence or absence of autocorrelation used Durbin Watson d- statistics and analysis of the results obtained d values of 1.697 and 1.917 level, region level farming households. Because the value of d is obtained located at 1.65 < DW < 2.35 it can be concluded that there is no autocorrelation, (3) heteroscedasticity test is done by using a scatterplot diagram. From the results of analysis show that there are points in the scatter diagram and do not form a particular pattern which means it does not happen heteroscedasticity. It can be interpreted that the errors have the same variance was going homoskedastisitas [12].

3.3. Main Factors Influencing the Dynamics of Farmers Household Food Security Levels in Great Solo

As presented in Table 1; partially significant variables on the dynamics of food security at the household level up to the Great Solo confidence level of 99% or α = 1% is; production of rice, rice consumption, amount of land tenure and family members.

Table 1. Effect of Variables on Household Food Security Farmers in Great Solo

Model	Unstandardized Coefficients		Standardized Coefficients	t	Sig.
	B	Std. Error	Beta		
(Constant)	-.064	.312	-.206	-.206	.838
Production	1.162***	.021	54.361	54.361	.000
Consumption	-.081**	.039	-2.087	-2.087	.043
Total Expenditure	-.054^{NS}	.044	-1.240	-1.240	.222
Farm income	-.003^{NS}	.026	-.131	-.131	.896
Total Land Tenure	.058*	.031	-1.853	-1.853	.071
Education of Household Head	.054^{NS}	.046	1.186	1.186	.243
Number of Family Members	.147***	.050	2.970	2.970	.005
Total Household income	.011^{NS}	.032	.345	.345	.732
Age	-.103^{NS}	.073	-1.413	-1.413	.165

Dependent Variable: Balance of Rice

Source: primary data analysis (2013)

Description : *** = significant to the 99 % confidence level (α = 1 %), ** = significant to the 95% confidence level (α = 5 %); * = significant to the 90 % confidence level (α = 10 %), NS = not significant until the 90 % confidence level (α = 10 %).

Standard regression coefficient indicate rank of variables that affect the household food security of farmers in Great Solo. The greater the value of the regression coefficient, the greater the influence of the independent variables.

Table 2. Variable Regression Standard koefisisen value (Key Factors) which Influential Against Household Food Security Dynamics Farmers Great Solo

Variable	Partial regression coefficient of variable	Rating
Rice production	1.3014	1
Rice consumption	-0.2045	2
Tenure	0.10417	3
Family size	0.48479	4

Source: Primary data analysis (2013)

Analysis of the main factors that affect farmers' household food security based on the order of magnitude in Great Solo coefficients are presented in Table 2. Based on the order of magnitude after the partial regression coefficients weighted by the standard regression coefficients then the greatest effect are the production of rice, rice consumption, land tenure, and number of family members. The four main factors that increase 1 unit will be significantly used effects on household food security status opportunities farmer respectively in Great Solo at 1.3014; -0.2045; 0.10417; 0.48479 units of surplus rice staple food commodities .

Rice production is the main factor of the most influential, so efforts to increase rice production, especially with intensification in Great Solo is still needed. From the description can be summarized that the main factors affecting food security at household level farmers in Great Solo are: (1) the number of family members (participation age to work in the agricultural sector), (needs improvement); (2) the production of rice, (need to increase productivity); (3) the acquisition and exploitation of land, (there needs to be institutional innovation and intensification of land). Of the main factors affecting consumption are : (1) consumption of rice, (need to decrease/diversification). In terms of the welfare of farmers , the main factors that influence are: (1) the production of rice, (need to increase productivity); (2) land tenure, (need to increase the intensity); (3) farmer exchange value, (needs improvement). These findings are in line with and confirms the main factors affecting food security at regional level aggregate in Great Solo, except exchange rates and household income farmers.

4. Conclusions

Conclusions of this study are first, consumption of rice per capita farming households in Great Solo average of 96.8 kg/capita/year. This number is much lower than the national rate of 136.7 kg/capita/year. Third, food balancing of Farmers household was surplus carbohydrate (rice was 676.33 kg and 2.13 kg maize) during the growing season. However, farmers exchange value is very low at 0.35. Value of surplus agricultural food production deficit more to meet non-food needs. This is the main source of farm household poverty. Fourth, the main factor in the level of farmer households in Great Solo will be influence on food self-sufficiency opportunities are: rice production, rice consumption, land tenure, and number of family members.

Recommendations from this research are first, farm household consumption needs improvement in terms of diversification of food sources so not depending on rice. Second, strengthening food security at the regional level with a reduced number of Great Solo family members (in the long run) , decrease in the level of per capita consumption of rice, accompanied by intensification of rice production through improved in institutional of expanse scale farming, fixes and improve the farmers exchange value to increase the farmer household welfare. Third, strengthening food security at the farmers household level in Great Solo with increased productivity of rice to meet the feasible scale, lowering the level of consumption per capita rice consumption, land tenure, and in the long term reduce the number of family members. Fourth, realizing food self-sufficiency at both the regional and farm households still need in Great Solo effectiveness of policy instruments menyakut rice production and consumption management.

IOP Conf. Series: Materials Science and Engineering **193** (2017) 012056 doi:10.1088/1757-899X/193/1/012056

References

[1] BPS 2010 *Food Consumption Statistic* (Jakarta : Statistics Indonesia)
[2] Darsono 2012 *Dimensions of Agricultural Development in the Global Challenge* (Surakarta: UNS Press)
[3] Darsono 2012 *SEPA Journal XV* **1** 1-21
[4] Hanani N A R 2011 *Food Security: Subsystem Availability* (Jakarta: National Food Security Council)
[5] Feng S 2008 *Journal Njas* **55** 363-378
[6] Saliem H P, M Ariani, Marisa Y, Purwantini T B, and Lokollo E M 2001 *Analysis of Household food security and regional level. Research Report. Research centers and the Social and Economic Development of Agriculture* (Bogor: Department of Agricultural Research and Development Agency)
[7] Little P D B, Cellarius C, Barrett and Coppock D L 1999 *Economic diversification and risk management among east African herders: A preliminary assessment and literature review* (Risk Management Project Technical Report)
[8] Darsono 2009 *Role of Government in Promoting Agriculture and Agro-Industry performance (Critical Analysis of the New Order and the Order of the Reformation)* (Surakarta: UNS Press in collaboration with the Institute of Education Development (LPP UNS) and the Research Center for Rural and Regional Development SBRC UNS)
[9] Darajati W 2011 Building a National Food Sovereignty (Yogyakarta: Gadjah Mada University)
[10] Singarimbun M and Effendi S 1997 *Research Methods Survey* (Jakarta: LP3ES)
[11] World Food Programme 2009 *Comprehensive Food Security and Vulnerability Analysis Guidelines edition 1*
[12] Gujarati D 2004 *Basic Econometric* Fourth Edition (New York: The Mc Graw−Hill Companies)
[13] Darsono 2004 *Cashew Nut Economy* (Surakarta: Caraka)
[14] Food Security Council 2012 *National Food Plan 2006* (Jakarta: Security Public Policy)
[15] Food Security Council 2012 *National Food Plan* (Jakarta: DKP RI)
[16] Council food resilience and the World Food Programme 2005 *Food Insecurity Map Indonesia (FIA)* (Jakarta: Bappenas)

IOP Conf. Series: Materials Science and Engineering **193** (2017) 012057 doi:10.1088/1757-899X/193/1/012057

The effect of cinnamon bark (*Cinnamomum burmanii*) essential oil microcapsules on vacuumed ground beef quality

I N Brilliana[1*], G J Manuhara[1], R Utami[1], L U Khasanah[1]
[1] Food Science and Technology, Universitas Sebelas Maret, Jl. Ir Sutami 36A, Surakarta, Central Java, Indonesia.

*E-mail: istiqomahniab@gmail.com

Abstract. Ground beef has a short shelf life because it is susceptible to damage due to microbial contamination and lipid oxidation. So some sort of preservation method such as refrigerated storage, vacuum packaging or natural preservative addition is needed to extend the shelf life of ground beef. A natural preservative that can be used as a food preservative is the cinnamon bark (*Cinnamomum burmanii)* essential oil microcapsules. The aim of the research was to determine the influence of a cinnamon bark essential oil microcapsules (0%;0.5% and 1% w/w of the ground beef) on the Total Plate Count (TPC), Thiobarbituric Acid (TBA), pH and color of ground beef during refrigerated storage (4±1°C). The result showed that cinnamon bark essential oil microcapsules affected the TPC, TBA, pH and color of ground beef. The addition of the cinnamon bark essential oil microcapsules on ground beef can inhibit microbial growth, inhibit lipid oxidation, inhibit discoloration and lowering pH of fresh ground beef during refrigerated storage compared to the control sample. The higher of the microcapsules were added, the higher the inhibition of microbial growth, lipid oxidation and discoloration of ground beef, indicating better preservation effects.

1. Introduction

Beef is a livestock product that is widely popular among consumers. The production of beef in Indonesia increase by 505,477 tons in 2012 with 6.76% rise each year [1]. Beef is a nutritious food. The nutritional values of beef were 18.8% protein, 14% lipid, 11 mg/g Ca, 170 mg/g phosphate, 2.8 mg/g Fe, 30 SI vitamin A and 0.08 mg/g vitamin B [2]. In addition to the form of chunks, beef is also sold in the form of ground beef. Fat and other tissue in beef were trimmed and size reduced by using a meat grinder [3,4]. Because of the size reduction, the surface of the ground beef will be wider and more susceptible to damage [5]. The deterioration of ground beef occurs due to microbial contamination and lipid oxidation. So, the shelf life of beef was 24 hours at room temperature storage and 1-4 days in refrigerated storage [6]. To extend ground beef shelf life, use of a preservation combination is a must. Preservation methods that can be used for ground beef are refrigerated storage, vacuum packaging [7] and addition of a natural preservative such as essential oil [8].

Essential oils can be extracted from Indonesian cinnamon bark (*C. burmanii*). The main component of cinnamon bark essential oil is cinnamaldehyde [9-11]. Cinnamaldehyde is an antimicrobial substance that acts against aerobic microbial (*Staphylococcus aureus)* and anaerobic facultative microbial such as *Escherichia coli, Bacillus cereus, Listeria monocytogenes,* and *Pseudomonas aeruginosa.* Those microbial are an example of spoilage flora in ground beef [6, 12, 13]. Beside the antimicrobial component, cinnamon bark essential oil also contains linalool [11] and cinnamic acid [14] as an antioxidant to prevent lipid oxidation. But essential oil has several weaknesses. There is evaporation of volatile components, it is oxidized easily, it is poorly dispersible in dried material, and is susceptible due to high temperatures, UV light and moisture [15,16]. So, essential oil needs further treatment by microencapsulation to protect the liquid core taking a solid form [17].

The cinnamon bark essential oil microcapsules with a coating material made of maltodextrin and gum arabic (2:3) had the best physical characteristics for cinnamon oil microcapsules [18]. The active components will be protected in microcapsules form and will be released rapidly in certain application because of the water solubility of the microcapsules [19]. Cinnamon bark essential oil microcapsules can be used for ground beef preservation. The aim of the research was to determine the effect of the cinnamon bark essential oil microcapsules (0%;0.5% and 1% w/w of the ground beef) on the Total Plate Count (TPC), Thio Barbituric Acid (TBA), pH and color of ground beef during refrigerated storage (4±1°C).

2. Experimental

2.1. Cinnamon Bark Essential Oil Preparation

Cinnamon bark (*C. burmanii*) from Bubakan, Girimarto, Wonogiri (Indonesia) was sun dried for four days until the water content was 10-12%. The dried cinnamon bark was milled and sieved detained a 10 mesh. The cinnamon powder was subjected to steam distillation for four hours [11].

2.2. Cinnamon Bark Essential Oil Microcapsules Preparation

The cinnamon bark essential oil microcapsules were made from maltodextrin and gum arabic coating materials in a 2:3 ratio. The ratio of essential oil and coating materials was 1:25. Then, the formula was dissolved in distilled water with a ratio of coating material and distilled water of 1:20 [18]. All of the materials were homogenized with an "Ultra Turax® Basic® Woke" homogenizer for 15 minutes at 8000 rpm. The next process was microencapsulation by a spray dryer "SD Basic Lab Plant" with an inlet temperature of 109°C, a feed flow rate 15 - 20 ml/minute [20] and exhaust temperature of 57°C.

2.3. Application of the cinnamon bark essential oil microcapsules to ground beef

Tenderloin beef from a local slaughter house Sapto Raharjo, Surakarta, Indonesia was obtained three hours after slaughter and kept in a cool box to prevent contamination. The fat and other tissue of the beef was cut off to make it easier to grind. Beef grinding and application of microcapsules was done 4 hours after slaughtered. 70 g of beef was grinding by "Miyako BL 151" for each sample. The microcapsules (0%, 0.5 % and 1% w/w of the ground beef) were mixed with ground beef in 250 ml beaker glass and stirred with a stirrer stick. The ground beef was packed with vacuum packaging "Maksipack Automatic Vacuum Packager DZ-260/PD", using vacuum plastic packaged "Salaman" nylon material 75 micrometers thick. The vacuumed ground beef was preserved at 4±1°C in a "Samsung" refrigerator. TPC [21], TBA [22], pH [22] and the color (using Lutron Electronic RGB-1002) of the vacuum ground beef was determined on days 0, 4, 8, 12 and 16 day.

3. Results and discussion

3.1. Total Plate Count (TPC)

Table 1. Effect of Cinnamon Bark Essential Oil Microcapsules in Total Plate Count of Vacuumed Ground Beef during refrigerated storage (4±1°C).

Microcapsules	TPC (log CFU/g)				
	Day-0	Day-4	Day-8	Day-12	Day-16
0%	5.04^{aA}±0.000	5.73^{aB}±0.014	5.84^{aC}±0.044	5.92^{bD}±0.019	6.56^{cE}±0.003
0.5 %	4.99^{aA}±0.084	5.37^{aB}±0.278	5.64^{aB}±0.139	5.69^{aBC}±0.046	6.07^{bC}±0.008
1 %	4.83^{aA}±0.040	5.14^{aB}±0.055	5.54^{aC}±0.046	5.60^{aC}±0.052	5.74^{aD}±0.075

The figures follow the same large superscript letters in the same row and small superscript letters in the same column showed no significant difference at the 5% significance level.

During refrigerated storage for 16 days, the TPC of all samples increased significantly. The initial TPC of the vacuumed ground beef was around 4.83 – 5.04 log CFU/g. This is similar to the result which related to minced beef with the addition of acetic acid, CO_2 and mustard over 15 days of storage, when the initial TPC was 5.72 – 5.82 log CFU/g [23]. The TPC of the vacuumed ground beef with the addition of cinnamon bark essential oil microcapsules was significantly lower than that of the control sample. At the end of the storage (day 16), the TPC of the vacuumed ground beef with addition of 1% cinnamon bark essential oil microcapsules was still below the maximum acceptable limit 6 log CFU/g. However, the TPC of control and sample with the addition of 0.5% cinnamon bark essential oil microcapsules exceeded the standard level. This indicates that the addition of cinnamon bark essential oil microcapsules could inhibit microbial growth. A similar result found that edible coating enriched with 2% cinnamon oil could inhibit bacterial growth on beef during refrigerated storage. The TPC of treated sample was 6.03 log CFU/g while, the control sample was 7.13 log CFU/g at day 20 [24]. Previous research also reported that TPC of chilled pork samples with the addition of chitosan nanoparticles loaded with cinnamon essential oil was 6.60 – 7.61 log CFU/g at day 15, but the control sample reached 6.19 log CFU/g just after six days [25].

The TPC of the vacuumed ground beef with the addition of cinnamon bark essential oil microcapsules was significantly lower than that of the control sample because cinnamon bark essential oil microcapsules has an active component that is cinnamaldehyde [9]. Cinnamaldehyde is an antibacterial [13] and antimicrobial component [26]. The active component in the microcapsules will be released when the microcapsules dissolves in the water [19] of ground beef, which has 66% water content [2].

The mechanism of microbial inhibition by cinnamon bark essential oil's active component is because of the active component of essential oil affects the lipid bilayer cell membrane and mitochondria, disturbs the structure and leakages of the cell [27]. Bioactive compounds will react with the proteins on the wall of microbial cell membranes or in the cytoplasm and cause denaturation. The damage to the cell wall and cell membrane is caused by the weakening of the wall structure and cell membranes that become abnormal and whose cell pores are enlarged. Thus the cell wall and cell membrane cannot selectively regulate the exchange of substances from and into the cell and ultimately cell lysis occurs [28]. Cinnamaldehyde inhibits ATPase activity. Cinnamaldehyde can increase membrane permeability, leakage of cytoplasm, and interact with enzymes located on the cell wall. Interaction of cinnamaldehyde with the cell membrane may cause leakage of small ions or it can inhibit the enzymes necessary for amino acid biosynthesis [29].

3.2. Thiobarbituric Acid (TBA)

Table 2. Effect of Cinnamon Bark Essential Oil Microcapsules on Thiobarbituric Acid (TBA) of Vacuumed Ground Beef during refrigerated storage (4±1°C).

Microcapsules	TBA (mg malonaldehide/kg)				
	Day-0	Day-4	Day-8	Day-12	Day-16
0%	$0.044^{bA}\pm0.002$	$0.061^{aB}\pm0.003$	$0.067^{bB}\pm0.001$	$0.094^{bC}\pm0.004$	$0.116^{aD}\pm0.005$
0.5 %	$0.046^{bA}\pm0.001$	$0.055^{aB}\pm0.005$	$0.060^{aB}\pm0.001$	$0.078^{aC}\pm0.004$	$0.111^{aD}\pm0.001$
1 %	$0.027^{aA}\pm0.002$	$0.054^{aB}\pm0.001$	$0.063^{aC}\pm0.002$	$0.083^{aD}\pm0.001$	$0.102^{aE}\pm0.005$

The figures follow the same large superscript letters in the same row and small superscript letters in the same column showed no significant difference at the 5% significance level.

The addition of cinnamon bark essential microcapsules causes inhibition of lipid oxidation on vacuumed ground beef. This is shown by the result for the TBA value of the treatment sample with the addition of 0.5% and 1% microcapsules, which was lower than the control sample at the end of the storage (day 16). The maximum acceptable limit for the TBA value was 2 malonaldehide/kg [30]. In this research, all samples had lower TBA value than the maximum acceptable limit by the end of the storage period. Previous research found that edible coating enriched with 2% cinnamon oil could inhibit lipid oxidation and the TBA value was lower than control during 20 days of storage [24]. Another research also reported that TBA of chilled pork samples with the addition of chitosan nanoparticles loaded with cinnamon essential oil reached 2.09 malonaldehide/kg on day 15 while the control exceeded the maximum acceptable limit on day six [25]. The TBA value of the treatment sample with the addition of 0.5% and 1% microcapsules which lower than the control happened because cinnamon bark essential oil contains cinnamaldehyde [14] and linalool [14]. Besides that, cinnamon bark extract also contains eugenol and cinnamic acid. Cinnamaldehyde, linalool, eugenol and cinnamic acid perform an antioxidant activity [31]. Antioxidants are free radical scavengers and hydrogen atom donors so can inhibit lipid oxidation [32].

3.3. pH

Table 3 Effect of Cinnamon Bark Essential Oil Microcapsules on pH of Vacuumed Ground Beef during refrigerated storage (4±1°C).

Microcapsules	pH				
	Day-0	Day-4	Day-8	Day-12	Day-16
0%	$5.8^{aC} \pm 0.1$	$5.7^{aBC} \pm 0.1$	$5.5^{aAB} \pm 0.0$	$5.5^{bAB} \pm 0.0$	$5.5^{aA} \pm 0.1$
0.5 %	$5.9^{aC} \pm 0.1$	$5.8^{aBC} \pm 0.1$	$5.8^{cBC} \pm 0.0$	$5.6^{bB} \pm 0.0$	$5.4^{aA} \pm 0.1$
1 %	$6.1^{aD} \pm 0.1$	$5.7^{aC} \pm 0.0$	$5.6^{bC} \pm 0.0$	$5.4^{aB} \pm 0.1$	$5.2^{aA} \pm 0.1$

The figures follow the same large superscript letters in the same row and small superscript letters in the same column showed no significant difference at the 5% significance level.

During refrigerated storage for 16 days, the pH of all samples decreased significantly. Previous research reported that the pH of ground beef packaged with vacuum packaging decreased from 6.04 at day 0 to 5.45 at day 14 [7]. The growth of lactic acid bacteria that produce lactic acid decreased pH of packaged meats. The pH levels of the treatment samples with the addition of cinnamon bark essential oil microcapsules were lower than that of the control sample. This happened because glucose, lactic acid and some of the amino acid in the beef will be degraded by all of microbial contaminant into alkaline metabolite. This alkaline metabolite could make the higher of pH [33]. The pH decreasing of the control sample was slower than that of the treatment samples because the alkaline metabolite in the sample control was higher than in the treatment samples so it could inhibit the decrease of pH. But in the treatment samples, the alkaline metabolite content was a little lower because of the cinnamaldehyde compound can inhibit microbial growth which can produce alkaline metabolites.

3.4. Color

Table 4 Effect of Cinnamon Bark Essential Oil Microcapsules in Color of Vacuumed Ground Beef during refrigerated storage (4±1°C).

	Microcapsules	Color				
		Day-0	Day-4	Day-8	Day-12	Day-16
Red	0%	$207.5^{aA} \pm 17.7$	$237.5^{cB} \pm 2.1$	$273.5^{cC} \pm 0.7$	$318^{bD} \pm 1.4$	$332^{bD} \pm 9{,}9$
	0.5 %	$182.3^{aA} \pm 17.3$	$222.5^{bB} \pm 0.7$	$231^{bB} \pm 5.7$	$226^{aB} \pm 5.7$	$225.5^{aB} \pm 7{,}8$
	1 %	$208.5^{aB} \pm 12.0$	$185.5^{aA} \pm 7.8$	$197.5^{aAB} \pm 9.2$	$213^{aB} \pm 7.1$	$248.5^{aC} \pm 3{,}5$
Green	0%	$100.5^{aA} \pm 7.8$	$121.5^{aB} \pm 0.7$	$146.5^{bC} \pm 0.7$	$164.5^{bD} \pm 0.7$	$166^{cD} \pm 2{,}8$
	0.5 %	$96.8^{aA} \pm 3.9$	$101.5^{aA} \pm 2.1$	$117{,}8^{aB} \pm 3.8$	$113.5^{aB} \pm 2.1$	$112.5^{aB} \pm 3{,}5$
	1 %	$103^{aA} \pm 5.7$	$107^{aA} \pm 8.5$	$113^{aAB} \pm 4.2$	$122^{aB} \pm 4.2$	$126^{bB} \pm 2{,}8$
Blue	0%	$84.5^{aA} \pm 7.8$	$97.5^{aB} \pm 0.7$	$124.5^{bC} \pm 3.5$	$134.5^{bC} \pm 2.1$	$134.5^{cC} \pm 0{,}7$
	0.5 %	$93.8^{aA} \pm 1.8$	$90^{aA} \pm 1.4$	$95{,}8^{aA} \pm 3.2$	$95^{aA} \pm 1.4$	$93{,}5^{aA} \pm 3{,}5$
	1 %	$87.5^{aA} \pm 4.9$	$90.5^{aAB} \pm 7.8$	$101^{aBC} \pm 2.8$	$101^{aBC} \pm 4.2$	$109.5^{bC} \pm 2{,}1$

The figures follow the same large superscript letters in the same row and small superscript letters in the same column showed no significant difference at the 5% significance level.

The redness of all vacuumed ground beef samples increased significantly over 16 days of refrigerated storage. The change of meat color happened because of the change of fresh ground beef pigment. Myoglobin is the main pigment associated with meat color. The oxygenated form of myoglobin is oximyoglobin that responsible for the bright-red color, while the oxidized form of myoglobin is metmyoglobin that responsible for browning [34]. The change of redness of the samples treated with cinnamon bark essential oil microcapsules was lower than the control sample because cinnamon bark essential oil has antioxidant compounds such as cinnamaldehyde, eugenol and linalool [11,14,31]. These antioxidants inhibit the discoloration of ground beef. Previous research stated that oregano extract that contains antioxidant can inhibit the discoloration of beef patties packaged in modified atmosphere storage [35].

The greenness of all vacuumed ground beef samples increased significantly over the 16 days of refrigerated storage. The change of greenness is because of oxidation of myoglobin to metmyoglobin, which then combines with H_2S which is a product of bacteria to form sulphmyoglobin (green pigment) and the growth of *Penicillium* [36]. *Pseudomonas fluorescens* produced slime and a green pigment on meat [6]. The samples treated with cinnamon bark essential oil microcapsules had lower greenness than the control sample. Antimicrobial compound such as cinnamaldehyde in the cinnamon bark essential oil can inhibit the growth of *Penicillium* [37] and *Pseudomonas fluorescens* [14] which form green pigment.

During 16 days of refrigerated storage, the blueness of the vacuumed ground beef control sample and the samples with a 1% addition of cinnamon bark essential oil microcapsules increased significantly. The blueness of the vacuumed ground beef samples with the addition of cinnamon bark essential oil microcapsules were lower than that of the control sample. The blueness is formed by *Penicillium* which produced a blue-green pigment [36]. Cinnamaldehyde as an antimicrobial compound in cinnamon bark essential oil can inhibit the growth of *Penicillium* [37].

4. Conclusion

The addition of the cinnamon bark essential oil microcapsules on ground beef can inhibit microbial growth, inhibit lipid oxidation, inhibit discoloration and lowering pH of fresh ground beef during refrigerated storage compared to the control sample. These results because of the active components of the cinnamon bark essential oil microcapsules that are cinnamaldehyde, linalool, eugenol. The higher of the microcapsules were added, the higher the inhibition of microbial growth, lipid oxidation and discoloration of ground beef, indicating better preservation effects.

Acknowledgements

This work was financially supported by a research project of PNBP UNS No. 632/UN27.21/LT/2016 from Universitas Sebelas Maret, Indonesia.

References

[1] Indonesian Food and Agriculture Directorate 2013 *Rencana Pembangunan Jangka Menengah Nasional (RPJMN) Bidang Pangan dan Pertanian 2015-2019* (Jakarta: Kementerian Perencanaan Pembangunan Nasional/Badan Perencanaan Pembangunan Nasional)

[2] Muchtadi T R, Sugiyono dan Fitriyono A 2010 *Ilmu Pengetahuan Bahan Pangan* (Bogor: Alfabeta)

[3] Butler F, Geraldine D, Dan E, Anna M L and R Bruce T 2006 *Case Study: Escherichia coli O157:H7 in Fresh Raw Ground Beef the use of microbial risk assessment outputs to develop practical risk management strategies: metrics to improve food safety* (Report) (Germany: Kiel)

[4] Heinz G, and P Hautzinger 2007 *Meat Processing Technology* (Bangkok: FAO Regional Office for Asia and the Pacific)

[5] Raharjo S 2010 *Aplikasi Madu Sebagai Pengawet Daging Sapi Giling Segar Selama Proses Penyimpanan* (Skripsi) (Surakarta: Universitas Sebelas Maret)

[6] Buckle K A 1985 *Ilmu Pangan* (Jakarta: UI Press)

[7] Irkin R, Esmer O K, Degirmencioglu N and Degirmencioglu A 2011 *Bulg. J. Agric. Sci.* **17** (5) 655-63

[8] Amany M S, Reham A, Amin, Gehan S A, dan Afifi 2010 *J. Am. Sci.* **6** (12) 691 – 700

[9] Khasanah L U, R Baskara K A, Rohula U, Dimas R A M and Mukhamad L 2013 *Procedia of Agribusiness National Seminar* (Surakarta Universitas Sebelas Maret)
[10] Prasetya N B A and Ngadiwiyana 2006 *JSKA* **9** (1) 1-4
[11] Yuliarto F T, Lia U K dan R Baskara K A 2012 *Jurnal Teknosains Pangan* **1** (1) 12– 23
[12] Amalaradjou M A R, Sangeetha A B, Ranjith R, Anup K J, Anu S C, Satyender R V, Tyler M, David S, Vijay K J, Ricard M and Kumar V 2010 *Food Microbiol.* **27** 841–4
[13] Gupta C, Amar PG, Ramesh C U and Archana K 2008 *Afr. J. Microbiol. Res.* **2** 258-61
[14] Prasetyaningrum, Rohula U dan R Baskara K A 2012 *Jurnal Teknosains Pangan* **1** (1) 24-31
[15] Koswara S 2007 *Teknologi Enkapsulasi Flavor Rempah-Rempah* ebookpangan.com
[16] Petrovic G M, Gordana SS and Niko S R 2010 *J. Med. Plant Res.* **4** (14) 1382-90
[17] Supriyadi and A Sakha R 2013 *Jurnal Teknologi dan Industri Pangan* **24** (2) 201–8
[18] Hermanto R F, L U Khasanah, Kawiji, W Atmaka, G J Manuhara and R Utami 2015 *IOP Conf. Series: Mater. Sci. Eng* **107**
[19] Khasanah L U, Baskara K A, Titiek R, Rohula U dan Godras J M 2015 *AGRITECH* **35** (4) 414–21
[20] Murti L W 2012 *Karakteristik Fisik dan Kimia Mikroenkapsulan Oleoresin Kayu Manis (Cinnamomum burmanii) dengan Variasi Ratio Gum Arab dan Maltodekstrin Sebagai Bahan Penyalut* (Skripsi) (Surakarta: Universitas Sebelas)
[21] Fardiaz S 1993 *Analisis Mikrobiologi Pangan* (Jakarta: PT. Raja Grafido Persada Jakarta)
[22] Apriyantono A, D Fardiaz, N L Puspitasari, Sedarwati dan S Budiyantono 1989 *Analisis Pangan: Petunjuk Laboratorium* (Bogor: PAU Pangan dan Gizi IPB)
[23] Elabbasy M T, Kamal I E and Alaa E M 2014 *Life Science Journal* **11** (9) 185 – 190
[24] Ayuningrum Z K 2015 *Pengaruh Minyak Atsiri Kulit Batang Kayu Manis (Cinnamomum burmanii) terhadap Karakteristik Edible Film dan Kualitas Daging Sapi Segar* (Skripsi) (Surakarta: Universitas Sebelas Maret)
[25] Hu J, X Wang, Z Xi and W Bi 2015 *Food Sci. Tech.* **63** 519–26
[26] Jayasena D D and Cheorun J 2013 *Trends in Food Sci. Tech.* **34** 96–108
[27] Burt S 2004 *Int. J. Food Microbiol.* **94** 223–53
[28] Naufalin R dan S R Herastuti 2013 *Animal Prod.* **15** (1) 8–14
[29] Negi, P S 2012 *International Journal of Food Microbiology* **156** 7–17
[30] Campo M M, G R Nute, S I Hughes, M Enser, J D Wood, R I Richardson 2005 *J. Meat Sci.* **72** 303–11
[31] Andriyanto A, M A M Andriani dan Esti W 2013 *Jurnal Teknosains Pangan* **2** (2) 13–20
[32] Aritonang S N, Yuherman dan Tiara N 2013 *Jurnal Ilmu Ternak* **13** (2) 22–26
[33] Dias F S, C L Ramos and R F Schwan 2013 *Food Sci. Tech. Camp.* **33** (3) 468–74
[34] Danijela S Z, Lazic, L V, Levic B L, Pezo L L, Tomovic M V and Hromis M N 2013 *Food and Feed Res.* **40** (1) 1-10
[35] Escalante A S, D Djenane, G Torrescano, J A Beltran and P Roncales 2003 *J. of Food Sci.* **68** (1) 339–45
[36] Lawrie R A 2006 *Meat Science.* (Cambridge: Woodhead Publishing Limited Abington)
[37] Tanackov S D K, and G R Dimic 2013 *Formatex* **2013** 838–49

IOP Conf. Series: Materials Science and Engineering **193** (2017) 012058 doi:10.1088/1757-899X/193/1/012058

Natrium dischargement from peripheral blood as a predominant factor influenced by the administration of banana (*Musa paradisiaca*) on elderly female hypertensive patient

A Pramono[1], N Noriko[1], S B Komara[1]
[1]Department of Biology, University of Al Azhar Indonesia, Sisingamangaraja Street, Jakarta, 12110, Indonesia

E-mail: apramesyanti@gmail.com

Abstract. Hypertension is more common in eldery female that triggered by diet and lifestyle changes. Bananas were not only useful for the food, but also for hypertension therapy and preserving life. Administration of bananas decreased blood pressure in hypertensive patients. This study aims to identify of factors that influenced by the administration of banana (*Musa paradisiaca*) on elderly female hypertensive patient. Twenty of eldery female patient were divided into 2 respondents group: control (11 patients) and treatment (9 patients). The treatment groups received banana twice a day during 2 weeks, but the control group didn't. Here, we showed the administration of banana significantly decreased blood pressure on elderly female hypertensive patient ($p = 0.00$) in both systole and diastole. There was a significant decrease in sodium levels ($p = 0.037$) in the blood, but potassium levels remained the same. Erythrocyte sedimentation level ($p = 0.136$) and trombocyte count ($p = 0.176$) in treatment group, were not affected by banana administration. Taken together, banana administration on elderly female hypertensive patient decreased the blood pressure significantly, greatly affected by the natrium dischargement from the blood. Thus, our findings contribute to preliminary comprehension of banana effect on hypertension reduction.

1. Introduction

Hypertension or high blood pressure is a health issue that predominantly occured in some developed countries. Cardiovascular diseases caused 17 million deaths annually, nearly a third of the world population. About 9.4 million deaths annually by cardiovascular disease, caused by hypertension [1]. Hypertension caused due to lifestyle such as diet and lifestyle changes. Changes of lifestyle is not easy to be done, therefore need a comprehensive approaches to achieve the expected results. [2]. Age factor is also one of the hypertension influence risk.

Hypertension is a medical condition which the blood pressure in the artery rises above normal limits [3]. Based on data in 2013 from World Health Organization (WHO), in 2025 there are one billion people that suffering hypertension. Two-thirds of that number will comes from developing countries, including Indonesia. Based on data from WHO World Health Statistics, hypertension accounted for 51% deaths from stroke and 45% deaths from coronary heart disease [1].

The hypertension prevalence in Indonesian was 31.7%, or 1 in 3 persons had hypertension. The prevalence of hypertension decreased from 31.7% (2007) to 25.8% (2013). However, the prevention and treatment management of hypertension were still needed. As many as 75% of hypertensive patients did not know realized that they had hypertension. They realized if there had been complications. In Indonesia, the threat of hypertension should not be ignored [4]. Government programs related to early disease detection and hypertension complications were desperately needed. Depok City Health Department issued data that showed the number of new cases of hypertension in 2013 as many as 19.275 cases (53,9%). Therefore, hypertension are the highest disease cases in the Depok area. From the 10 diseases of patients hospitalized in Depok City in 2013, hypertension occupies the 2nd position (13,63%) [5]. Most hypertensive patients who comes to the clinic in the Depok City in 2010 are women [6]. Those data showed that hypertension in women has been a serious problem which must be addressed, especially in Depok City.

A wide variety of plants has diuretic and potassium content. Diuretics has an antihypertensive effect by increasing the release of water and natrium. Potassium maintained stable in body through the electrolyte sodium potassium pump, reduced the amount of water and salt in the body, and loosen up blood vessels. The conditions made blood pressure became normal. Potassium intake from fruit were added in the diet of hypertensive patient can decreased of systolic and diastolic blood pressure [7].

A study diet therapy on hypertensive patients blood pressure, proved that bananas decreased high blood pressure due to the high content of potassium. The work mechanism of banana to decrease of hypertension is similar with antihypertensive drug [8,9]. In this study, we collected preliminary information of banana effect on hypertension patient, as well as to identify the factors in the blood and urine, that influenced by banana administration.

2. Methods

2.1 Time and Place Research

This research was conducted in Posbindu Gandul Village, located Cinere Subdistrict, Depok City, Indonesia from February till May 2016. The blood and urine examination were performed in biology laboratory Department of Biology University of Al Azhar Indonesia.

2.2 Study Design

This study involved 20 hypertensive patients (9 control, 11 treatment) with the inclusion criteria were women aged 45-70 years which has a history of blood pressure $\geq$140 / 90 mmHg (seen from the records of the Posbindu per month), has no history of other diseases that caused hypertension (diabetes, coronary heart disease, kidney disorders, etc.). Beside that, the respondents were not undergoing drug therapy of hypertension and willing to cooperate in this study. All respondents had knowingly signed informed consent. All examination procedures of blood samples were conducted in Posbindu Gandul Village, located Cinere Subdistrict, Depok City.

The study performed in four stages; banana administration, daily nutritional intake documentation, daily blood pressure examination, blood and urine samples examination. The administration of banana gave to treatment group only at a dose twice a day (morning and afternoon) during 2 weeks. Blood pressure of both respondents (control and treatment group) checked routinely during 2 weeks using digital sphygmomanometer. The blood pressure was checked using digital sphygmomanometer. Daily nutritional intake data was recorded as additional reference data. Blood and urine tests conducted to all respondents from day 0 (before study) and day 14 (after study).

Blood examination parameters were erythrocyte sedimentation rate (ESR), number of thrombocyte, and electrolyte concentration (potassium, natrium and chlorin). Urine examination parameter were leucocyte, nitrit, urobilinogen, protein, pH, blood content, specific gravity (SG), keton, bilirubin, and glucose level. The urine data transformed into score data from 0 (negative) to 6 (highest expression), respectively. The score data from urine was averaged to facilitate analysis.

This study involving two groups: the control group and the treatment group. The patients were taken in accordance with the inclusion criteria. Data were analyzed using bivariate analysis independent T test. As a normality prerequisite test on systolic and diastolic, we transformed the data into a logarithm due to abnormaly data distribution, then continued with Mann Whitney U test.

3. Result and Discussion

3.1 Result

During two weeks of the study, blood pressure values of all respondents were fluctuated (figure 1). The trendline data showed decreased in systolic and diastolic treatment group, while the systolic and diastolic trendline of control remain increased in value. Both, systolic and diastolic value daily were fluctuated. The average decline of blood pressure was 6.8/6.3 mmHg per day. Banana administration during 2 weeks at a dose of twice a day had able reduced the systolic ($p = 0.00$) and diastolic ($p = 0.00$) of blood pressure significantly. This was confirmed by Mariani (2007) which proved that the fruits are high in potassium (including bananas) maintained homeostasis diastolic blood pressure [5]. In other hand, erythrocyte sedimentation rate and thrombocyte count were not influenced by banana administration (figure 2).

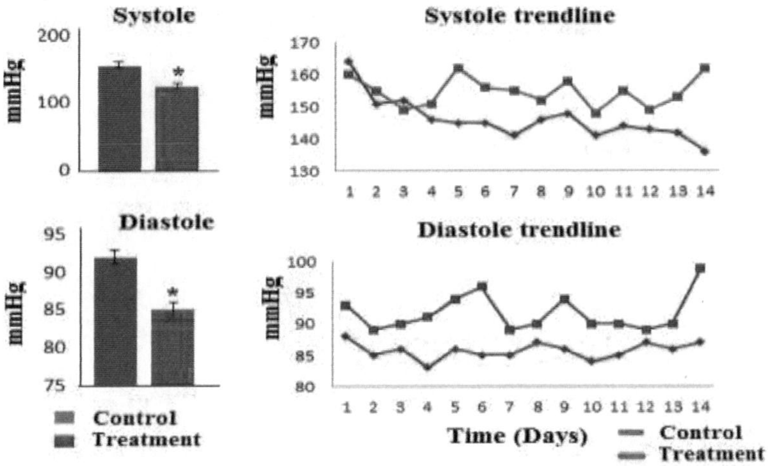

Figure 1. Blood pressure average and trendline from systole and diastole.

Banana administration reduced systolic and diastolic level daily. (A-B) Systole data showed significant reduction of blood pressure ($p = 0.00$). The systole trendline showed the treatment group was decreased daily, compared to control group. (C-D) Diastole data showed significant reduction of blood pressure ($p = 0.00$). The diastole trendline showed the treatment group was decreased daily, compared to control group. Data A and C represent means ± SD and are derived from two times independent experiments (*, $P < 0.05$).

Banana administration did not affected ESR and thrombocyte count significantly. (A) ESR data showed reduction, but the alteration was not significant ($p = 0.136$). (C) Thrombocyte count data showed reduction, but the alteration was not sinificant ($p = 0.176$). Data represent means ± SD and are derived from two times independent experiments (*, $P < 0.05$).

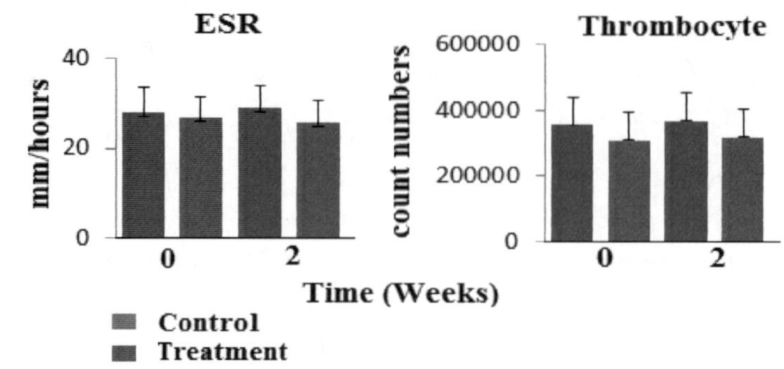

Figure 2. ESR and thrombocyte count average in the peripheral blood.

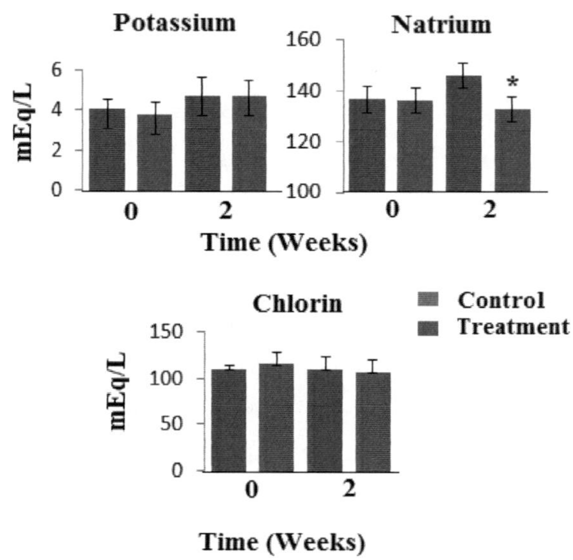

Figure 3. Potassium, natrium and chlorin average in the peripheral blood.

Banana administration did not affected to potassium and chlorin level significantly, but affected to natrium level in the blood significantly. (A) Potassium level showed no alteration ($p = 0.621$). (B) Natrium level showed significant reduction ($p = 0.037$). (C) Chlorin level showed no alteration ($p = 0.518$). Data represent means ± SD and are derived from two times independent experiments (*, $P < 0.05$). The average value of blood potassium content showed no difference between control group and treatment group, either on weeks 0 or weeks 2. It means that potassium content in the blood were not influenced by banana administration (figure 3). It's the same case as potassium blood, the clorin content in the blood showed no difference between control and treatment group, either on weeks 0 or weeks 2. In the other hand, blood natrium content on treatments group showed significant decline. On weeks 2, the blood natrium content of treatment group (132.81 mEq / L) compare to control group (146.06 mEq / L), was decreased significantly ($p = 0.037$, $p<0.05$). The natrium blood content of control group was only small exceeding normal limits on weeks 2 (figure 3). For adult, the normal range of natrium content in the blood was between 135-145 mEq / L [6]. This occurred because a very high content of potassium in bananas can depleted natrium in extracellular space and increased natrium excretion in the urine (natriuresis) [10]. Detailed mechanism needs further investigation.

Not all of the urine analysis data showed difference (supplementary). The urine analysis data showed that there was a decrease in urine leukocyte esterase in treatment group at week 0 and week two. This result were in line with study in interstitial cystitis patient which showed that a subset patient with positive PST/ Potassium Sensitivity Test, were caused by abnormal epithelial permeability showed that potassium administration would lead to epithelial permeability improvement, reduced the leucocyte esterase level and urinary infection level [11].

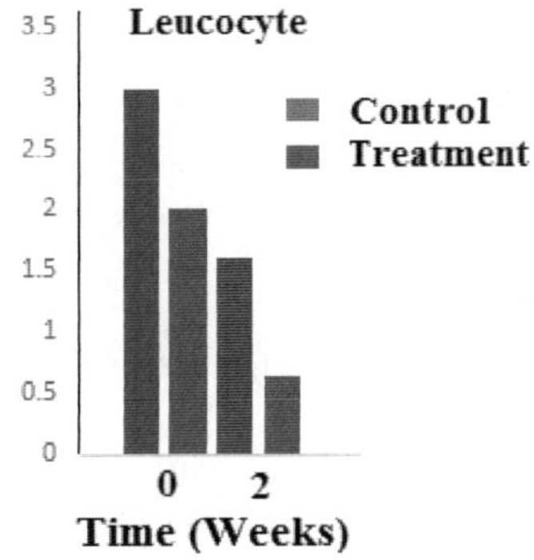

Figure 4. Average score for leucocyte esterase in urine on week 0 and week 2.

3.2 Discussion

Bananas administration to eldery female at a dose of 2 times a day during 2 weeks shown to decreased of blood pressure based on the results of statistical tests that had been carried out. This was also confirmed by the results of research conducted by Amran et al. (2010), which has been reported that the nutritional intake of fruits (not limited to bananas) for 2 weeks in the elderly can lowered blood pressure significantly [12]. Another study proved that the combination of banana juice and coconut water effectively lowered blood pressure in 25 elderly people in a nursing home within 5 days [7].

The mechanism of banana fruit in lowering blood pressure associated with one of dominant component in banana, which is potassium. The concentration of potassium in the banana fruit is estimated at 422 mg or 10% of daily potassium requirement for adults [13]. Potassium homeostasis levels are influenced by the kidneys that work to filter, reabsorption and excreted it under the influence of aldosterone [14]. Mechanism action of potassium to reducing blood pressure have been widely studied by researchers. Some researchers inferred that potassium works in several ways, which are affects the central nervous system and peripheral, change the activity of the renin-angiotensin and aldosterone, change directly peripheral resistance and has diuretic properties as well as the relationship physiological reciprocal with sodium, potassium can act as natriuresis [15].

Potassium affects the activity of the renin-angiotensin with working like ACE-inhibitor, which is an enzyme that may inhibit angiotensin II causes vasoconstriction of the blood vessels thus increasing blood pressure. Some researchers in India found the ACE-inhibitor in 6 different types of bananas [16]. In addition, potassium which administered either acute or chronic can suppressed the level of secretion or renin activity. Arteriolar smooth muscle response against angiotensin II and III could be reduced in animals when the diet is high of potassium. The increasing of potassium intake resulting of decrease in angiotensin circulation in the vascular [17].

In other hand, potassium can also initiate the process of diuretics and natriuresis in the body. One of study reported that dose of potassium to lead natriuresis ie with up to 300 mM per day which give for 4 days. Inspite of initiation and diuretic natriuresis occurs due to potassium intake, sodium balance are generally returned to normal in the first week. The study that conducted by Meneely support the natriuretic effect of potassium which revealed total body sodium levels declined during the administration of a combination of high-potassium salt diet in rats. This indicates that under these conditions, potassium effectively leading diuretic, minimal in the short term and even though the effect on the volume of body fluids is not clear, the mechanism is possible which can cause changes in blood pressure [15].

4. Conclusion

Banana administration to elderly female hypertensive patient decreased the blood pressure significantly. The natrium blood levels were greatly affected by banana administration. Our findings contribute to preliminary comprehension of banana effect on hypertension reduction.

Acknowledgement
We thanks to Mrs. Pudji Parwati Pramono and Mrs. Vitri Agustini for facility support.

References

[1] World Health Organization 2013 *A Global Brief of Hypertension: Silent Killer Global Health Crisis* (Switzerland)
[2] Rini A, Wahiddudin, Rismayanti 2013 *Risk Factors of Physical Activity, Smoking, and Alcohol Consumption on the Incidence of Hypertension in Elderly in the Region of Pattingalloang Health Center in Makassar City* (Makassar: University of Hasanuddin)
[3] Rahajeng E, Tuminah S 2009 *Maj. Kedokt. Indo* **59**
[4] Depkes RI 2013 *Riset Kesehatan Dasar* (Jakarta: Balitbangkes)
[5] Dinkes Kota Depok 2013 *Profil Kesehatan Kota Depok Tahun 2013* (Depok: Dinkes Depok)
[6] Sefni G 2010 *Evaluasi Penggunaan Antihipertensi Konvensional dan Kombinasi Konvensional Bahan Alam pada Pasie Hipertensi di Puskesmas Wilayah Depok* (Depok: Department of Pharmacy University of Indonesia)
[7] Yuli A, Febrianti, Lies I 2010 *J. Kes. Masy. Nas.* **5** 125
[8] Peni T, Sulisdiana 2015 *Hosp. Majap.* **7**
[9] Lizel R T, Sonny K, Gresty M 2013 *e. J. Kep.* **1**
[10] F.J Haddy. 1991. *Roles of sodium, potassium, calcium, and natriuretic factors in hypertension. Hypertension*. Nov: 18(5 Suppl):III179-83.
[11] Gupta S K, Pidcock L, Parr N J 2005 *BJU Internat. J.* **96**
[12] Yuli A, Febrianti, Lies I 2010 *J. Kes. Masy. Nas.* **5** 125
[13] Schmidi D *International Food Information Council Foundation* (Whasington DC)
[14] Maria G, Puspita D R, Sulistyowati Y 2012 *Hubungan Asupan Natrium dan Kalium dengan Tekanan Darah pada Pasien Hipertensi di Unit Rawat Jalan di RS Guido Valadares Dili Timor Leste* (Yogyakarta: Department of Nutrition Science University of Respati)
[15] Janet T, David P 1983 *Hypert* **5** 864
[16] Sharrock S and Lusty C 2000 *Nutritive Value of Banana* (Montpellier: INIBAP)
[17] Wein R M and Rolfe M 2010 *Clin. J. Am. Soc. Nephrol* **5** 531